Vegetable Brassicas and Related Crucifers

CROP PRODUCTION SCIENCE IN HORTICULTURE SERIES

This series examines economically important horticultural crops selected from the major production systems in temperate, subtropical and tropical climatic areas. Systems represented range from open field and plantation sites to protected plastic and glass houses, growing rooms and laboratories. Emphasis is placed on the scientific principles underlying crop production practices rather than on providing empirical recipes for uncritical acceptance. Scientific understanding provides the key to both reasoned choice of practice and the solution of future problems.

Students and staff at universities and colleges throughout the world involved in courses in horticulture, as well as in agriculture, plant science, food science and applied biology at degree, diploma or certificate level will welcome this series as a succinct and readable source of information. The books will also be invaluable to progressive growers, advisers and end-product users requiring an authoritative, but brief, scientific introduction to particular crops or systems. Keen gardeners wishing to understand the scientific basis of recommended practices will also find the series very useful.

The authors are all internationally renowned experts with extensive experience of their subjects. Each volume follows a common format covering all aspects of production, from background physiology and breeding, to propagation and planting, through husbandry and crop protection, to harvesting, handling and storage. Selective references are included to direct the reader to further information on specific topics.

Titles Available:

1. **Ornamental Bulbs, Corms and Tubers** A.R. Rees
2. **Citrus** F.S. Davies and L.G. Albrigo
3. **Onions and Other Vegetable Alliums** J.L. Brewster
4. **Ornamental Bedding Plants** A.M. Armitage
5. **Bananas and Plantains** J.C. Robinson
6. **Cucurbits** R.W. Robinson and D.S. Decker-Walters
7. **Tropical Fruits** H.Y. Nakasone and R.E. Paull
8. **Coffee, Cocoa and Tea** K.C. Willson
9. **Lettuce, Endive and Chicory** E.J. Ryder
10. **Carrots and Related Vegetable Umbelliferae** V.E. Rubatzky, C.F. Quiros and P.W. Simon
11. **Strawberries** J.F. Hancock
12. **Peppers: Vegetable and Spice Capsicums** P.W. Bosland and E.J. Votava
13. **Tomatoes** E. Heuvelink
14. **Vegetable Brassicas and Related Crucifers** G. Dixon
15. **Onions and Other Vegetable Alliums, 2nd Edition** J.L. Brewster
16. **Grapes** G.L. Creasy and L.L. Creasy
17. **Tropical Root and Tuber Crops: Cassava, Sweet Potato, Yams and Aroids** V. Lebot

VEGETABLE BRASSICAS AND RELATED CRUCIFERS

2nd Edition

Geoffrey R. Dixon

School of Agriculture, Policy and Development, Earley Gate, Whiteknights Road, PO Box 237, University of Reading, Reading, Berkshire RG6 6EU and GreenGene International, Hill Rising, Horsecastles Lane, Sherborne, Dorset DT9 6BH, UK

Rachel Wells

Department of Crop Genetics, John Innes Centre, Norwich Research Park, Colney Lane, Norwich, Norfolk NR4 7UH, UK

CAB Publishing
CAB INTERNATIONAL
Wallingford, Oxfordshire OX10 8DE, UK

CABI is a trading name of CAB International

CABI
Nosworthy Way
Wallingford
Oxfordshire OX10 8DE
UK

CABI
200 Portland Street
Boston
MA 02114
USA

Tel: +44 (0)1491 832111
E-mail: info@cabi.org
Website: www.cabi.org

T: +1 (617)682-9015
E-mail: cabi-nao@cabi.org

A catalogue record for this book is available from the British Library, London, UK.

Library of Congress Cataloging-in-Publication Data

Names: Dixon, Geoffrey R., author. | Wells, Rachel (Scientist), author.
Title: Vegetable brassicas and related crucifers / Geoffrey R. Dixon and
  Rachel Wells.
Other titles: Crop production science in horticulture ; 40.
Description: Second edition | Boston, MA : CAB International, [2023] |
  Series: Crop production science in horticulture ; 40 | Includes
  bibliographical references and index. | Summary: "Vegetable brassicas
  crops include broccoli, cauliflower, cabbage, kale and Brussel sprouts.
  This is an update of this popular title in the Crop Production Science
  in Horticulture series, originally published in 2006"-- Provided by
  publisher.
Identifiers: LCCN 2023019375 (print) | LCCN 2023019376 (ebook) | ISBN
  9781789249156 (paperback) | ISBN 9781789249163 (ebook) | ISBN
  9781789249170 (epub)
Subjects: LCSH: Brassica. | Cole crops.
Classification: LCC SB317.B65 D59 2023 (print) | LCC SB317.B65 (ebook) |
  DDC 635/.34--dc23/eng/20230825
LC record available at https://lccn.loc.gov/2023019375
LC ebook record available at https://lccn.loc.gov/2023019376

ISBN-13: 9781789249156 (paperback)
        9781789249163 (ePDF)
        9781789249170 (ePub)

DOI: 10.1079/9781789249170.0000

Commissioning Editor: Rebecca Stubbs
Editorial Assistant: Emma McCann
Production Editor: Rosie Hayden

Typeset by Exeter Premedia Services Pvt Ltd, Chennai, India
Printed and bound in the UK by CPI Group (UK) Ltd, Croydon, CR0 4YY

This book is dedicated to the memory of
Dr Michael Dickson, College of Agriculture and Life Sciences, Cornell
University, Geneva, NY, USA, who provided inspirational help and
support for the First Edition.

Contents

Preface

Brassicas are highly valuable sources of food, fodder, forage, condiments and ornamentals. They demonstrate significant genotypic and phenotypic diversity and flexibility; showing convergent evolution in the European and Asian forms of *Brassica oleracea* and *Brassica rapa*, respectively. A resultant rich array of fresh foodstuffs has evolved though selection and directed breeding developed initially by satisfying regional preferences. Since the first edition of *Vegetable Brassicas and Related Crucifers* plant breeders' hybrids have significantly amplified crop diversity, responding to changing market demands. In parallel, have come significant changes in crop science and technology. These now ensure high-quality crop production is achieved by sustainable means, which safeguard local biodiversity and conserve its environment. As a result, the use of synthetic crop protection formulations is diminishing, replaced by biostimulants, biofertilizers and active systems of biological pest and pathogen control. Regrettably, the spectrum of parasites, herbivores and microbes attacking crops is increasing, encouraged by climatic warming, moister, warmer environments and added virulence and aggressiveness. Physical change is happening in methods by which crops are monitored, measured and resource use calculated. Electronics are powering change in crop husbandry and the resultant opportunities are described and discussed. Automatic and autonomous systems are rapidly replacing muscle power for brassica production, harvesting, storage, cold-chain handling and delivery to the ultimate consumer as safe and trusted foods.

Wider arrays of brassica forms now offer higher quality fresh and more easily consumed brassicas. These trends encourage brassica use in diets that help reduce the impact of the diseases of affluence, such as cancers, coronary distress, and strokes. Evidence associating longer term brassica consumption with healthiness and greater welfare is now largely irrefutable. Rising consumption of brassica baby leaves, watercress and kale indicate public acceptance of this association. Each of these changes is supported by the vastly increased scientific literature reporting research concerning brassicas. Aspects of this research are absorbed into this revision. Massive growth in the worldwide importance of industrial brassicas, principally *Brassica napus* as oilseed rape or canola and the associated scientific research has in turn boosted the literature relevant to vegetable brassicas. The enormous scientific importance of *Arabidopsis thaliana*, the first plant whose genome was sequenced, adds additional weight and volume into the brassica literature. The first edition of *Vegetable Brassicas and Related Crucifers* retained relevance for a generation. Hopefully, this revision will similarly provide knowledge for researchers, students, industrialists, growers, advisors, consultants and those with an intelligent general interest in these fascinating plants.

Views and interpretation expressed in the second edition of *Vegetable Brassicas and Related Crucifers* are solely those of the two authors, Geoffrey R. Dixon and Rachel Wells. We accept responsibility for all errors, omissions and unconventional thinking.

Geoffrey Dixon's roots for this book are set deep in a lifetime of scientific fascination in the biology of *Brassica* allied to the husbandry associated with commercial production. His knowledge of brassica science, technology and husbandry reflects working and travelling nationally and internationally, epitomised by the concept of *'One foot in the furrow and one hand on the laboratory bench'*. Rachel Wells is a Brassica geneticist with a keen interest in translating scientific research into commercial reality. This includes working on anything from producing biolubricants in oilseed rape to the architectural development of a cauliflower or the feeding habits of cabbage stem flea beetle. She can just as often be found in a field with farmers, agronomists and plant breeders as in a controlled environment room supporting students in their research.

Geoffrey Dixon expresses deep gratitude for decades of support and encouragement from numerous friends, students and professional colleagues. Nationally, these encompass studying at Wye College (University of London), working in the National Institute of Agricultural Botany, Cambridge, Aberdeen School of Agriculture, the Scottish Agricultural College, Strathclyde University and, most recently, the School of Agriculture, Policy and Development of Reading University.

Particular friends and mentors are Professors Paul Hadley and Helmut van Emden who have offered much valued friendships and intellectual stimulation at Reading University. Professor Paul Williams of Wisconsin University at Madison, USA is a source of abiding friendship, deep intellectual sustenance and a common view of science over many decades. Professor Stephen Strelkov, University of Alberta, Canada brought a vibrant window into the opportunities resulting from molecular science. My friends in the Vegetable Consultants Association have moulded my views on the interactions between industry and science.

My dearest wife, Kathy, continues with her unstinting and unfailing support of my life and interests aided by Lucy, Dougal, Isabella, Hector and Richard, Amanda and James.

Rachel gives thanks to all those that have worked with her as part of 'Team Brassica' at the John Innes Centre and beyond; colleagues, mentors, and students alike. Deepest thanks go to Ian Bancroft and Judith Irwin, who supported her during her early career and most recently Lars Ostergaard, Richard Morris and Steve Penfield as amazing research collaborators and friends.

Last, to my wonderful, unconventional, family, that have supported me in my work. My daughter Charlie, who believes so much in her mother the scientist, and my partner Gary, for bringing light to my life and whose art graces the pages of this book.

Geoffrey R. Dixon, Sherborne, UK and Rachel Wells, Norwich, UK; 2024.

NAMING CONVENTIONS

Naming of plants, pests and pathogens used in this book has been very carefully assessed in an attempt to achieve consistency and conformity. There may, however, be failures and inconsistencies, since this is a complex and intricate topic, for which the authors accept responsibility.

SOURCE ACKNOWLEDGEMENTS

The use of 'after' indicates that permissions were granted for use in the first edition of *Vegetable Brassicas and Related Crucifers* and this has been carried forward into this second edition. Sincere apologies are offered to any collaborators who feel this is an unwarranted action. Significant attempts have been made to trace the sources of Figures 7.24 and 8.6 without success; sincere apologies are offered to the collaborators who created these illustrations for the failure to acknowledge their excellent illustrations.

Origins and Diversity of Brassica and Its Relatives

RACHEL WELLS*

*Department of Crop Genetics, John Innes Centre, Norwich
Research Park, Colney Lane, Norwich, Norfolk NR4 7UH, UK*

Abstract

The *Brassica* genus comprises an abundance of phenotypically diverse species that have been adapted during domestication into an array of vegetable, oilseed and condiment crops. Understanding *Brassica*[1] vegetables involves a fascinating, biological journey through evolutionary time, witnessing wild plant populations interbreeding and forming stable hybrids. Humankind took both the wild parents and their progeny, refined them by selection and further combination, and over time produced crops that are, together with the cereals, the mainstay of world food supplies. This, in part, is down to the complex nature of *Brassica* genome evolution: ancient genome duplications, speciation, gene loss, hybridization and polyploidization events. This complexity and variation provides the flexibility for speciation, adaptation and selection that drives crop development. Modern genetic marker technologies have vastly improved the resolution of population structure and phylogenetic analyses, greatly enhancing our previous understanding of *Brassica* crop evolution. Here we discuss the origins, evolution and vast levels of diversity that are observed in today's wild, feral and cultivated *Brassica* species.

ORIGINS AND DIVERSITY OF *BRASSICA* CROPS

Genetic diversity and flexibility are characteristic features of all members of the family Brassicaceae (previously Cruciferae). Possibly, these traits encouraged

*rachel.wells@jic.ac.uk

© Geoffrey R. Dixon and Rachel Wells 2024. *Vegetable Brassicas and Related Crucifers,* 2nd edition. (G.R. Dixon and R. Wells)
DOI: 10.1079/9781789249170.0001

their domestication by Neolithic people. *Brassica* crops were first described in a Chinese almanac from around 3000 BCE and ancient Indian texts from around 1500 BCE (Keng, 1974; Prakash *et al.*, 2011). Records show that the Ancient Greeks, Romans, Indians and Chinese all valued and used them greatly. The etymology of *Brassica* has been contested since Herman Boerhaave suggested in 1727 that it might come from the Greek αποτουβραξειν, Latin *vorare* (both meaning 'to devour') (Henslow, 1908). An alternative derivation from *Bresic* or *Bresych*, the Celtic name for cabbage, was suggested by Hegi (1919). This is a contraction of *praesecare* (to cut off early) since the leaves were harvested for autumn and early winter fodder. Another suggested origin is from the Greek βρασσω (crackle), coming from the sound made when the leaves are detached from the stem (Gates, 1953). A further suggestion is a Latin derivation from 'to cut off the head' and was first recorded in a comedy by Plautus in the 3rd century BCE. Aristotle (384–322 BCE), Theophrastus (371–286 BCE), Cato (234–149 BCE), Columella (1st century CE) and Pliny (23–79 CE) all mention the importance of brassicas.

Further east, the ancient Sanskrit literature *Upanishads* and *Brahmanas*, originating around 1500 BCE, mention brassicas and the Chinese *Shijing*, possibly edited by Confucius (551–479 BCE), refers to the turnip (Prakash and Hinata, 1980). European herbal and botanical treatises of the Middle Ages clearly illustrate several *Brassica* types and Dutch paintings of the 16th and 17th centuries show many examples of brassicas. In the 18th century, species of coles, cabbages, rapes and mustards were described in the genera *Brassica* and *Sinapis* in *Institutiones Rei Herbariae* (de Tournefort, 1700) and *Species Plantarum* (Linnaeus, 1735). Probably the most important early, formal classifications of *Brassica* were made by Otto Eugen Schultz (1874–1936) and published in *Das Pflanzenreich* and *Die Natürlichen Pflanzenfamilien* (Schulz, 1919 and 1936, respectively). These classifications were supported broadly by the great American botanist and horticulturist Liberty H. Bailey (1922, 1930).

Brassica crops worldwide provide the greatest diversity of products used by humans derived from a single genus. Other members of the family Brassicaceae extend this diversity. Overall, brassicas deliver: leaf, flower and root vegetables that are eaten fresh, cooked and processed; fodder and forage, contributing especially as an overwintering feed supply for meat- and milk-producing domesticated animals; sources of protein and oil used in low-fat edible products, fuel for illumination and industrial lubricants; condiments such as mustard, herbs and other flavourings; flowering and variegated ornamentals; and soil conditioners as green manure and composting crops.

Wild diploid *Brassica* and related hybrid amphidiploids (Greek: *amphi* = both; *diploos* = double; possessing the diploid genomes from both parents) evolved naturally in inhospitable places with abilities to withstand drought, heat and salt stresses (Gómez-Campo and Prakash, 1999). The Korean botanist Woo Jang-choon (known in scientific literature as Nagaharu U)

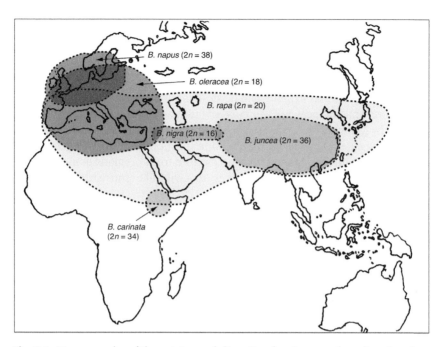

Fig. 1.1. Biogeography of the origins and diversity of major crop-founding *Brassica* species. (After UNFAO, courtesy Garry Breeze).

(Nagaharu, 1935) deduced that three basic diploid *Brassica* forms were probably the parents of subsequent amphidiploid crops. *Brassica nigra* (black mustard), itself the ancestor of culinary mustards, is found widely distributed as annual herbs growing in shallow soils around most rocky Mediterranean coasts. Natural populations of *Brassica oleracea* and associated types are seen historically as potential progenitors of many European cole vegetables. These populations inhabit rocky cliffs in cool damp coastal habitats. They have slow, steady growth rates and are capable of conserving water and nutrients. Domestication of *B. oleracea* occurred in a Mediterranean location, with the crop reaching the Atlantic coast through the movement of people and cultures (Maggioni, 2015). The putative ancestor of many Asian brassica vegetables, *Brassica rapa*, originates from the 'Fertile Crescent' in the high-plateau regions of today's Iran, Iraq and Turkey. Here, these plants grow rapidly in the hot, dry conditions forming copious seed. Other family members evolved as semi-xerophytes in the Saharo-Sindian regions in steppe and desert climates. Early hunter-gatherers and farmers discovered that the leaves and roots of these plants provided food and possessed medicinal and purgative properties when eaten either raw or boiled. Some types supplied lighting oil, extracted from the seed, and others were simply used for animal feed. These simple herbs have developed into a massive array of essential crops grown all around the world (see Fig. 1.1).

BIODIVERSITY

Wild diploid *Brassica* species still cling to survival in inhospitable habitats and thus are indicative of the natural diversity of this genus. Such species can be seen in Table 1.1.

Such diversity has expanded in domestication and the service of human-kind. However, wild species of our current cultivated brassicas are rare. Recent analyses by Mabry *et al.* (2021) suggest that many, previously considered wild, C genome species are feral, containing a proportion of cultivated germplasm within the genome, and therefore have escaped from previous domestication. This suggests cultivated forms may revert to a wild-like state with relative ease (Mabry *et al.*, 2021).

Wild hybrids

Wild *Brassica* and its close relatives hybridized naturally to form polyploids. These amphidiploids and their parental wild diploids were key building blocks from which our domesticated brassica crops have evolved. Three hybrid species are of especial interest as ancestors of the crop brassicas as described by Nagaharu (1935). The relationships between the hybrid amphidiploids and their parental species are summarized in the gene flow 'Triangle of U' (Nagaharu, 1935) (see Fig. 1.2).

Brassica carinata (BBCC, $n = 17$, genome size ~1300 Mb) is proposed to have evolved through spontaneous hybridization between the wild kale form of *B. oleracea* (CC, $n = 9$, genome size ~490 Mb) and *B. nigra* (BB, $n = 8$, genome size ~515 Mb) in the adjoining regions of the highlands of Ethiopia, East Africa and the Mediterranean coast (Seepaul *et al.*, 2021). This hypothesis is supported by evidence of the presence of these progenitor species in the region during the emergence and domestication of *B. carinata* (Alemayehu and Becker, 2002) and that *B. carinata* shares the chloroplast genome with the hybridization donor, *B. nigra* (Li *et al.*, 2017). This species is characterized by the slow, steady growth of *B. oleracea* and the mustard oil content of *B. nigra*. Wild forms of *B. carinata* are not known but primitive domesticated types are cultivated in upland areas of Ethiopia and further south into Kenya. *Brassica carinata* has been traditionally cultivated as both an oilseed and leafy vegetable in the Ethiopian Highlands (Ojiewo *et al.*, 2013). Carinata crops themselves are locally referred to as *gomen-zer* in the Amharic language (Hagos *et al.*, 2020), Abyssinian mustard, Ethiopian mustard or Ethiopian cabbage, though this is not necessarily synonymous with the sophisticated heads seen on today's supermarket shelves. It is one of the most drought- and heat-tolerant species within the Brassicaceae. However, both kale and carinata crops thrive in the cool environments that local farmers term 'kale gardens', typical of the Ethiopian Highlands.

Brassica juncea (AABB, $n = 18$, genome size ~930 Mb) is a hybrid between *B. rapa* (AA, $n = 10$, genome size ~350 Mb) × *B. nigra* (BB, $n = 8$, genome size ~515 Mb) and can be divided into four recognized subspecies. These

Table 1.1. Examples of the diversity of some wild *Brassica* species. (After Tsunoda *et al.*, 1984).

Name	Chromosome Count	Geographical Distribution	Habitat
Brassica amplexicaulis	$n = 11$	Intermountain area south-east of Algiers	Small plant, colonizes colluvial slopes, especially medium-sized gravel
Brassica barrelieri	$n = 9$	Iberian Peninsula, extending to Morocco and Algeria	Common on wastelands, especially areas of more dense vegetation
Brassica elongata	$n = 11$	Plateau steppe – lands of south-eastern Europe and western Asia, as far as Iran	Semi-arid areas
Brassica fruticulosa ssp. *fruticulose*	$n = 8$	Found around the Mediterranean coasts, especially among pine trees; the ssp. *cossoniana* extends beyond the coastal zone and is found on the Saharan side of the Middle Atlas mountains in Morocco	Biennial or perennial Annual
Brassica fruticulosa ssp. *cossoniana*	$n = 8$		Both subspecies require well-drained sites; on inland sites they characteristically colonize stony mountain slopes and alluvial areas
Brassica maurorum	$n = 8$	Endemic to North Africa; inhabits arable land in Morocco and Algeria	Colonizes stony pastures in semi-arid areas from the coast to low mountainous zones; on arable land grows to 2 m high in dense clumps similar to *Brassica nigra*
Brassica oxyrrhina	$n = 9$	Southern Portugal, Spain and north-western Morocco	Coastal sandy habitats
Brassica repanda and *Brassica gravinae*	$n = 10$	Inland rocky areas	These species grow together in the lithosol in the crevices of rocky outcrops A polypoid ($n = 20$) form has been found north of Biskra in Algeria
Brassica spinescens	$n = 8$	Endemic to North Africa	Coastal calcareous or siliceous cliffs Diminutive growth habit with small, thick glabrous leaves; *B. fruticulosa*, *B. maurorum* and *B. spinescens* possibly form a single cytodeme
Brassica tournefortii	$n = 10$	Coastal areas of the Mediterranean extending to western Asia as far as India	Capable of colonizing arid alluvial sand where other vegetation is sparse

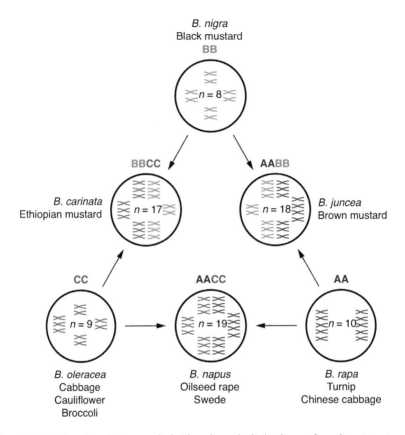

Fig. 1.2. Relationships between diploid and amphidiploid crop-founding *Brassica* species. (The 'Triangle of U': Nagaharu, 1935, courtesy Garry Breeze).

include *juncea* (seed mustard) that is used as oilseed and condiment, *integrifolia* (leaf mustard) with a diverse variation of leaf morphology, *napiformis* (root mustard) with a tuberous root and *tumida* (stem mustard) with an enlarged edible stem (Kang *et al.*, 2022). Yang *et al.* (2016, 2018) demonstrated that *B. juncea* first diversified into root mustard, followed by seed mustard, leaf mustard and stem mustard. They determined a monophyletic origin for *B. juncea* based on phylogenetic analyses of the A subgenome. This was confirmed by analysis of variation within cytoplasmic DNA by Kang *et al.* (2021). *Brassica juncea* is used as a source of vegetable oil in India and throughout Asia, especially in China and Japan, while vegetable forms are of immense dietary importance. Feral forms are classed as weeds in cropping systems throughout China and Japan (Sun *et al.*, 2018). Reputedly, wild forms are still found on the Anatolian Plateau and in southern Iran.

The third hybrid, *Brassica napus* (AACC, *n* = 19, genome size ~1130 Mb), developed from *B. rapa* (AA, *n* = 10, genome size ~350 Mb) × *B. oleracea* (CC, *n* = 9, genome size ~490 Mb). There are three recognized subspecies:

rapeseed/oilseed rape (*B. napus* ssp. *oleifera*), swede or rutabaga (*B. napus* ssp. *rapifera*) and Siberian kale or leaf rape (*B. napus* ssp. *pabularia*). Wild populations do not exist, and the true species progenitors are unknown. *Brassica napus* may have Mediterranean origins or this hybrid may have formed as *B. oleracea* types expanded into agricultural regions along the coasts of northern Europe and *B. rapa* extended from the Irano–Turanian regions. Lu *et al.* (2019) determined the *B. napus* A subgenome evolved from the ancestor of the European turnip; and hypothesized the *B. napus* C subgenome evolved from the common ancestor of kohlrabi, cauliflower, broccoli and Chinese kale. However, it is believed that within the last 1000 years, further gene flow has occurred from the two progenitor species. Feral populations of *B. napus* have acquired major scientific significance as a means of determining the potential for gene flow to and from genetically modified cultivars of oilseed rape.

Diversity within the amphidiploids

Considerable genetic diversity is present within the three amphidiploid species. This is hypothesized to be due to two major factors: multiple hybridizations with different diploid parents and genome modifications following polyploidization. Evidence of multiple hybridizations was reported by Song *et al.* (1996) showing four cytoplasmic types were present within *B. napus* accessions that matched different parental diploid cytoplasm (see Fig. 1.3).

In more recent analysis, Li *et al.* (2017) performed *de novo* assembly of 60 complete chloroplast genomes of *Brassica* genotypes for all six species within U's triangle. Chloroplast genome sequences, which are maternally inherited and therefore only represent the maternal lineage, have been used extensively for inferring plant phylogenies. Phylogenetic analysis separated the *Brassica* species into four clades: Clade I contained *B. juncea*, *B. rapa* and *B. napus*; Clade II *B. oleracea*; Clade III *B. rapa* and *B. napus*; and Clade IV *B. nigra* and *B. carinata*. *Brassica rapa* showed evidence of two types of chloroplast genomes, with the Clade IV type specific to some Italian broccoletto accessions, while *B. oleracea* and *B. nigra* were only represented by a single clade. No amphidiploid hybrids were grouped with *B. oleracea*, suggesting that *B. oleracea* is not the maternal parent to any amphidiploid species. This fits with the observation that in interspecific crosses, *B. oleracea* can only be used as a male parent. *Brassica carinata* and *B. juncea* share their chloroplast genome with one of their hybridization donors, *B. nigra* and *B. rapa*, respectively, fitting with U's model. Chloroplast genomes of all eight *B. juncea* accessions clustered with the chloroplast genomes of *B. rapa* accessions in subclade 'I-a' (Japanese leafy types and turnips, plus one broccoletto), supporting the hypothesis of Palmer *et al.* (1983) that *B. rapa* is the ancestral maternal parent of the amphidiploid. *Brassica napus* clearly had evidence for two independent hybridization events, as accessions were either within Clade I *B. rapa* subclade 'e' (sarson-like morphotype) or *B. rapa* Clade IV (Italian broccoletto morphotype).

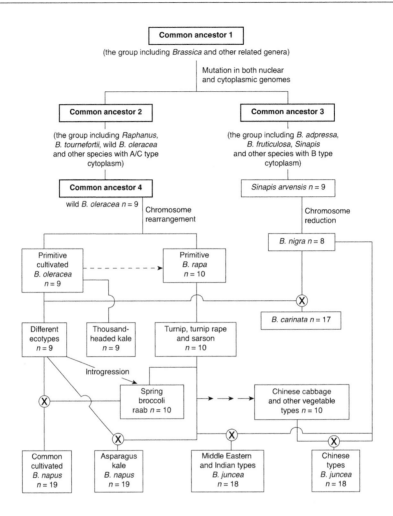

Fig. 1.3. Hypothetical scheme for genome evolution of *Brassica* species and cultivated forms. (After Song *et al.*, 1988, courtesy Garry Breeze).

Direct evidence for genome changes after polyploidization can be obtained by studying synthetic amphidiploids developed by resynthesis via interspecific hybridization of the diploid progenitor species. The production of synthetics is a recognized methodology used to increase genetic variation and introduce desired phenotypic characters. However, it is also associated with genome instability and rapid genomic change. Analysis for morphological traits, chromosome numbers and restriction fragment length polymorphisms (RFLPs) in chloroplast, mitochondrial and nuclear DNA clones in artificially synthesized analogues of *B. napus, B. juncea* and *B. carinata*, in comparison with the natural amphidiploids, showed the synthetic hybrids were closer to their diploid parents when compared with natural polyploids. Genome changes,

mainly involving either loss or gain of parental fragments and novel fragments, were seen in the early generations of synthetic amphidiploids (Song
et al., 1993). It has also been shown, that within the first generations after
interspecific hybridization, *Brassica* hybrids experience altered gene expression
patterns (Lloyd *et al.*, 2018), changes in transposable element activity (Sarilar
et al., 2013) and altered gene methylation patterns (Lukens *et al.*, 2005; Gaeta
et al., 2007). Samans *et al.* (2017) showed the size and number of rearrange-
ments per generation is much lower in natural populations of *B. napus* than in
resynthesized *B. napus*. This indicates that genome stability increases across
evolutionary time, as many individuals with detrimental chromosomal rear-
rangements are lost, and suggests natural *B. napus* must have mechanisms
that prevent non-homologous chromosome pairing.

The frequency of genome change, and the direction of its evolution in
the synthetic hybrids, were associated with divergence between the parental
diploid species. Quantitative analysis of phylogenetic trees based on RFLP
data from *Brassica* by Song *et al.* (1988), and more recently by Li *et al.* (2017),
suggest that:

1. *Brassica nigra* originated from one evolutionary pathway with *Sinapis
   arvensis*, or a close relative, as the likely progenitor, whereas *B. rapa* and
   B. oleracea came from another pathway with a possible common ancestor in
   wild *B. oleracea*, or a closely related species possessing nine chromosomes.
2. The estimated divergence of *B. rapa* and *B. oleracea* from *B. nigra* varies
   between studies – 7.9 million years ago (Mya) (Lysak *et al.*, 2005), 13.7 Mya
   (Li *et al.*, 2017), and 20 Mya (Arias *et al.*, 2014).
3. The estimated divergence of the A genome *B. rapa* (*B. juncea* and *B. napus*)
   from *B. oleracea* is between 0.12 and 1.37 Mya (Cheung *et al.*, 2009),
   2.18 Mya (Li *et al.*, 2017) and 3.7 Mya (Inaba and Nishio, 2002).
4. The amphidiploid species *B. napus* and *B. juncea* have evolved through
   different combinations of the diploid morphotypes and thus polyphyletic
   origins may be a common mechanism generating the natural occurrence
   of amphidiploids in *Brassica*.
5. The cytoplasm has played an important role in the nuclear genome evolution of amphidiploid species when the parental diploid species contain
   highly differentiated cytoplasms.

Contrasting the physiology and morphology of wild and cultivated brassicas

It will be evident to the reader by now that many of the wild and feral *Brassica*
spp., and their close allies, inhabit dry coastal, arid rocky or desert habitats.
These wild plants have very thick leaves containing less chlorophyll and many
more cell wall components compared with cultivated plants. Typically, they
have well-developed xylem vessels and small leaf areas. These characteristics

increase the efficiency of water conservation in plants. The foliage of wild xerophyllous plants has evolved high photosynthetic rates per unit leaf area (or per quantum of light received), even in dry air conditions.

Conversely, cultivated brassicas have broadly expanded, thin leaves that are well supplied with chlorophyll. These characteristics are advantageous for receiving, absorbing and utilizing solar radiation when there are ample supplies of water and nutrients available. Typically, these are mesophyllous environments found in fertile, cultivated fields.

Similar contrasts between the ecology of wild progenitors of crops and cultivated plants are found between the wild allies of wheat and artificial cultivars. Wild forms possess small, thick leaves whereas wheat cultivars have large, thin leaves. Both wild *Brassica* and *Hordeum* spp. evolved strategies for successful growth under arid conditions involving the restriction of transpiration, intensification of water movement to sites of photosynthesis, restriction of light absorption and efficient fixation of the absorbed solar radiation. Such traits became redundant in cultivation and consequently were removed by generations of artificial field selection and more latterly by directed plant breeding.

Other brassica relatives illustrating the biodiversity of this family

Several other members of the Brassicaceae illustrate the diversity of this family and their evolution in cultivation. Examples are summarized in Tables 1.2 and 1.3.

CULTIVATED *BRASSICA* SPECIES AND FORMS

Brassica oleracea group (*n* = 18) – the European group

The European brassica vegetables originate from *B. oleracea* and some probably closely related Mediterranean species. They can be divided into subordinate groups often at the variety (var.), subvariety (subvar.) or cultivar (cv.) levels. Much of the basis of current understanding of diversity within this group comes from the detailed studies of American horticultural botanist Bailey (1922, 1930, 1940).

Multiple origins and parents

One school of thought suggests that the constituent crops within the *B. oleracea* group have multiple origins derived from cross-breeding between closely related *Brassica* species living in geographical proximity to each other. In consequence, the taxonomy of parents and progeny is confused and clouded still further by millennia of horticultural domestication. For example, the progenitors of headed cabbages and kales were postulated by Netroufal (1927)

Table 1.2. Genera allied to *Brassica* that form crop plants: *Eruca, Sinapis* and *Raphanus.* (After Tsunoda *et al.,* 1984).

Eruca sativa (*n* = 11, syn. *Eruca vesicaria* ssp. *sativa*), commonly known as 'garden or salad rocket', is a salad vegetable in southern Europe. In India it forms an oilseed (taramira) and is used for fodder. *Eruca sativa* has recently become a popular salad vegetable in western Europe. Wide distribution occurs across southern Europe, North Africa, western Asia and India. Variation within the cytodeme *Eruca* is substantial with ecotypes evolved for several habitats in relation to available water.

Sinapis alba (*n* = 12) (white mustard) grows wild in Mediterranean areas with abundant moisture and ample soil nutrients, often coexisting with *Sinapis arvensis* (*n* = 9) (charlock or wild mustard) although the latter prefers lower soil moisture levels. In dry areas, *Sinapis turgida* (*n* = 9) is common. In wet habitats, *S. alba* reaches 2 m in height with a high mustard oil content. It is grown in Europe as 'white' or 'yellow' mustard. Cultivation is increasing, particularly in Canada, in combination with 'brown' mustard (*Brassica juncea*).

Raphanus (radish) – there are about 18 genera in the subtribe Raphininae including *Rapistrum, Cakile* and *Crambe.* Possibly all *Raphanus* spp. with *n* = 9 form a single cytodeme. The genus is found along Mediterranean coasts forming the dominant plant on coastal areas of the Sea of Marmara and the Bosporus Strait. Around field margins, *Raphanus raphanistrum* and *Raphanus rapistrum rugosum* tend to coexist. Culinary radishes (*Raphanus sativum*) possibly evolved in southern Asia and provide a wide diversity of root forms and flavours. Fodder radishes (*Raphanus sativus* var. *oleiferus*) are a source of animal fodder obtained from both roots and foliage. *Crambe* is a source of seed oil and is beginning to be used more extensively in North America.

as *Brassica montana* and of kohlrabi as *Brassica rupestris*. Later, Schiemann (1932) realised that several of the Mediterranean wild types had formed the origins for locally cultivated landraces. Schulz (1936) supported this view and identified *Brassica cretica* as a progenitor of cauliflower and broccoli.

Lizgunova (1959) grouped cultivars into five different species and proposed a multiple origin from wild forms. Helm (1963), in devising a triple origin, combined cauliflower, broccoli and sprouting broccoli into one line, thousand-headed kale and Brussels sprouts in another, and all other crop forms in a third. Further analysis was completed by Toxopeus (1974) and Toxopeus *et al.* (1984) suggested that for simplicity a horticulturally based taxonomy was preferable to attempted botanical versions.

Potential wild species contributing to the *Brassica oleracea* group

Populations of wild relatives, which cross-fertilize with *B. oleracea* and form interbreeding groups, are found on isolated cliffs and rocky islets. They form distinct units that often display phenotypic differentiation leading to several layers of variation superimposed on each other.

Table 1.3. Selected examples of wild relatives of *Brassica*. (After Tsunoda *et al.*, 1984).

Name	Chromosome count	Geographical Distribution	Habitat
Diplotaxis acris	*n* = 7	Saharo–Sindian area	Dry, desert regions
Diplotaxis harra	*n* = 13		
Diplotaxis tenuisiliqua	*n* = 9	North Africa	Moist areas, sometimes associated with *Sinapis alba*
Erucastrum cardaminoides	*n* = 9	Endemic to the Canary Islands	
Erucastrum laevigatum	*n* = 28	Southern Italy, Sicily and North Africa	Probably an autotetraploid of *Erucastrum virginatum* (*n* = 7)
Erucastrum leucanthum	*n* = 8	Endemic to North Africa, particularly the seaward side of the Middle Atlas mountains	Coexists with *Hirschfeldia incana*
Hirschfeldia incana	*n* = 7	Dominant component of Mediterranean flora	Large roadside colonies especially favoured by fine textured soils
Hutera spp.	*n* = 12	Iberian Peninsula	Rocky outcrops and colluvial sites; *Hutera* and *Rhynchosinapis* form a cytodeme complex found in the Spanish Sierra Morena regions
Sinapidendron spp.	*n* = 10	Madeira	Rocky cliffs; perennial habit

Maggioni (2015) detailed the domestication of *B. oleracea*. This identified the 11 accepted wild species of the *B. oleracea* group, including three subspecies of *B. cretica*, from the National Plant Germplasm System (NGPS)/ Germplasm Resources Information Network (GRIN) taxonomy for plants (USDA, Agricultural Research Service, National Plant Germplasm System, 2023). Some separate species detailed by Maggioni (2015) can be grouped into the *Brassica rupestris–incana* complex as they demonstrate distinct regional variation. These potential progenitors of European *B. oleracea* are detailed in Table 1.4.

A detailed study into the evolutionary history of wild, domesticated and feral *B. oleracea* was published by Mabry *et al.* (2021). This provided new genetic evidence combined with knowledge of archaeology, literature and environmental niche modelling to support the hypothesis of a single eastern Mediterranean domestication origin for *B. oleracea*, agreeing with the conclusions of Maggioni *et al.* (2018). Using population structure modelling Mabry *et al.* (2021) identified *B. cretica* and *Brassica hilarionis* as likely progenitor

Table 1.4. Examples of potential progenitors of European (*Brassica oleracea*) brassica vegetables. (After Tsunoda *et al.*, 1984).

Brassica cretica: Populations occur around the Aegean, including Crete, southern Greece and in south-western Turkey. The plant has a branching, woody habit, carrying glabrous, fleshy leaves and persisting in perennial form for 5–8 years but flowering in the first year under favourable environments. The inflorescence axis elongates between buds prior to opening with light yellow to white flowers. Three subspecies are known: *Brassica cretica* ssp. *cretica*, which grows on the mountainous cliffs and gorges of the Peleponnese and Crete, and central to south Lebanon (probably introduced); *Brassica cretica* ssp. *aegaea* growing in southern Greece, south-west Turkey and Israel (Mount Carmel) (probably introduced); and *Brassica cretica* ssp. *laconica* from south Peloponnese.

The *Brassica rupestris–incana* complex: This group has developed a number of distinct regional variants. The complex is characterized by forming a strong central stem, and a large apical inflorescence succeeded by further branching. The top of each partial inflorescence forms a tight grouping of buds which may open before the axis between them elongates. Foliage consists of large petiolate leaves which are poorly structured since they shrivel when past maturity, in contrast to *B. cretica* which retains mature leaves with a recognizable morphology. Hairs are characteristically found on both seedlings and adults. Distribution includes Sicily, southern and central Italy, and parts of Croatia, Bosnia and Montenegro. Species that have been separated but which are included within the complex include: *Brassica incana* Ten., *Brassica villosa*, *Brassica rupestris* Rafin., *Brassica tinei* Lojac., *Brassica drepanensis* (Car.) Dam., *Brassica botteri* Vis., *Brassica mollis* and *Brassica cazzae* Ginzb. and Teyb.

Brassica macrocarpa Guss.: A species restricted to Aegadian Islands to the west of Sicily with a habit similar to *B. rupestris–incana* with which it forms fertile hybrids and characterized by possessing smooth-surfaced, thick fruits containing seeds produced in two rows within the loculus. There are questions as to whether this species should be placed in the genus *Brassica* in view of the seed capsule. Conceivably, this species is a subordinate of the *B. rupestris–incana* complex.

Brassica insularis Moris.: Found in Corsica, Sardinia and Tunisia, the plants are similar in low-branching habit to *B. cretica* but the stiff, glabrous leaves, with pointed lobes and large white fragrant flowers and seed glucosinolates, differ from other *Brassica* spp. Species rank has been given to the Tunisian form as *B. atlantica* (Coss.) Schultz, but the morphology and intercrossing indicate clear inclusion in *B. insularis* without subdivision.

Brassica montana Pourr. (syn. *B. robertiana* Gay): Found growing in north-eastern coastal areas of Spain, southern France and northern Italy, these plants are shrubby perennials with lobed green, glabrous or possibly hairy leaves. An intermediate status between the *B. rupestris–incana* complex and *B. oleracea* is suggested by biogeographical and morphological evidence.

Continued

Table 1.4. Continued

Brassica oleracea: Found on coasts of northern Spain, western France and southern and south-western UK; as a stout perennial forming strong vegetative stocks which then flower and branch. The greyish coloration and glabrous nature of the leaf surfaces distinguish *B. oleracea*. Morphological differences between *B. bourgeaui* Kuntz, endemic to the Canary Islands, and *B. oleracea* are limited, therefore its status as a separate species is unclear.

Brassica hilarionis: Located solely in the Kyrenia Mountains of Cyprus, *B. hilarionis* has fruit reminiscent of the size and texture of *B. macrocarpa* and vegetatively a habit and leaf morphology close to *B. cretica*.

species of *B. oleracea* cultivars. This was further refined to support *B. cretica* as the progenitor species.

Crops developed within *Brassica oleracea* and allies

The vast array of crop types that have developed within *B. oleracea* (and also *B. rapa*) is probably unique within economic botany (Nieuwhof, 1969). This has led to acceptance at the subspecies and variety (cultivar) levels of descriptions based around the specialized morphology of the edible parts and habits of growth within the crop types (Wellington and Quartley, 1972). The nomenclature and common names of cultivated *B. oleracea* are given in Table 1.5.

Brassica oleracea (the cole or cabbage brassicas)

BRUSSELS SPROUTS: *B. OLERACEA* VAR. *GEMMIFERA*.   Brussels sprouts may have emerged in the Low Countries (coastal Rhine–Meuse–Scheldt delta) in the medieval period and risen to prominence in the 18th century around the city of Brussels. Subsequently, they became established as an important vegetable crop in north-eastern Europe, especially the northern Netherlands and parts of the UK. Local open-pollinated landraces were developed that were suited to specific forms of husbandry but usually subdivided into early, mid-season and late maturity groups. Often, they would be capable of resisting pests and pathogens common within their locality and have morphologies adapted to prevailing climatic conditions. In the 1930s, early maturing types were developed in Japan where many of the original F_1 hybrids were produced. These hybrids, and their derivatives, formed the basis for cultivars ideally suited to the emerging 'quick-freeze' vegetable processing industry. The entire worldwide crop of Brussels sprouts is now dominated by an F_1 germplasm derived by American, Dutch and Japanese breeders and originating from the initial crosses.

Table 1.5. The nomenclature and common names of cultivated *Brassica oleracea*. (After Wellington and Quartley, 1972).

B. oleracea L.	Common synonym	Common name
var. *alboglabra* (L.H. Bailey) Musil.	*Brassica alboglabra* L.H. Bailey	Chinese kale, Kailan
var. *botrytis* L.		Cauliflower
var. *capitata* L.		Red/white/Shetland cabbage
var. *costata*		Portuguese kale/Tronchuda kale
var. *gemmifera*		Brussels sprouts
var. *gongylodes* L.	*Brassica caulorapa* (DC.) Pasq.	Kohlrabi
var. *italica* Plenck		Broccoli
var. *medullullosa* Thell.		Marrow-stem kale
var. *palmifolia*		Palm kale/Jersey kale
var. *ramosa* CD.	ssp. *fruticosa* Metzg.	Thousand-headed kale, Branching bush kale
var. *sabauda* L.		Savoy cabbage
var. *sabellica* L.		Curly kale
var. *viridis* L.	var. *acephala* DC.	Kale, collard

In the 1990s, trials revealed that the content of the glucosinolates, sinigrin and progoitrin, was found to be correlated with bitterness ($r^2_{multiple}$ = 0.67 and 0.93, respectively) (van Doorn *et al.*, 1998). Later studies showed these traits were under strong genetic control with high heritability (van Doorn *et al.*, 1999). This knowledge was used for the selection of sweeter-tasting lines with low levels of sinigrin and progoitrin from the natural variation in historical varieties. Botanically, the plants are biennial with simple erect stems up to 1 m tall. Axillary buds develop into compact miniature cabbage heads or 'sprouts' that are up to 30 mm in diameter. At the top of the stem is a rosette of leaves; the leaves are generally petiolate and rather small, with a subcircular leaf blade (see Chapter 2 section, Floral Biology as Related to Controlled Pollination).

CAULIFLOWER: *B. OLERACEA* VAR. *BOTRYTIS*; BROCCOLI: *B. OLERACEA* VAR. *ITALICA*. A remarkable diversity of cauliflower- and broccoli-like vegetables developed in Europe, probably emanating from Italy, and possibly evolved from germplasm introduced in Roman times from the eastern Mediterranean. A classification of the colloquial names used to describe these crops was proposed by Gray (1982) and is shown in Table 1.6.

Over the past 400 years, white-headed cauliflowers (derived from the Latin *caulis* (stem) and *floris* (flower)) have spread from Italy to central and northern Europe, which became important secondary centres of diversity for

Table 1.6. Classification of *Brassica oleracea* var. *botrytis* and var. *italica* with associated colloquial crop names. (After Gray, 1982).

Brassica oleracea L. var. *botrytis* DC.	Cauliflower
	Heading broccoli
	Perennial broccoli
	Bouquet broccoli
	White-sprouting broccoli[a]
Brassica oleracea L. var. *italica* Plenck	Purple-sprouting broccoli
	Cape broccoli
	Purple cauliflower
	Calabrese and other green-sprouting forms (broccoli in North America)
	White-sprouting broccoli[a]

[a]White-sprouting broccolis are thought to have evolved independently in northern Europe. Their close affinity to winter-hardy cauliflower suggests that the late form may be more correctly regarded as a form of *B. oleracea* var. *botrytis*.

the annual and biennial cauliflowers now cultivated worldwide in temperate climates. Cauliflowers adapted to hot humid tropical conditions have evolved in India during the past 200 years from biennial cauliflowers, mainly of British origins.

Crisp (1982) proposed a taxonomic basis for grouping the various types of cauliflower found in cultivation. He admits this has limitations but at least it gives order where little previously existed (see Table 1.7).

Cauliflower is a biennial or annual herb, 50–80 cm tall at the mature vegetative stage and 90–150 cm when flowering. The root system is strongly ramified, concentrating in the top 30 cm of soil with thick laterals penetrating to deeper layers. The stem is unbranched, 20–30 cm long and thickened upwards. There is a rosette (frame) of 15–25 large, oblong, erect leaves surrounding the compact terminal flower head (curd). Usually, lateral buds do not develop in the leaf axils. The glabrous leaves are almost sessile and coated with a layer of wax; the leaf blade is grey to blue–green in colour with whitish main and lateral veins. Leaves vary in shape from short and wide (40–50 cm × 30–40 cm) with curly edges to long and narrow (70–80 cm × 20–30 cm) with smooth edges. The curd consists of a dome of proliferated floral meristems that are white to cream or yellow in colour, growing on numerous short and fleshy peduncles. The curd varies from a rather loose to a very solid structure, with a flattened to deeply globular shape from 10 to 40 cm in diameter. Young leaves may envelop the curd until a very advanced stage of development is reached. Bolting cauliflower plants often have several flower stalks (see Chapter 2 section, Floral Biology as Related to Controlled Pollination).

Broccoli is an Italian word from the Latin *brachium*, meaning an arm of a branch. In Italy, the term is used for the edible floral shoots on brassica plants,

Table 1.7. Groups of cauliflower as determined by their phylogeny. (After Crisp, 1982).

Group name	Characteristics	Common types
Italian	Very diverse, includes annuals and biennials and types with peculiar curd conformations and colours.	Jezi Naples (= Autumn Giant) Romanesco Flora Blanca
North-west European biennials	Derived within the last 300 years from Italian material.	Old English Walcheren Roscoff Angers St Malo
Northern European Annuals	Developed in northern Europe for at least 400 years. Origin unknown, perhaps Italian or possibly eastern Mediterranean.	Le Cerf Alpha Mechelse Erfurt Danish
Asian	Recombinants of European annuals and biennials, developed within the last 250 years. Adapted to tropical climates.	Four maturity groups are recognized by Swarup and Chatterjee (1972)
Australian	Recombinants of European annuals and biennials, and perhaps Italian stocks; developed during the last 200 years.	Not yet been categorized

including cabbages and turnips, and was originally applied to sprouting forms, but now includes heading types that develop a large, single, terminal inflorescence. Broccoli with multiple green, purple or white flower heads (sprouting broccoli) became popular in northern Europe in the 18th century. Broccoli with a single, main, green head (calabrese – the name has been taken from the Calabria region of Italy) was introduced into the USA by Italian immigrants during the early 20th century (see Fig. 1.4). It has become a popular 'convenience' vegetable, spreading back into Europe from the USA and into Japan and other parts of the Pacific Rim over the past 50 years.

The white-heading forms are also colloquially referred to as cauliflower. Broccoli is often used to describe certain forms of cauliflower, notably in the UK where the term heading or winter broccoli is traditionally reserved for biennial types. The term broccoli, without qualification, is also generally applied in North America to the annual green-sprouting form known in the UK and Italy as calabrese. The term 'sprouting' as used in sprouting broccoli refers to the branching habit of this type, the young edible inflorescences often being referred to as sprouts. The term 'Cape' used in conjunction with broccoli, or as a noun, is traditionally reserved for certain colour-heading forms of *B. oleracea* var. *italica*. A classification of broccoli is given in Tables 1.8 and 1.9.

Green broccoli (the single-headed or calabrese type) differs from cauliflower in the following respects: the leaves are more divided and petiolate,

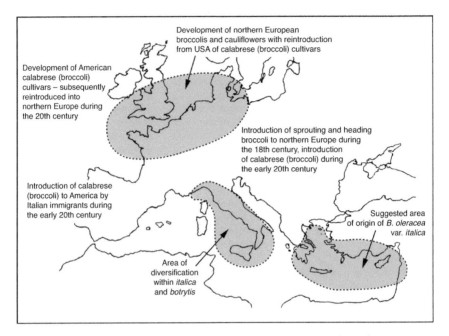

Development of northern European
broccolis and cauliflowers with reintroduction
from USA of calabrese (broccoli) cultivars

Development of American
calabrese (broccoli)
cultivars – subsequently
reintroduced into
northern Europe during
the 20th century

Introduction of sprouting and heading
broccoli to northern Europe during
the 18th century, introduction
of calabrese (broccoli) during
the early 20th century

Introduction of calabrese
(broccoli) to America by
Italian immigrants during
the early 20th century

Suggested area
of origin of *B. oleracea*
var. *italica*

Area of
diversification
within *italica*
and *botrytis*

Fig. 1.4. The evolution of broccoli and cauliflower in Europe and North America. (After Gray, 1982, courtesy Garry Breeze).

Table 1.8. Classification of colour-heading and sprouting broccoli. (After Giles, 1941).

Coloured-heading types	
Dark purple heading	
Copper-coloured or purplish-brown heading	Early, intermediate and late maturing cultivars
Green heading	
Sulfur-coloured or yellowish-green heading	
Sprouting types	
Green sprouting	
Purple sprouting	

and the main head consists of clusters of fully differentiated green or purple flower buds, which are less densely arranged with longer peduncles. Axillary shoots with smaller flower heads usually develop after removal of the dormant terminal shoot. The flower head is fully exposed from an early stage of development. Green broccoli plants carry inflorescences from the lateral branches as well. Sprouting forms of broccoli bear many, more-or-less uniform and relatively small flower heads instead of the single large head of the calabrese type.

Table 1.9. Classification of Italian sprouting broccoli by morphological types. (After Giles, 1944).

Type	Description
Green-sprouting of Naples	Small shoots produced on long stems, considered to be a counterpart to white-sprouting broccoli.
Early summer broccoli of Naples	Shows aggregation of smaller shoots with fewer larger shoots. Stems shorter than the Naples green-sprouting type.
Calabrese	Further reduction in stem length, which gives the plant a heading appearance.

CHINESE KALE: *B. OLERACEA* SSP. *ALBOGLABRA*.   Chinese kale (*B. oleracea* ssp. *alboglabra*) has formed a cultivated stock since ancient times without apparent wild progenitors, but there are possible similarities to *Brassica cretica* ssp. *nivea*. Following early cultivation in the eastern Mediterranean trade centres, it could have been taken to China. The lines cultivated in Europe may have lost their identity through uncontrolled hybridization. Recently, much horticultural attention has focused on Chinese kale. Chinese kale is now a cultigen native to southern and central China. It is popular and widely cultivated throughout China and South-east Asia and is used as leaves in salads and other dishes. The flower bud, flower stalk and young leaves are consumed. A classification into five groups, which vary in flower colour from white to yellow and in depth of green coloration in the leaves and their shape, was produced by Okuda and Fujime (1996) using cultivars from Japan, Taiwan, China and Thailand as examples.

It is an annual herb, up to 0.4 m tall during the vegetative stage and reaching up to 1–2 m at the end of flowering. All the vegetative organs are glabrous and glaucous. The narrow single stem forks at the top. Leaves are alternate, thick, firm and petiolate and leaf blades ovate to orbicular–ovate in shape. The margins are irregularly dentate and often undulate and characteristically auriculate at the base or on the petiole. The basal leaves are smaller and sessile without auricles. The inflorescence is a terminal or axillary raceme 30–40 cm long, with pedicels 1–2 cm long (see Chapter 2 section, Floral Biology as Related to Controlled Pollination). The taproot is strongly branched.

OTHER KALES: *B. OLERACEA* SSP. *ACEPHALA*.   Many groups are distinguished: borecole or curly kale, collard, marrow-stem kale, palm tree kale, Portuguese kale and thousand-headed kale. Kales are ancient cole crops, closely related to the wild forms of *B. oleracea*, and many distinctive types were developed in Europe. There are residual populations of the original progenitors, such as the wild kale of Crimea, variously ascribed to *B. cretica* and *B. sylvestris* but now identified as a hairy form of *B. rupestris–incana*. It is suggested that, as a consequence of trade around the Mediterranean, this form was transferred to the

Crimea and is evidence of early widespread cultivation of *B. rupestris–incana* types. A similar relic population exists in the wild kale of Lebanon, inhabiting the cliffs near Beirut, which is morphologically similar to *B. cretica* ssp. *nivea*. Both are possible evidence for widespread trade by the earliest Mediterranean civilizations that moved the botanical types around. This allowed interbreeding resulting in the widening diversity of horticultural crops that were artificially segregated from the botanical populations.

Brussels sprouts, the kales and kohlrabi are part of a similar group of polymorphous, annual or biennial erect herbs growing up to 1.5 m tall, glabrous and often much branched in the upper parts. In particular, kales are extremely variable morphologically, most closely resembling their wild cabbage progenitors. The stem is coarse, neither branched nor markedly thickened and 0.3–1 m tall. At the apex is a rosette of generally oblong, sometimes red-coloured leaves. Sometimes, the leaves are curled. This is caused by disproportionately rapid growth of leaf tissues along the margins. In borecole or curly kale, the leaves are crinkled and more-or-less finely divided. Often green or brownish-purple and they are used as vegetables. Collards have smooth leaves, usually green, and they are most important as forage in western Europe. Marrow-stem kale has a succulent stem up to 2 m tall and is used as animal forage. Palm tree kale is up to 2 m tall with a rosette of leaves at the apex – it is mainly used as an ornamental. Portuguese kale has leaves with succulent midribs that are used widely as a vegetable. Thousand-headed kale carries a whorl of young shoots at some distance above the soil. Together they are more-or-less globular in outline, and this type of kale is mainly used as forage.

KOHLRABI: *B. OLERACEA* VAR. *GONGYLODES*.   Kohlrabi first appeared in the Middle Ages in central and southern Europe. The crop has become well-established in parts of Asia over the last two centuries and is economically important in China and Vietnam. Kohlrabi are biennials in which secondary thickening of the short stem produces the spherical edible portion, 5–10 cm in diameter and coloured green or purple. The leaves are glaucous with slender petioles arranged in compressed spirals on a swollen stem.

WHITE-HEADED CABBAGE: *B. OLERACEA* VAR. *CAPITATA* F. *ALBA*; RED-HEADED CABBAGE: *B. OLERACEA* VAR. *CAPITATA* F. *RUBRA*; SAVOY-HEADED CABBAGE: *B. OLERACEA* VAR. *SABAUDA*.   These varieties were defined by Nieuwhof (1969). Heading cabbage are the popular definitive image of vegetable brassicas in Europe, indeed the terms 'cabbage garden' and 'vegetable garden' were synonymous in some literature.

Early civilizations used several forms of 'cabbage' and these were probably refined in domestication in the early Middle Ages in north-western Europe as important parts of the human diet and medicine and as animal fodder. It is suggested that their progenitors were the wild cabbage (*B. oleracea*) feral forms, which are now found on the coastal margins of western Europe, especially England and France, and leafy, unbranched and thick-stemmed kales that had been disseminated by the Romans. Pliny described methods for the preservation of cabbage and sauerkraut was of major importance as a source of vitamins in winter and on long sea journeys. In most cabbages it is chiefly the leaves that are used. Selection pressure in cultivation has encouraged the development of closely overlapping leaves forming tight compact heads, the heart or centre of which is a central undeveloped shoot surrounded by young leaves. Head shape varies from spherical to flattened to conical. The leaves are either smooth, curled or savoyed (Milan type). Seed propagation of cabbage is relatively straightforward and in consequence large numbers of localized regional varieties, or landraces, were selected with traits that suited them to particular climatic and husbandry niches such as: Aubervilliers, Brunswick, de Bonneuil, Saint Denis, Strasbourg, Ulm and York. From the 16th century onwards, European colonists spread cabbages worldwide. Scandinavian and German migrants introduced cabbages to North America, especially the mid-western states such as Wisconsin. In the tropics, cultivation is usually restricted to highland areas and to cooler seasons. White-heading cabbage is especially important in Asia and India. The majority of cultivars are now F_1 hybrids coming from a circumscribed group of breeders using similar parental genotypes. Forms derived originally from the Dutch White Langedijk dominated the market for storage cabbage and more recently fresh white cabbage in supermarkets. Refinement of Savoy types through breeding of F_1 hybrids has expanded the range now on offer. Large-headed cabbages with ample anthocyanin pigmentation are found in Shetland, used as winter sheep fodder or part of the human diet in harsh conditions.

Cabbages are biennial herbs that are 0.4–0.6 m tall at the mature vegetative stage and 1.5–2.0 m tall when flowering in the second year. Mature plants have a ramified system of thin roots, 90% in the upper 0.2–0.3 m of the soil, but some laterals penetrate down to 1.5–2 m deep. Stems are unbranched, 20–30 cm long, gradually thickening upward. The basal leaves form in a rosette of 7–15 sessile outer leaves each 25–35 cm × 20–30 cm in size. The upper leaves form in a compact, flattened, globose to ellipsoidal head, 10–30 cm in diameter, composed of a large number of overlapping fleshy leaves around the single growing point. These leaves are grey to blue–green, glabrous and coated with a layer of wax, on the outside of the rosette, and light green to creamy white inside the head, especially with white-headed cabbage. The leaves are red–purple in red-headed cabbage and green to yellow–green and puckered in Savoy-headed cabbage. The inflorescence is a 50–100 cm bractless long raceme on the main stem and on axillary branches of bolted plants. Germination is epigeal and the seedlings have a thin taproot and cordate cotyledons; the first true leaves are ovate with a lobed petiole (see Chapter 2 section, Floral Biology as Related to Controlled Pollination).

Novel crop types

Breeders continuously select for novel characteristics desired by consumers. These include miniature cauliflower, sprouting broccoli and cauliflower and exploit the vast variation in colour available within brassicas (Dixon, 2017). Crossing between different morphological types also offers the development of novel forms. Crosses between stable and uniform kale and Brussels sprout parent lines, resulted in the development of the kalette (also known as kale sprout or flower sprout). Varieties were selected that combine the best characters of both crop types while eliminating undesirable characters. The kalette is a tender Asian-type vegetable without the tough stalk of kale, but with improved flavour due to removing the bitterness sometimes found in both kale and Brussels sprouts. The open rosette of the kalette allows for quick cooking. Kalettes grow very much like Brussels sprouts and are particularly winter hardy without many cultivation issues.

Hybridization between taxa

Crossings occur even between distant taxa of the Brassicaceae, giving at least semi-fertile hybrids, and this may be analogous to the means by which genetic mixing between wild forms led to the horticultural types grown commercially today. Meiotic pairing is normal and indicates close identity particularly throughout the $2n = 18$ forms. Although pollen fertility and seed set are variable there is usually enough to provide for the survival of further generations.

These characteristics indicate that where races, varieties or species are cultivated in close proximity, crossings will occur. Self-sterility is found in many of the taxa. It is far from absolute but sufficiently robust to ensure high proportions of outbreeding. Outbreeding normally results from a high frequency of similar S genes between individuals belonging to the same population (see Chapter 2 section, Hybrid Production: Self-incompatibilty). It is concluded that present day cultivars include much introgressed genetic material derived from other cultivated or wild forms. Consequently, it is important to understand and use the historical literature that describes crops derived from *B. oleracea* alongside that derived from genetical and taxonomic sources in order to interpret the status of modern forms and hybrids.

Comparative taxonomy using ancient and medieval literature and science

Some syntheses of the literature have been attempted specifically for the Brassicaceae, notably that of Toxopeus (1974). Greek writers, especially Theophrastos (370–285 BCE), discussed cole crops. It is evident that 'branching' types were known at that date, and these may have resembled bushy kales that were also found in uncultivated ground. Possibly, this indicates the domestication of *B. cretica*. Comments are found describing bushy kales with curled leaves, thus both forms may have been present and undergoing hybridization before spreading to other parts of Europe. The Romans (Cato, 234–149 BCE; Pliny the Elder, 23–79 CE) knew of both stem kales and heading cabbage which were cultivated together. Since seed would be produced locally, hybrids could form and the best selected for further improvement thereby developing local cultivars. Highly prized types might then have spread further as items of trade. Zeven (1996) suggested that the 'perpetual kale' (*B. oleracea* var. *ramosa*) was the Tritian kale of the Romans (referred to by Pliny in 70 CE), which they took throughout their empire. Some relic populations are still grown in various parts of Europe (Belgium, England, France, Ireland, Netherlands, Portugal, Scotland) and in Brazil, Ethiopia and Haiti. The crop is known as 'Hungary Gap' in England and 'Cut and Come Again' in Scotland. Plants reach up to 3 m in height, some forms appearing to have lost the capacity to flower, as in those found in the Dutch province of Limburg, resulting from long selection pressure for leafiness and multiple branching habit. The patchwork of dissemination in Europe suggests previously widespread distribution by traders. A picture of a 'kail stock' (cabbage stalk) by Richard Waite painted in 1732 and entitled 'The Cromartie Fool' (the Earl of Cromartie's Jester) is owned by the Scottish National Portrait Gallery.

Cole and neep crops were grown throughout Europe (Sangers, 1952, 1953). Analysis of archives indicates that cole crops were well recognized in the early 14th century to the extent that the name *Coolman* or *Coelman* (cabbage-man) were common surnames while the term *coeltwn* (modern Dutch = *kooltuin*) indicated a cabbage garden. Trade was established between the Low Countries and England for the export of cabbages by the 1390s. Dodenaeus had, by 1554 (Zeven and Brandenburg, 1986), classified cole crops as white cabbage, Savoy cabbage, red cabbage and curly kale and had recognized the turnip which, in 1608, he had differentiated into flat-rooted and long-rooted forms.

Useful evidence of the forms of brassicas in cultivation comes from studies of the Dutch and Flemish painters of the 15th and 16th centuries where red and white cabbages and cauliflowers figure prominently. Only turnips (*B. rapa*) are seen in these paintings with an apparent absence of swede (*B. napus*) at this time (Toxopeus, 1974, 1979, 1993). Some evidence is available for the presence of radishes in these paintings, but unfortunately there is also conflict with similarities to turnips. However, it is probable that the French 'icicle' radish can be distinguished. It is likely that all subgroups of *Brassica*: kohlrabi, cauliflower and sprouting broccoli were developed by medieval times and spread westwards and northwards. While further south and south-east other *Brassica* groups were developed for cultivation but generally with the exception of *B. oleracea* ssp. *alboglabra*.

Diversification of *Brassica* crops is well demonstrated in Portugal where original cole crops were introduced by Celtic tribes over several centuries before the Common Era (CE) began and in advance of the Roman conquests. These developed into the Galega kale (*B. oleracea* ssp. *acephala*), Tronchuda cabbage

(*B. oleracea* var. *tronchuda*) and Algarve cabbage (*B. oleracea* var. *capitata*). The Tronchuda types are vigorous growing collard-type plants with a small, loose head and large, thick leaves; while the Galega types are leafy and headless plants, with large leaves having long petioles and a single indeterminate stem which can attain 2–3 m before bolting (Monteiro and Williams, 1989). These crops, together with vegetable rape (*B. napus* var. *napus*), Nabo (turnip) (*B. rapa* var. *rapa*), Nabica (turnip greens) (*B. rapa* var. *rapa*) and Grelos (turnip tops) (*B. rapa* var. *rapa*), form an essential part of the rural diet in Portugal. There are numerous landraces of these crops distributed throughout Portugal which have very low within-population uniformity due to the allogamic pollination mechanism, associated with the poor isolation used by farmers for seed production. The high levels of variability in shape, size, colour, taste, earliness and pest and pathogen resistance in these populations constitutes an immense reservoir of diversity for breeding purposes.

Comparative morphology gives further information on the origins of brassicas, an important character is the greyish surface texture of the west European *B. oleracea* found principally in headed cabbage and Brussels sprouts. The strong, dominating, central structure of stem kales is found in the *B. rupestris–incana* from which primary origin could be inferred. *Brassica cretica* is probably the origin of the bushy kales since they share common branching, shrubby habit and fleshy leaves. The presence of white flowers may have been derived from *B. oleracea* ssp. *alboglabra* and *B. cretica* or combinations between them.

Local stocks (landraces)

Various localities still utilize old cultivars and landraces although there is great economic pressure for these to be supplanted by high-yielding, standardized and often hybrid cultivars. Information concerning the older open-pollinated types is fragmentary but of great value in understanding the history of *Brassica* in cultivation. Around the Aegean, primitive kales, similar to *B. cretica*, are still cultivated, some with branching inflorescences similar to sprouting broccoli. Even wild types may be utilized in some island villages as salad vegetables.

In Bosnia, Croatia, Montenegro and Serbia, wild type kales grow on field margins, waste areas and building sites and are still used as animal fodder. Two forms are apparent: a tall single-stem type, similar to marrow-stem kale, and a more branching type with high anthocyanin content, often with a habit similar to cabbage. Neither of these produce heads but are possibly early kale types similar to the wild progenitors.

In general, it is inferred that the west European headed-cabbage types derived from *B. oleracea* on the grounds of morphology with cross-fertilization with Roman kales. The Savoy type is possibly a result of further introgression between other coles. The branching bushy kales possibly originate from *B. cretica* with perhaps 2000 years of hybridization with other forms. Stem kales may well originate from the *B. rupestris–incana* complex in the Adriatic or more southerly parts of Italy. Hybridization to cabbages and perhaps *B. cretica* will have added to variation and type differentiation. Origins for the inflorescence kales, cauliflower and broccoli are still unresolved. Their rapid growth and morphology may have suggested that *B. cretica* is involved but the leaf characteristics would also indicate that *B. oleracea* is a progenitor. The rapid flowering *B. oleracea* ssp. *alboglabra* possibly segregated from *B. cretica* ssp *nivea*. Then, via cultivation and trade, it spread from ancient Greece into the eastern Mediterranean and then further eastwards.

The relationships between regional groups of Italian landrace cauliflower (*B. oleracea* var. *botrytis*) and broccoli (*B. oleracea* var. *italica*) have recently been unravelled (Massie *et al.*, 1996). A large pool of genetic diversity exists within cauliflower and broccoli grown throughout Italy following centuries of selection for local conditions and preferences. Different provinces of Italy have been associated with specific variant types, for example: Romanesco cauliflower in the Lazio region; Di Jesi, Macerata and Tardivo di Fano varieties in the Marche region; Cavolfiore Violetto di Sicilia (Sicilian purple cauliflower) in Sicily, except for the Palermo province where a green cauliflower is typical. Intermediate forms are also reported, such as between the Macerata and Sicilian purples (Gray and Crisp, 1985).

Brassica rapa group (*n* = 18) – the Asian group

The International Code of Nomenclature for Algae, Fungi, and Plants (previously known as the International Code of Botanical Nomenclature: Gilmour *et al.*, 1969; Stafleu *et al.*, 1972) rules that the author who first combines taxa of similar rank bearing epithets of the same date chooses one of them for the combined taxon. Metzger (1833) first united *B. rapa* and *Brassica campestris* of 1753 under *B. rapa* as used in this text.

Brassica rapa (AA, *n* = 10, genome size ~350 Mb) and the amphidiploids *B. carinata* (BBCC, *n* = 17, genome size ~1300 Mb), *B. juncea* (AABB, *n* = 18, genome size ~930 Mb) and *B. napus* (AACC, *n* = 19, genome size ~1130 Mb), and *R. sativus* (RR, *n* = 9) are grown extensively throughout Asia with a huge number of distinct varieties. Differentiation results from selection both in nature and by the forces of cultivation; this is a two-way process with great intermingling and recombination. Several distinct groupings are distinguished, especially in the headed Chinese cabbages and Japanese radishes.

Initially, the centre of origin of *B. rapa* is postulated as the Mediterranean from where it spread northwards and eastwards to Germany and into central Europe, and eventually towards Asia (Mizushima and Tsunoda, 1967). Along the way, great local variation in cultivation developed. The plant reached China via Mongolia as an agricultural crop and was introduced to Japan either via China to the western part of the country or via Siberia and into the eastern part of the country. In India, *B. rapa* and *B. rapa* var. *sarson* (derived from the former species) are used as oil plants, with no records of wild progenitors having been found. Phylogenetic analysis by Qi *et al.* (2017), using 126 global accessions with more than 31,000 genome-wide single nucleotide polymorphism (SNP) markers, suggested five distinct genetic groups, while also supporting a European–Central Asian origin followed by an eastward expansion. Evidence suggested that pak choi, Chinese cabbage and yellow sarson are monophyletic groups while the oil-type *B. rapa* ssp. *oleifera* and brown sarson were polyphyletic (see Fig. 1.5).

Based on morphology, seven vegetable groups can be differentiated: var. *campestris*, var. *pekinensis*, var. *chinensis*, var. *parachinensis*, var. *narinosa*, var. *japonica* and var. *rapa*. The leaf vegetables are thought to have developed after they entered China, with the exception of the var. *japonica* group which has a common ancestor in the oil rapes. Parallels are frequently drawn by the development of *B. oleracea* vegetables in Europe and *B. rapa* vegetables in Asia. The headed Chinese cabbages tend to dominate the use for cultivation, but there are parts of China itself where *B. juncea* is prominent and in central China the forms var. *chinensis* and var. *narinosa* are of importance largely for climatic reasons.

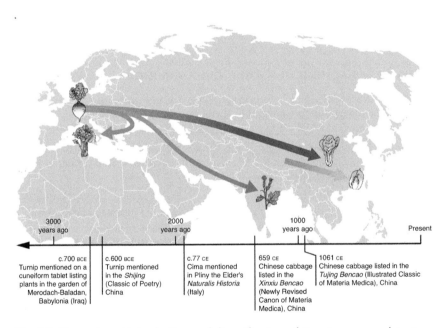

Fig. 1.5. The eastward introduction and diversification of *Brassica rapa* as shown by the coloured arrows. The *Brassica rapa* genetic groups are represented by: European *Brassica rapa* – green (on the left side of the map); rapini (ssp. *sylvestris*) – orange; yellow sarson (ssp. *trilocularis*) – blue; pak choi (ssp. *chinensis*) – red; and Chinese cabbage (ssp. *pekinensis*) – yellow. A timescale of occurrence recorded within historical written records is shown on the arrow below. (After Qi *et al.*, 2017, courtesy Garry Breeze).

There are close parallels between the manner of differentiation and selection in cultivation of *B. oleracea* in Europe and of *B. rapa* and *B. juncea* in Asia. Species may have developed along similar lines in a single region as, for example, the trend towards entire glabrous leaves in *B. rapa*, *B. juncea* and *R. sativus*. It is possible to find atypical characters, such as thick stems, which are poorly represented in the putative parents but which have formed extensively in the progeny. Thus neither *B. rapa* (A genome) nor *B. nigra* (B genome) possess thick stems but there are forms of *B. juncea* (AB genome) which are far more developed than the parents and are similar to *B. oleracea* (C genome) (see Table 1.2).

Crop plants of *B. rapa*, *B. juncea* and *R. sativus* are used as leaf or root vegetables in Japan and China, whereas in India they are developed as oil plants and *R. sativus*, in particular, has produced very impressive siliquae. Thus, differentiation has arisen over time in the respective directions of cultivation for either vegetative or reproductive organs. In China, differentiation of headed Chinese cabbages has produced types adapted to several climatic zones. In the north, cold-tolerant forms are used in the summer with a similar but separate segregation in the south for types suited to winter culture. Distinct types have been developed separately in the two areas capable of accomplishing similar tasks.

Leaf greens types for crops have been developed in the main from *B. rapa* and *B. juncea* to the exclusion of radishes and turnips, which could form the same product but with a lower level of efficacy. Vegetables which are eaten in high volume as fresh forms tend to have been differentiated into the greatest numbers of improved types, whereas those used for preserves and processing have less variation produced by segregation in cultivation. Fig. 1.6 summarizes leaf greens developed from both *B. oleracea* ssp. *alboglabra* and *B. rapa*.

Crops developed within *Brassica rapa* and allies

CAISIN: SYN. VAR. *B. PARACHINENSIS*; *B. CHINENSIS* VAR. *PARACHINENSIS*.　　　Caisin (also known as choy sum) is generally thought to have differentiated along with the leaf neeps (Chinese cabbage, pak choi) from oil-yielding turnip rapes, which were introduced into China from the Mediterranean area through western Asia or Mongolia. Caisin originated in middle China where it was selected and popularized for its inflorescences. It may be seen as a parallel variation in *B. rapa* comparable with Chinese kale (*B. oleracea* cv. group Chinese kale) in *B. oleracea*.

Where headed Chinese cabbage is grown, caisin and the non-heading leaf neeps (e.g. pak choi) are also indispensable vegetables. Caisin is cultivated in southern and central China, in South-east Asian countries, such as Indonesia, Malaysia, Thailand, Vietnam and in other parts of the region, and in areas of western India.

The var. *parachinensis* is probably a derivative of var. *chinensis*. Used for the flower stalk and very popular in central China, the plants will bolt readily and the time from seeding to harvest can be 40–80 days depending on the cultivar

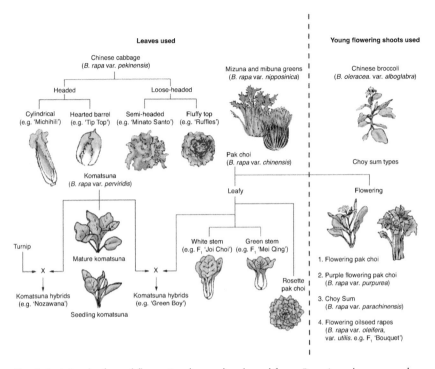

Fig. 1.6. Asian leafy and flowering forms developed from *Brassica oleracea* and *Brassica rapa*. (After Larkcom, 1991, courtesy Garry Breeze).

used (Herklots, 1972). The var. *campestris* is the most primitive of vegetables, closely resembling oil rape and used for flower stalks and rosette leaves.

Caisin is an annual, taprooted herb, 0.2–0.6 mm tall with a usually open, erect or sometimes prostrate growth habit. The stems are normally less than 1 cm diameter, small in comparison with other leafy cabbages, and usually profusely branched. There are few leaves in the rosette, with only one or two leaf layers. There are long petiolate, spathulate or oblong, bright green stem leaves. These may be glabrescent to glabrous, green to purple–red and finely toothed when young. Lower stem leaves are ovate to nearly orbicular and central stem leaves ovate to lanceolate to oblong, with long and narrow-grooved petioles that are sometimes obscurely winged. Upper stem leaves gradually pass into narrow bracts. Inflorescences form as a terminal raceme, elongating when in fruit.

CHINESE CABBAGE: SYN. *B. PEKINENSIS*; *B. CAMPESTRIS* SSP. *PEKINENSIS*; *B. RAPA* SSP. *PEKINENSIS*.   Chinese cabbage is a native of China. It probably evolved from the natural crossing of pak choi (non-headed Chinese cabbage), which was cultivated in southern China for more than 1600 years, and turnip, which was grown in northern China. Much of its variety differentiation took place in China during the past 600 years. Its derivatives were introduced into Korea in the 13th century, the countries of South-east Asia in the 15th century and Japan in the 19th century. An illustration of the headed shape of Chinese cabbage with wrapping leaves was first recorded in China in 1753. Chinese cabbage is now grown worldwide.

The origins of var. *pekinensis* may correlate with the oil rapes of northern China, developing first as a headed type where the lower parts of the plant were swollen and latterly where the entire plant developed a headed form. Several variants exist with head forms distinguished as either 'wrapped-over' or 'joined- up'. In the former, the leaves overlap at the top of the conical head which does not happen in the 'joined-up' or multi-leaved types. The 'wrapped-over' forms are heavy-leaved, early maturing and round-headed with an adaptation to warmer climates.

The 'joined-up' types are late maturing with firm texture and adaptation to cooler climates. It is possible that both forms originated first in the Shandong Peninsula and the 'joined-up' types spread northwards from there, while the 'wrapped-over' forms developed in a southerly direction. The latter eventually differentiated into a more 'southerly' type which has very early maturity, a small head and heat tolerance (see Fig. 1.7).

Shapes of Chinese cabbage are classified on the degree of heading, as non-, half- and completely headed types with further refinements to give long, short, tapered, round-topped, wrapped-over and joined-up forms. The Chinese largely use these cabbages in the autumn for preserving and processing. In Japan, they are freshly boiled or salt-pickled with early and late types but there is insufficient production to meet consumer demands. Consequently, crops are distributed between regions in Japan according to seasonal availability. The crop probably entered Japan from Shandong in the Meiji era (1868–1912), being well adapted to the Japanese climate.

Chinese cabbage is a biennial herb, cultivated as an annual, 0.2–0.5 m tall during the vegetative stage and reaching up to 1.5 m in the reproductive state. The taproot and lateral roots are prominent in older plants forming an extensive, fibrous, finely branched system. During vegetative stages the leaves are arranged in an enlarged rosette. This forms a short, conical, more-or-less compact head, with ill-defined nodes and internodes and alternate heading and non-heading leaves; leaves are 20–90 cm × 15–35 cm in size. Leaves vary in shape with different growth stages, the dark-green outer leaves are narrowly ovate with long with winged petioles. Inner heading leaves are broad, subcircular and whitish-green. The flowering stem carries lanceolate leaves, much smaller than heading leaves, with broad, compressed petioles and blades clasping the stem.

The origin of *B. rapa* is not known. The wide variation of neep crops evolved in different parts of the Eurasian continent. Besides Chinese cabbage, pak choi and caisin, the leafy vegetable types or leaf neeps comprise cultivar

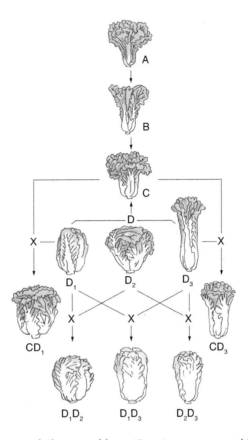

Fig. 1.7. The evolution of Chinese cabbage (*Brassica rapa* ssp. *pekinensis*), where:
A = var. *dissoluta*, B = var. *infarcta*, C = var. *laxa*, D = var. *cephalata*, D_1 = f. *ovata*,
D_2 = f. *depressa*, D_3 = f. *cylindrica*, CD_1 = var. *laxa* × f. *ovata*, CD_3 = var. *laxa* ×
f. *cylindrica*, D_1D_2 = f. *ovata* × f. *depressa*, D_1D_3 = f. *ovata* × f. *cylindrica*,
D_2D_3 = f. *depressa* × f. *cylindrica*. (After Dickson, courtesy Garry Breeze).

groups Mizuna, Neep Greens and Tatsoi, developed in temperate regions of
Asia.

MIZUNA: SYN. *B. JAPONICA*; *B. CAMPESTRIS* SSP. *NIPPOSINICA*; *B. RAPA* SSP. *NIPPOSINICA*.
The var. *japonica* is a vegetable unique to Japan, carrying many basal
branches and leaves. This plant resembles *B. juncea* since the inflorescence
stalk has no leaves, the petioles and siliquae are slender, and the seeds are
small (Matsumura, 1954). Possibly, this variety is derived from hybridization
between *B. rapa* and *B. juncea*. Grouping of cultivars rests on the level of
dissection of the leaves.

NEEP GREENS: SYN. *B. PERVIRIDIS*.   See Bailey (1940) for certain forms
(komatsuna).

TATSOI: SYN. *B. NARINOSA*: *B. CAMPESTRIS* SSP. *NARINOSA*; *B. RAPA* SSP. *NARINOSA*. The var. *narinosa* developed in middle China. It is very cold-tolerant with thick leaves and crisp petioles that are used as boiled vegetables.

VEGETABLE TURNIP: SYN. *B. CAMPESTRIS* SSP. *RAPA*; *B. CAMPESTRIS* SSP. *RAPIFERA*; *B. RAPA* SSP. *RAPA*.   The many forms of vegetable turnip are highly regarded in Japan as well as in Europe where also fodder turnip used to be a very popular crop (root neeps). Oilseed types (seed neeps), grown for rape oil, are important in India and Canada.

The turnip is the oldest *B. rapa* crop on record. It was described in the ancient Greek times of Alexander the Great, whose empire included the Middle East and Persia, from where it may have found its way to eastern Asia. Quite independently of each other in Europe and in Japan, well-defined polymorphic groups of vegetable turnips had been created by the 18th century.

Although turnips (var. *rapa*) are grown around the world, Japan is one of the major areas of variety development, while in eastern Asia radishes have predominated. Turnips grow best in cool climates. Worldwide, turnips can be classified into: Teltou turnips, West European turnips with dissected leaves, Anatolia and Palestine turnips, Russian turnips of the Petrovskij type, Asiatic turnips with dissected leaves of the Afghan type and a subgroup of Asiatic turnips of the Afghan type with entire leaves, Japanese entire-leaved turnips with entire glabrous leaves, and European entire-leaved turnips with pubescent leaves. (Sinskaia, 1928).

The Afghan types are thought to be close to the original var. *rapa* with glabrous dissected leaves, ascending rosette and small taproots. Japanese forms possibly developed from these with entire leaves, no hairs and well-formed taproots demonstrating the effects of selection in cultivation. A further classification (Shebalina, 1968) identified Anatolia, Central Asia, Europe, Iraq and Japan types. The latter includes types derived either indigenously in Japan or developed from European types with several intermediates. Characters used to differentiate types include the form of the seed coat, which either swell (type A) or form a thick layer of epidermal cells (type B) (Shibutani and Okamura, 1954).

The cultivar groups, Mizuna, Neep Greens, Tatsoi and vegetable turnips, are annual or biennial herbs with stout taproots, often becoming fusiform to tuberous (turnips). Stems are erect, branched and up to 1.5 m tall. Leaves are very variable, depending on the cultivated type. They grow in a rosette during the vegetative phase. The basal leaves are petioled, bright green and lyrate–pinnatipartite, dentate, crenate or sinuate, bearing large terminal lobes and up to five pairs of smaller lateral lobes. Lower cauline leaves are sessile, clasping and pinnatifid, and upper cauline leaves are sessile, clasping, undivided, glaucous, and entire to dentate.

Cultivar group Mizuna consists of spontaneously tillering plants with pinnate leaves (mizuna cultivars) or entire leaves (mibuna cultivars). The cultivar group Neep Greens comprises essentially non-heading plants including crops such as komatsuna, zairainatane, kabuna and turnip greens. The cultivar group Tatsoi typically grows a flat rosette of many, small dark-green leaves. The cultivar group vegetable turnip consists of forms in which the storage organ (swollen hypocotyl and root (i.e. the turnip)) is used as a vegetable as well. Turnip roots vary widely in shape, from flat through globose to ellipsoid and cylindrical, blunt or sharply pointed, flesh white, pink or yellow, apex white, green, red, pink or bronze. All these characteristics may occur in cultivars in any imaginable combination.

PAK CHOI: SYN. *B. CHINENSIS*; *B. CAMPESTRIS* SSP. *CHINENSIS*; *B. RAPA* SSP. *CHINENSIS*. Pak choi evolved in China and its cultivation was recorded as far back as the 5th century CE. It is widely grown in southern and central China, and Taiwan. This group is a relatively new introduction in Japan where it is still referred to as 'Chinese vegetable'. It was introduced into South-east Asia in the Malacca Sultanate in the 15th century. It is now widely cultivated in the Philippines and Malaysia, and to a lesser extent in Indonesia and Thailand. In recent years, it has gained popularity in North America, Australia and Europe.

The var. *chinensis* is the 'large white cabbage' of China, while var. *narinosa* and the like are 'small white cabbage' (Wu, 1957). These varieties are particularly important where headed forms are not grown and are extensively used in Malaysia and Indonesia. As with var. *pekinensis*, these forms are differentiated in China from oilseed rape with distinctions between type related to the width and cross-section of the petioles. Those with narrow, flat petioles are nearest to the parental origin forming both branching and non-branching types. Petiole colour varies from white to green, the former being used in autumn while the latter is more cold-tolerant and resistant to bolting. Frequently, the Chinese forms are larger than those grown in Japan.

Pak choi is an erect biennial herb, cultivated as an annual. In the vegetative state it is glabrous, dull green and 0.15–0.3 m tall and in the generative stage reaches 0.7 m The leaves are arranged spirally, not forming a compact head but spreading in groups of 15–30. Petioles are enlarged, terete or flattened, 1.5–4 cm wide and 0.5–1 cm thick, growing in an upright manner forming a subcylindrical bundle. Each white, greenish-white to green leaf blade is orbicular to obovate, 7–20 cm × 7–20 cm. Stem leaves are entire, tender, smooth or blistering, shiny green to dark green and auriculate-clasping.

Brassica juncea

BRASSICA JUNCEA: SYN. *SINAPIS JUNCEA*; *S. TIMORIANA*; *B. INTEGRIFOLIA*.   *Brassica juncea* crops are grown worldwide, from India to North Africa, to Central Asia, to Europe and North America. The exact origin is unknown, but as an amphidiploid it seems logical that it originated in an area where the parental species *B. nigra* and *B. rapa* overlap in their distribution (e.g. Central Asia). It is generally agreed that the primary centre of diversity of *B. juncea* is north-west

India, including the Punjab and Kashmir, with secondary centres in central and western China, eastern India and Myanmar, and Anatolia (via Iran). In *B. juncea*, two types of mustards with varying uses have evolved – these are oilseed types and vegetable types. The oilseed types (oilseed mustard) are particularly important in India, Bangladesh and China. The vegetable types contain forms with edible leaves (leaf mustard), stems (stem mustard) and roots (root mustard).

The vegetable mustards are widely cultivated in Asian countries. The highest degree of variation occurs in China, which is regarded as the primary centre of varietal differentiation. The early Chinese traders might well have carried the crop into South-east Asia, whereas the appearance of *B. juncea* near European ports suggests a connection with grain imports. It has also been suggested that Indian contract labourers brought it to the Caribbean. In South-east Asia it is the leaf mustards which are the most common.

Brassica juncea is an erect annual to biennial herb, 0.3–1.6 m tall. Normally unbranched, sometimes with long, ascending branches in the upper parts, in appearance it is subglabrous and subglaucose. The taproot is sometimes enlarged (root mustard). Leaves are highly variable in shape and size, either pinnate or entire with petioles that are pale to dark green, smooth or pubescent, and heading or non-heading.

Oilseed mustards are grown as crops worldwide but are chiefly found in a belt from India to Europe. As vegetables, they appear mainly in Asia, and especially China, with a great diversity of types. In India, they are spice plants used in curries while the young leaves may be eaten as greens. Most of the *B. juncea* vegetables of Asia have been developed as pickles hence their use in Europe is limited.

In China, vast amounts of different types of mustard are consumed. Where leaf vegetables of *B. rapa* and *R. sativus* are difficult to grow, as in southern China, *B. juncea* is also used as a leaf vegetable. Growth is slow with tolerance to high temperatures and humidity. Long days are required to stimulate flowering with some varieties being days neutral. This is in contrast to *B. rapa* and explains the distribution of crop types. The greatest differentiation of *B. juncea* for greens is found in China, especially in Sichuan Province. Sinskaia (1928) differentiated on leaf characters in relation to geographical occurrence with an eastern Asian group with bipinnate leaves, a Central Asian Group with entire leaves and a Chinese Group with crisp leaves. Variations of stem characters were used to differentiate forms in Taiwan (Kumazawa and Akiya, 1936), some with compact stems (Kari-t'sai) and others with enlarged stems (Ta-sin-t'sai). Indian types are characterized by their content of the oil 3-butenyl isothiocyanate, which is absent from Chinese types (Narain, 1974), and with the recognition of early bolting north-eastern, late bolting north-western and intermediate types. (Singh *et al.*, 1974). In Japan, succulent leafy types were reputedly introduced in the Meiji era (1868–1912).

Seven types may be distinguished by geographical and botanical characters:

Hakarashina group (*B. juncea*) with pinnate leaves, distributed in India, Central Asia and Europe. They have course-textured, dissected leaves like those of radishes. The majority are cultivated for oil but may form greens when young.

Nekarashina group (*B. juncea* var. *napiformis*) with enlarged roots, distributed in Mongolia, Manchuria and north China. The taproot is well-enlarged like those of turnips. The crop is absent in Japan. Variation in root character is simple as the root shape is solely conical and root colour is equally upper green and lower white. This group is acclimatized to cold unlike most *B. juncea* crops and hence is an important Chinese crop.

Hsueh li hung (*B. juncea* var. *foliosa*) and Nagan sz kaai (*B. juncea* var. *japonica*) group with glabrous leaves and many branches, distributed in middle and northern China. They are characterized by bipinnate leaves dissected in thread-like segments like those of carrots which form a cluster of rosettes. Dissection is deeper in Hseuh li hung and rosettes more vigorous in Nagan sz kaai. Cold tolerance is relatively high, the leaves are pubescent, dark green and pungent to taste.

Azamina group (*B. juncea* var. *crispifolia*), ecologically similar to Takana, but the leaves are firmly dissected, like parsley, with a general appearance of curly kale. They are used for salads and as ornamentals in the USA, possibly as progeny of the introduced cultivar Fordhook Fancy.

Takana group (*B. juncea* var. *integrifolia*) with entire succulent leaves distributed in southern and middle China, and South-east Asia towards the Himalayas (Kumazawa, 1965). They have either dissected or entire leaves, the latter being more frequent, forming vigorous large plants up to 1 m high. The leaves are soft in texture with midribs varying from narrow to wide and crescent shaped in transverse section, internodes are extended. The leaves are also glabrous or crisp and green to dark green. Cultivated as specialized regional products in Japan.

Takana group (*B. juncea* var. *rugosa*), large plants with leaves that have wide, flat, entire midribs and are extremely succulent. The Paan sum of Guangdong Province has a moderately wrapped head and the Chau chiu tai kai tsai forms a completely wrapped head. The Liu tsu chieh of Zhejiang Province bears protuberances on the midrib.

Ta hsin tsai group (*B. juncea* var. *bulbifolia*), plants with succulent stems and elongated internodes. The leaves are not eaten. Those used for Za-tsai have enlarged stems up to 6–7 cm with protuberances on the petioles; stem enlargement is encouraged by low temperature.

Raphanus sativus – radish

Radish (*R. sativus*) is an anciently annual or biennial cultivated vegetable (Frandsen, 1943). The radish crop can be classified into five morphotypes: European small (*R. sativus* var. *sativus*), Asian or daikon (*R. sativus* var.

longipinnatus), black (*R. sativus* var. *niger*), rat-tail (*R. sativus var. mougri*) and oil (*R. sativus* var. *oleiformis*) (Yamagishi, 2017). In addition to cultivated varieties, four wild taxa are currently recognized: *R. raphanistrum, R. raphanistrum* ssp. *landra, R. raphanistrum* ssp. *raphanistrum* and *R. raphanistrum* ssp. *rostratus*.

World radish production is estimated at 7 million t/year, which is 2% of total vegetable production. They are particularly prominent in Japan, Korea and Taiwan. The mild-flavoured East Asian radish produces large roots that can be sliced, diced and cooked. The Daikon form is increasing in popularity worldwide, especially as it is reputed to be tolerant to clubroot disease caused by *Plasmodiophora brassicae*. The small-rooted European radish is used for relishes, appetizers and to add variety and colour to green leaf salads. Other forms are used as leaf greens, green manure crops and sources of oilseeds.

The origins of the culinary radish are not known since there is no immediately obvious wild progenitor. However, the zone of maximum diversity runs from the eastern Mediterranean to the Caspian Sea and eastwards to China and Japan. Radish crops were cultivated around the Mediterranean before 2000 BCE and are reported in China in 500 BCE and Japan in 700 CE. New varieties such as var. *caudatus* and var. *hortensis* probably originated during migration. These would then have been domesticated in South Asia, South-east Asia and East Asia. *Raphanus sativus* var. *hortensis* is thought to have been transported to Japan via southern China. They are now found worldwide.

Until recently, little has been known about the origin and spread of radishes. However, recent studies have demonstrated that cultivated radishes have multiple origins from their wild progenitors. Yang *et al.* (2002) hypothesized that *Raphanus* was derived from a hybridization between the *B. nigra* and *B. rapa* lineage, the latter as the female parent, with the production of *Raphanus* inferred to be about 4.7–8.8 Mya. Other studies suggest *Raphanus* diverged from *B. rapa* 13–19 Mya (Moghe *et al.*, 2014) and 16.7 Mya (Mitsui *et al.*, 2015).

Kobayashi *et al.* (2020) surveyed genome-wide SNPs in 520 radish accessions. Principle component analysis grouped the accessions into four populations and showed a strong association with their geographic distribution: (i) Japan, (ii) China and Korea, (iii) South and South-east Asia, and (iv) Europe and the eastern Mediterranean. Phylogenetic analysis identified five clades – four clades comprised cultivated radish accessions, while one clade comprised wild radish species such as *R. raphanistrum* and *R. maritimus*.

Note

[1] Throughout this book '*Brassica*' is used in botanical contexts and 'brassicas' is used in horticultural contexts.

REFERENCES

Alemayehu, N. and Becker, H. (2002) Genotypic diversity and patterns of variation in a germplasm material of Ethiopian mustard (*Brassica carinata* A. Braun). *Genetic Resources and Crop Evolution* 49(6), 573–582. DOI: 10.1023/A:1021204412404.

Arias, T., Beilstein, M.A., Tang, M., McKain, M.R. and Pires, J.C. (2014) Diversification times among *Brassica* (Brassicaceae) crops suggest hybrid formation after 20 million years of divergence. *American Journal of Botany* 101(1), 86–91. DOI: 10.3732/ajb.1300312.

Bailey, L.H. (1922) The cultivated Brassicas. *Gentes Herbarum* 1(2), 53–108.

Bailey, L.H. (1930) The cultivated Brassicas, second paper. *Gentes Herbarum* 2(5), 209–267.

Bailey, L.H. (1940) Certain noteworthy Brassicas. *Gentes Herbarum* 4, 319–330.

Cheung, F., Trick, M., Drou, N., Lim, Y.P., Park, J.-Y. *et al.* (2009) Comparative analysis between homoeologous genome segments of *Brassica napus* and its progenitor species reveals extensive sequence-level divergence. *Plant Cell* 21(7), 1912–1928. DOI: 10.1105/tpc.108.060376.

Crisp, P. (1982) The use of an evolutionary scheme for cauliflowers in the screening of genetic resources. *Euphytica* 31(3), 725–734. DOI: 10.1007/BF00039211.

de Tournefort, J.P. (1700) *Institutiones Rei Herbariae*, Editio Altera., Vol. 1. Typographia Regia, Paris, pp. 219–227. DOI: 10.5962/bhl.title.713.

Dixon, G.R. (2017) The origins of edible brassicas. *Plantsman* 16(3), 180–185.

Frandsen, K.J. (1943) The experimental formation of *Brassica juncea* Czern et Coss. *Dansk Botanisk Arkiv* 11(4), 1–17.

Gaeta, R.T., Pires, J.C., Iniguez-Luy, F., Leon, E. and Osborn, T.C. (2007) Genomic changes in resynthesized *Brassica napus* and their effect on gene expression and phenotype. *Plant Cell* 19(11), 3403–3417. DOI: 10.1105/tpc.107.054346.

Gates, R.R. (1953) Wild cabbages and the effects of cultivation. *Journal of Genetics* 51(2), 363–372. DOI: 10.1007/BF03023303.

Giles, W.F. (1941) Cauliflower and broccoli. What they are and where they came from. *Journal of the Royal Horticultural Society* 66, 265–278.

Giles, W.F. (1944) Our vegetables: whence they came. *Journal of the Royal Horticultural Society* 69, 167–173.

Gilmour, J.S.L., Horne, F.R., Little Jr, E.L., Stafleu, F.A. and Richens, R.H. (eds) (1969) International Code of Nomenclature of Cultivated Plants. *Regnum Vegetabile* 64.

Gómez-Campo, C. and Prakash, S. (1999) Origin and domestication. In: Gómez-Campo, C. (ed.) *Biology of Brassica Coenospecies*. Elsevier, Amsterdam, pp. 33–58.

Gray, A.R. (1982) Taxonomy and evolution of broccoli (*Brassica oleracea* L.var. *italica* Plenck). *Economic Botany* 36(4), 397–410. DOI: 10.1007/BF02862698.

Gray, A.R. and Crisp, P. (1985) Breeding improved green-curded cauliflower. *Cruciferae Newsletter* 10, 66–67.

Hagos, R., Shaibu, A.S., Zhang, L., Cai, X., Liang, J. *et al.* (2020) Ethiopian mustard (*Brassica carinata* A. Braun) as an alternative energy source and sustainable crop. *Sustainability* 12(18), 7492. DOI: 10.3390/su12187492.

Hegi, G. (1919) *Illustrierte Flora von Mittel-Europa*. NBd IV/1. Lehmann Verlag, Munich.

Helm, J. (1963) Morphologisch-taxonomische gliederung der kultursippen von *Brassica oleracea* L. *Journal Die Kulturpflanze* 11(1), 92–210. DOI: 10.1007/BF02136113.

Henslow, G. (1908) History of the cabbage tribe. *Journal of the Royal Horticultural Society* 34, 15–23.

Herklots, G.A.C. (1972) *Vegetables in South-East Asia*. George Allen & Unwin, London.

Inaba, R. and Nishio, T. (2002) Phylogenetic analysis of Brassiceae based on the nucleotide sequences of the *S*-locus related gene, *SLR1*. *Theoretical and Applied Genetics* 105(8), 1159–1165. DOI: 10.1007/s00122-002-0968-3.

Kang, L., Qian, L., Zheng, M., Chen, L., Chen, H. *et al.* (2021) Genomic insights into the origin, domestication and diversification of *Brassica juncea*. *Nature Genetics* 53(9), 1392–1402. DOI: 10.1038/s41588-021-00922-y.

Kang, L., Qian, L., Chen, H., Yang, L. and Liu, Z. (2022) Resequencing in *Brassica juncea* for elucidation of origin and diversity. In: Kole, C. and Mohapatra, T. (eds) *The Brassica Juncea Genome (Compendium of Plant Genomes)*. Springer, Cham, pp. 257–267. DOI: 10.1007/978-3-030-91507-0.

Keng, H. (1974) Economic plants of ancient North China as mentioned in 'Shih Ching' (Book of Poetry). *Economic Botany* 28(4), 391–410. DOI: 10.1007/BF02862856.

Kobayashi, H., Shirasawa, K., Fukino, N., Hirakawa, H., Akanuma, T. *et al.* (2020) Identification of genome-wide single-nucleotide polymorphisms among geographically diverse radish accessions. *DNA Research* 27(1), dsaa001. DOI: 10.1093/dnares/dsaa001.

Kumazawa, S. (1965) *Vegetable Gardening*. Yokendo, Tokyo.

Kumazawa, S. and Akiya, R. (1936) Study of vegetables in Taiwan and the South area of China. 2. Takana. *Agriculture and Horticulture* 11, 1741–1748.

Larkcom, J. (1991) *Oriental Vegetables: The Complete Guide for Garden and Kitchen*. John Murray, London.

Li, P., Zhang, S., Li, F., Zhang, S., Zhang, H. *et al.* (2017) A phylogenetic analysis of chloroplast genomes elucidates the relationships of the six economically important *Brassica* species comprising the Triangle of U. *Frontiers in Plant Science* 8, 111. DOI: 10.3389/fpls.2017.00111.

Linnaeus, C. (1735) *Species Plantarum*. Laurentius Salvus, Stockholm.

Lizgunova, T.V. (1959) The history of botanical studies of the cabbage *Brassica oleracea* L. *Bulletin of Applied Botany, Genetics and Plant Breeding* 32, 37–70 (Russian with English summary).

Lloyd, A., Blary, A., Charif, D., Charpentier, C., Tran, J. *et al.* (2018) Homoeologous exchanges cause extensive dosage-dependent gene expression changes in an allopolyploid crop. *New Phytologist* 217(1), 367–377. DOI: 10.1111/nph.14836.

Lu, K., Wei, L., Li, X., Wang, Y., Wu, J. *et al.* (2019) Whole-genome resequencing reveals *Brassica napus* origin and genetic loci involved in its improvement. *Nature Communications* 10(1), 1154. DOI: 10.1038/s41467-019-09134-9.

Lukens, L.N., Pires, J.C., Leon, E., Vogelzang, R., Oslach, L. *et al.* (2005) Patterns of sequence loss and cytosine methylation within a population of newly resynthesized *Brassica napus* allopolyploids. *Plant Physiology* 140(1), 336–348. DOI: 10.1104/pp.105.066308.

Lysak, M.A., Koch, M.A., Pecinka, A. and Schubert, I. (2005) Chromosome triplication found across the tribe Brassiceae. *Genome Research* 15(4), 516–525. DOI: 10.1101/gr.3531105.

Mabry, M.E., Turner-Hissong, S.D., Gallagher, E.Y., McAlvay, A.C., An, H. *et al.* (2021) The evolutionary history of wild, domesticated, and feral *Brassica oleracea* (Brassicaceae). *Molecular Biology and Evolution* 38(10), 4419–4434. DOI: 10.1093/molbev/msab183.

Maggioni, L. (2015) The domestication of *Brassica oleracea* L. Doctoral Thesis, Swedish University of Agricultural Sciences, Alnarp, Sweden.

Maggioni, L., von Bothmer, R., Poulsen, G. and Lipman, E. (2018) Domestication, diversity and use of *Brassica oleracea* L., based on ancient Greek and Latin texts. *Genetic Resources and Crop Evolution* 65(1), 137–159. DOI: 10.1007/s10722-017-0516-2.

Massie, I.H., Astley, D. and King, G.J. (1996) Patterns of genetic diversity and relationships between regional groups and populations of italian landrace cauliflower and broccoli (*Brassica oleracea* L. var. *botrytis* L. and var. *italica* Plenck.). *Acta Horticulturae* 407(3), 45–54. DOI: 10.17660/ActaHortic.1996.407.3.

Matsumura, T. (1954) On the silique types of n=10 group Brassicas of Japan. *Japanese Journal of Breeding* 4(3), 179–182. DOI: 10.1270/jsbbs1951.4.179.

Metzger, J. (1833) *Systematische Beschreibung Der Kultivirten Kohlarten*, Vol. 2. August Oßwald, Heidelberg, pp. 1–68.

Mitsui, Y., Shimomura, M., Komatsu, K., Namiki, N., Shibata-Hatta, M. *et al.* (2015) The radish genome and comprehensive gene expression profile of tuberous root formation and development. *Scientific Reports* 5, 10835. DOI: 10.1038/srep10835.

Mizushima, U. and Tsunoda, S. (1967) A plant exploration in Brassica and allied genera. *Tohoku Journal of Agricultural Research* 17, 249–276.

Moghe, G.D., Hufnagel, D.E., Tang, H., Xiao, Y., Dworkin, I. *et al.* (2014) Consequences of whole-genome triplication as revealed by comparative genomic analyses of the wild radish *Raphanus raphanistrum* and three other Brassicaceae species. *The Plant Cell* 26(5), 1925–1937. DOI: 10.1105/tpc.114.124297.

Monteiro, A.A. and Williams, P.H. (1989) The exploration of genetic resources of Portuguese cabbage and kale for resistance to several *Brassica* diseases. *Euphytica* 41(3), 215–225. DOI: 10.1007/BF00021588.

Nagaharu, U. (1935) Genome analysis in *Brassica* with special reference to the experimental formation of *B. napus* and peculiar mode of fertilization. *Japanese Journal of Botany* 7, 389–452.

Narain, A. (1974) Oilseeds rape and mustard. In: Hutchinson, J. (ed.) *Evolutionary Studies in World Crops*. Cambridge University Press, Cambridge, pp. 67–78.

Netroufal, F. (1927) Zytologische studien über die kulturrassen von *Brassica oleracea*. *Österreichische Botanische Zeitschrift* 76(2), 101–115. DOI: 10.1007/BF01246243.

Nieuwhof, M. (1969) *Cole Crops*. Leonard Hill, London.

Ojiewo, C., Teklewold, A., Weyesa, B., Tesfaye, M., Wakjira, A. *et al.* (2013) Good agricultural practices for production of Ethiopian mustard (*Brassica carinata* A. Braun) in sub-Saharan Africa. *Scripta Horticulturae* 15, 103–114.

Okuda, N. and Fujime, Y. (1996) Plant growth characters of chinese kale (*Brassica oleracea* L. var. *alboglabra*). *Acta Horticulturae* 407, 55–60. DOI: 10.17660/ActaHortic.1996.407.4.

Palmer, J.D., Shields, C.R., Cohen, D.B. and Orton, T.J. (1983) Chloroplast DNA evolution and the origin of amphidiploid *Brassica* species. *Theoretical and Applied Genetics* 65(3), 181–189. DOI: 10.1007/BF00308062.

Prakash, S. and Hinata, K. (1980) Taxonomy, cytogenetics and origin of crop *Brassica*, a review. *Opera Botanica* 55, 1–57.

Prakash, S., Wu, X.-M. and Bhat, S.R. (2011) History, evolution, and domestication of *Brassica* crops. In: Janick, J. (ed.) *Plant Breeding Reviews*, Vol. 35. Wiley-Blackwell, Oxford, pp. 19–84. DOI: 10.1002/9781118100509.

Qi, X., An, H., Ragsdale, A.P., Hall, T.E., Gutenkunst, R.N. *et al.* (2017) Genomic inferences of domestication events are corroborated by written records in *Brassica rapa*. *Molecular Ecology* 26(13), 3373–3388. DOI: 10.1111/mec.14131.

Samans, B., Chalhoub, B. and Snowdon, R.J. (2017) Surviving a genome collision: genomic signatures of allopolyploidization in the recent crop species *Brassica napus*. *The Plant Genome* 10(3), plantgenome2017.02.0013. DOI: 10.3835/plantgenome2017.02.0013.

Sangers, W.J. (1952) *De ontwikkeling van de Nederlandse tuinbouw (tot het jaar 1930)*. W.E.J. Tjeenk Willink, Zwolle, Netherlands.

Sangers, W.J. (1953) *Gegevens betreffende de ontwikkeling van de Nederlandse tuinbouw (tot het jaar 1800)*. W.E.J. Tjeenk Willink, Zwolle, Netherlands.

Sarilar, V., Palacios, P.M., Rousselet, A., Ridel, C., Falque, M. *et al.* (2013) Allopolyploidy has a moderate impact on restructuring at three contrasting transposable element insertion sites in resynthesized *Brassica napus* allotetraploids. *The New Phytologist* 198(2), 593–604. DOI: 10.1111/nph.12156.

Schiemann, E. (1932) *Entstehung der Kulturpflanzen. Handbuch der Vererbungwissenschaft*, Vol. 3. Gebrüder Borntraeger, Stuttgart.

Schulz, O.E. (1919) Cruciferae Brassiceae, Pars prima. In: Engler, H.G.A. (ed.) *Das Pflanzenreich*, Vol. IV 105 (Heft 70). Engelmann, Leipzig, pp. 1–290.

Schulz, O.E. (1936) Cruciferae. In: Engler, A. and Harms, H. (eds) *Die Natürlichen Pflanzenfamilien*, Vol. 17b. Engelmann, Leipzig, pp. 227–658.

Seepaul, R., Kumar, S., Iboyi, J.E., Bashyal, M., Stansly, T.L. *et al.* (2021) *Brassica carinata*: biology and agronomy as a biofuel crop. *GCB Bioenergy* 13(3), 582–599. DOI: 10.1111/gcbb.12804.

Shebalina, M.A. (1968) The history of the botanical investigation and classification of turnip. *Bulletin of Applied Botany Genetics and Plant Breeding* 38, 44–87.

Shibutani, S. and Okamura, T. (1954) On the classification of turnips in Japan with regard to the types of epidermal layers of the seed. *Journal of the Japanese Society for Horticultural Science* 22(4), 235–238. DOI: 10.2503/jjshs.22.235.

Singh, C.B., Asthana, A.N. and Mehra, K.L. (1974) Evolution of *Brassica juncea* under domestication and natural selection in India. *Genetica Agraria* 28, 111–135.

Sinskaia, E.N. (1928) The oleiferous plants and root crops of the family Cruciferae. *Bulletin of Applied Botany, Genetics and Plant Breeding* 19(3), 555–626.

Song, K.M., Osborn, T.C. and Williams, P.H. (1988) Brassica taxonomy based on nuclear restriction fragment length polymorphisms (RFLPs): 1. Genome evolution of diploid and amphidiploid species. *Theoretical and Applied Genetics* 75, 784–794. DOI: 10.1007/BF00265606.

Song, K., Tang, K. and Osborn, T.C. (1993) Development of synthetic *Brassica* amphidiploids by reciprocal hybridization and comparison to natural amphidiploids. *Theoretical and Applied Genetics* 86(7), 811–821. DOI: 10.1007/BF00212606.

Song, K., Tang, K., Osborn, T.C. and Lu, P. (1996) Genome variation and evolution of *Brassica* amphidiploids. *Acta Horticulturae* 407(2), 35–44. DOI: 10.17660/ActaHortic.1996.407.2.

Stafleu, F.A., Bonner, C.E.B., McVaugh, R., Meikle, R.D., Rollins, R.C. *et al.* (1972) International Code of Nomenclature adopted by the Eleventh International Botanical Congress, Seattle, August 1969. *Regnum Vegetabile* 82. A. Oosthok, Utrecht, Netherlands.

Sun, X.Q., Qu, Y.Q., Li, M.M., Song, X.L. and Hang, Y.Y. (2018) Genetic diversity, genetic structure and migration routes of wild *Brassica juncea* in China assessed by SSR markers. *Genetic Resources and Crop Evolution* 65(3), 1581–1590. DOI: 10.1007/s10722-018-0628-3.

Swarup, V. and Chatterjee, S.S. (1972) Origin and genetic improvement of Indian cauliflower. *Economic Botany* 26(4), 381–393. DOI: 10.1007/BF02860710.

Toxopeus, H. (1974) Outline of the evolution of turnips and coles in Europe and the origin of winter rape, swede-turnips and rape kales. In: *Proceedings of Eucarpia 'Cruciferae 1974' Meeting, 25–27 September 1974*. Dundee, UK, pp. 1–7.

Toxopeus, H. (1979) The domestication of Brassica crops in Europe: evidence from the herbal books of the 16th and 17th centuries. In: *Proceedings of Eucarpia "Cruciferae 1979" Conference, 1–3 October 1979*. Wageningen, Netherlands, pp. 29–37.

Toxopeus, H. (1993) *Brassica rapa* L. In: Jansen, P.C.M., Westphal, E., Siemonsma, J.S. and Piluek, K. (eds) *Plant Resources of South-East Asia. No. 8. Vegetables*. PROSEA, Bogor, Indonesia/Pudoc, Wageningen, Netherlands, pp. 121–123.

Toxopeus, H., Oost, E.H. and Reuling, G. (1984) Current aspects of the taxonomy of cultivated brassica species. *Cruciferae Newsletter* 9, 55–58.

Tsunoda, S., Hinata, K. and Gómez-Campo, C. (eds) (1984) *Brassica Crops and Wild Allies: Biology and Breeding*. Japan Scientific Societies Press, Tokyo.

USDA, Agricultural Research Service, National Plant Germplasm System (2023) Germplasm Resources Information Network (GRIN Taxonomy). National Germplasm Resources Laboratory. Available at: https://www.ars.usda.gov/northeast-area/beltsville-md-barc/beltsville-agricultural-research-center/national-germplasm-resources-laboratory/ (accessed 16 August 2023).

van Doorn, H.E., van der Kruk, G.C., van Holst, G.-J., Raaijmakers-Ruijs, N.C.M.E., Postma, E. *et al.* (1998) The glucosinolates sinigrin and progoitrin are important determinants for taste preference and bitterness of Brussels sprouts. *Journal of the Science of Food and Agriculture* 78(1), 30–38. DOI: 10.1002/(SICI)1097-0010(199809)78:1<30::AID-JSFA79>3.0.CO;2-N.

van Doorn, J.E., van der Kruk, G.C., van Holst, G.J., Schoofs, M., Broer, J.B. *et al.* (1999) Quantitative inheritance of the progoitrin and sinigrin content in Brussels sprouts. *Euphytica* 108(1), 41–52. DOI: 10.1023/A:1003600227319.

Wellington, P.S. and Quartley, C.E. (1972) A practical system for classifying, naming and identifying some cultivated Brassicas. *Journal of the National Institute of Agricultural Botany* 12(3), 413–432.

Wu, G.M. (1957) *Vegetable Gardening in China*. Scientific Publisher, Beijing.

Yamagishi, H. (2017) Speciation and diversification of radish. In: Nishio, T. and Kitashiba, H. (eds) *The Radish Genome*. Springer, Dordrecht, Netherlands, pp. 11–30. DOI: 10.1007/978-3-319-59253-4.

Yang, J., Liu, D., Wang, X., Ji, C., Cheng, F. *et al.* (2016) The genome sequence of allopolyploid *Brassica juncea* and analysis of differential homoeolog gene expression influencing selection. *Nature Genetics* 48(10), 1225–1232. DOI: 10.1038/ng.3657.

Yang, J., Zhang, C., Zhao, N., Zhang, L., Hu, Z. *et al.* (2018) Chinese root-type mustard provides phylogenomic insights into the evolution of the multi-use diversified allopolyploid *Brassica juncea*. *Molecular Plant* 11(3), 512–514. DOI: 10.1016/j.molp.2017.11.007.

Yang, Y.-W., Tai, P.-Y., Chen, Y. and Li, W.-H. (2002) A study of the phylogeny of *Brassica rapa, B. nigra, Raphanus sativus,* and their related genera using noncoding regions of chloroplast DNA. *Molecular Phylogenetics and Evolution* 13(3), 455–462. DOI: 10.1016/S1055-7903(02)00026-X.

Zeven, A.C. (1996) Sixteenth to eighteenth century depictions of cole crops, (*Brassica oleracea* L.), turnips (*B. rapa* L. cultivar group vegetable turnip) and radish (*Raphanus sativus* L.) from Flanders and the present-day Netherlands. *Acta Horticulturae* 407(1), 29–33. DOI: 10.17660/ActaHortic.1996.407.1.

Zeven, A.C. and Brandenburg, W.A. (1986) Use of paintings from the 16th to 19th centuries to study the history of domesticated plants. *Economic Botany* 40(4), 397–408. DOI: 10.1007/BF02859650.

Breeding, Genetics and Models

2

Rachel Wells*

Department of Crop Genetics, John Innes Centre, Norwich Research Park, Colney Lane, Norwich, Norfolk NR4 7UH, UK

Abstract

The Brassicaceae is one of the most flexible plant families in terms of inter-specifc and intergenomic crosses. Here, we discuss breeding methodologies, such as intercrossing using embryo rescue and fusion and the use of less well-known species in Brassicaceae to extract valuable genes for resistance to pathogens and pests, and other economic characters. Recent advances in new breeding technologies, such as genetic modification and genome editing, offer exciting potential for targeted trait development.

In addition, we consider studies of model Brassicas, such as thale cress (*Arabidopsis thaliana*), and forward and reverse genetic approaches in model diploid Brassicas, for the identification of genes which can be applied in crop improvement. High-throughput sequencing methodologies have produced a step change in progress towards understanding component genes, genomic interactions, protein products and resultant phenotypic characteristics. Genome sequences, genetic markers and linkage maps are available for the major species, and these genomic studies have shown conservation of genes across species, supporting the transfer of knowledge and traits between models and crops.

Finally, we focus on breeding for key traits in brassica vegetables.

*rachel.wells@jic.ac.uk

© Geoffrey R. Dixon and Rachel Wells 2024. *Vegetable Brassicas and Related Crucifers,* 2nd edition. (G.R. Dixon and R. Wells)
DOI: 10.1079/9781789249170.0002

FLORAL BIOLOGY AS RELATED TO CONTROLLED POLLINATION

The Brassicaceae are characterized by the common floral form of four-petalled, cross-shaped flowers. The flower (see Fig. 2.1) differentiates by the successive development of four sepals, six stamens, two carpels (comprising stigma and style) and the four petals. The carpels form a superior ovary with a 'false' septum and two rows of campylotropous ovules. The nucellar tissue is largely displaced by the embryo sac and when the buds open, the ovules mainly consist of the two integuments and the ripe embryo sac. The buds open under pressure from the rapidly growing petals. Opening begins in the afternoon, with anthesis, where the flowers become fully expanded and viable pollen becomes available the following morning. Petals can range from bright yellow, to peach, primrose, cream and even white, growing to 10–25 mm long and 6–10 mm wide. The sepals are erect. Nectar is secreted by four nectaries situated between bases of the short stamens and the ovary. The flowers are borne in racemes on the main stem and its axillary branches. The inflorescences may attain lengths of 1–2 m. The slender pedicels are 15–20 mm long.

Pollination of the flowers is usually performed by insects, particularly bees, which collect pollen and nectar. Self-pollination can occur with plant movement. However, many species can exhibit self-incompatibility (SI) and

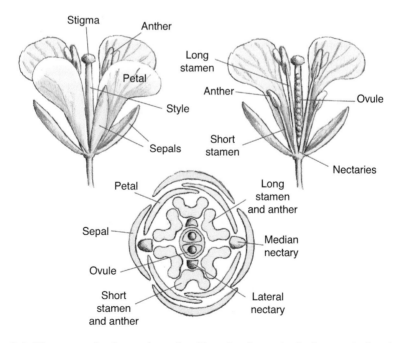

Fig. 2.1. Diagrammatic views of a stylized brassica flower including vertical and horizontal half-sections. (Gary Breeze).

so are obligate out-crossers. Hand pollination is best performed with freshly released pollen in the morning.

SEED DEVELOPMENT

Progression through seed development following pollination can be seen in Fig. 2.2. The embryo develops from a fertilized egg cell via the globular, heart, linear and bent cotyledon-shaped forms through a series of asymmetric cell divisions. As embryo maturation occurs, cell expansion and differentiation replace active cell division, and seed storage products such as proteins and oils accumulate. The embryo fills most of the seed coat after 3–5 weeks, by which time the endosperm has been almost completely absorbed. Nutrient reserves for germination are stored in the cotyledons, which are folded together with the embryo radicle lying between them. Upon maturation, seed desiccation takes place allowing the embryo to enter a quiescent state of seed dormancy.

The fruits of cole crops are glabrous siliquae, 4–5 mm wide and 40–100 mm long. The siliqua contains an ovary, comprising two chambers containing rows of seeds, separated by the pseudoseptum (see Fig. 2.3). The seeds are enclosed by two linear valves, separated by a replum, and a valve margin with a dehiscence zone. One siliqua contains 10–30 seeds.

The siliqua elongates as the seeds develop. Between 3–4 weeks after the opening of a flower it reaches its full length and diameter. When it is ripe the two valves dehisce. Separation begins at the attached base and works towards the unattached distal end, leaving the seeds attached to the placentas. Physical force ultimately separates the seeds, usually by the pushing of the dehisced siliquae against other plant parts either by wind or in threshing operations.

BREEDING METHODOLOGIES

Self-pollination

Self-pollination of any of the brassicas can be obtained by brushing or shaking the open flowers if the plant is self-compatible. If the plant is self-incompatible,

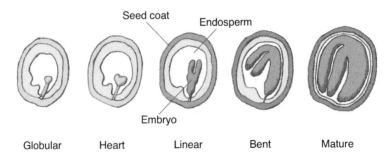

Fig. 2.2. Brassica seed developmental time series. (Gary Breeze).

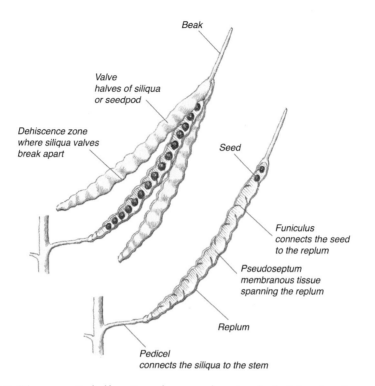

Fig. 2.3. Diagrammatic half-section of an open brassica dry fruit (seed pod or siliqua) displaying the seeds. (Gary Breeze).

the buds can be opened 1–4 days before they would open naturally and be bud-pollinated. If the buds are too small then bud pollination is usually more difficult and likely to be unsuccessful, and if left too late then the incompatibility factor may have developed and few seeds will be produced. A young bud, 3–4 days prior to natural opening, is preferred as the SI factor will not have developed, but the bud is still large enough for the stigma to be receptive.

Usually, six or eight buds can be opened and pollinated at one time on a raceme, with a pointed object such as a toothpick or forceps. Pollen from older open flowers can be transferred to the stigma with a brush or finger, or by using the open flower itself to brush the anthers across the stigmatic surface to transfer the pollen. Williams (1980) suggested the use of 'bee sticks', constructed by using the hairy abdomen of bees mounted on a toothpick, as pollen carriers.

Cross-pollination

When the pollination has a breeding objective other than as a study for SI, then the bud should be emasculated using those expected to open in 1 or 2 days, eliminating the possibility of selfing. The desired pollen is then transferred to the stigma in the same manner as for bud pollination. If a male sterile or self-incompatible plant is used as a female parent, then an open flower or one with a protruding pistil can be used. When the cross-incompatibility between two plants is being assayed, the open flowers need not be emasculated. The major advantages of hybrids with vegetable brassicas are the uniformity of plant size, shape and maturity, and the protection that hybrids give to a company that produces the seed. Vigorous open-pollinated cultivars can be just as productive as hybrid ones, although they may lack desirable ancillary benefits such as uniformity of maturity.

SEED INCREASES

Hand pollination is best performed in the greenhouse or in a large, screened cage to eliminate insects. If cross- or self-pollination are desired in the field, cheesecloth bags can be used to enclose the blossoms of one or two plants. It is preferable is to enclose several plants in a screen cage 2 m high.

The plants should be held away from the cage walls to avoid pollination through them by visiting bees and other insects. Bees are the best pollinators, but flies can also be used. If large-scale increases are to be performed outdoors, then the minimum isolation between lots should be 2000 m or more. Greater isolation is needed if one lot is downwind from another or the location of the hive results in bees crossing one lot of plants to reach the second.

All brassicas can be dug after growing in the field, potted and moved to the greenhouse after a period of vernalization that is required, except for green broccoli (calabrese) and summer cauliflower and some Chinese cabbage. Digging American-type cauliflowers (Snowball forms) is more difficult, but success can be high if the plants are dug with a root ball and potted with minimum root disturbance. If the temperatures are high, then placing the plants in a refrigerator or controlled environment room (5–10°C) for a couple of weeks while they are re-rooting helps survival. European and Australian cauliflower survive transplanting much more readily, with more than 95% success rates. Increasing cauliflower by tissue culturing the curd is also used, but this slows seed production by at least 6 months. However, it should eliminate the loss of selected plants.

Asexual propagation can be used for cabbage by cutting off the head and allowing the lateral buds to develop. The buds can then be excised, rooted on a moist medium, vernalized and allowed to flower in small pots. The buds from Brussels sprouts can be treated in a similar way. For green broccoli, cuttings can be taken from the lateral branches and rooted in a mist bed after being dipped in a hormone rooting powder. Rooting starts in 10–14 days and the plants develop rapidly thereafter.

INTERSPECIFIC HYBRIDIZATION

The close genetic relationship between *Brassica* species as demonstrated by the 'Triangle of U' (see Fig. 1.2) illustrates the potential for hybridization between species. Interspecific hybridization between these six *Brassica* species; *Brassica rapa* (A genome), turnip and Chinese cabbage/leaf types, $2n = 20$; *Brassica nigra* (B genome), black mustard, $2n = 16$; *Brassica oleracea* (C genome), cole crops (e.g. cabbage, cauliflower, green broccoli), $2n = 18$; *Brassica juncea* (AB genome), brown mustard, $2n = 36$; *Brassica carinata* (BC genome), Ethiopian mustard, $2n = 34$; *Brassica napus* (AC genome), swedes or rutabagas, rape, or oilseed rape (canola) $2n = 38$; has been performed to create synthetic allopolyploids. The designations of cytoplasmic and nuclear genomes of *Brassica* and *Raphanus* species are shown in Table 2.1.

Interspecific hybridization is difficult due to pre- and post-fertilization barriers and the subsequent abortion of the hybrid embryo. However, hybrids

Table 2.1. Designations of cytoplasmic and nuclear genomes of *Brassica* and *Raphanus* species. (After Williams and Heyn, 1980).

Species	Subspecies or variety	Cytoplasm	$2n$ genome	Common name
Brassica nigra		B	*Bb*	black mustard
Brassica oleracea		C	*Cc*	cole vegetables
Brassica rapa		A	*Aa*	
	chinensis		aa.c	pak choi
	nipposinica		aa.n	mizuna
	oleifera		aa.o	turnip rape
	parachinensis		aa.pa	caisin (choy sum)
	pekinensis		aa.p	Chinese cabbage, petsai
	rapifera		aa.r	vegetable turnip
	trilocularis		aa.t	sarson
Brassica carinata		BC	*Bbcc*	Ethiopean mustard
Brassica juncea		AB	*Aabb*	vegetable mustard
Brassica napus		AC	*Aacc*	fodder rape, oilseed rape (canola), swede
Raphanus sativus		R	*Rr*	vegetable radish, daikon

Williams and Heyn (1980) suggested using the single capitalized letter representing the genome descriptor to designate the cytoplasm in which the nuclear genes are functioning.

can be regenerated via siliqua culture, ovule culture and embryo culture on nutrient medium. For siliqua culture, young, fertilized pistils are excised 4–6 days after pollination (DAP). These pistils are cultured on Murashige and Skoog medium before dissecting out the developing ovule for ovule culture. Ovules can also be cultured directly, from the day after pollination (Akmal, 2021). The success of embryo rescue, the direct culture of the embryo, is highly dependent on the developmental stage of the embryo. This will vary between genotypes and species. Peng *et al.* (2018) determined that 15 days after pollination was the best time for embryo rescue in the diploid–diploid (*B. rapa* × *B. oleracea*) crosses, while 20 days after pollination was optimum for amphidiploid–diploid (*B. napus/B. juncea* × *B. rapa*) crosses. The embryo is generally still small 2 weeks after pollination and at this stage embryo rescue can be first attempted. Peng *et al.* (2018) identified cotyledonary embryos were best for regeneration. Cultured embryos are haploid, containing one genome from each progenitor. However, chromosome doubling, to give viable diploid plants, can occur spontaneously during culture or be induced by treatment with colchicine. *Raphanus sativus* (R genome), radish $2n = 18$, genome has also been used to create synthetic allotetraploids (Karpechenko, 1928; Lange *et al.*, 1989). Hybridization between the seven species is easier than to the other cruciferous species. Newly resynthesized hybrids provide an excellent source of genetic variation. However, they often suffer from chromosome mis-segregation and homoeologous exchange, resulting in phenotypic instability and sterility.

Protoplast fusion

Hybridization can also be performed via protoplast fusion, a technique which allows the setting up of hybrid and cybrid combinations of species that are sexually incompatible, facilitating the gene transfer in sexually incompatible species, without genetic transformation.

Several vegetable lines with new combinations of nuclear and cytoplasmic genomes have been created by protoplast fusion. The usual procedure is to fuse protoplasts of regenerable lines, such as *B. oleracea*, with protoplasts from a cytoplasm donor irradiated to eliminate the nuclear DNA. Some plants recovered from such experiments are cybrids, containing the nuclear genome of the *B. oleracea* partner and the mitochondria and/or chloroplasts of the other partner.

The initial fusion product contains the mitochondria and chloroplasts of both fusion partners, in contrast to sexually derived zygotes which almost always contain only the chloroplasts of the maternal parent. Sorting out a mixed population of organelles during subsequent cell division and regeneration has made it possible to overcome the chlorosis seen when the radish-derived Ogura type (Ogura, 1968) of cytoplasmic male sterile (CMS) *Brassica* lines are grown at low temperatures (Walters *et al.*, 1992). This chlorosis results from the interaction of radish chloroplasts with a *Brassica* nucleus, so fusion-mediated replacement of the radish chloroplasts with *Brassica* ones can produce CMS vegetable material suitable for hybrid production. Further fusions allow transfer of a cold-tolerant Ogura CMS from green broccoli to cabbage in one step, eliminating the need for many generations of back-crossing (Sigareva and Earle, 1997a). Other types of male sterile cytoplasms (e.g. Polima and Anand) have also been transferred to *B. oleracea* from other *Brassica* species by protoplast fusion.

Cauliflower and green broccoli cybrids, with chloroplast-encoded resistance to atrazine herbicides, have been created through fusions using *B. rapa* or *B. napus* as the sources of the resistant chloroplasts. The green broccoli cybrids combined two organelle phenotypes (atrazine resistance and the '*nigra*' type of CMS) (Jourdan *et al.*, 1989). Atrazine-resistant *B. oleracea* did not show the negative effects seen in atrazine-resistant oilseed rape (*B. napus*).

Protoplast fusion has been used successfully in *B. rapa* and *B. oleracea* for the production of somatic hybrids to transfer biotic stress resistance genes involved in the bacterial soft rot disease caused by *Erwinia carotovora* ssp. *carotovora* (Ren *et al.*, 2000). A dominant gene for resistance to *Xanthomonas campestris* pv. *campestris* (black rot) has also been transferred from Ethiopian mustard (*B. carinata*) to green broccoli by fusion, followed by back-crossing to green broccoli via embryo rescue and conventional sexual crosses. But somatic hybrids combining *B. oleracea* with more distantly related brassicas often suffer from poor fertility, making back-crossing more difficult.

HYBRID PRODUCTION

Brassica vegetables exhibit strong heterosis and a high degree of hybrid vigour. F_1 hybrid cultivars provide many advantages, especially when breeding for traits such as uniform maturity, high early and high total yield, quality, and resistance to insect pests, pathogens and abiotic stresses (Kucera *et al.*, 2006). F_1 hybrid production is, therefore, an important objective within brassica crop improvement and the development of efficient pollination methodologies is key for hybrid seed production. Two naturally available pollination control mechanisms, sporophytic SI and male sterility (primarily CMS), are being commercially exploited for hybrid seed production in brassica crops (Singh *et al.*, 2019).

Self-incompatibility

Self-incompatibility is defined as the prevention of fusion of fertile (functional) male and female gametes of the same plant (Gowers, 1989). The recognition and following rejection of self, results in dependency on cross-pollination. Historically, the use of SI was the accepted and only successful method of production of hybrid vegetable brassicas, until the very recent development of CMS systems.

Previously, genetic studies had hypothesized the incompatibility specificities of brassicas were controlled by one locus, called the S gene. However, molecular analysis has identified at least two genes within the S locus; one gene, *SP11*, functioning as the male and the other, *SRK*, as the female determinant. SRK acts in the plasma membrane of papilla cells of stigma while SP11 is expressed in the anther tapetum during the maturation of the pollen grains. In an SI reaction, when a pollen grain lands on the stigma SP11 binds with SRK resulting in the prevention of pollen-tube growth (Narayanapur *et al.*, 2018).

Pollen is viable and will achieve fertilization in most cross-pollinations, excluding a small percentage for which the incompatibility specificity of the plant functioning as the female is the same as the incompatibility specificity of the pollen.

Assaying self-incompatibility

Self-incompatibility can be assessed either by pollination or cytological methods. Using a pollination assay, the level of SI can be assessed by the number of seeds produced 60 days after a self- or cross-pollination. Self-compatible lines result in successful pollination and the formation of a greater number of seeds. Assessing SI using this method is slow and the number of seeds formed can be influenced by many other factors such as temperature, humidity, pests and pathogens. Alternatively, using fluorescent microscopy techniques, that readily display those pollen tubes that have penetrated the style, provides a direct measure of incompatibility which can be completed within 12–15 h. Aniline blue stain accumulates in the pollen tubes and fluoresces when irradiated with ultraviolet light. With appropriate light filters under a fluorescent microscope, the tubes are visible, whereas the background of stylar tissue is largely unseen. Penetration by a few tubes indicates incompatibility, while penetration by many tubes indicates compatibility. The rapid results greatly accelerate the evaluation compatibility levels and the planning of breeding programmes.

Prior to assessing SI, breeding for pest and pathogen resistance and for other horticultural characteristics is necessary. Then the evaluation of compatibility should follow. The selected plants can be tested by bud self-pollination and self-pollination of open buds. The resultant seed set or pollen-tube penetration from the open flowers will measure the intensity of SI. Where an interesting family is established, 11 plants should be selected and cross-pollinated in all combinations to establish the types of dominance and relative degrees of dominance within the family. The details of testing and evaluation of SI were reviewed by Dickson and Wallace (1986).

Self-incompatibility is not always stable and may be broken down by high temperature (Gonai and Hinata, 1971), therefore, it poses a risk of selfing within hybrid seed production. Maintenance and multiplication of parental S-allele lines must be performed using methods to break down SI such as bud pollination or treatment with a high concentration of carbon dioxide (CO_2) and sodium chloride spray (Singh *et al.*, 2019). For these reasons, alternative approaches for F_1 production are also required.

Male sterility

Male sterility refers to the degeneration or loss of function of male organs in bisexual plants and can be divided into genetic or CMS (Ji *et al.*, 2020). Genetic male sterility can be divided into either dominant or recessive genes both of which have been used for hybrid production. Cytoplasmic male sterility has been used widely used in hybrid brassica production. Most CMS in *B. oleracea* was transferred from other cruciferous crops (e.g. *R. sativus* and *B. napus*). The main types of CMS include Nigra CMS, Polima CMS, Ogura CMS and Anand CMS (Ji *et al.*, 2020).

The first *B. oleracea* cytosterile was developed by Pearson (1972) when he crossed *B. nigra* with green broccoli, and this was developed further in cabbage as the Nigra CMS system. Unfortunately, this system was complicated by petaloidy and lack of development of the nectaries giving a male sterile that was unattractive to pollinating bees. Polima CMS was discovered in oilseed rape bred in Poland and is used in *B. napus*, *B. juncea* and *B. rapa* (Ji *et al.*, 2020). Polima CMS was transferred from *B. napus* to *B. oleracea* via protoplast fusion. However, the male sterile plants showed abnormal flowers and siliqua development and incomplete pollen abortion, which made the system unsuitable for use in *B. oleracea* (Yarrow *et al.*, 1990).

A better male sterile was developed when Bannerot *et al.* (1974) crossed a CMS radish (R cytoplasm) 'Ogura' with cabbage. In the fourth backcross generation he obtained normal plants with $2n = 18$ that were totally male sterile, the flowers having only empty pollen grains or vestigial anthers. Male sterile plants had the problem of pale or white cotyledons and pale-yellow leaves during development, accentuated at low temperatures. The cold temperature chlorosis has been overcome, however, by replacing the *Raphanus* chloroplasts with *Brassica* ones via protoplast fusion (Pelletier *et al.*, 1983; Yarrow *et al.*, 1986; Jourdan *et al.*, 1989). Initially, however, there was reduced female fertility (Pellan-Delourme and Renard, 1988) which resulted in limited acceptance of this material. Walters *et al.* (1992) transferred the resistance to cauliflower and subsequently into green broccoli. This, and similar material, have resulted in lines with good seed production and no apparent horticultural problems after vigorous selection for seed yield in early generations (see Fig. 2.4). So far, Ogura CMS is the most widely studied and widely used type of male sterility in cruciferous vegetable breeding (Ji *et al.*, 2020).

Some plant breeders developing similar male sterility apparently did not select sufficiently rigorously for seed-setting in early generations following

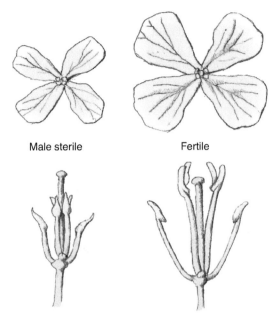

Male sterile   Fertile

Fig. 2.4. Flowers from cytoplasmic male sterile and fertile *Brassica* plants. (Gary Breeze).

fusion. This resulted in subsequent problems with a failure of seed production. Apart from the academically produced CMS, various seed companies have patented their own CMS in *Brassica*. This inhibited use of the Ogura type of CMS due to high patent charges for the right to use this male sterile. Cold-tolerant Ogura male sterile cytoplasm in pak choi and other types of Chinese cabbage have been produced.

Cardi and Earle (1997) transferred the 'Anand' cytoplasm – originally from *B. juncea*, but later attributed to *B. tournefortii* from *B. rapa* 'rapid cycling' lines – to *B. oleracea* and economic forms of *B. rapa*. After vigorous screening in the field, green broccoli lines with good seed yield were released and also cauliflower, cabbage and Chinese cabbage. Both the cytosteriles, the Ogura and Anand, have the typical male sterile genes inherited in a similar manner to that seen in onions and carrots. This means that the sterile requires the appropriate 'S' sterile cytoplasm plus male sterile genes. All *B. oleracea*, *B. napus* and *B. rapa* have male sterile genes but no fertility restoring genes, and act as maintainers, while *Raphanus* does have fertility restoring genes. Obtaining the male sterile version of an inbred line only requires back-crossing to obtain the desired inbred female parent. Incompatibility can, however, upset this programme, if the recurrent parent has not previously been selected for self-fertility, which is not always so easy to find since most breeders have been selecting for SI. Sigareva and Earle (1997b) have found they can substitute the nucleus of inbred lines into the appropriate sterile cytoplasm by fusion and avoid the need for the slow back-crossing procedure.

Bud pollination

Bud pollination to overcome SI is accomplished by opening the bud and transferring pollen from an open flower of the same plant. Fertilization will not occur where this is done when the bud is very small. At this stage the style is not receptive, but about 3–4 days before the flower has opened the style and stigma are fully receptive and the SI factor has not yet developed. Therefore, self-fertilization is possible and SI can be bypassed.

The bud is opened with a pointed object, such as a mounted needle, and pollen transferred from an open flower. In some large breeding programmes, many pollinators can be used to produce selfed seed on self-incompatible plants.

There are two other methods now widely used to overcome SI. The first involves spraying 3–4% sodium chloride solution on to the open flower. Leaving it for 20–30 min, removing the excess salt solution by blowing it off the flower or blotting with a damp cloth or paper towel, and then allowing self-pollination. The salt removes the inhibitor from the stigma, permitting self-pollination. A second and more efficient method if many plants are involved, is to self-pollinate the open flowers, and then place the plants in a closed unit or room into which 5% CO_2 can be admitted. Then the SI factor is overcome, and this allows the production of seed following self-pollination. Single plants can be self-pollinated: the flowers are enclosed in a plastic bag, the bag inflated by exhaling into it, using a straw, and then the bag is closed. This may be repeated, later in the day. The effect is similar to enclosing the plant in a room and adding CO_2.

When the only method of hybrid production was using SI, self-fertile plants were usually discarded. Now. with the interest in male sterility. the self-fertile plants are desired and the self-incompatible plants are discarded. It is equally important, however, to establish that the plants are homozygous for self-compatibility and not just setting seed because they are heterozygous for SI genes.

Production of F_1 hybrids

Inbred brassicas, as with most other out-crossing crops, tend to lose vigour and although a single cross will restore vigour in the F_1, the seed production on the inbred parents may be uneconomically low. For this reason, whether hybrid production involves SI or male sterility, three-way crosses are often used. If SI is used, and if both parents are self-incompatible ($S_1S_1 \times S_2S_2$) then seed from both parents can be harvested. Or an open pollinated line can be used as the pollen parent and seed only saved from the self-incompatible line. Where the original S_1S_2 hybrid is self-incompatible then it can be used as a male or female line together with a self-incompatible C line and again all the seed may be harvested. But that is more complicated and requires deeper understanding of the levels of SI and the relationships of each inbred line to the others involved.

With male sterility, the A or male sterile line *Smsms* is maintained by back-crossing to a self-fertile B line *Nmsms* (S = sterile and N = normal or fertile cytoplasm). After several generations of back-crossing, the A line will resemble the B line, but there will be some inbreeding depression. For production of hybrid seed, the A line is crossed to a further unrelated C line. Only the seed on the A line can be saved. Three-way crossing could be made by crossing the A × C hybrid to a further line, D, and this may result in higher seed production levels since the female parent will now be a hybrid.

Seed production success depends on issues such as whether the two parents flower at the same time. This can easily become a problem where the parents may require different degrees of vernalization and exposure times to produce flower stalks and flowers.

MICROPROPAGATION

The development of fixed homozygous lines for research or parental lines for F_1 hybrid production requires seven to eight rounds of recurrent selfing. This is labour intensive and difficult due to SI. Therefore, the development of homozygous lines via microspore culture is a valuable breeding tool. Microspore-derived embryogenesis was first completed successfully on green broccoli by Keller *et al.* (1975) and has now been implemented in multiple crop types. Now, embryogenesis of microspores is routine in large-scale crucifer breeding programmes. Multiple factors affect the success of microspore culture. These include donor plant growth conditions, which will affect pollen fitness viability, plant genotype, bud size and the resulting developmental stage of the pollen (Sharma *et al.*, 2020). Bhatia *et al.* (2016) determined that in cauliflowers, late uninucleate to early binucleate pollen contains the highest percentage of viable microspores. This mirrors results obtained in *B. napus* (Pechan and Keller, 1988). Other factors that affect the success of microspore culture include microspore culture density, osmotic pressure and culture nutrient conditions, and additives such as activated charcoal, plant growth regulators and plant hormones (Sharma *et al.*, 2020).

In most cases, a high percentage of haploid embryos will double spontaneously providing homozygous plants. However, chemical doubling agents, such as colchicine, can also be applied (Ahmadi and Ebrahimzadeh, 2020). Embryogenesis is an expensive procedure, but the time saved in obtaining homozygous lines compared with using seven generations of single-seed descent for conventional inbreeding can be well worthwhile.

CALLUS CULTURE AND TRANSFORMATION

Tissue culture techniques have been applied to *B. oleracea* vegetables, either for clonal propagation or for development of novel and sometimes improved plant types. Plants have been regenerated from diverse multicellular explants including: immature embryos; seedling parts such as hypocotyls or cotyledons; stem pieces; leaves; roots; and floral tissues such as flower stalks or cauliflower curds. Plant regeneration from somatic tissues makes it possible to maintain populations that fail to produce seeds and are difficult to propagate from standard methods.

Plants can be recovered from single wall-free protoplasts, usually isolated from leaves or hypocotyls of plantlets grown *in vitro*. Plant regeneration from protoplasts has been reported in all brassica species, although protoplast regeneration efficiency remains low. Strong genotype specificity is usually noted in studies comparing regeneration from various lines of a given vegetable, with some performing very well and others showing little or no response. Efficiency will also be affected by a combination of protoplast isolation method, protoplast density for culture, culture conditions and media ingredients, and the developmental stage of protoplast calli capable of shoot induction (Li *et al.*, 2021).

Genetic transformation offers a significant advance in the ability to introduce advantageous genetic variation without negatively affecting the elite genetic background by introducing unwanted, deleterious variation. It also provides the means to introduce variation that may not be present within the genetic background of a species, or closely related species. Transformation systems are available for most agriculturally important brassica species. Transformation can be achieved via a number of methodologies, including direct transformation using biolistics, electroporation, microinjection and polyethylene glycol (PEG) treatment or indirect methods such as *Agrobacterium*-mediated transformation.

Transgenic green broccoli, cauliflower and cabbage have been created, primarily through procedures involving the vectors *Agrobacterium tumefaciens* or *Agrobacterium rhizogenes*. This approach has previously been used in conferring resistance to insects (from *Bacillus thuringiensis* (Bt)), herbicides and viruses. Floral phenotypes have also been altered through transfer of genes preventing pollen development or modifying the SI response. As with regeneration, susceptibility to *Agrobacterium* varies between *B. oleracea* genotypes. Detailed protocols and video guides are available from the John Innes Centre for the self-compatible green broccoli type line AG DH1012, a doubled haploid genotype from the *Brassica oleracea* ssp. *alboglabra* (A12DHd) crossed with *B. oleracea* var. *italica* (Green Duke GDDH33) used as a model *Brassica* (Hundleby and Chhetry, 2020).

More recently, advances have been made in the area of genome editing, a group of technologies that allow genetic material to be added (knock-in), removed (knock-out) or altered at a specific point within the genome.

A number of genome-editing technologies have been developed, including zinc-finger nucleases and transcription activator-like effector nucleases (TALENs), which use endonuclease catalytic domains attached to modular DNA-binding proteins to induce targeted DNA double-stranded breaks (DSBs) at specific genomic loci. The most well-known editing system, the CRISPR (clustered regularly interspersed short palindromic repeats)/Cas9 system, has been adapted from a naturally occurring system that bacteria use as an immune defence. The Cas9 nuclease is targeted to genomic DNA by a small 20-nucleotide guide RNA and mediates a DSB. Upon cleavage, the target locus undergoes DNA damage repair. However, re-ligation leaves errors in the form of insertion/deletion (indel) mutations, resulting in a potential loss of gene function. Larger deletions can also be created by inducing multiple DSBs (Ran *et al.*, 2013). First published in 2013, CRISPR offers a faster and simpler approach to gene knockout in both single and multiple genetic locations, within a single or small number of generations, in a way that has not been possible through alternative breeding methods. In green broccoli, Neequaye *et al.* (2021) performed the first CRISPR-edited field trial to knock out *MYB28*, resulting in down-regulation of aliphatic glucosinolate biosynthesis genes and a reduction in accumulation of the methionine-derived glucosinolate, glucoraphanin, in leaves and florets. This proof-of-concept trial highlights the potential of genome editing within brassica research and breeding.

SELECTION AND PREPARATION OF PLANTS FOR BREEDING FOLLOWING SELECTION

Most vegetable brassicas are relatively easy to transplant and will withstand considerable abuse during selection and evaluation phases prior to being allowed to grow to maturity and flower. All the vegetable brassicas can be transplanted in a bare-rooted state without damage except for cauliflower. Cauliflower does not regenerate roots very readily. If, however, a cauliflower at market stage is dug carefully, preferably from wet or moist soil, then a root ball can be maintained. The plant is then placed in a large pot with moist potting soil and very few selections will be lost. About half of the leaves should be removed to reduce desiccation by transpiration. Plant breeders at Cornell University, New York State, USA, rarely lose over 5% of the plants. If it is hot (over 27°C) in the glasshouse, then it is desirable to place the freshly potted plants in a cool place at 10–20°C for 2–3 weeks. If the glasshouse temperature will not exceed 27°C then the plants should develop and produce seed. If the cauliflower is a late summer or winter type it may need varying lengths of cold storage before it will develop flowers from the curd.

Whole green broccoli plants can be dug and potted, or have cuttings taken from them, dipped into an auxin-based hormonal rooting medium and placed in a mist chamber for 10–14 days, by which time they should be rooted and ready for potting. Cabbage, Brussels sprouts and kale can be dug, some of the leaves removed, and planted in pots, prior to being vernalized, or the cabbage head can be removed, and adventitious buds allowed to develop. These cabbage buds and buds of Brussels sprouts can be removed and placed in moist soil where they will root while being vernalized.

BIOLOGICAL MODELS

When research attempts to understand the fundamental processes of nature irrespective of whether these are biological, chemical, mathematical or physical the initial vehicle used to translate theory into experimentation must be as simple as possible. Simplicity of operation allows the principles that are being investigated to be identified and analysed in manners that are uncluttered by other extraneous and unrelated processes. In biology, the complexities associated with crop plants can, for example, be added at later stages when research moves downstream into applied areas and results in useful applications. Initially, the fundamental biologist seeks plant or animal systems that are easily amenable to being cultured in confined and controlled environments with a rapid, simple, ephemeral life cycle and small mass. These organisms become 'model forms' around which enormous amounts of scientific knowledge are assembled and theories developed. Zoologists have used the fruit fly, *Drosophila melanogaster*, since the early 1900s as a convenient vehicle for genetic studies. Similarly detailed studies have been made with the yeast fungus, *Saccharomyces cerevisiae*, and the nematode, *Caenorhabditis elegans*.

Whyte (1960) identified what he termed 'Botanical Drosophilas'. These did, however, include several complex crop plants such as cereals, soybeans, peas, *Kalenchoe blossfeldiana*, *Perilla* spp., and the simpler weeds thale cress (also known as mouse-ear cress) (*Arabidopsis thaliana*) and cocklebur (*Xanthium pennsylvanicum*). Where crop plants are used in the early stages of the translation of theory into experimentation, it is because a particular plant has some specific trait that is amenable to laboratory studies. Thus, much use is made of spinach (*Spinacea oleracea*) in photosynthetic research because it has large and relatively accessible chloroplasts and photoperiodic studies have centred on the *Chrysanthemum × morifolium* because its responses are clear cut and amenable to manipulation. Much of our knowledge of plant morphogenesis has come from studies of carrot (*Daucus carota* var. *carota*) cells in culture because they are suited to this form of husbandry where single root cells exhibit clear totipotency. *Antirrhinum* spp. and *Petunia* spp. have formed admirable vehicles for the analysis of primordial differentiation and growth in flowers.

The family Brassicaceae contains several models, of which *Arabidopsis thaliana* and rapid-cycling genotypes of both *B. oleracea* and *B. rapa* are most commonly used.

Arabidopsis thaliana – botany and ecology

The genus *Arabidopsis* contains a small number of species of slender annual or perennial herbs. Thale cress (*A. thaliana*) is common throughout northern Europe living on dry soils in walls, banks, hedgerows and waste places. It is believed to be native to Africa and Eurasia although its exact geographical origin in unclear. It has been collected or reported as present in many different regions and climates ranging from high elevations in the tropics to the cold climate of northern Scandinavia and including locations in Europe, Asia, Australia and North America. A related species, *Arabidopsis suecica*, a post-glacial allopolyploid species formed via hybridization of *A. thaliana* and *Arabidopsis arenosa*, is restricted to Scandinavia and related genera in the tribe Sisymbrieae include *Sisymbrium* species themselves (the rockets) and *Isatis* (woad) (Stace, 2001).

Botanically, *A. thaliana* is an ephemeral with simple erect stems varying between 5 and 50 cm tall. Usually, the stems are rough and hairy bearing simple hairs. The basal leaves form into a rosette, each having an elliptical to spathulate shape, and each leaf is stalked with a toothed edge, especially at the distal end. The flowers are small, usually less than 3 mm diameter, and formed of white petals in the characteristic cruciform pattern. Each inflorescence is composed of 4–10 flowers, and the fruits are dry siliquae. In miniature, therefore, *A. thaliana* resembles the floral structure of many other much larger brassicas described in this book.

Once mature, the dry seed may be stored for several years with little loss of germination capacity, but once moistened they germinate rapidly. In optimal conditions, each plantlet produces a rosette of leaves, forms a flower stalk and blossoms within 4 weeks of sowing. The rate of development and time to flowering are controlled by the prevailing environmental conditions. A shortage of nutrients, for example, leads to a decrease in regeneration time but decreases the seed set. Temperature and day length may also have an effect (*A. thaliana* flowers more rapidly in long days or in continuous light, while short days retard flowering significantly; it is also dependent on the genetic constitution as dictated by various morphotypes) (see Table 2.2). The commonly used ecotypes such as 'Columbia' and 'Landsberg' have a regeneration time of approximately 6 weeks. Since most types of *A. thaliana* continue forming flowers and seeds over several months it is possible to harvest in excess of 10,000 seeds from each plant. The plant grows satisfactorily in either soil or compost and can be cultured on nutrient agar or liquid sterile media. Consequently, it is admirably suited to growth in artificially lit, controlled environment chambers and 'growth cabinets' or 'growth rooms' well divorced from traditional field husbandry.

Use of *Arabidopsis thaliana* in genetic and molecular studies

This small cruciferous plant has taken on the mantle of being a ubiquitous research tool in studies of many aspects of plant, and some animal, biology. Proof-of-concept studies almost invariably utilize *A. thaliana* and, in

Table 2.2. Examples of morphological mutants of *Arabidopsis thaliana*.

Character	Effect
Angustifolia and asymmetric leaves	Narrow or asymmetric leaves when homozygous
Glabra and distorted trichomes	Removes or changes the shape of the leaf hairs
Erecta and *compacta*	Alters the disposition and size of stems
Brevipedicellus	Reduces the pedicels when homozygous
Apetala-1	Reduces petals
Eceriferum	Changes the morphology of the epidermal wax

consequence, there is an enormous literature describing every facet of its germination, growth and reproduction in genetic and molecular terms. It was originally identified as an experimental tool by Laibach (1943) in Germany following detailed investigations of its genetics that started in 1907. Its haploid chromosome number was identified as $n = 5$, a collection of ecotypes was established and mutant forms resulting from exposure to chemical and physical mutagens were catalogued.

There is a set of recessive mutants that cause homeotic variations that can be used in studies of developmental biology (Pruitt *et al.*, 1987) and a group of colour variants make useful visible markers. Lethal recessive mutants can be employed for developmental, biochemical and physiological studies. Resistance to herbicides and a range of plant pathogens offer practical agricultural and horticultural applications. *Arabidopsis thaliana* is thus well characterized genetically – a wealth of useful and interesting mutations have been induced, analysed and mapped.

It is the small size, short generation time, high seed set, ease of mutagenesis and transformation of *A. thaliana*, making it easier and faster to induce, select and characterize new mutations, that attract scientists of many biological disciplines. Added to this are very attractive cellular and nuclear characteristics. *Arabidopsis thaliana* is attractive for both quantitative and qualitative genetics because of its small genome size and flexible genomic organization (Somerville, 1989; Griffing and Scholl, 1991; Meyerowitz and Somerville, 1994; Meyerowitz, 1997). It has the honour of being the first plant where the entire genome has been sequenced.

The advantages offered by *A. thaliana* include:

- Small size – hence the ease and simplicity of cultivation in controlled environments where space is at a premium in terms of availability and cost. More than 1000 plants can be grown in the space occupied by this open book.
- Short life cycle – regeneration from seed to seed takes between 4 and 6 weeks which offers short, repeatable and reproducible experiments.

- Perfect flowers – these are self-fertile so that pollen transfer is simplified with no requirement for complex procedures to ensure cross-pollination and the flowers tend not to open-pollinate. Self-fertilization exposes recessive mutations in homozygous form in the M_2 generation (the second generation following the application of mutagens).
- Ability to produce large amounts of seed – up to 10,000 seeds can be harvested from one plant. This provides large populations very quickly from a single cross. In turn, this minimizes the space requirements and makes the statistical analysis of populations easier with the large numbers of individuals available.
- A very compact genome – the genome is among the smallest of the higher plants with a haploid size of 135 Mb of DNA. This makes *A. thaliana* an ideal organism for genetic and molecular studies such as mutant analysis, molecular cloning of genes and the detailed construction of a physical map of the genome.
- Easy genetic engineering – *Arabidopsis thaliana* can be genetically engineered with relative simplicity using either *Agrobacterium tumefaciens*-mediated transformation or direct particle bombardment of the tissues.

Arabidopsis genomics

Arabidopsis thaliana has the smallest known genome among the higher plants of ~135 Mb. It is more resistant to ionizing radiations than many other angiosperms. This property correlates with its genome size. The genome size is 11 times larger than the yeast fungus (*S. cerevisiae*) and 30-fold larger than the common intestinal bacterium, *Escherichia coli*. The contrast of the *A. thaliana* genome with those of other higher plants frequently used in molecular and genetic studies is striking (see Table 2.3). The genome size of many higher plants is a result of species hybridization and genome duplication and thought to be an evolutionary insurance against environmental changes. Less than 25% of a complex organism's genes may be required for growth and reproduction. Possession of multiple copies increases the chances that it will possess a variant (allele) that will make the difference between surviving or succumbing to an environmental stress such as drought.

The significance of this small DNA content for molecular genetics is that *A. thaliana* is easier, simpler and more economical to sequence and assemble. *Arabidopsis thaliana* has a low content of repeated sequences. Those elements that are repeated are set at far distances from each other, greatly simplifying genome assembly and genetic studies.

The entire genome of *A. thaliana* was the first angiosperm to be sequenced (The Arabidopsis Genome Initiative, 2000). This presented unlimited opportunities to mine new knowledge for functional and metabolic processes controlled by the about 25,000 genes present in *A. thaliana* (The Arabidopsis Genome Initiative, 2000). With the advent of next-generation sequencing technologies, the

Table 2.3. Comparison of genome sizes.

Species	Common name	Assembled genome size (~Mb)	
Brassicaceae			
Arabidopsis thaliana	Thale cress	120	a
Brassica rapa	Chinese cabbage	347	a
Brassica nigra	Black mustard	515	b
Brassica oleracea	Cole crops	489	a
Brassica juncea	Brown mustard	933	a
Brassica carinata	Ethiopian mustard	1310	c
Brassica napus	Oilseed rape	848 (1130)	a
Gramineae			
Oryza sativa ssp. *japonica*	Rice	427	a
Hordeum vulgare	Barley	4226	a
Triticum aestivum	Bread wheat	14,547	a
Zea mays	Maize (corn)	2182	a
Leguminosae			
Pisum sativum	Pea	3920	a
Glycine max	Soybean	978	a
Phaseolus vulgaris	Common bean	521	a
Solanaceae			
Solanum tuberosum	Potato	810	a
Solanum lycopersicum	Tomato	828	a
Hominidae			
Homo sapiens	Human	3097	d

[a]https://plants.ensembl.org (accessed 15 August 2023).
[b]Paritosh *et al.* (2020).
[c]Song *et al.* (2021).
[d]https://www.ensembl.org (accessed 15 August 2023).

Arabidopsis 1001 Genomes project has provided sequences for 1135 worldwide *Arabidopsis* accessions. This has allowed the identification of genome variants in a larger and more representative sample of accessions to empower genome-wide association studies and investigations into the species demographic history. It has also enabled the identification of features that make specific geographic or genetic subsets particularly well-suited for forward genetics, field experiments and selection scans (1001 Genomes Consortium, 2016).

Information from the analysis of these genomes is available by online databases, such as Ensembl Plants, The Arabidopsis Information Resource (TAIR), 1001 Genomes Database and the European Nucleotide Archive, along with details of the proteins that are coded for mutant phenotype, gene expression data, publication data and linked to germplasm collections. Similarities between plant and animal biology are highlighted by the information obtained from the *Arabidopsis* projects (Leitch and Bennett, 2003). The boundaries between these sectors of biology have largely disappeared and there are distinct possibilities that knowledge obtained from *Arabidopsis* can be applied to the study of animal and human genomes and to understanding their nuclear and cellular processes in health and disease (Sanderfoot and Raikhel, 2001). Recently, comparative genomics between *Arabidopsis* and cultivated *Brassica* species has progressed from studies of large-scale synteny using restriction fragment length polymorphism (RFLP) probe comparison to whole-genome sequencing-based studies.

Use of *Arabidopsis thaliana* to understand gene function

Mutagenesis is a key tool for studying the genetic control of biological traits. The phenotypic analysis of mutants compared to the wild-type parental genotype provides a direct measure of a gene's contributions to a plant's characteristics at a biochemical, cellular, tissue, organ or developmental level. It allows the biological assignment of gene function.

In the late 1990s and early 2000s, the ability to genetically transform *A. thaliana* allowed the development of *Agrobacterium tumefaciens* transfer DNA (T-DNA)-induced collections. The insertion of the T-DNA results in mutation of a gene by disrupting gene function. Large populations of T-DNA transformant lines have resulted in mutant alleles being available for the majority of *Arabidopsis* genes. Within Colombia (Col-0) these include the SALK, GABI-KAT, SAIL and WISC lines, which can be ordered from stock centres such as The European Arabidopsis Stock Centre (NASC: http://arabidopsis.info/) making mutants for functional studies readily available (O'Malley *et al.*, 2015).

Transformation also permits the complementation of a mutant phenotype with cloned genes which is the critical final stage in confirming gene function. Many *A. thaliana* genes have been cloned and characterized. This underpins many translational crop studies and comparative analysis of genome structure and evolution.

BRASSICA GENOMICS AND GENETIC MAPPING

Crucifers enjoy a pivotal role in the developing our understanding of plant genomics and mapping, mainly through the extensive study of *A. thaliana* which shares 86% sequence identity with *Brassica* species. *Arabidopsis* has a simple five-chromosome genome with minimal levels of duplication, which made it the ideal candidate for the first sequenced plant genome, which was completed in 2000. Since then, advances in sequencing technology have introduced a step change in genomics. The Arabidopsis 1001 Genomes project published 1135 sequenced accessions in 2016, opening up the potential for detailed studies using natural variation. In addition to these data, we now have detailed knowledge on specific biological processes underlying *Arabidopsis* growth and development and know the function of many of the individual genes.

Parkin *et al.*, (2005) demonstrated the segmental structure of the whole *B. napus* genome based on comparative analysis with *Arabidopsis*, using over 1000 RFLP markers mapped to homologous regions between the model and the crop. This identified three colinear segments in each of the diploid *Brassica* genomes for every region of *Arabidopsis*, thus suggesting genome triplication followed by a small number of insertions/deletions/translocations would provide the simplest explanation for the present *Brassica* diploid genome structure.

This relationship underpins the comparative analysis between species. *Arabidopsis* is an excellent resource for gene sequence and structure, potential markers and predicting the location of genetic regions of interest. However, due to gene loss and rearrangement it is impossible to predict the number of homologous gene copies that have been retained based on this relationship alone.

The first *Brassica* genome, the *B. rapa* Chinese cabbage Chiifu, was published in 2011 by the *Brassica rapa* Genome Sequencing Project Consortium of partners from China, Korea, UK, Canada, Australia, France, USA and Germany using a combination of BAC by BAC (bacterial artificial chromosome) and Illumina GA II sequencing. Consisting of 284 Mb of gene space and 41,174 gene models, *B. rapa* was shown to contain five times more transposon-related sequences in the *B. rapa* gene space than in the *A. thaliana* genome, with long terminal repeat (LTR)-type transposon-related sequences estimated to occupy over a quarter of the *B. rapa* genome (Wang *et al.*, 2011). The *B. oleracea* genome (var. *capitata*, line 02-12), produced with a combination of Illumina, Roche 454 and Sanger sequence data, was published in 2014. The identified genome size was 540 Mb, 90% bigger than that of *B. rapa*. However, despite this large genome expansion, *B. oleracea* was only estimated to contain 44,940 gene models, with expansion the result of a greater level of transposons (Liu *et al.*, 2014). Chalhoub *et al.* (2014) sequenced the genome of polyploid *B. napus* (oilseed rape, napus kale, swede), which originated approximately 7500 years ago from the combination of the *B. rapa* (A genome) and *B. oleracea* (C genome) genomes. The *B. napus* A subgenome, at 314.2 Mb, and C subgenome, at 525.8 Mb, are equivalent in size to genomes of the diploid progenitors.

Genome sequencing is now routine within *Brassica* species and the combination of long-read technology, such as PacBio and Nanopore sequencing, with chromosome conformation capture, such as Hi-C, results in high-quality genome assemblies. With the release of the *B. juncea* genome (Yang *et al.*, 2016), the *B. nigra* genome sequence (Perumal *et al.*, 2020) and the *B. carinata* genome (Song *et al.*, 2021), complete genomes are now available for all six species in the Brassica 'Triangle of U', providing extensive resources for comparative and functional genomics analyses of Brassicaceae species.

The increase in genome sequencing capability has allowed the identification of variation within genomes. Structural variants, namely presence/absence variants (PAVs) and copy number variants, mean reference genomes often cannot capture the entire gene contents of a species and, within crops, have been shown to lack agronomically important genes. This is particularly true in *B. oleracea* which demonstrates a vast diversity of crop types and morphology. Pangenomes, which combine sequence from multiple accessions, allow the capture of not only the core genome, containing sequences shared between all individuals of the species, but also those genes which are present within individual crop types or accessions. The *B. oleracea* pangenome comprises nine morphologically diverse *B. oleracea* varieties: two cabbage, a kale, a Brussels sprout, a kohlrabi, two cauliflower, a green broccoli and a wild relative, *Brassica macrocarpa*. The pangenome comprises 61,379 genes, 18.7% of which demonstrate PAVs in the varieties analysed (Golicz *et al.*, 2016). Agronomically important genes showing PAVs include flowering time and metabolite-related genes. In 2021, Bayer *et al.* published pangenomes comprising 87, 77 and 59 individuals for *B. oleracea*, *B. rapa* and *B. napus* respectively. This number of individuals is sufficient to capture the majority of gene content in the population (Bayer *et al.*, 2021).

Molecular markers and genetic maps are now commonly used in research and breeding. Historical markers such as isozyme, RFLP and random amplified polymorphic DNA markers have been replaced by simple sequence repeats and more abundant single nucleotide polymorphism (SNP) markers which can be mined from the genome sequence. These can be assayed using technologies such as genotyping by targeted sequencing (GBTS), the Brassica 60k Illumina Infinium SNP array (Clarke *et al.*, 2016) and Kompetitive Allele Specific PCR (KASP) (Semagn *et al.*, 2014). Shen *et al.* (2021) identified millions of SNPs via whole-genome sequencing of 23 representative green broccoli genotypes. A core set of 1167 SNPs was used to genotype 372 accessions of green broccoli representing most of the variability of green broccoli in China via GBTS and KASP to assay diversity and population structure. Peng *et al.* (2018) mapped quantitative trait loci (QTL) for clubroot resistance within an F_2 population of 94 individuals using 3218 SNPs assayed using the Infinium array.

Multiple studies to identify gene loci controlling traits of interest have been performed using QTL analysis. However, brassica vegetables have also been the subject of mapping methodologies made possible or more effective by the ease of sequencing. Bulked segregant analysis, now also known as QTL-seq, was first described by Michelmore *et al.* (1991) as the process by which common alleles must be present when DNA is bulked from a group of plants sharing the same phenotype. Consequently, two bulked pools of segregating individuals differing for a trait will differ only at the loci that harbour that trait. Modern genome sequencing and SNP marker identification increase the resolution, facilitating the identification of candidate variation associated with the trait of interest. Ce *et al.* (2021), used QTL-seq to identify two candidate genes controlling for clubroot resistance in a *B. oleracea* F_2 population.

A functional marker was developed exhibiting around 95% accuracy in identifying resistance within $56 F_2$ lines. This could be used for marker-assisted selection in breeding.

Genome-wide association studies (GWAS), based on the statistical association between a genetic variant and a trait of interest in a diverse panel of lines, also benefit from increased marker density. Associative transcriptomics (AT) is a form of genome-wide association analysis where transcriptome sequence is used to identify genetic variation for marker association. The transcriptome also provides gene expression levels from the tissue sequenced, therefore association analysis with gene expression level (Gene Expression Marker analysis) can also be performed. Woodhouse *et al.* (2021), developed an AT pipeline from a diverse panel of *B. oleracea*, identifying associations for heading and flowering traits under different environmental conditions. Gene expression marker association identified variation in expression of *BoFLC.C2* as a candidate for vernalization response. This pipeline is now a resource available for the study of further traits.

New aspects of brassicas

The wide genetic diversity and flexibility of *Brassica* taxa discussed in Chapter 1, and referred to in this chapter, has provided many opportunities for both natural and human-driven evolution. In domestication, these characteristics have resulted in a vast range of morphologically different brassica crop types, often individually tailored to particular locality and regional husbandry systems. In Europe, *B. oleracea*, and in Asia, *B. rapa*, especially have been intensively selected over many centuries of cultivation, producing plant forms that are markedly different in their horticultural and culinary traits. Botanically, however, they are closely similar and frequently difficult to characterize on strictly taxonomic criteria. Frequently, they may be divided solely by horticultural characteristics (Toxopeus, 1979).

In the 20th century, members of the family Brassicaceae have taken on new roles as ideal model forms for use in basic scientific research and more latterly as educational tools. The very genetic flexibility and diversity that allowed generations of horticulturists to hone new crop types as divergent as Brussels sprouts and cauliflower offers new sources of benefits to humankind as a route to fundamental genetic and evolutionary information. Indeed, it could be argued that the value of this family to humankind is set to increase substantially because of the roles of its members in the molecular and genetic biological revolutions which will be a central part of scientific endeavour in the 21st century. It might not stretch credulity too far to compare the emerging opportunities for the Brassicaceae as the plant science equivalent to stem cell research in human and animal biology.

Tools and resources for studying brassica genetics

By comparison with *Arabidopsis*, the long life cycle of *Brassica* species slows the rate of progress in understanding gene function. However, many tools and resources are now available for the study of *Brassica* gene function and the development of germplasm.

Mutagenized populations have been produced for the model diploid *Brassicas*, *B. rapa* R-o-18 and *B. oleracea* DH1012. Unlike the *Arabidopsis* T-DNA insertion mutants, to create the populations, *Brassica* seeds were chemically mutagenized by treatment with ethyl methane sulfonate to primarily produce single G:C to A:T base-pair transitions via by guanine alkylation. Tissue was collected from the M_2 generation to produce DNA for screening for mutations within genes of interest and M_3 seed collected for the studies of these mutants. Mutational screens can be performed using techniques such as Targeting Induced Local Lesions IN Genomes (TILLING) and are available as a service via the John Innes Centre Molecular Genetics platform (https://www.jic.ac.uk/research-impact/technology-research-platforms/molecular-genetics/ (accessed 15 August 2023)). A *B. napus* population is also available. The *B. rapa* R-o-18 population has been subject to exome capture amplicon sequencing. Therefore, just as for *Arabidopsis* mutants, it is possible to select and order mutants within genes of interest online.

Diverse germplasm collections represent natural variation within genes of interest as opposed to that induced by chemical mutation but can be screened for variation in a similar way, known as EcoTILLING. Multiple germplasm collections for *Brassica* species are now available.

Generation speed can be increased using a method known as 'speed breeding' which utilizes an extended day length of 22 h, combined with suitable light spectra and levels, and elevated temperature. Generation times in *B. rapa* R-o-18 and *B. oleracea* DH1012 were reduced from 41 to 36.5 days and 61.2 to 49.2 days, respectively, when grown under speed-breeding conditions as compared to control 16 h supplementary light (Ghosh *et al.*, 2018). Both lines used within this study are rapid-cycling varieties and do not require vernalization. It is currently unknown if speed breeding can override vernalization requirements or can be combined with vernalization treatment to reduce generation time in brassica vegetables.

Rapidly reproducing forms of *Brassica* named 'Fast Plants®' have been bred for use in association with inexpensive growing systems developed from recycled plastic containers at the University of Wisconsin–Madison, WI, USA. This combination offers an effective and efficient set of tools for biological studies at all levels of complexity from the research bench to school students. Depending on the genotype used, Fast Plants® will germinate in 1 day, grow and flower in 2 weeks and produce viable seed in little more than 28 days. Changes in the patterns of growth and development can be studied over 24-h periods with physiological changes monitored hourly. The rapid life cycle, small size and ease of growth of these plants makes them ideal for investigations into genetics, reproduction, physiology, ecology and growth (Greenler and Williams, 1990).

Fast Plants® resulted from a project that screened over 2000 *Brassica* accessions obtained from the United States Department of Agriculture (USDA) National Germplasm System (Williams and Hill, 1986). A few plants of each species screened flowered in significantly shorter times compared with the average. These faster flowering genotypes were developed further providing populations tailored for use in experiments made under controlled conditions. Combining genes from several early flowering forms provided material with even greater levels of accelerated flowering.

Rapid reproduction traits were combined with other characteristics such as diminutive size that allowed large numbers of plants to be grown in stand-ardized laboratory and classroom conditions. In use, the fast-flowering types of *Brassica*, and more recently *Raphanus*, are cultured at 24°C with continuous illumination of 250 μm/s/m² at high population densities.

Test populations were bred by interpollinating several early flowering types within each *Brassica* and *Raphanus* species. The criteria used in the selection process included characteristics such as: minimum time from sowing to flowering, rapidity of seed maturation, absence of seed dormancy, small plant size and high levels of female fertility. Populations of approximately 300 plants were used in each cycle of reproduction and 10% of the population that flowered most quickly was selected and mass pollinated to produce the next generation. Once the reduction in the average number of days to flowering was stabilized and when more than 50% of any individual population flowered within a 2–3-day period, the selection process for that population was stopped. Each population fulfilling these criteria was then increased by mass pollina-tion and designated as a 'rapid-cycling base population' (RCBP). The flowering and growth characteristics of each base population are given in Table 2.4. Curves comparing when six *Brassica* species flower for the first time are shown in Fig. 2.5. This transition from germination through to flowering and seed formation is shown pictorially in Fig. 2.6 for *B. rapa*.

The RCBPs are relatively homogeneous for plant habit and time to flower-ing. They retain considerable variation for other traits such as their response to pathogens causing disease. They respond strongly to changes in the cultural environment. Where greater space for foliage and root growth is provided the plants become much larger forming abundant quantities of seed.

These genotypes may be used as the safe repositories of genes for future use in molecular and conventional genetic breeding studies. Self-compatible stocks of the three diploid species are available together with tetraploid forms of *B. oleracea*, *B. rapa* and *R. sativus*, obtained by colchicine treatment of diploids, the latter being used to form triploids and subsequent trisomic and aneuploid forms. A range of mutant forms with traits, such as gibberellin responders, dark-green dwarfs, elongated internodes, chlorophyll deficien-cies and anthocyanin suppressors, offer characteristics for studies of inherit-ance and as molecular markers. As an aid for plant breeders a range of tester incompatibilities have been generated. Cell and protoplast lines of RCBPs of *B. oleracea* and *B. napus* regenerate simply and easily. Hence, they can be used

Table 2.4. Phenotypic characters of rapid-cycling *Brassica* and *Raphanus* base populations. (After Williams and Hill, 1986).

Crucifer Genetics Cooperative Stock Number	Species	Genome	Days to flowering (SD)	Length (cm) to first flower (SD)	Seeds per plant (SD)	Days per cycle	Cycles per year
1	*Brassica rapa*	Aaa	16 (1)	11.9 (3.1)	78 (54)	36	10
2	*Brassica nigra*	Bbb	20 (2)	27.1 (4.9)	69 (49)	40	9
3	*Brassica oleracea*	Ccc	30 (3)	22.6 (5.3)	18 (21)	60	6
4	*Brassica juncea*	Abaabb	19 (1)	29.6 (4.0)	107 (46)	39	9
5	*Brassica napus*	Acaacc	25 (2)	35.3 (7.1)	76 (53)	55	6
6	*Brassica carinata*	BCbbcc	26 (2)	41.7 (6.6)	67 (46)	56	6
7	*Raphanus sativus*	Rrr	19	n/a	n/a	48	7

Cultural conditions: 24°C and continuous light; data are expressed as mean values: (SD) = standard deviation; the cytoplasmic genome is indicated by the upper case and nuclear genome by the lower case: a = 10 chromosomes, b = 8 chromosomes, c and r = 9 chromosomes.

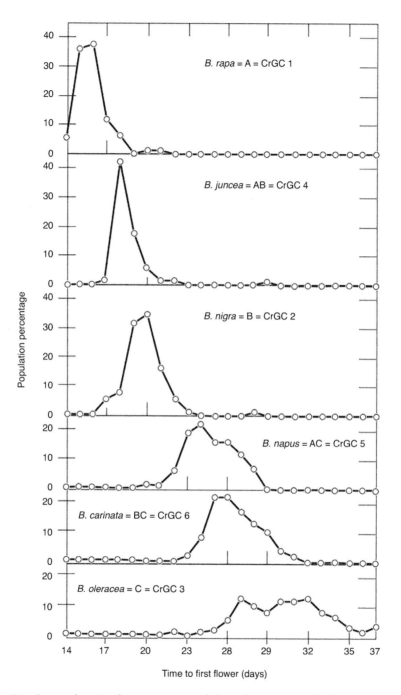

Fig. 2.5. Curves showing the percentages of plants flowering for the first time in rapid-cycling base populations of six *Brassica* species grown under standard conditions. (After Williams and Hill, 1986).

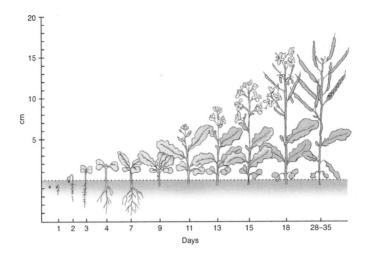

Fig. 2.6. Pictorial representation of germination, growth, flowering and seed set in rapid-cycling base populations for *Brassica rapa*. (After Williams and Hill, 1986; courtesy Garry Breeze).

with haploid embryo cultures of the natural polyploid species (*B. carinata*, *B. juncea* and *B. napus*) and offer great potential for use in breeding and genetic transformation studies. In addition to serving as repositories for nuclear genes, the RCPBs have shown especially useful properties for introducing the nuclei of various species into cytoplasms, expressing distinct phenotypes such as male sterility and triazine herbicide resistance.

Because the cytoplasmic traits are transmitted only through the female line, the nuclei of a species can be introduced into the cytoplasm of any other species by making crosses between selected parents. Normally, after making interspecific or intergeneric crosses the amount of seed set becomes very low and floral morphology and nectary functions become abnormal. In the RCBPs, normal levels of seed production and floral characteristics can be restored. An example of the potential of rapid-cycling *Brassica* stocks may be seen in breeding lines carrying multiple resistance to fungal, bacterial and viral pathogens. Resistance to *Plasmodiophora brassicae* (the causal agent of clubroot disease) from *B. rapa*, to powdery mildew (*Erysiphe cruciferarum*) and viral pathogens from Chinese cabbage (*B. rapa* ssp. *pekinensis*), and CMS from radish (*Raphanus* spp.) were combined. This provided a source for multiple pathogen resistance and CMS for superior Chinese cabbage lines that was constructed in less than 2 years.

Educational tools

The RCBPs offer a source of stimulating and interesting living material for use in classroom education at all levels of age and attainment. Most biology courses make use of model animal materials such as fruit flies (*Drosophila* spp.).

There is little plant material that can be used for courses in botany, horticulture, agriculture, forestry and general science education that matures sufficiently quickly so as to allow experiments to be completed within the 6–8 weeks required by many school and higher educational curricula. Use of RCBPs in such classes allows students to explore aspects of plant growth, development and reproduction, physiology, genetics, evolution and ecology. The RCBPs are well suited to this purpose since their hallmark is remarkably rapid development.

The plants can be induced to flower in 14–18 days, they are small and compact, and may be grown at high population densities (e.g. 2500 plants/m²) under continuous illumination in the classroom. The ease of cultivation and array of interesting morphologies makes these plants attractive models for both teachers and students. FastPlants® are becoming integrated into national and international educational programmes, particularly in the USA and UK, offering 'hands-on' experiences of the principles of plant biology.

The Crucifer Genetics Cooperative and international collaboration

In 1982, the development of rapid-cycling *Brassica* genotypes prompted the formation of the Crucifer Genetics Cooperative (CrGC) in the USA, as a vehicle for distributing seed and information among research workers and others around the world. Since then, CrGC has held regular meetings to discuss research findings. In 1994, the 9th CrGC Workshop was held in Lisbon, Portugal in collaboration with the 1st International *Brassica* Symposium, sponsored by the International Society for Horticultural Science (ISHS). This arrangement of combined International *Brassica* Symposia and CrGC Workshops has successfully led to further joint meetings. While considerations of oilseed rape (*B. napus*) are largely beyond the scope of this book, it is worth noting that there is also a series of International Rapeseed Congresses and International Plant Pathology Congresses which run parallel with the International *Brassica* Symposia, details of which are available from ISHS headquarters in Leuven, Belgium (www.ishs.org (accessed 15 August 2023)).

VERNALIZATION IN CROP SCHEDULING AND BREEDING

Ensuring a synchronous transition from the vegetative to the reproductive phase is important for maximizing the harvestable produce from brassica vegetables. For many brassica varieties, a period of prolonged cold exposure, known as vernalization, is required to induce flowering and as such it is a key trait of interest. The requirement for vernalization, or lack of, is determined by the plant's life history strategy, e.g. if the plant is a winter annual, perennial or biennial or whether it is rapid-cycling or a summer annual (Woodhouse *et al.*, 2021). In *B. oleracea*, vernalization-requiring biennials are represented by cabbage and Brussels sprouts, with summer annual crops including some green broccoli and cauliflower cultivars.

Much research has been performed on the control of flowering in the model plant *Arabidopsis*, with a core set of 306 genes implicated in the control of flowering time and available via the FLOR-ID database (Bouché *et al.*, 2016). However, most of the natural variation in *Arabidopsis* flowering time maps to two loci, *FRIGIDA* (*AtFRI*) and *FLOWERING LOCUS C* (*AtFLC*). In *Arabidopsis*, *FRI* confers a vernalization requirement and thus a winter annual habit, by increasing the expression of *FLC*, a repressor of flowering. There are two *B. oleracea FRI* orthologues (*BolC.FRI.a* and *BolC.FRI.b*). Analysis of the genetic variation at *BolC.FRI.a* within cultivated *B. oleracea* germplasm identifies two major alleles, which appear equally functional (Irwin *et al.*, 2012). Two indels have been found to be related to winter annual or biennial habit of cauliflower and cabbage. Guo *et al.* (2021) identified a deletion in the promoter region of *BolC.FRI.b* present within 98% of the cauliflowers studied, while the majority of cabbages (87%) showed the alternate allele.

Genetic variation at *FLC* is a major determinant of heading date by altering vernalization response. There are five *FLC* orthologues in *B. oleracea*. A deletion/loss-of-function mutation at *BolFLC.C2* in cauliflower, kale and green broccoli has been associated with an early flowering phenotype (Ridge *et al.*, 2015; Woodhouse *et al.*, 2021), indicating that *BoFLC.C2* has an equivalent role in *B. oleracea* to *FLC* in *Arabidopsis*. Irwin *et al.* (2016) identified two functional alleles of *BoFLC.C2* that both conferred a requirement for vernalization, but showed variation for vernalization sensitivity, and thus the length of vernalization required.

Under most conditions, summer cauliflower and green broccoli will flower without any artificial vernalization. However, under hot summer conditions in the eastern USA, some Dutch cauliflower cultivars can be so late that they only mature after the first frosts, which can destroy them. This indicates that these cauliflowers require a cooler temperature for vernalization and if dug with the curds in a prime market stage, flowering may be accelerated by storage for a month at 5°C. Biennial crops, such as some cauliflower types, cabbage, kale, kohlrabi, collards and Brussels sprouts, all require 2–3 months of vernalization below 10°C.

Insufficient vernalization extends the time between removal from the cool conditions to flowering by as much as 2–3 months. This results in an actual delay in time of flowering and often in incomplete or poor flowering or even devernalization after the seed stalk has been initiated. Consequently, an adequate duration of vernalization is essential. In most cases, 5°C is the ideal temperature for vernalization. Lower or higher temperatures result in the need for a longer vernalization period. Within *B. oleracea* crosses the more annual habit is dominant to the more biennial, but there is a continuous range in this variation indicating polygenic inheritance. Some *B. oleracea* genotypes also possess a juvenile phase during which it is not possible to trigger flowering by cold exposure. Plants must be pre-grown for 10 weeks or to the 8–10 leaf stage and 10–15 mm stem diameter before they will respond to vernalization. Prior to that stage the plant may not respond to reduced temperatures. Seed of *B. rapa and B. napus* can also be vernalized.

BREEDING RESISTANCE TO PATHOGENS

Comprehensive details of methods used to screen brassicas for pathogen resistance are provided by Williams (1981). The author also lists some of the rapid-cycling accessions of *Brassica* and their specific genes for use by breeders. Plant breeders from across the world have been active in incorporating resistance to the many pathogens attacking vegetable brassicas. However, resistance in most individual cultivars is limited to a few pathogens. Consideration of the successfulness of resistance breeding to pathogens, pests and disorders is given in Chapters 7 and 8.

Resistance and dominance

As with some other crops, such as the cucurbits, many resistances are dominant, although this is often due to major genes at specific loci. This holds true for white rust (*Cystopus candida*), black leg (*Leptosphaeria maculans*), black rot (*Xanthomonas campestris*), dark leaf spot (*Alternaria* spp.), downy mildew (*Hyaloperonospora parasitica*), powdery mildew (*Erysiphe cruciferarum*), cabbage yellows (*Fusarium oxysporum* f. sp. *conglutinans*), soft rot (*Erwinia carotovora* ssp. *carotovora*) and turnip mosaic virus. As such, these can be used in hybrid production, where one parent can be resistant to some pathogens and the other resistant to alternate pathogens, but the hybrid can have good tolerance to all pathogens (see Chapter 7 section, Strategies for Pest and Pathogen Control).

Juvenile vs mature plant resistance

In some cases, breeders have assumed both seedling and mature plant resistance resulted in resistance throughout the life of the plant and in all environments. It has become accepted that with many pathogens of brassicas, and other crops, that there may be both juvenile and mature plant resistance and these traits may be expressed separately. If this fact is not taken into account, the results can be confusing and misleading. For example, resistance to downy mildew, black rot, dark leaf spot and white rust all respond differently with plant age. It is also common for the expression of resistance to be affected by the environment in which a crop is grown (see Chapter 7 section, Strategies for Pest and Pathogen Control) and the emergence of populations compatible with individual resistances (see Fig. 7.1).

SELECTION TECHNIQUES FOR SPECIFIC CHARACTERS IN VEGETABLE BRASSICAS

Cabbage head shape

The desirable cabbage head shape in commerce has changed from pointed, flat or round to almost exclusively round. Pointed head is genetically dominant to round. Many genetic factors, however, influence head shape. Selection can best be made by cutting the head vertically through the core. This allows selection for head shape, core length, diameter and solidity, leaf toughness, leaf configuration within the head and leaf or rib size. The cut core will heal and lateral buds develop from it so that selected plants can be saved for seed production.

Heading vs non-heading

A distinguishing character of cabbage is the development of several wrapper leaves surrounding the terminal bud. As these leaves develop the head will become solid and develop a head or heart. Heading is recessive to non-heading. The F_1 between a non-heading or raceme type and a heading cabbage will be intermediate.

Depending how wide the cross, two genes (n_1 and n_2) or more are involved in heading. Likewise, the loose head of a Savoy type is recessive to the hard heading of a smooth-leaf cabbage. By the time a second backcross is reached within *B. oleracea*, the head type can be regained, provided reasonably large populations are used.

Head leaves

In a cross between lines with few and many wrapper leaves, few is dominant. There are modifying factors. The number of head leaves is also confounded by whether the cabbage is an early or late maturing type, the early maturing type having fewer leaves. The evaluation and selection are made following vertical splitting of mature heads with a knife.

Size of head

Head size is a quantitative trait and may be related to hybrid vigour, but not necessarily as open-pollinated lines can be just as large. Cabbage is responsive to day length and in northern latitudes (such as Alaska) the cabbages can become very large, weighing as much as 20 kg. Likewise, the shorter the day length as in winter, the smaller the head. Agronomic issues such as crop spacing and row spacing confound head size. Dense cropping naturally results in smaller heads.

Plant height

There is a major gene, 'T', for plant height. Most cultivars are recessive in this respect. But again, there are genetic modifiers. A long stem is an undesirable characteristic as the plant may fall over from the weight of the head. For machine harvesting (see Chapter 5 section, Introduction), however, the stem should not be too short and it is important that the head stands upright. Generally, in *B. oleracea* tall is dominant to short, but plant height involves several genes acting additively.

Core width

Wide core appears to be dominant to narrow. A narrow core is desirable, but less important than a shorter core. A wide core is usually softer or more tender than a thin core. Tough cores result in 'woodiness' caused by the deposition of lignified tissues in the fresh and processed products and are not desirable.

Core length

Core length is controlled by two incompletely dominant genes for short core. A short core of less than 25% of head diameter would be desirable. Fig. 2.7 demonstrates the undesirability of long cores.

Frame size

Older cultivars generally had large frames and basal leaves. In adapting cabbage to mechanized harvesting and in the effort to develop high-yielding cultivars, the trend has been towards smaller frames. The compact fresh market cultivars generally have smaller frames than processing types. An adequate frame is needed, however, for photosynthesis, but expression is influenced also by factors such as season, spacing and fertility. Adequate water and nutrition early in the plant

Fig. 2.7. Extended core in landrace cabbage Ladozskaja. (Geoff Dixon).

establishment and development will result in a larger frame, which also may cause the plant to develop tipburn (see Chapter 8 section, Physiological Disorders: Tipburn) if water is withheld or is in short supply during maturity.

Head splitting

Chiang (1972) identified three genes that control head splitting and act additively with partial dominance for early splitting. Narrow sense heritability for splitting was 47%. In 2015, Su *et al.* (2015) identified QTLs for head splitting resistance on chromosomes C3, C4, C7 and C9 based on 2 years of phenotypic data using both multiple QTL mapping and inclusive composite-interval mapping that collectively explained 39.4–59.1% of phenotypic variation. Three major QTLs (*Hsr 3.2, 4.2, 9.2*) showing a relatively large effect were robustly detected in different years. To evaluate cultivars for splitting they must be allowed to grow to full maturity. Long-cored cultivars usually split at the top of the head, while short-cored ones tend to split at the base.

Storability

Late, slow-growing cultivars are most suitable for storing. Dry matter of the better storage types is higher (9–10%) than standard and early cultivars (6–7%). The best storing cultivars usually have finely veined leaves. Eating quality is usually inferior because the leaves are tougher and harder. Selection for breeding for long-term storage is made most effectively by cutting the head from the stem as for storage.

At the end of the storage period, the selected heads can be placed on moist potting mix and roots will grow out from the base of the head, or the lateral buds can be allowed to develop and then be removed and rooted and allowed to produce an inflorescence and flowers. Cabbage used for coleslaw should be white internally, while cabbage for fresh market should preferably be green.

Winter cultivars

The best cultivars for growth in more southern regions of the UK and Europe in the winter, and those that may be subject to considerable cold, should be bolting resistant and require a long, cold induction before flower stems will develop. They are often slightly savoyed. Selection can best be done in the areas where they will be grown and subject to cold induction. They should be planted earlier than the commercial crop so that they are larger when the cold develops and are subject to longer periods of cold. Plants that do not bolt rapidly under such conditions must be bolting resistant and can be saved for seed production.

GREEN BROCCOLI BREEDING

Advances in the genetics and molecular breeding of green broccoli are reviewed by Han *et al.* (2021). Multiple studies using transcriptomics, next-generation sequencing, genetic mapping and comprehensive phenotyping have greatly increased the number of QTL and candidate regions controlling traits of interest. Despite this, further work is required before exploitation in breeding.

Heat tolerance

Susceptibility to high-temperature damage is one of the major problems in green broccoli. This makes the crop agronomically more suitable for cultivation in cool, moist climates. Heat at harvest time is not critical, except that high temperatures will reduce the period over which the crop is marketable. About 3–4 weeks prior to the head being marketable is the critical period when the growing point is differentiating to become reproductive (see Chapter 4 section, Climate Change). Bjorkman and Pearson (1998) studied the development stage at which the temperature response is most critical. The developing reproductive structure is most sensitive at the straightened and bowed stages which is a relatively short period of 4–5 days. Leaf bracts can enlarge and grow through the head.

When the plant is sensitive to heat the buds will become enlarged and grow through the head, due to elongation of the sepals. The degree of enlargement is quantitative and influenced by the amount of heat. Additionally, the secondary inner whorls of buds may turn yellow, causing a condition called 'yellow eye', 'pheasant eye' or 'starring' (see Fig. 2.8). Yellow eye tends to be dominant to normal bud formation. Also, leaflet extension may be encouraged in the head. Non-leafiness tends to be dominant to leafy. All three defects are enhanced or triggered by heat and can render the crop unmarketable. The best screening method is to grow the plants under normal ideal conditions. They are then moved to a hot environment (35–25°C day/night) for 10 days when the growing point is at the critical 'straightened and bowed' stage before returning the plants to a normal growing regime of 20–25°C. Multiple QTL for heat tolerance or susceptibility to permit breeding for heat tolerance have been identified.

The plants can be grown in the field in a location where high temperatures are expected when the plants are maturing and subsequently field selections are made for resistance to heat injury. Some seed companies have bred vigorously for heat tolerance and selections show considerable tolerance compared with most green broccoli hybrids. Yang *et al.* (1998) developed hybrid 'Ching-Long 45' with heat tolerance from the cross of green broccoli and an inbred involving Chinese kale (*B. oleracea* ssp. *alboglabra*).

Branching

Branching is also influenced by temperature, and a hot summer increases the degree of branching. Branching is dominant to non-branching. In the

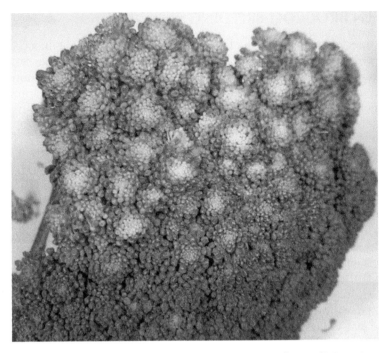

Fig. 2.8. 'Yellow eye', 'pheasant eye' or 'starring' in a green broccoli (*Brassica oleracea* var. *italica*) head. (Geoff Dixon).

most extreme non-branching type of green broccoli, the plant is similar to a cauliflower in that no lateral buds develop in the leaf axils and if the head is removed, no lateral buds will develop. Under these conditions, if the head is removed no seed can be produced on the saved plant. The degree of branching varies, however, depending on the cultivar, temperature and vernalization treatment. High temperature increases the degree of branching. Moderate branching is probably desirable, but excessive branching is not.

Storability

The need for improved harvest efficiency in green broccoli has resulted in screening of cultivars for concentration of yield and uniformity of harvest, as well as postharvest storage quality. McCall *et al.* (1996) stored green broccoli at 2°C for 14 days and thereafter at 12°C for a further 19 days. The cultivars kept well at 2°C, but deteriorated rapidly at 12°C. Modifications of this system have been used to select individual heads in segregating populations for quality and then for storability and considerable variation was apparent (see Chapter 8 section, Storage Environment). However, many handling factors support postharvest storage such as end of day harvest, controlled atmosphere and postharvest low-level light. Much more selection and breeding for shelf life is needed and preliminary observations show there is opportunity to develop more superior cultivars with potentially extended shelf lives.

Other characteristics

Solid stem is desirable in green broccoli and cutting the terminal head off the plant reveals the internal stem structure, and will permit selection. The plant should not be too tall, but a very short plant is not desirable. Most very tall plants do not have solid or large heads. Tallness is dominant to intermediate and short stem but can be stabilized at any desired height. Cabbage produces the tallest plants at flowering stage, compared with the other common vegetable brassicas. In green broccoli and cauliflower elongation is more limited, with little or no elongation of green broccoli beyond the natural flower position in the head at market stage.

The green broccoli head should be dome shaped and heavy at maturity. The trend is for new cultivars to have small buds or beads. Large bud is dominant to small bud. So far, small buds and heat tolerance have been hard to combine. Plant breeders are now developing hybrids which will be suitable for automated, robotic harvesting as compared with current manual processes.

CAULIFLOWER BREEDING

Cauliflower can be divided into several subgroups (see Chapter 1 section, *Brassica oleracea* (the cole or cabbage brassicas)) dependent on their origin from Australia, Europe (especially Italy), India and the USA. There are further subgroupings of cauliflower from different regions within European countries, such as France, Germany, Italy, Netherlands and Cornwall in south-west England. In general, the early cauliflowers, such as the Snowball types, are self-compatible while, progressively, the later cultivars are more self-incompatible. The Indian cauliflowers that are mostly early are quite self-incompatible. The Australian cauliflowers that are of intermediate to late maturity are quite self-compatible. The Australian cauliflowers are usually very large.

Careful handling of selected cauliflower plants when being dug for seed production is essential. Digging mature plants with root balls disturbed as little as possible, along with the removal of about half of the leaves, usually resulted in over 95% survival in the Cornell University breeding programmes. The curd will eventually elongate and produce flower stalks and plenty of flowers. If, however, the seed crop produced is not too heavy, and if the plant remains healthy, a plant will continue to develop flowers from undeveloped sections of the curd for several months.

Some breeders have had trouble with saving field selections. Grafting curd portions on to young stock plants has been fairly successful and does not delay seed production appreciably. Alternatively, tissue culture has been a means of vegetative propagation. Small pieces of the curd are removed from the parent plant, the surface sterilized and small pieces, 1–2 mm in diameter, are placed on paper wicks in nutrient solution. Roots and shoots develop and can be transferred to pots and grown to produce new plants (Crisp and Gray, 1975a, b, c).

Heat tolerance

Cauliflower is heat susceptible, but not to the extreme degree of green broccoli. Some of the Indian cauliflowers, such as Pusa Katki, have the greatest heat tolerance, but these are early and have relatively poor-quality curds. If the temperature is high then the curd may develop bracts (green bracts corresponding to axillary leaves are usually present in the curd, which make it unmarketable). Bracting is recessive, and heritability was reported at 73%. The curd may also produce true green leaves in the head (see Fig. 2.9).

The curd may also turn purple if exposed to the sun. This may be due to development of small, purple buds, which are similar to the small, white, velvety buds which develop in some curds and is called 'ricey'.

Growing the plants under hotter than ideal conditions will allow for selection for heat tolerance. Solidity of curd is also desirable and continued selection for this trait is required as soft-curded plants will bolt more readily and have a selective advantage. Solid tends to be recessive, although it is a quantitative factor, and there may be more than one genetic route to solidity, so that crosses of two solid curds does not necessarily mean the hybrid will be solid.

Curd colour

The most desired colour in cauliflower is a white or cream curd. In many cauliflowers, if the curd is exposed to the sun it will turn brownish-yellow. This is especially so for the Snowball type. Cauliflowers are protected from the sun either by wrapping the leaves over the head and tying them or developing plants with long leaves, which protect the curd from exposure. The upright leaf character is partially dominant with a heritability of 67% and three loci. This is particularly true for some of the later autumn and winter type cauliflowers but has been bred into some early types. The other alternative is to breed persistent whiteness into the plant. A cauliflower, Plant Introduction (PI) 183214, from Egypt had this character and remained completely white even when exposed to full sun (Dickson and Lee, 1980). Several recessive genes are involved and unless large F_2s are grown (over 250) there is little chance of obtaining persistent white curds. The curd of persistently white plants will remain white even after extensive elongation.

There are several curd colours, especially in Italian cultivars 'Autumn Giant' and 'Flora Blanca'. Italy is regarded as a centre of origin for the cauliflower (see Chapter 1 section, *Brassica oleracea* (the cole or cabbage brassicas)). Colours such as bright green (due to two genes), orange (*Or*, β-carotene gene), purple (*Pr*, anthocyanin genes) and yellow or golden are also available. The

Fig. 2.9. Bracting in Romanesco cauliflower observed under high light conditions. (Rachel Wells).

purple can be due to precociously developed flower buds that have developed on the curd and can assume the purple pigmentation of the buds. The green curds are more inherently frost tolerant than white curds (Gray, 1989).

The orange colour is due to β-carotene pigments and these can range from over 400 μg/100 g of tissue in a deep orange curd to about 3 μg/100 g in a very white curd (Dickson *et al.*, 1988). Orange curd is dominant to white (Crisp *et al.*, 1975) but the degree of colour is quantitative and influenced by the whiteness of the normal white curd parent. Therefore, crossing white curd parents with orange curd types results in an orange curd hybrid that is pale. Introgression of the '*Or*' gene led to the release of the first biofortified cauliflower in India, Pusa Beta-Kesari with 8–10 ppm β-carotene content. Introgression of the '*Pr*' gene into Indian cauliflowers led to the development of the 'Pusa' purple cauliflower, with intense purple curd colour and average anthocyanin concentration of 43.7 mg/100 g (Singh *et al.*, 2020).

SWEDE BREEDING

Brassica napus is a cool-season vegetable crop primarily grown in the northern countries of Europe and Canada, where it is also called rutabaga. The swollen root is pale orange-fleshed with the upper half being off-white or slightly yellow and the lower half brown. Desirable characters are uniform interior colour, smooth round exterior shape, smaller rather than larger leaf attachment and good storability. They have been sources of resistance to some populations of *P. brassicae* (clubroot) and these have been used in interspecific crosses. Also, the interspecific progeny from crosses to *B. oleracea* have been partially male sterile. In the latter, however, flowers are often fertile and the level of fertility is inconsistent from year to year, so that this has not been a satisfactory source of male sterility.

Turnip mosaic virus can be a major problem (see Chapter 7 section, Virus Pathogens) and there are wide differences in cultivar reaction. There are also different reactions among resistant lines, which are inherited independently. These are symptomless for necrosis or mosaic development. Various isolates differed in virulence but caused similar host reactions.

Cabbage maggot or root fly (*Delia brassica*) is a major pest problem in swede (*B. napus* ssp. *rapifera*) and breeding efforts to control it have had varying success. There does not appear to be immunity to this pest in either *B. napus* or *B. oleracea*, but some selections are much less susceptible than others.

BRUSSELS SPROUT BREEDING

Sprouts vary from early maturity (90 days) to late maturity (120 days or more) depending on vernalization requirements. Dark-green sprouts are preferred over pale sprouts. Buds can vary from small to large at maturity, but mid-sized ones are preferred. Likewise, the plants can vary in height. Taller plants have higher yield potential, but unless the crop is hand-harvested it is difficult to obtain uniform sprout maturity along the whole length of the stem. Sprouts that are planted too close together are difficult to harvest, while wide spacing will reduce yield. Easy detachment of the sprouts themselves is a very desirable trait, as is easy removal of the leaves.

Flavour varies widely from quite bitter to sweet, with the latter being preferred. Frost, however, results in improvement in flavour just as it does in root crops such as parsnip. Some new hybrids have much improved flavour without being chilled. Almost all sprouts now grown are hybrids, produced using SI.

Sprouts suffer from a number of pest and disease problems. Therefore, breeding for pathogen resistance is of key interest. Sprouts are susceptible to all the pathogens causing problems with other *B. oleracea*, including *Alternaria, Pseudomonas, Xanthomonas, Mycosphaerella, Verticillium, Albugo* and downy mildew (*Hyaloperonospora parasitica*). In the USA, the major disease problem is cabbage yellows (*F. oxysporum* f. sp. *conglutinans*) and some of the newer hybrids are resistant. Susceptible hybrids can develop serious rotting, especially of the lower buds. Pests include aphids, armyworms, cutworms, diamondback moths, caterpillars and flea beetles. However, research into resistance is ongoing. Ellis *et al.* (2000) tried to select for aphid resistance. Some glossy-leaved sprouts are resistant to aphids and caterpillars, but the glossy leaf is not acceptable to the consumer. Hondelmann *et al.* (2020) screened 16 commercialized Brussels sprout cultivars, identifying lines showing antixenosis resistance with significantly less infestation from cabbage whitefly (*Aleyrodes proletella*). Certain cultivars showed antibiosis resistance with significantly increased mortality, prolonged developmental times and reduced weights.

CHINESE CABBAGE BREEDING

Chinese cabbage (*B. rapa* ssp. *pekinensis*) types are as diverse as *B. oleracea*. Chinese cabbage is susceptible to almost all the pathogens, pests and environmental stresses affecting *B. oleracea*. Telekar and Griggs (1981) discuss issues of specific importance in the breeding of Chinese cabbage. The evolution of Chinese cabbage forms is shown in Fig. 1.7. Most types are readily cross compatible. As with *B. oleracea*, hybrids have been made using SI, but recently there has been an increasing interest in using male sterility, of both genic and cytoplasmic types.

REFERENCES

1001 Genomes Consortium (2016) 1,135 genomes reveal the global pattern of polymorphism in *Arabidopsis thaliana*. *Cell* 166(2), 481–491. DOI: 10.1016/j. cell.2016.05.063.

Ahmadi, B. and Ebrahimzadeh, H. (2020) In vitro androgenesis: spontaneous vs. artificial genome doubling and characterization of regenerants. *Plant Cell Reports* 39(3), 299–316. DOI: 10.1007/s00299-020-02509-z.

Akmal, M. (2021) Embryo culture and embryo rescue in *Brassica*. In: Aminul Islam, A.K.M., Hossain, M.A. and Mominul Islam, A.K.M. (eds) *Brassica Breeding and Biotechnology*. IntechOpen. DOI: 10.5772/intechopen.96058.

Arabidopsis Genome Initiative (2000) Analysis of the genome sequence of the flowering plant *Arabidopsis thaliana*. *Nature* 408(6814), 796–815. DOI: 10.1038/35048692.

Bannerot, H., Boulilard, L. and Chupeau, Y. (1974) Cytoplasmic male sterility transfer from *Raphanus* to *Brassica*. *Cruciferae Newsletter* 1, 52–54.

Bayer, P.E., Scheben, A., Golicz, A.A., Yuan, Y., Faure, S. *et al.* (2021) Modelling of gene loss propensity in the pangenomes of three *Brassica* species suggests different mechanisms between polyploids and diploids. *Plant Biotechnology Journal* 19(12), 2488–2500. DOI: 10.1111/pbi.13674.

Bhatia, R., Dey, S.S., Sood, S., Sharma, K., Sharma, V.K. *et al.* (2016) Optimizing protocol for efficient microspore embryogenesis and doubled haploid development in different maturity groups of cauliflower (*B. oleracea* var. *botrytis* L.) in India. *Euphytica* 212(3), 439–454. DOI: 10.1007/s10681-016-1775-2.

Bjorkman, T. and Pearson, K.J. (1998) High temperature arrest of inflorescence development in broccoli (*Brassica oleracea* var. *italica* L.). *Journal of Experimental Botany* 49(318), 101–106. DOI: 10.1093/jxb/49.318.101.

Bouché, F., Lobet, G., Tocquin, P. and Périlleux, C. (2016) FLOR-ID: an interactive database of flowering-time gene networks in *Arabidopsis thaliana*. *Nucleic Acids Research* 44(D1), D1167–D1171. DOI: 10.1093/nar/gkv1054.

Cardi, T. and Earle, E.D. (1997) Production of new CMS *Brassica oleracea* by transfer of 'Anand' cytoplasm from *B. rapa* through protoplast fusion. *Theoretical and Applied Genetics* 94(2), 204–212. DOI: 10.1007/s001220050401.

Ce, F., Mei, J., He, H., Zhao, Y., Hu, W. *et al.* (2021) Identification of candidate genes for clubroot-resistance in *Brassica oleracea* using quantitative trait loci-sequencing. *Frontiers in Plant Science* 12, 703520. DOI: 10.3389/fpls.2021.703520.

Chalhoub, B., Denoeud, F., Liu, S., Parkin, I.A.P., Tang, H. *et al.* (2014) Early allopolyploid evolution in the post-Neolithic *Brassica napus* oilseed genome. *Science* 345(6199), 950–953. DOI: 10.1126/science.1253435.

Chiang, M.S. (1972) Inheritance of head splitting in cabbage (*Brassica oleracea* L. var. *capitata* L.). *Euphytica* 21(3), 507–509. DOI: 10.1007/BF00039347.

Clarke, W.E., Higgins, E.E., Plieske, J., Wieseke, R., Sidebottom, C. *et al.* (2016) A high-density SNP genotyping array for *Brassica napus* and its ancestral diploid species based on optimised selection of single-locus markers in the allotetraploid genome. *Theoretical and Applied Genetics* 129(10), 1887–1899. DOI: 10.1007/s00122-016-2746-7.

Crisp, P. and Gray, A. (1975a) Propagating cauliflowers vegetatively. *The Grower* 83(17), 860–861.

Crisp, P. and Gray, A. (1975b) Cauliflower breeding by grafting curd portions on to stock plants. *The Grower* 83(18), 915–916.

Crisp, P. and Gray, A. (1975c) Tissue culture in cauliflowers. *The Grower* 83(19), 966–967.

Crisp, P., Walkey, D.G.A., Bellman, E. and Roberts, E. (1975) A mutation affecting curd colour in cauliflower (*Brassica oleracea* L. var. *botrytis* DC). *Euphytica* 24(1), 173–176. DOI: 10.1007/BF00147182.

Dickson, M.H. and Lee, C.Y. (1980) Persistent white curd and other curd characters of cauliflower. *Journal of the American Society for Horticultural Science* 105(4), 533–535. DOI: 10.21273/JASHS.105.4.533.

Dickson, M.H. and Wallace, D.H. (1986) Cabbage breeding. In: Basset, M.J. (ed.) *Breeding Vegetable Crops*. AVI Publishing, Westport, CT, pp. 395–432.

Dickson, M.H., Lee, C.Y. and Blamble, A.E. (1988) Orange-curd high carotene cauliflower inbreds, NY 156, NY 163, and NY 165. *HortScience* 23(4), 778–779. DOI: 10.21273/HORTSCI.23.4.778.

Ellis, P.R., Kift, N.B., Pink, D.A.C., Jukes, P.L., Lynn, J. *et al.* (2000) Variation in resistance to the cabbage aphid (*Brevicoryne brassicae*) between and within wild and cultivated *Brassica* species. *Genetic Resources and Crop Evolution* 47(4), 395–401. DOI: 10.1023/A:1008755411053. (accessed 8 August 2023).

Ghosh, S., Watson, A., Gonzalez-Navarro, O.E., Ramirez-Gonzalez, R.H., Yanes, L. *et al.* (2018) Speed breeding in growth chambers and glasshouses for crop breeding and model plant research. *Nature Protocols* 13(12), 2944–2963. DOI: 10.1038/s41596-018-0072-z.

Golicz, A.A., Bayer, P.E., Barker, G.C., Edger, P.P., Kim, H. *et al.* (2016) The pangenome of an agronomically important crop plant *Brassica oleracea*. *Nature Communications* 7, 13390. DOI: 10.1038/ncomms13390.

Gonai, H. and Hinata, K. (1971) Effect of temperature on pistil growth and phenotypic expression of self-incompatibility in *Brassica oleracea* L. *Japanese Journal of Breeding* 21(4), 195–198. DOI: 10.1270/jsbbs1951.21.195.

Gowers, S. (1989) Self-incompatibility interactions in *Brassica napus*. *Euphytica* 42(1), 99–103. DOI: 10.1007/BF00042620.

Gray, A. (1989) Green curded cauliflowers. *Journal of the Royal Horticultural Society* 114(1), 31–33.

Greenler, J. and Williams, P.H. (1990) Rapid-cycling *Brassica rapa* (Fast Plants) for hands-on teaching of plant science. In: McFerson, J.R., Kresovich, S. and Dwyer, S.G. (eds). *Proceedings of the 6th Crucifer Genetics Workshop*. Plant Genetic Resources Unit, Cornell University, Geneva, NY, p. 24.

Griffing, B. and Scholl, R.L. (1991) Qualitative and quantitative genetic studies of *Arabidopsis thaliana*. *Genetics* 129(3), 605–609. DOI: 10.1093/genetics/129.3.605.

Guo, N., Wang, S., Gao, L., Liu, Y., Wang, X. *et al.* (2021) Genome sequencing sheds light on the contribution of structural variants to *Brassica oleracea* diversification. *BMC Biology* 19(1), 93. DOI: 10.1186/s12915-021-01031-2.

Han, F., Liu, Y., Fang, Z., Yang, L., Zhuang, M. *et al.* (2021) Advances in genetics and molecular breeding of broccoli. *Horticulturae* 7(9), 280. DOI: 10.3390/horticulturae7090280.

Hondelmann, P., Paul, C., Schreiner, M. and Meyhöfer, R. (2020) Importance of antixenosis and antibiosis resistance to the cabbage whitefly (*Aleyrodes proletella*) in brussels sprout cultivars. *Insects* 11(1), 56. DOI: 10.3390/insects11010056.

Hundleby, P. and Chhetry, M. (2020) *Brassica oleracea* transformation. In: To, K.-Y. (ed.) *Genetic Transformation in Crops*. IntechOpen. DOI: 10.5772/intechopen.93570. (accessed 24 May 2023).

Irwin, J.A., Lister, C., Soumpourou, E., Zhang, Y., Howell, E.C. *et al.* (2012) Functional alleles of the flowering time regulator *FRIGIDA* in the *Brassica oleracea* genome. *BMC Plant Biology* 12(1), 21. DOI: 10.1186/1471-2229-12-21.

Irwin, J.A., Soumpourou, E., Lister, C., Ligthart, J.-D., Kennedy, S. *et al.* (2016) Nucleotide polymorphism affecting *FLC* expression underpins heading date variation in horticultural brassicas. *The Plant Journal* 87(6), 597–605. DOI: 10.1111/tpj.13221. (accessed 8 August 2023).

Ji, J., Huang, J., Yang, L., Fang, Z., Zhang, Y. *et al.* (2020) Advances in research and application of male sterility in *Brassica oleracea*. *Horticulturae* 6(4), 101. DOI: 10.3390/horticulturae6040101.

Jourdan, P.S., Earle, E.D. and Mutschler, M.A. (1989) Synthesis of male sterile, triazine-resistant *Brassica napus* by somatic hybridization between cytoplasmic male sterile *B. oleracea* and atrazine-resistant *B. campestris*. *Theoretical and Applied Genetics* 78(3), 445–455. DOI: 10.1007/BF00265310.

Karpechenko, G.D. (1928) Polyploid hybrids of *Raphanus sativus* L. × *Brassica oleracea* L.: on the problem of experimental species formation. *Zeitschrift fur Induktive Abstammungs- und Vererbungslehre* 48(1), 1–85. DOI: 10.1007/BF01740955.

Keller, W.A., Rajhathy, T. and Lacapra, J. (1975) *In vitro* production of plants from pollen in *Brassica campestris*. *Canadian Journal Genetics and Cytology* 17(4), 655–666. DOI: 10.1139/g75-081.

Kucera, V., Chytilova, V., Vyvadilova, M. and Klíma, M. (2006) Hybrid breeding of cauliflower using self-incompatibility and cytoplasmic male sterility. *Horticultural Science* 33(4), 148–152. DOI: 10.17221/3754-HORTSCI.

Laibach, F. (1943) *Arabidopsis thaliana* (L.) Heynh. als Objekt fürgenetische und entwicklungs-physiologische untersuchungen. *Botanisches Archiv* 44, 439–455.

Lange, W., Toxopeus, H., Lubberts, J.H., Dolstra, O. and Harrewijn, J.L. (1989) The development of Raparadish (× Brassicoraphanus, $2n = 38$), a new crop in agriculture. *Euphytica* 40(1–2), 1–14. DOI: 10.1007/BF00023291.

Leitch, I.J. and Bennett, M.D. (2003) Integrating genomic characters for a holistic approach to understanding plant genomes. *Biology International* 4, 18–29.

Liu, S., Liu, Y., Yang, X., Tong, C., Edwards, D. *et al.* (2014) The *Brassica oleracea* genome reveals the asymmetrical evolution of polyploid genomes. *Nature Communications* 5, 3930. DOI: 10.1038/ncomms4930.

Li, X., Sandgrind, S., Moss, O., Guan, R., Ivarson, E. *et al.* (2021) Efficient protoplast regeneration protocol and CRISPR/Cas9-mediated editing of glucosinolate transporter (*GTR*) genes in rapeseed (*Brassica napus* L.). *Frontiers in Plant Science* 12, 680859. DOI: 10.3389/fpls.2021.680859.

McCall, D., Sorensen, L. and Jensen, B.D. (1996) *Broccoli Varieties*. S.P. Rapport-Status Planteaulsforsog No. 8, 32 pp.

Meyerowitz, E.M. (1997) Plants and the logic of development. *Genetics* 145(1), 5–9. DOI: 10.1093/genetics/145.1.5.

Meyerowitz, E.M. and Somerville, C.R. (eds) (1994) *Arabidopsis*. Cold Spring Harbor Laboratory Press, Cold Spring Harbor, NY.

Michelmore, R.W., Paran, I. and Kesseli, R.V. (1991) Identification of markers linked to disease-resistance genes by bulked segregant analysis: a rapid method to detect markers in specific genomic regions by using segregating populations. *Proceedings of the National Academy of Sciences USA* 88(21), 9828–9832. DOI: 10.1073/pnas.88.21.9828.

Narayanapur, V.B., Suma, B. and Minimol, J.S. (2018) Self-incompatibility: a pollination control mechanism in plants. *International Journal of Plant Sciences* 13(1), 201–212. DOI: 10.15740/HAS/IJPS/13.1/201-212.

Neequaye, M., Stavnstrup, S., Harwood, W., Lawrenson, T., Hundleby, P. *et al.* (2021) CRISPR-Cas9-mediated gene editing of *MYB28* genes impair glucoraphanin accumulation of *Brassica oleracea* in the field. *The CRISPR Journal* 4(3), 416–426. DOI: 10.1089/crispr.2021.0007.

Ogura, H. (1968) Studies on the new male sterility in Japanese radish, special reference to the utilization of this sterility towards the practical raising of hybrid seeds. *Memoirs of the Faculty of Agriculture, Kagoshima University* 6(2), 39–78.

O'Malley, R.C., Barragan, C.C. and Ecker, J.R. (2015) A user's guide to the Arabidopsis T-DNA insertion mutant collections. *Methods in Molecular Biology* 1284, 323–342. DOI: 10.1007/978-1-4939-2444-8_16.

Paritosh, K., Pradhan, A.K. and Pental, D. (2020) A highly contiguous genome assembly of *Brassica nigra* (BB) and revised nomenclature for the pseudochromosomes. *BMC Genomics* 21, 887. DOI: 10.1186/s12864-020-07271-w.

Parkin, I.A.P., Gulden, S.M., Sharpe, A.G., Lukens, L., Trick, M. *et al.* (2005) Segmental structure of the *Brassica napus* genome based on comparative analysis with *Arabidopsis thaliana*. *Genetics* 171(2), 765–781. DOI: 10.1534/genetics.105.042093.

Pearson, O.H. (1972) Cytoplasmically inherited male sterility characters and flavor components from the species cross *Brassica nigra* (L) Koch × *B. oleracea* L. *Journal of the American Society for Horticultural Science* 97(3), 397–402. DOI: 10.21273/JASHS.97.3.397.

Pechan, P.M. and Keller, W.A. (1988) Identification of potentially embryogenic microspores in *Brassica napus*. *Physiologia Plantarum* 74(2), 377–384. DOI: 10.1111/j.1399-3054.1988.tb00646.x.

Pellan-Delourme, R. and Renard, M. (1988) Cytoplasmic male sterility in rapeseed (*Brassica napus* L.): female fertility of restored rapeseed with "Ogura" and cybrids cytoplasms. *Genome* 30(2), 234–238. DOI: 10.1139/g88-040.

Pelletier, G., Primard, C., Vedel, F., Chetrit, P., Remy, R. *et al.* (1983) Intergeneric cytoplasmic hybridization in cruciferae by protoplast fusion. *Molecular and General Genetics* 191(2), 244–250. DOI: 10.1007/BF00334821.

Peng, S., Nath, U.K., Song, S., Goswami, G., Lee, J.-H. *et al.* (2018) Developmental stage and shape of embryo determine the efficacy of embryo rescue in introgressing orange/yellow color and anthocyanin genes of *Brassica* species. *Plants* 7(4), 99. DOI: 10.3390/plants7040099.

Perumal, S., Koh, C.S., Jin, L., Buchwaldt, M., Higgins, E.E. *et al.* (2020) A high-contiguity *Brassica nigra* genome localizes active centromeres and defines the ancestral *Brassica* genome. *Nature Plants* 6(8), 929–941. DOI: 10.1038/s41477-020-0735-y.

Pruitt, R.E., Chang, C., Pang, P.P.-Y. and Meyerowitz, E.M. (1987) Molecular genetics and development of *Arabidopsis*. In: Loomis, W. (ed.) *Genetic Regulation of Development. 45th Symposium of the Society for Developmental Biology*. A.R. Liss, New York, pp. 327–338.

Ran, F.A., Hsu, P.D., Wright, J., Agarwala, V., Scott, D.A. *et al.* (2013) Genome engineering using the CRISPR-Cas9 system. *Nature Protocols* 8(11), 2281–2308. DOI: 10.1038/nprot.2013.143.

Ren, J.P., Dickson, M.H. and Earle, E.D. (2000) Improved resistance to bacterial soft rot by protoplast fusion between *Brassica rapa* and *B. oleracea*. *Theoretical and Applied Genetics* 100(5), 810–819. DOI: 10.1007/s001220051356.

Ridge, S., Brown, P.H., Hecht, V., Driessen, R.G. and Weller, J.L. (2015) The role of *BoFLC2* in cauliflower (*Brassica oleracea* var. *botrytis* L.) reproductive development. *Journal of Experimental Botany* 66(1), 125–135. DOI: 10.1093/jxb/eru408.

Sanderfoot, A.A. and Raikhel, N.V. (2001) *Arabidopsis* could shed light on human genome. *Nature* 410(6826), 299. DOI: 10.1038/35066726.

Semagn, K., Babu, R., Hearne, S. and Olsen, M. (2014) Single nucleotide polymorphism genotyping using Kompetitive Allele Specific PCR (KASP): overview of the technology and its application in crop improvement. *Molecular Breeding* 33(1), 1–14. DOI: 10.1007/s11032-013-9917-x.

Sharma, R.K., Choudhary, P.K. and Agarwal, A. (2020) Doubled haploid production in *Brassica oleracea* L.: a review. *Plant Archives* 20(2), 6293–6300.

Shen, Y., Wang, J., Shaw, R.K., Yu, H., Sheng, X. *et al.* (2021) Development of GBTS and KASP panels for genetic diversity, population structure, and fingerprinting of a large collection of broccoli (*Brassica oleracea* L. var. *italica*) in China. *Frontiers in Plant Science* 12, 655254. DOI: 10.3389/fpls.2021.655254. (accessed 24 May 2023).

Sigareva, M. and Earle, E.D. (1997a) Intertribal somatic hybrids between *Camalina sativa* and rapid cycling *Brassica oleracea*. *Cruciferae Newsletter* 19, 49–50.

Sigareva, M. and Earle, E.D. (1997b) *Capsella bursa-pastoris*: regeneration of plants from protoplasts and somatic hybridization with rapid cycling *Brassica oleracea*. *Cruciferae Newsletter* 19, 57–58.

Singh, S., Dey, S.S., Bhatia, R., Kumar, R. and Behera, T.K. (2019) Current understanding of male sterility systems in vegetable *Brassicas* and their exploitation in hybrid breeding. *Plant Reproduction* 32(3), 231–256. DOI: 10.1007/s00497-019-00371-y.

Singh, S., Singh, S.P., Singh, A. and Yadav, S. (2020) Molecular mapping and marker assisted selection for development edible colour, β-carotene and anthocyanin biofortification in cole and root crops. *Advances in Crop Science and Technology* 8(6), 1000457.

Somerville, C. (1989) *Arabidopsis* blooms. *The Plant Cell* 1(12), 1131–1135. DOI: 10.1105/tpc.1.12.1131.

Song, X., Wei, Y., Xiao, D., Gong, K., Sun, P. *et al.* (2021) *Brassica carinata* genome characterization clarifies U's triangle model of evolution and polyploidy in *Brassica*. *Plant Physiology* 186(1), 388–406. DOI: 10.1093/plphys/kiab048.

Stace, C. (2001) *New Flora of the British Isles*. Cambridge University Press, Cambridge.

Su, Y., Liu, Y., Li, Z., Fang, Z., Yang, L. *et al.* (2015) QTL analysis of head splitting resistance in cabbage (*Brassica oleracea* L. var. *capitata*) using SSR and InDel makers based on whole-genome re-sequencing. *PLoS ONE* 10(9), e0138073. DOI: 10.1371/journal.pone.0138073.

Telekar, N.S. and Griggs, T.D. (eds) (1981) *Chinese Cabbage, Proceedings of the First International Symposium*. Asian Vegetable Research and Development Centre, Shanhau, Taiwan.

Toxopeus, H. (1979) The domestication of Brassica crops in Europe: evidence from the herbal books of the 16th and 17th centuries. In: *Proceedings of the Eucarpia 'Cruciferae 1979' Conference, 1–3 October 1979*. Wageningen, Netherlands, pp. 29–37.

Walters, T.W., Mutschler, M.A. and Earle, E.D. (1992) Protoplast fusion-derived Ogura male sterile cauliflower with cold tolerance. *Plant Cell Reports* 10(12), 624–628. DOI: 10.1007/BF00232384.

Wang, X., Wang, H., Wang, J., Sun, R., Wu, J. *et al.* (2011) The genome of the mesopolyploid crop species *Brassica rapa*. *Nature Genetics* 43(10), 1035–1039. DOI: 10.1038/ng.919.

Whyte, R.O. (1960) *Crop Production and Environment*. Faber & Faber, London.

Williams, P.H. (1980) Bee-sticks, an aid in pollinating Cruciferae. *HortScience* 15(6), 802–803. DOI: 10.21273/HORTSCI.15.6.802.

Williams, P.H. (1981) *Screening Crucifers for Multiple Disease Resistance*. Department of Plant Pathology, University of Wisconsin, Madison, WI.

Williams, P.H. and Heyn, F.W. (1980) The origins and development of cytoplasmic male sterility in Chinese cabbage. In: Telekar, N.S. and Griggs, T.D. (eds) *Chinese Cabbage, Proceedings of the First International Symposium*. Asian Vegetable Research and Development Centre, Shanhau, Taiwan, pp. 293–300.

Williams, P.H. and Hill, C.B. (1986) Rapid-cycling populations of *Brassica*. *Science* 232(4756), 1385–1389. DOI: 10.1126/science.232.4756.1385.

Woodhouse, S., He, Z., Woolfenden, H., Steuernagel, B., Haerty, W. *et al.* (2021) Validation of a novel associative transcriptomics pipeline in *Brassica oleracea*: identifying candidates for vernalisation response. *BMC Genomics* 22, 539. DOI: 10.1186/s12864-021-07805-w.

Yang, J., Liu, D., Wang, X., Ji, C., Cheng, F. *et al.* (2016) The genome sequence of allopolyploid *Brassica juncea* and analysis of differential homoeolog gene expression influencing selection. *Nature Genetics* 48(10), 1225–1232. DOI: 10.1038/ng.3657.

Yang, Y.W., Tsai, C.C. and Wang, T.T. (1998) A heat-tolerant broccoli F_1 hybrid, 'Ching-Long 45'. *HortScience* 33(6), 1090–1091. DOI: 10.21273/HORTSCI.33.6.1090.

Yarrow, S.A., Wu, S.C., Barsby, T.L., Kemble, R.J. and Shepard, J.F. (1986) The introduction of CMS mitochondria to triazine tolerant *Brassica napus* L., var. 'Regent', by micromanipulation of individual heterokaryons. *Plant Cell Reports* 5(6), 415–418. DOI: 10.1007/BF00269630.

Yarrow, S.A., Burnett, L.A., Wildeman, R.P. and Kemble, R.J. (1990) The transfer of 'Polima' cytoplasmic male sterility from oilseed rape (*Brassica napus*) to broccoli (*B. oleracea*) by protoplast fusion. *Plant Cell Reports* 9(4), 185–188. DOI: 10.1007/BF00232176.

SEED AND SEEDLING MANAGEMENT

3

GEOFFREY R. DIXON*

School of Agriculture, Policy and Development, Earley Gate, Whiteknights Road, PO Box 237, University of Reading, Reading, Berkshire RG6 6EU and GreenGene International, Hill Rising, Horsecastles Lane, Sherborne, Dorset DT9 6BH, UK

Abstract

Factors controlling seed development, physiological maturation, genetic control of seed development and size, purity and vigour are identified as having prime importance for eventual sustainable crop production. There follow discussions of germination, seed enhancement, priming, coating and conditioning. Subsequently, the technology of seed placement in the field or alternatively its protected propagation in advance of transplantation is reviewed. Robotics and automation are having beneficial impacts on the efficiency of field placement of both seed and transplants. Monitoring subsequent growth characteristics is increasingly achieved by remote sensing using unmanned aerial vehicles or satellites.

Seed position within the siliqua is now recognized as having considerable effect on the performance of individual grains. Raising the biological performance of seed and subsequent seedlings via greater understanding of interactions between genotype and environment from parent plants to germination and subsequent field crops is a thread of knowledge and research running throughout this entire chapter.

*geoffrdixon@gmail.com

© Geoffrey R. Dixon and Rachel Wells 2024. *Vegetable Brassicas and Related Crucifers,* 2nd edition. (G.R. Dixon and R. Wells) DOI: 10.1079/9781789249170.0003

INTRODUCTION

Seed and its subsequent development into robust, reliable seedings and vigorous transplants are the fundamental elements on which profitable brassica crops are established. The biology of seed structure, formation, maturation, ripening and the processes of germination is discussed in this chapter, highlighting how viable plantlets are produced. Vitally important aspects include the distinctness, uniformity and stability of seed and seedling phenotypes which result from the work of plant breeders (see Chapter 2, Breeding Methodologies). Aspects of seed quality and its control, involving cultivar identification, establishing purity by laboratory testing, field trials, seed crop inspection, and appropriate handling and storage are discussed in detail by Kelly (1988), Basra (2006), Black *et al.* (2006) and Leskovar *et al.* (2014). Protecting the intellectual property rights of plant breeders for their new cultivars is governed by the International Union for the Protection of New Varieties of Plants (UPOV) based in Geneva, Switzerland and the UPOV convention first adopted in 1961. It was last revised in 1991, with the USA becoming a member.

SEED DEVELOPMENT

Seed is the most fundamental input in vegetable production on which the effectiveness of other inputs and outputs depends (Deleuran *et al.*, 2018). Viability, vigour, uniformity, purity and health of the seeds are quality parameters important for all production systems. The ideal would be seeds that germinate uniformly at a high percentage and subsequently grow relatively quickly and produce high-quality crops (see Fig. 3.1).

The important crop genus *Brassica* contains self-incompatible outbreeding species where *Brassica oleracea*, for example, displays strong parent-of-origin effects on seed development (Kaminski *et al.*, 2020). The development of brassica seed can be divided into three phases: cell division and early expansion in the first 14 days after anthesis (DAA), followed by reserve accumulation in the following 14–49 days and finally a period of dehydration over the final 7 days (Norton and Harris, 1975; Dasgupta and Mandal, 1993; Gurusamy and Thiagarajan, 1998). See also Chapter 2, Seed Development.

The origin of the parent plants producing the seed influences the detailed timing of each of these phases. Ancestral parents of cauliflower (*B. oleracea* var. *botrytis*), for example, originated in temperate, cool climates, hence its seed requires longer periods for development from flowering to maturity.

Seed coat (testa) morphology is a useful character for taxonomic and evolutionary studies. Zeng *et al.* (2004) proposed that variation in the patterns of seed coat development could provide a means for analysing the relationships among amphidiploids and their ancestral parents. Seed pods (siliquae) change colour from green to pinkish-yellow and brown by about day 49. As the seeds mature, their internal cellular membranes strengthen and become more capable of retaining solutes within the cell. Pro-anthocyanidins are the

Fig. 3.1. Comparison of growth stages – seed, seedling, juvenile plant and mature green broccoli (*Brassica oleracea* var. *italica*). (Geoff Dixon).

main flavonoids which affect the seed coat colour in *Brassica* species (Ren *et al.*, 2017). Based on studies of the *Brassica rapa* genome, these authors suggested that *TRANSPARENT TESTS GLABRA 1*, is a candidate gene responsible for controlling seed coat colour.

The quantities of free sugars and amino acids in cells increases from the start of seed formation up to 21 DAA. Thereafter, free solutes decline with accelerating speed as the seeds mature. The decrease in free, soluble compounds in the seed results from their transformation to bound, insoluble forms over the period of development (see Figs 3.2 and 3.3). The activity of dehydrogenase enzymes can be equated with seed quality and subsequent seedling vigour and relates inversely with seed moisture content (Khattra *et al.*, 1993). Enzyme activity increases slowly with advancing seed maturity reaching a maximum at 56 DAA (see Fig. 3.4).

In dry seeds of *Brassica napus*, *Nasturtium officinale*, *Lepidium sativum*, *Camelina sativa* and *B. oleracea* iron localizes in vacuoles of cells surrounding pro-vasculature in cotyledons and hypocotyl (Ibeas *et al.*, 2017). At the torpedo development stage, iron localizes in the nuclei including integument, free cell endosperm and almost all embryo cells. Later, iron is detected in cytoplasmic structures in different embryo cell types. Iron accumulates in nuclei in specific stages of embryo maturation before localization in vacuoles of cells surrounding pro-vasculature in mature seeds. Studies of the ultrastructure

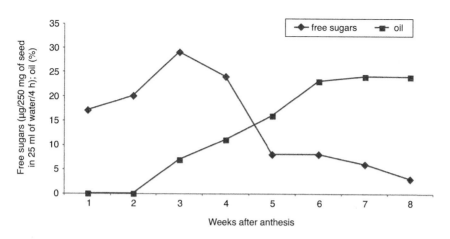

Fig. 3.2. Changes in free sugars and oil content of cauliflower (*Brassica oleracea* var. *botrytis*) seeds during development and maturation. (After Gurusamy and Thiagarajan, 1998).

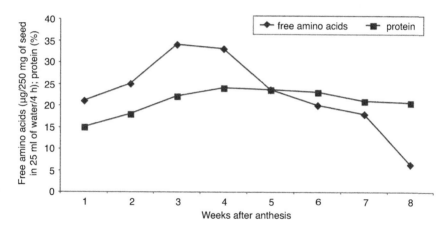

Fig. 3.3. Changes in free amino acids and protein in cauliflower (*Brassica oleracea* var. *botrytis*) seeds during development and maturation. (After Gurusamy and Thiagarajan, 1998).

of endosperm and embryo in *Eruca sativa* cv. Nemat demonstrated the presence of large amounts of lipids and glucosinolates (Alessio *et al.*, 2010). This suggested that the thick and abundant micropylar endosperm, completely surrounding the suspensor, may be an active source of nutrients for the embryo. Protein content increases linearly in the first 28 DAA and slowly thereafter. Oil content increases steadily up to seed harvest.

The activities of cytosolic and leucoplastic glycolytic pathway enzymes and oxidative pentose phosphate pathway enzymes increased by an average of 60% and 90%, respectively, between early (20 DAA) and mid-early (35 DAA)

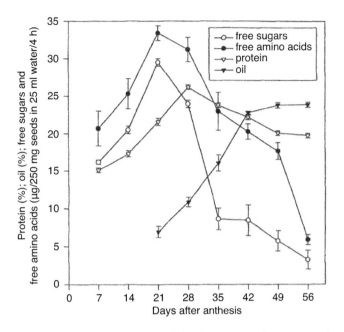

Fig. 3.4. Biochemical changes during seed development and maturation in cauliflower (*Brassica oleracea* var. *botrytis*). (After Gurusamy and Thiagarajan, 1998).

stages of seed development, and by an average of 40% and 70%, respectively, between early (20 DAA) and mid-late (50 DAA) stages (Yadav and Singh, 2005). This suggested the need for an exchange of metabolites in the two compartments through various translocators acting in cooperation and producing energy, reductants and carbon skeletons for different biosynthetic activities.

PHYSIOLOGICAL MATURITY

Length and width of maturing brassica pods increases towards maturity coinciding with seed swelling as they accumulate mass and storage compounds and as the moisture content falls. Physiological maturity is reached when the seed achieves maximum dry weight and this marks the end of the swelling phase of the seed (Shaw and Loomis, 1950). In red-headed cabbage (*B. oleracea* var. *capitata*) seeds, for example, physiological maturity occurred at or after maximum seed dry mass (Still and Bradford, 1998). In cauliflower (*B. oleracea* var. *botrytis*), physiological maturity coincides with the reduction of seed moisture content to about 40% (see Fig. 3.5). This may represent a normal part of maturation (McIlrath *et al.*, 1963) and again is influenced by the environment in which the seed-producing parents are grown (Sreeramulu *et al.*, 1992).

The precise timing of these phases depends on the particular *Brassica* species, its cultivars and the environments in which the seed crops are grown.

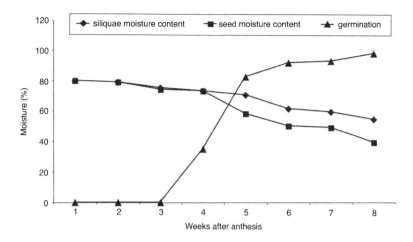

Fig. 3.5. Changes in moisture content and germinability of cauliflower (*Brassica oleracea* var. *botrytis*) seeds during development. (After Gurusamy and Thiagarajan, 1998).

Studies of green broccoli (calabrese) cv. Waltham-29, an open-pollinated form, showed, for example, that it matures more rapidly than cauliflower. Maximum seed weight was reached 42 days after pollination (DAP). Abscission of the seed from the placenta (funiculus) started 42–49 days after pollination and coincided with a sharp decline in fresh weight. Chlorophyll content of both seed testa and siliquae declined from 42 days after pollination and they acquired a tan and reddish-brown coloration, respectively, by 56 days. This is a common phenomenon with dry fruits, such as pods, where desiccation coincides with maximum dry weight accumulation. In mustard, for instance, there is a complete loss of chlorophyll from the seed coat by 42 days after pollination, while in green broccoli the majority of seed pods have shed (dehisced) their seed by 84 days after pollination (Jett and Welbaum, 1996). Typical changes in the appearance of seed pods during development are shown in Table 3.1, using red-headed cabbage as an example.

The maturity status and quality of seeds can be determined non-destructively (Jalink *et al.*, 1999) using measurements of chlorophyll fluorescence. A seed sample of white cabbage (*B. oleracea* var. *capitata*) with high chlorophyll fluorescence measurements equates with low quality.

Seed maturity is not an absolute prerequisite for the onset of germination. The *Brassica* embryo has the capability of germinating during development (Bewley and Black, 1994). Cauliflower germination may behave in a precocious manner and the speed of germination increases as seeds approach maturity (see Fig. 3.6).

Maximum germination rate for cabbage was reported at 48 days after pollination (Still and Bradford, 1994). Mature seeds require sufficient nutrients and other stored reserves to support germination and post-germination

Table 3.1. Moisture content, siliquae and seed characters of red-headed cabbage (*Brassica oleracea* var. *capitata*) during seed development. (After Still and Bradford, 1998).

Days after full bloom	Moisture content (%)	Siliquae characters	Seed characters
27	85	Waxy, dark green	Light green, liquid endosperm
33	80	Waxy, green, moist inside siliquae	Light green
44	60	Light green–yellow, slightly moist	Dark green, shiny, still moist inside, fibre beginning to form along the inside length of the siliquae
50	55	Light green–yellow, more fibrous	Dark dull green, not moist inside
54	50	Yellow–green, fibrous	Brown and black
56	40	Yellow, fibre covering entire inside of siliquae	Black, beginning to wrinkle
61	10	Yellow–white, brittle, dehiscent	All seeds black

activity until the seedling is capable of independent growth. Generally, the speed at which seedlings grow and increase in dry weight correlates with the maturity of the parent seed (see also Ellis *et al.*, 1987). Harrington (1972) suggested that physiological maturity of seed is defined as the developmental stage at which they achieve maximum viability and vigour. At this point nutrients cease entering the seed from the parent plant and ageing begins.

Generally, there may be trade-offs between fertility and offspring viability which underpins plant reproductive responses in suboptimal environmental conditions (Dogra and Dani, 2019). Senescence involves internal resource limitation, and it is a suboptimal physiological condition. Possibly senescence affects age-specific fertility and seed viability (quality) in indeterminate plants. Pod position can be an indicator of parental age. It then becomes a predictor of seed viability or seed quality. Seeds formed in low-fertility pods during late-senescence phases correlate with weakening of maternal control over seed-size optimization (see also Samec *et al.*, 2019).

Physiological maturity is the most appropriate time to harvest seed (Ellis *et al.*, 1987). In cauliflower, the period between 49 and 56 DAA is the point of physiological and harvest maturity. After this time, there are reductions in seed and siliquae mass and dimensions with deterioration commencing due to oxidation, volatilization, desiccation and loss of nutrients. By comparison, white mustard (*Sinapis alba*) reached physiological maturity 60 days after pollination (Fischer *et al.*, 1988). Immature mustard seed harvested without

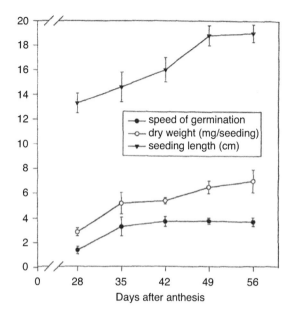

Fig. 3.6. Changes in the speed of germination, seedling dry weight and seedling length of cauliflower (*Brassica oleracea* var. *botrytis*) during development. (After Gurusamy and Thiagarajan, 1998).

drying germinated as early as 14 days after pollination, although desiccation tolerance was not attained until 35 days after pollination (Ahuja *et al.*, 1981).

GENETIC CONTROL OF SEED DEVELOPMENT

Interploidy crosses fail in many plant species, not least *Brassica* spp., due to abnormalities in endosperm development. Crossing within *Brassica* species, particularly *B. oleracea*, may fail due to the strong incompatibility and high mortality of embryos (Kaminski *et al.*, 2020; Stoute *et al.*, 2012). The pattern of physiological events in seed development follows identifiable sequences of gene expression. Proteins, carbohydrates and lipids build up in seeds during embryogenesis resulting in increased dry mass. The accumulation phase is terminated by ovule abscission and marked by maximum dry mass. Over 20,000 distinct genes are expressed at specific times during embryogenesis (Goldberg *et al.*, 1989), but there are a number of recognizable patterns. Expression continues after ovule abscission controlling the synthesis of storage compounds, preparation for desiccation, prevention of premature germination and the establishment of dormancy. It is suggested that a post-abscission programme of gene expression is completed before subsequent maximum seed vigour is attained.

SEED SIZE AND MATURITY

Botanically, dispersal and establishment ability influence evolutionary processes such as geographic isolation, adaptive divergence and extinction probability (see also Tyc *et al.*, 2020). There is evidence that the evolution of increased dispersal ability is associated with larger seed size, which may increase establishment ability and can also influence macro-evolutionary processes, possibly by increasing the propensity for long-distance dispersal (Willis *et al.*, 2014). As a result, speciation and consequent diversification rates are increased.

In *Brassica* species, determination of seed maturity is complicated by the indeterminate growth and extended flowering periods of the parent plants (Jett and Welbaum, 1996; see also Martinez-Laborde *et al.*, 2016). Overall inflorescence development continues for a considerable period of time within individual racemes and between different racemes. The processes of flower opening, pollination and fertilization each require an extended period of time, allowing the fruits and seeds to mature in rotation. Harvesting too early results in poor-quality, immature seed while delayed harvesting leads to possibly 50% seed losses due to shattering. Seeds are released when the ripe fruits dehisce or 'shatter' (Dickson and Wallace, 1987). Maturity is not, however, an absolute prerequisite for germinability – the embryo may be capable of germinating during development (Bewley and Black, 1994). In their study of cauliflower (*B. oleracea* var. *botrytis*), these workers found that germination occurred precociously but germination capacity and speed increased as the seeds matured. Importantly, mature seeds contain sufficient nutrients and other stored reserves to support germination and post-germination growth. Coincidentally, various curd sizes of the cultivar Pusa Snowball K-1 had a direct effect on seed yield and quality parameters in cauliflower (Khulbe *et al.*, 2010).

Where the entire raceme is harvested mechanically, seeds at different stages of maturity and differing in viability and vigour are inevitably mixed together. Seed maturity is associated with the presence of growth regulators, particularly abscisic acid (ABA). This compound suppresses germination early in development, stimulates the accumulation of storage reserves, imposes dormancy and determines the sensitivity of germination to the seed water potential (ψ). Also see Gowtham *et al.* (2021).

Each green broccoli (*B. oleracea* var. *italica*) seed pod, for example, contains 10–30 seeds and all are at slightly different stages of maturity. In some *Brassica* species, up to one third of the carbon in developing embryos was fixed by photosynthesis in the siliquae walls. But it is not known whether assimilates are supplied uniformly to developing ovules (Opeña *et al.*, 1988). Little is known of the effect of seed position within the *Brassica* pod and its subsequent impact on vigour. In other crops, such as soybean (*Glycine max*) and runner bean (*Phaseolus coccineus*), seeds produced at the stylar end of the pod (i.e. those formed first) are heavier than those formed at the peduncular end (formed last and closest to the flower stem). Generally, it appears that where fruits contain less than three seeds then their position has little or no influence on subsequent seedling vigour. Where there are more than five seeds per fruit then position in the pod begins to influence substantially subsequent seed size and the vigour of growing seedlings.

Senescence may affect age-specific fertility and seed viability (quality) in indeterminate annuals (Dogra and Dani, 2019). Maturation post-fertilization may be divided into three phases: (i) early-senescence (initial flowers), (ii) mid-senescence and (iii) late-senescence (wilting leaves). Seed viability probability as a function of pod position on the inflorescence (a proxy for parent's age) and seed position within pod was verified by germination tests in *Brassica* and then analysed using a binomial logistic regression model. This revealed age-specific fertility increased gradually, peaked and then declined significantly during senescence in *Arabidopsis* and *Brassica*. Seeds positioned closest to stigma tended to be heavier and more viable than others in highly fertile pods, characteristic of the mid-senescence phase in *Brassica*. Pod position (parent's age) was a significant predictor of seed viability probability or seed quality, which improved in old and senescing brassica. Proximity to stigma can increase seed quality. The unexpected increase in fertility and seed viability during the early-senescence phase is likely to be due to highly conserved developmental constraints on leaf and pod phenotype.

Studies of the cauliflower cv. Kibo Giant (Gurusamy and Thiagarajan, 1998) demonstrated that patterns of seed abortion and development within developing siliquae are non-random. Ovules located in the middle zone of a siliqua produced more mature seeds than ovules in stylar and basal zones in that order. The tetralocular form of *B. rapa* has a wide siliqua, containing more seeds than in the bilocular type (Lee *et al.*, 2018). The expression of genes coding for oleosins and fatty acid synthesis and elongation markedly increased at 28 through 35 days after flowering, respectively, remaining high thereafter. The expression of major storage protein genes increased at 28 or 35 days after flowering. The accumulation of seed storage compounds in tetralocular *B. rapa* was highly dependent on photo-assimilates in seeds and the siliqua wall. Overall, dynamic changes to transcript abundance of most genes in relation to seed-storage products occurred between 28 and 35 days after flowering. Lee *et al.* (2018) proposed that the tetralocular ovary in *B. rapa* may be a useful trait for improving seed yield and yield in *Brassica* species.

Brassica seeds are frequently sorted commercially and sold on the basis of size; large seeds are perceived to be more vigorous and have greater financial and agronomic value compared with small ones (Liou, 1987). The effects of seed size on emergence, growth and yield of many vegetables are well documented (McDonald, 1975). In green broccoli, for example, seed size has been correlated positively with seedling growth rate. Seedling dry weight, stand establishment and final crop head yield are improved by using larger, potentially more vigorous seeds (Heather and Sieczka, 1991). General rules concerning seed and seedling vigour are difficult to establish. For instance, no differences were detected in the germination rate of large cabbage seed compared with small ones (Jett and Welbaum, 1996). But there is a tendency for larger seed to produce larger seedlings at least in the early stages of growth (Komba *et al.*, 2007). Six commercial seed lots of two cultivars of kale (*B. oleracea* var. *acephala*) which varied in germination and vigour were size-graded into four categories: results did not support the hypothesis that large seeds have superior performance to small seeds, or that small seeds have lower vigour.

Vision-based, auto-sorting systems provide estimates of the shape, colour and textural features of seeds (Huang and Cheng, 2017) and efficiently sorted Chinese cabbage (*B. rapa* ssp. *pekinensis*) seeds. Chlorophyll fluorescence may also be used for sorting seeds (Yadav *et al.*, 2015). Tests with 14-year-old seeds of cabbage (*B. oleracea* var. *capitata*) showed that the germination of aged cabbage seeds could be improved by 16%. This technique may be used commercially for sorting seeds and upgrading quality for seeds where chlorophyll is present in the seed coat and persists during storage.

Seed shape may also influence subsequent performance. Image analysis has been used in the study of imbibition in white cabbage seed (Dell'Aquila *et al.*, 2000) describing changes in seed size during swelling. Subsequently, measurements were made for each seed using an area and a roundness factor as descriptors of seed size and shape, respectively (Dell'Aquila, 2004). Close correlations were established between the normalized values of the roundness factor and subsequent radicle length. This suggests that seed shape could provide estimates of potential radicle growth at the start of the visible germination. Additionally, seed grading may offer means for increasing the seed quality and consequently the transplant performance and stand establishment in the field (Silva *et al.*, 2018). Grading increased seed weight and especially vigour and large seeds resulted in increased transplant quality. This reinforced studies by Cardoso *et al.* (2002) who showed that small seeds produced lighter seedlings. There were increased numbers of plants not producing curds in cauliflower hybrid Shiromaru I as seed size decreased (Cardoso and da Silva, 2009). Inflorescence thinning is sometimes used as a means for improving seed quality and seed yield; however, Sukthong (2008) found this of little value for green broccoli–Chinese kale (*B. oleracea* ssp. *alboglabra*) hybrids.

SEED QUALITY

Seed that attains maximum viability and vigour is physiologically mature and thereafter deterioration commences (Powell and Matthews, 1984: Bhat *et al.*, 2017). Seed should be harvested, ideally, when it has attained physiological maturity and has not started to deteriorate. Reaching greatest dry mass is generally correlated with optimum maturity and quality as quantified by maximum germination in controlled tests. This state may not correlate, however, with the maximum seedling vigour under field conditions. For growers of vegetable brassica crops this difference has crucial significance. Seed is very expensive and forms a considerable element in the variable costs of growing each crop. The costs of propagating excess plants where the seed merchant makes an erroneously low estimate of viability forms a substantial financial burden for the grower. Simply expressing the quality of seed by germination and purity values does not reflect the potential field performance and the cost–benefits of the products from one seed house compared with another. Consequently, attempts have been made to establish laboratory tests that quantify the future potential vigour of seedlings in the field and how this translates into 'useable plants' that are placed in the field. Good quality seeds are the first line of security for producing high-quality crops (Vieira *et al.*, 2019). This requires information on the physiological and sanitary quality of seeds before sowing, complemented by germination tests. Combining germination with accelerated ageing tests permitted the ranking of seed lots of kale (*B. oleracea* var. *acephala*) seeds with transplants emergence.

A seed-quality test developed by Kataki and Taylor (1997) measured the anaerobic-to-aerobic (ANA) ratio of ethanol production and the relationship of the ANA ratio index to seed quality. Expanding the usefulness of this technique, Rutzke *et al.* (2008) tested cabbage (*B. oleracea* var. *capitata*) as a model determining whether ethanol production was influenced by ageing treatment, hydration level, seed integrity (grinding) and oxygen availability. Results suggested that lower ethanol production in aged ground seeds was caused by hypoxia following rapid hydration damage as a consequence of lost seed coat integrity.

The predictability of the seed germination was, for example, assessed using radish (*Raphanus sativus*), Chinese cabbage (*B. rapa* ssp. *pekinensis*) and European cabbage (*B. oleracea* var. *capitata*) using a logistic regression model with amino-acid leakage parameters, such as colour intensity and absorbance of seed leachate treated with ninhydrin reagent, as predictors. Seed germination could be predicted from both the colour intensity and absorbance of the seed leachate (Min *et al.*, 2013). Seeds of Chinese cabbage and European cabbage provided more robust results compared with those obtained from radish samples.

SEED VIGOUR

Seed vigour is a key trait essential during the production of sustainable and profitable brassica crops (Awan *et al.*, 2018). The genetic basis of variation in seed vigour has recently been determined in *B. oleracea*, but the relative importance of interactions with parental environments is unknown. These authors found, however, a consistent effect of beneficial alleles across production environments; but abiotic stresses during seed production had a large impact that enhanced the genetic differences in performance, measured as germination speed. The genotype–environment interaction revealed provided an evolutionary mechanism which minimized risk during subsequent germination, and might have resulted in development of seed traits such as reduced dormancy and more rapid germination observed during crop domestication. Earlier, Bettey *et al.* (2000) studied 105 doubled haploid *B. oleracea* lines derived from crossing Chinese kale (*B. oleracea* var. *alboglabra* line A12DHd) and green broccoli (*B. oleracea* var. *italica* line GDDH33). Results indicated that germination and pre-emergence seedling growth are under separate genetic controls. Quantitative trait loci (QTL) analyses revealed significant loci on linkage groups O1, O3, O6, O7 and O9. It is suggested that genes at these loci are important in determining predictable seed germination and seedling establishment.

Seed vigour has been defined as 'the sum total of the properties of seed which determine its potential activity and performance during germination and seedling emergence' (adapted from Perry, 1978). Seeds are subjected to varying degrees of water and temperature stress during their development, maturation, harvest and storage during seed production, and then again during rehydration and germination following sowing. Those with high vigour are capable of rapid germination and the establishment of healthy seedlings, resulting in more efficient crop production. Three key traits were identified as defining vigour: rapid germination, rapid initial downward growth of the radicle and a high potential for upward shoot growth in soil of increasing impedance (Finch-Savage *et al.*, 2010).

Plants respond to environmental stresses by forming a series of specific proteins. These include 'heat shock proteins' (HSPs) and 'late embryogenesis abundant proteins' (LEA proteins) (Bettey and Finch-Savage, 1998). These proteins also form in the absence of external stress as part of the normal developmental processes. They are found at specific stages in the growth cycle, especially during dehydration as occurs in pollen and seed formation. The HSPs prevent protein aggregation and act as molecular chaperones ensuring that protein folding is correctly completed. Late embryogenesis abundant proteins have conserved elements existing as amphiphilic α-helices that may be associated with protection from desiccation.

Both types of protein increase during the later stages of seed development; this is when seed vigour also increases. Hence, there can be a positive direct correlation between the occurrence of these two characteristics. It has yet to be established, however, whether stress protein content contributes directly to seed vigour. This question was investigated with batches of the F_1 hybrid white cabbage, *B. oleracea* var. *capitata* cv. Bartolo, using seed lots grown in different years, representing the product of genotype interactions with a range of environmental conditions and possessing high levels of viability. Differences between them were twofold, relating to vigour characters and abilities to withstand stressful conditions. HSP17.6 contributed to the performance of cabbage seed under stressful conditions and, therefore, is an important component of vigour. When the seed was subjected to 'rapid ageing' processes at 4°C for 10 days with 10% moisture content, HSP17.6 was unaffected and seed with the highest concentrations withstood subsequent stresses.

Since vigour results from a combination of characters, seed batches with high vigour relative to others as determined by one set of tests may have lower vigour when this is tested using a different set of environmental conditions. As a result of this variability a combination of tests is essential for the reliable prediction of seedling vigour. Conditions must be precisely controlled when testing for seed vigour and this is especially important when test periods are short, such as used for radicle emergence testing (Bowden and Landais, 2018). As an example of variations, they found results from radicle emergence tests of radish (*Raphanus* spp.) seeds were higher when tested in the dark compared with those made in high light.

The cabbage cv. Varazdin was monitored for germination and swelling during imbibition by combining a standard germination and a salt stress integrated germination test using a computer-aided image analysis system (Dell'Aquila, 2003). Their findings suggest that rapid image processing and recording of seed germination and size may represent an innovative technique for more accurate determination of any variation of seed hydration status. In addition, it was concluded that the measure of seed area may be taken as a good marker of seed vigour and tolerance to both advancing deterioration and salt stress. The use of computerized image analysis of seedlings potentially provides an efficient technique for the evaluation of the quality of seed lots. It is claimed to be quick in responding, simple in execution and reproduceable. Abud *et al.* (2017) concluded that it is possible to detect vigour differences among green broccoli seed lots by the computerized image analysis of seedlings with a seed-vigour imaging system (SVIS®).

Seed-storage conditions affect two key seed-quality traits: seed viability (ability to germinate and produce normal seedlings) and vigour (germination performance) (Schausberger *et al.*, 2019). Accumulated oxidative damage accompanies the loss of seed vigour and viability during ageing, indicating that redox control is key to longevity. Controlled deterioration in two *B. oleracea* genotypes with allelic differences at two QTLs that result in differences in ABA signalling and seed vigour were compared. Both ABA signalling and seed ageing affected seed vigour but not necessarily through the same biochemical mechanisms. Morris *et al.* (2016) used natural variation and fine mapping in the crop *B. oleracea* showing that allelic variation at three loci influences the key vigour trait of rapid germination. They proposed a mechanism by which both seed ABA content and sensitivity to it determined the speed of germination.

The genetic basis of variation in seed vigour has recently been identified in *B. oleracea* (Awan *et al.*, 2018). Seeds were produced in a range of maternal environments, including global warming scenarios, comparing lines with similar genetic background but different alleles (for high and low vigour) at the QTL responsible for determining seed vigour by altering ABA content and sensitivity. The genetic–environmental interaction revealed provides a robust mechanism of bet-hedging which minimizes environmental risk during subsequent germination (Gols *et al.*, 2015).

EXAMPLES OF GERMINATION AND VIGOUR TESTS

Standard germination tests

These are performed according to internationally accepted rules established by the International Seed Testing Association (ISTA) using prescribed temperature regimes. Detailed methodology established for testing seed by analysis was in accordance with the International Rules (Wellington, 1970). Example regimes could be either continuous 20°C or alternating 20°C and 30°C during dark and illuminated periods. Lighting regimes are also standardized. Radicle emergence is recorded daily for the first 5 days of incubation and counts of normal and abnormal seedlings are made after 10 days by suitably qualified seed analysts.

For research purposes, gathering cumulative germination data is a very laborious task that is often prohibitive for large experiments (Joosen *et al.*, 2010). The GERMINATOR package contains three modules: (i) design of experimental setup with various options to replicate and randomize samples; (ii) automatic scoring of germination based on the colour contrast between the protruding radicle and seed coat on a single image; and (iii) curve fitting of cumulative germination data and the extraction, recap and visualization of the various germination parameters. The curve-fitting module enables analysis of general cumulative germination data and can be used for all plant species. Results showed that the automatic scoring system works for *Arabidopsis thaliana* and *Brassica* spp. seeds, and is likely to be applicable for other species, as well. GERMINATOR is a low-cost package that allows the monitoring of several thousands of germination tests, several times a day, by one person.

Controlled deterioration test

The moisture content of seed is adjusted to 24% and they are sealed in laminated polyethylene–aluminium foil pouches and held at 1°C overnight allowing moisture to equilibrate throughout the seed sample. The sealed pouches are placed in a water bath at 45°C for 24 h (Matthews and Powell, 1987). The standard germination test is then made at 20°C and normal seedlings counted. Internationally agreed procedures for the validation of a controlled deterioration test for small-seeded vegetable species such as *Brassica* spp. are described by Powell and Matthews (2005). The controlled deterioration test involves the deterioration of seed samples from seed lots in a precise and controlled manner at an elevated moisture content (Chamling *et al.*, 2017).

Conductivity tests

Staining seed samples with 2,3,5-triphenyl tetrazolium chloride, which turns red when in contact with actively metabolizing tissues, provides a further means of assessing seed quality and subsequent potential seedling vigour (Burris *et al.*, 1969). Conductivity tests are completed in 24 h, which is significantly shorter than some other procedures (Hampton *et al.*, 2009). These authors tested seed lots of *B. napus*, *B. oleracea* var. *alboglabra*, *B. rapa* ssp. *pekinensis* and *B. rapa*. Standard germination of all 26 seed lots was related to subsequent field emergence.

Determination of initial seed viability (K_i)

Seed is brought to 18% moisture content, sealed in laminated polyethylene–aluminium foil pouches and held at 1°C for 3 days, allowing moisture to equilibrate throughout the sample. The pouches are placed in a germination cabinet at 40°C and sampled daily, for 10 days, to perform a standard germination test at 20°C. A germination test is also made directly after moisture equilibration (Ellis and Roberts, 1981).

Emergence tests

Trays are filled with commercial compost and each sown with 100 seeds at 40 mm depth and covered by a standard weight of compost. Subsequent seedlings are grown either under ambient conditions, at a standardized temperature (e.g. 20°C) or at an elevated temperature (e.g. 30°C). Emerged seedlings are removed and counted on a daily basis. Temperature profiles are monitored with thermistors inserted directly into the compost.

SEED PURITY

Control of seed purity is vital for all crops but is of especial importance with vegetable brassicas where the uniformity of plant stand is essential for the future profitability of the crop. Seeds are often contaminated by seeds with non-hybrid properties (Baranek and Raddova, 2015). The detection and elimination of these siblings is of particular importance where F_1 hybrids are predominantly used as cultivars, as with most brassica crops. The sibling problem arises because *B. oleracea* possesses a single locus, multi-allelic, sporophytic incompatibility system. Plant breeders usually maintain their lines by bud pollination (see Chapter 2, Assaying self-incompatibility). All incompatibility alleles are not equally effective, however, and varying amounts of self-fertilization may take place within inbred lines that are homozygous for S alleles. Additionally, the incompatibility reaction may be weakened by environmental factors, and the ratio of selfing to crossing can be affected by the behaviour of pollinating insects and the availability of alien pollen (Wills *et al.*, 1979).

Sibling plants are, at best, far less productive for the grower than the hybrid cultivar and represent a failure by the seed-producing company. Elimination is achieved by testing hybrid populations or by producing complex multi-parent hybrids where any siblings may also be productive. This may not always be feasible and increases the cost and complexity of a breeding programme. Quality control has generally been achieved by sampling hybrid populations as seed, cotyledons or leaves followed by visual inspection or by biochemical testing such as the use of iso enzyme markers. Visual inspection is not satisfactory since the morphological differences may require a long time period over which they emerge. Iso enzyme studies are suggested as quick, simple and accurate but again may not always be applicable since the high level of inbreeding within the parent lines of a hybrid cultivar means that they may express similar iso enzyme bands. Use of DNA probes could circumvent this problem. Image analysis also offers a further solution.

Image analysis of seeds, cotyledons and leaves in two dimensions and seeds in three dimensions, using populations of Brussels sprouts (*B. oleracea* var. *gemmifera*) and cabbage, determined the amount of sibling contamination in F_1 hybrid seed (Fitzgerald *et al.*, 1997). Image analysis has the advantages of being non-invasive, open to automation and is increasingly interactive and user-friendly. Studies so far have shown that use of image analysis with cotyledons or early adult leaves gives a more accurate evaluation of the number of siblings when compared with iso enzyme analysis. Development of image analysis offers the seed industry opportunities for much improved quality control. Sibling detection and vigour estimations may be made based on mean seed size and population variance. Commercial development requires refinement to the image analysis techniques, subsequent statistical analysis, greater precision in the selection of biological parameters to be analysed (seed part, organ or tissue of the developing seedling) and enhanced user-friendly computerized interfaces.

Several simple methods of DNA preparation can be used for the identification of S (self-incompatibility) haplotypes of breeding lines in green broccoli (*B. oleracea* var. *italica*) and cabbage (*B. oleracea*) and in purity tests of F_1 hybrid seeds (Sakamoto *et al.*, 2000). Culinary radish (*R. sativus*) crops are important worldwide and hybrid seed purity is crucially important. Random amplified polymorphic DNA (RAPD) markers will identify the presence of siblings (Huh and Choi, 2009). Potentially, three types of polymerase chain reaction-based markers, RAPD, inter simple sequence repeat and simple sequence repeat, are tools for cabbage hybrid seed purity determination (Liu *et al.*, 2007).

SEED ENHANCEMENT

'Seed enhancement' is an industrial term that has recently acquired scientific use and covers beneficial techniques applied to seeds between harvesting and sowing. The objective is to improve germination, seedling growth and raise the efficiency of seed delivery and associated materials at the time of sowing. Enhancement covers three aspects: pre-sowing hydration (priming), seed coating and seed conditioning (Taylor *et al.*, 1998; Halmer, 2008). Pre-sowing hydration treatments include non-controlled water uptake systems; methods in which water is freely available and not restricted by the environment and controlled systems; and methods that regulate seed moisture content preventing the completion of germination. Three techniques are used for controlled water uptake: priming with solutions, priming with solid particulate systems or by controlled hydration with water. Seed conditioning equipment upgrades seed quality by physical criteria. Integration of these methods can be performed to upgrade seed quality in brassicas that combines hydration, coating and conditioning. Upgrading is achieved by detecting sinapine leakage from non-viable seeds in a coating material surrounding the seeds. Seed coat permeability directly influences leakage rate, and seeds of many species have a semipermeable layer. The semipermeable layer restricts solute diffusion through the seed coat, while water movement is not impeded.

Seed priming

Seed priming benefits the speed, synchronization and uniformity of germination, often leading to improved stand establishment (Kikuti and Marcos-Filho, 2008; Baenas *et al.*, 2016; Hassini *et al.*, 2017). Seed priming is a commercially used technique for improving seed germination and vigour. It involves imbibition of seeds in water under controlled conditions which initiates the early events of germination, followed by drying the seed back to its initial moisture content (Varier *et al.*, 2010). Seeds require water, oxygen and a suitable temperature for germination. Water uptake follows a three-phase pattern with an initial rapid uptake or imbibition (Phase 1), followed by a lag period (Phase 2) and then a second increase in water uptake associated with seedling growth (Phase 3). Seeds are tolerant of desiccation during Phases 1 and 2 but frequently intolerant of it in Phase 3 (Taylor *et al.*, 1998). Water uptake may be either uncontrolled or controlled. In the former, water is freely available and not restricted by the environment. The seeds may be soaked or placed on moistened blotters. Soaking can involve the total immersion of seed in water with or without artificial aeration. Provided the seeds are viable and not dormant with a sufficient oxygen supply available and a suitable temperature, germination will proceed. The process is arrested at a specific time inhibiting the onset of Phase 3.

Homogeneity of field establishment from either direct-drilled crops or emergence from seed sown into modules increases the ultimate uniformity of crop growth and concentrates the harvest into shorter periods of time, increasing efficiency and decreasing costs. Where brassicas are grown below or above their optimal temperatures for germination and establishment, direct-seeded crops tend to be drilled at high rates leading to the need for subsequent thinning to achieve efficient final stand densities. Alternatively, seed priming techniques may be employed.

Priming is a controlled hydration process followed by redrying that allows the metabolic activities of germination to commence but not reach the stage of radicle emergence. Priming using polyethylene glycol (PEG) increased the germination rate of smaller seeds compared with larger ones. When seeds are rehydrated after priming, germination rate is increased and, in some cases, the temperature range for continued germination may be expanded. Two forms of priming treatments are available – either osmotic or matric. In the former, seeds are incubated in either solutions of low-molecular-weight inorganic salts, such as potassium nitrate, sodium chloride or potassium phosphate, or high-molecular-weight non-penetrating solutes, like PEG, at a water potential that is low enough to inhibit germination.

Priming has been applied successfully to several vegetable brassica crops. In particular, it will increase seedling vigour in cold, moist soils. Matric priming utilizes moistened solid carriers, such as the clay mineral, vermiculite or calcium silicate, to hydrate seeds. In several vegetable crops, matric-primed seeds germinate more rapidly compared with those that have been osmotically treated. To achieve osmotic priming, seeds are placed on two layers of filter paper saturated with the priming agent and sealed in containers, such as Petri dishes, for up to 7 days in darkness. For matric priming, water and calcium silicate are mixed thoroughly and placed in a sealed container for 24 h before adding the seeds. The containers are rotated every 12 h to ensure the uniform mixing of the carrier and seed for 7 days. After priming by either method, the seeds are washed for 2 min in tap water to remove either the osmoticum or the solid carriers, rinsed in distilled water and blotted dry. The seed is then dried with forced air at 37°C for 20 min and finally reduced to a moisture content of 5–6% on a dry weight basis by being placed over silica gel in a desiccator. Seeds used in Europe were often then coated with thiram (tetramethylthiuram disulfide) or other approved fungicides and insecticides, practices now discontinued on environmental and public health grounds. Subsequently, it was sealed into tinfoil packages or held in glass or plastic bottles at 4°C. The use of agrochemicals as seed dressings may continue elsewhere worldwide. The effectiveness of protecting soil from degradation, compaction and erosion over winter by growing cover crops is improved by priming the seed (Snapp *et al.*, 2008). Priming treatment increased the specific volume of the cabbage (*B. oleracea* var. *capitata*) seeds (Sakata and Tagawa, 2009). For organic crops, thermotherapy may be useful (Soriano *et al.*, 2019).

Seed coating

Seeds vary greatly in size, shape and colour (Costa and Taranto, 2005). Many seeds are small and irregular making singularization and precision placement difficult. Additionally, seeds require protection from pests and pathogens. Seed coating allows mechanical sowing in precise patterns to achieve uniformity of plant spacing and provides carriers for plant protectants and increasingly biostimulants. Seeds may be both pelleted and film-coated. Pelleting is defined as the deposition of a layer of inert material that transforms the original shape and size of the seed (see Figs 3.7 and 3.8). The shape is changed to spherical with increased weight and improved accuracy of placement. Film coating retains the original shape and size of the seed with minimal increase in weight. Since brassica seeds are naturally spherical film coating is more frequently used. Coatings may contain polymers, pesticides, biological agents such as bacteria, coloured markers or dyes and other additives depending on the regulations prevailing where the crops are grown and market acceptance.

Film coating – multi-layering options

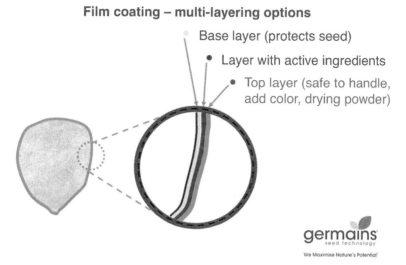

Base layer (protects seed)

Layer with active ingredients

Top layer (safe to handle, add color, drying powder)

Fig. 3.7. Film coating – multi-layers. (Courtesy Jordan Long, Germains Seed Technology).

Fig. 3.8. Film-coated brassica seeds. (Courtesy Jordan Long, Germains Seed Technology).

There are some indications that the European Union may require changes in the formulation of seed coatings, dispensing with the use of microplastics. In some instances, seed coating increases green broccoli (*B. oleracea* var. *italica*) nutrient content (Carvajal *et al.*, 2015). This is significant since green broccoli is a rich source of minerals which play an important role in human health. Capsulation treatment improves germination and vegetative growth and production. This may result from greater absorption of nitrogen, iron and higher carbon fixation.

Seed conditioning

Harvested seed is seldom pure and contains undesirable materials including poor-quality seed that requires removal. Conditioning has two objectives. First, removal of contaminants such as other crop or weed seeds and inert materials and results in pure samples of the desired cultivar. Second, conditioning eliminates poor-quality seed that may be immature, damaged or an undesirable size. Conditioning is usually carried through with a series of automated-grading stages controlled by microprocessors. Further improvement is achieved by exploiting specific physical characteristics such as seed colour. Colour-based sorting has improved greatly over the past couple of decades using optical systems, microprocessors and detection systems. Colour sorters that sense light from 360° have been developed. Light detectors can also quantify reflectance in the ultraviolet (UV), near infrared (NIR) and fluorescence wavelengths and have expanded the capacity of colour sorters.

Ultrasound technology is now being added to these systems, providing sorters that can discriminate between seeds by differences of size, shape, texture, colour, lustre, mass, hardness and mechanical damage. Ultimately, such systems will identify the presence of diseased seeds before symptoms are expressed visually. It has been demonstrated that while the peak value of the ultrasound waves decreased, the slope and bandwidth values increased where seeds were invaded by a pathogen.

Seed enhancement offers systems for the detection and removal of poor-quality seed from seed lots. Pre-sowing hydration elevates seed moisture content and reactivates cellular functions. Cell membrane integrity may be assessed indirectly by measurement of solute leakage. Seed coatings offer systems that hold compounds in close proximity to single seeds and colour sorting separates high- and low-quality seed. In sequence, pre-sowing hydration, seed coating and conditioning are applied to enhance seed-lot quality in brassicas.

A projected quality enhancement system can be developed for brassica seeds based around the presence of sinapine. This fluorescent alkaloid forms naturally during brassica seed development. It has a yellow colour at high alkaline pH (>10) and leaks from non-viable but not from viable seeds during germination. Measurement of sinapine leakage from brassica seed has proved more accurate at predicting germination success than electrolyte conductivity tests. Single-seed tests for sinapine leakage can form the basis for quality control in brassica seed lots. First, the seeds are hydrated by soaking in water for 4 h or primed in PEG solutions for 24 h. Sinapine leakage is greatest from the non-viable seeds. Freshly hydrated seeds are then coated with a filler containing finely ground cellulose. This acts as an adsorbent trapping the sinapine leachate in the coating. Coated seeds are dried and sorted by UV light into non-fluorescent and fluorescent individuals. Seed conditioning improved germination for cabbage (*B. oleracea* var. *capitata*), green broccoli (*B. oleracea* var. *italica*) and cauliflower (*B. oleracea* var. *botrytis*).

The efficiency of upgrading seed lots by exploiting sinapine leakage is compromised if the seed coat restricts leakage from non-viable seed. Non-viable seeds were undetected causing false negatives where the solute leakage failed to diffuse through the seed coat into the coating. The cutin content of the seed coat may affect permeability, retarding sinapine diffusion. Soaking the seed in a dilute solution of sodium hypochlorite (NaClO) enhanced leakage. This characteristic can be used for detecting poor-quality brassica seeds by sinapine leakage as demonstrated by Taylor *et al.* (1991). These authors used a range of brassica types: cabbages (*B. oleracea* var. *capitata* cvs Danish Ballhead and King Cole), cauliflowers (*B. oleracea* var. *botrytis* cv. Snowball), green broccoli (*B. oleracea* var. *italica* cv. Citation F_1), oilseed rape (*B. napus* cv. Westar) and the ornamental wallflower (*Erysimum hieraciifolium* cv. Orange Bedder) with deteriorated and non-deteriorated seed. Sinapine leakage from deteriorating seeds was 42–300% greater than from non-deteriorating seeds. As a result, a system was developed for upgrading seed quality by exploiting sinapine leakage. Seeds were hydrated and then coated with an absorbent (10% Pelgel) which traps the leaking sinapine.

Seed germination at low temperatures can be predicted using thermal time models. In green broccoli, for example, priming lowered the mean thermal time to germination but had little effect on the minimum temperature required for actual germination. Primed seeds germinated more quickly because of their lower thermal time requirement. Priming advanced germination more rapidly per unit of thermal time compared with non-primed seeds but did not reduce the minimum temperature for germination.

Similar effects are found with tomato (*Lycopersicon esculentum*) and onion (*Allium cepa*) which have no dormancy requirement and hence the minimum temperature for germination is genetically controlled, showing little variation within a particular species. Where priming lowers the minimum temperature for germination, this is a response to the substitution of priming for after-ripening, as in musk melon (*Cucumis melo*) overcoming some part of the dormancy requirement and expanding the temperature range for germination.

Root growth is more sensitive to variations above and below the optimal growing temperature compared with radicle emergence. Hence, poor stands of direct-drilled crops may be due to a lack of root growth rather than an impairment of radicle growth. Matrically primed seed tends to germinate more quickly than osmotically primed seed. It has been noted that the calcium content of matrically primed seed increases development (Jett *et al.*, 1996).

Induced defence allows plants to manage energy reserves more efficiently by synthesizing defence compounds only when needed. A risk of induced defence occurs when plants are challenged by herbivores – they may suffer considerable damage before the defence is mounted (Haas *et al.*, 2018). Priming can cause a state of readiness for the induction of the defence response, leading to a reduction in the damage received in an energy-efficient and less costly manner. In these studies, an objective was to verify whether seed coating with jasmonic acid and chitosan could prime plants against chewing and sap-feeding herbivores by affecting the herbivory of treated plants. *Brassica oleracea* var. *capitata* cv. Derby Day seeds were treated with jasmonic acid and chitosan, and colonized by diamondback moth (*Plutella xylostella*) second-instar and newborn nymphs of the green peach-potato aphid (*Myzus persicae*). Jasmonic acid reduced the mean relative growth rate of *P. xylostella* and led to 84% pre-imaginal mortality, whereas chitosan reduced oviposition. The intrinsic rate of increase in the green peach-potato aphid was raised by jasmonic acid. Seed coating containing these natural products could induce long-term defence priming in *B. oleracea* against chewing and sap-feeding insects.

GERMINATION

Studies of cellular and molecular events during seed germination provide information on the processes that are activated in the cell nucleus during the transition from a quiescent to an active state. Upon imbibition, the initial events are DNA, RNA and protein synthesis (Bewley and Black, 1994). The major component of the early DNA synthesis comprises a DNA repair process taking place in the first hours of germination (Osborne, 1983). The processes of DNA replication and accumulation of β-tubulin are two parallel but independent events during seed germination (Górnik *et al.*, 1997). Both RNA and protein synthesis have been demonstrated in leek seed (*Allium porrum*) at higher rates in germinating primed seed compared with untreated controls, while the rate of synthesis correlated positively with seed vigour (Bray *et al.*, 1989). Work with white cabbage (*B. oleracea* var. *capitata* cv. Bartolo) showed that DNA replication is not a prerequisite for radicle protrusion and the initial extension growth. The embryo in cabbage is sufficiently differentiated in dry, quiescent seed in order to produce a plantlet without further DNA replication or cell division following imbibition. But further seedling development, including root growth and root hair formation, appears to be dependent upon DNA replication. Respiration increases rapidly during imbibition corresponding with developing mitochondrial activities. Oxygen uptake has been correlated with rates of germination, seedling growth, emergence in the field and ultimately with crop yield. In white cabbage (*B. oleracea* var. *capitata* cv. Bartolo) oxygen consumption increases at imbibition and at germination (Bettey and Finch-Savage, 1996). These changes reflected increasing oxidation of carbohydrate reserves via respiratory pathways. Relative differences in the activity of key enzymes in these pathways correlated with the germination rates (T_{50} = time to 50% germination) where there were large differences in seed vigour. Enzyme activity equated with differences in the flux through glycolysis where seed lots differed substantially in vigour, but they were not determinants of seed vigour. Increases in the rate of germination (decreased mean germination time) were observed after aerated hydration of all seeds. Cauliflower (*B. oleracea* var. *botrytis*) seeds that underwent aerated hydration for 12 and 28 h at 20°C either improved or reduced storage potential of low- or high-vigour seeds, respectively (Powell *et al.*, 2000).

Epigenetic modifications to DNA can be inherited and may play a key role in evolution, with epigenetic influences on life history traits such as the timing of germination and flowering (Kottler *et al.*, 2018). They found that 5-azacytidine treatment affected the timing of germination and that this effect differed across populations. This treatment delayed germination in *B. rapa* Fast Plants®, which have been artificially selected for rapid cycling, flowering and quicker germination. Consequently, epigenetic modifications can influence phenotypic traits in ways that are dependent on genetic identity, life history and light availability (see also Perez-Balibrea *et al.*, 2011; Lema *et al.*, 2019).

Non-heading Chinese cabbage (*B. rapa* var. *chinensis*), when subjected to heat stress, loses growth and yield (Yarra and Xue, 2020). Numerous regulatory genes in various crops have been shown to contribute thermotolerance. Heat-stress-responsive genes were up-regulated in the transgenic plants subjected to high-temperature or heat-shock treatment, demonstrating the role of DAED-box RNA helicases in improving heat-stress tolerance of transgenic plants.

SEED DORMANCY

Some *Brassica* species exhibit postharvest dormancy. This can vary in time between 0 and 140 days and is prolonged by extremely dry or humid storage. Storage at 10–70% relative humidity (RH) promoted the release from dormancy during storage (Watanabe, 1953; Tokumasu *et al.*, 1975). Dormancy in brassica is thought to be very limited and removed after a brief period of dry storage (Opeña *et al.*, 1988).

In green broccoli (*B. oleracea* var. *italica*), germinability was first apparent in 10% of fresh (undried) seed by 28 days after pollination (Jett and Welbaum, 1996). Germination percentages increased in days 42–56 after pollination when essentially all fresh seeds were germinable. Seed drying increased germination at 42 days after pollination. In some species, it has been proposed that dehydration provided a switch, changing gene expression from a developmental programme to a germinative one. In green broccoli this did not appear to happen since dehydration was not a prerequisite for germination.

Physical sanitation methods are used by the seed industry as a means of preventing transmission of seed-borne diseases, but sensitivity varies between seed lots (Groot *et al.*, 2006). Harvesting seeds as mature as possible and removing less mature seeds during seed processing is an advisable policy. Chlorophyll fluorescence analyses provide a useful method of sorting *B. oleracea*. This would result in more efficient physical sanitation of seed lots.

SEED PLACEMENT

Ultimately, seed may be placed in the soil with the expectation of producing reliable and robust plants capable of providing a positive return on investment. Vegetable seed placement is achieved using the following methods:

- Drills – which sow small to large seeds closely in a row. Seed spouts into the row with little regulation for spacing adjustment, so thinning is required where close spacing is not desired. Drills are unsuited for brassicas because of the high price of seed and need for regular spacing.
- Plate drills – which pick up seeds individually placing them into a cell of the plate from where they drop into a furrow. Plate drills are suitable for medium- to large-sized seed and space out in the row by changing the spacing gears.
- Precision drills – which operate with a high degree of accuracy, picking up one seed at a time and spacing it in a furrow. This is achieved using a punched belt, vacuum or a seed-cup system for singling out one seed at a time. Precision drills are the most suitable and efficient drills for most brassica crops.

Drills of all types are mounted on a toolbar which provides for multi-row planting. Inter-row spacing is changed by adjusting the position of drill units on the toolbar. Mechanical precision seed drilling, whereby individual seeds are spaced regularly and accurately, revolutionized field brassica growing. Precision drills were introduced from the 1960s onwards replacing the arduous but highly skilled manual task of singling and spacing crops. Brassica seed, because of its ovoid, bead-like shape, is ideal for precision drilling. A seed bulk is placed in a hopper above a continuous rotating belt or disc (see Fig. 3.9) which has holes equivalent in diameter to individual seeds or the cups into which they are placed.

Metered volumes of seed are dropped out into a furrow formed ahead by a share and closed over by a following share. Precision belts or discs sow seed in single or multiple rows depending on the brassica crop and the density required for efficient cropping. Where legislation permits, these machines are capable of autonomous working (see Fig. 3.10).

Multiple rows placed on raised beds are ideally suited for quickly maturing brassicas, such as the baby leaf salads, salad rocket (*E. sativa*) and wall rocket (*Diplotaxis tenuifolia*) or white mustard (*S. alba*). Detailed manuals on the structure, operation and maintenance of drills are available from Stanhay Webb Ltd, Bourne, UK.

Fig. 3.9. Seed positioned on a rotating disc. (Courtesy of Stanhay Webb Ltd).

Fig. 3.10. Stanhay triple-bed seeder, capable of autonomous working. (Courtesy of Stanhay Webb Ltd).

TRANSPLANTING

Increasing demand for uniformity in the growth and maturity of specialized crops, such as green broccoli (*B. oleracea* var. *italica*), cauliflower (*B. oleracea* var. *botrytis*), Brussels sprouts (*B. oleracea* var. *gemmifera*) or Chinese cabbage (*B. rapa* ssp. *pekinensis*) is satisfied by using hybrid cultivars. Seed of these cultivars is expensive and not suitable, therefore, for direct field drilling. Transplants are produced at high density in modularized trays until they reach a growth stage suitable for field placement. Early studies of transplant production were discussed by Anon (1981). The market for transplants is now satisfied by highly automated, environmentally controlled husbandry. The structure of the brassica industry now effectively divides between two business sectors:

- *Plant propagators* grow the transplants from seed using protected glasshouse cultivation involving similar techniques to those used for producing flowering ornamental bedding plants. Technological developments in compost design that improved air-fill porosity, water and nutrient retention and pathogen control underpin this process (Baker, 1957; Anon, 2018).
- *Ware crop growers* negotiate contracts with the propagators for transplant raising and use the products for fresh vegetable production in the field.

Larger growers now frequently combine both businesses, simplifyingtheir logistics chains and reducing costs. This business structure alsoremoves difficulties which can arise when weather windows make immediatetransplanting an imperative.

Producing transplants

Supplies of compost for transplant raising, formulated for specific nutrient content, aeration and water-holding capacities and constituent peat or alternative bulk medium, are delivered by container lorries from the manufacturers. Aliquots of compost are elevated into a hopper above a conveyor belt on which are mechanically placed modular growing trays. These hold several hundred individual units and are filled automatically with precise portions of compost. The conveyor belt passes these filled modular trays under a seeding unit (see Fig. 3.11).

A reservoir of seed is held above this machine. Individual seeds pass down into injection units and are placed singly into each module. Seeded modular trays are transferred automatically into controlled environment germination chambers. These provide standardized temperatures encouraging uniform germination in a predicted time. Once germinated (see Fig. 3.12), the trays of seedlings are automatically moved into glasshouses maintained at regulated temperatures and illumination (see Fig. 3.13).

As the plants reach a required growth stage they are automatically removed from the glasshouse and taken for transplanting in the field, conforming with previously established production and harvesting schedules.

Until recently, the transplanting stage required manual plant handling and field placement – laborious tasks often performed in inclement weather. Transplanting machinery was trailed behind tractors and field staff fed individual plants, taken from the modules, into rotating shares which lightly gripped each plant and placed it at set intervals in the row. The row was opened out ahead of the plants and closed around them afterwards. Transplanting machines provided for several rows of plants being handled at once, forming crops growing in single or multiple rows in beds.

Fig. 3.11. Trays of modules filled with compost passing under an automatic seeder. (Courtesy of G's Growers Ltd).

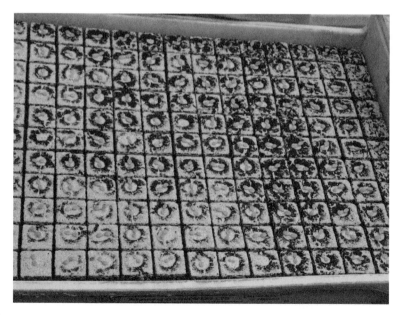

Fig. 3.12. Uniformly germinated seedings moving from controlled environment cabinets to glasshouse. (Courtesy of G's Growers Ltd).

Fig. 3.13. Glasshouse production of transplants prior to field placement. (Courtesy of G's Growers Ltd).

Automated and autonomous field placement

Automatic and autonomous transplanting machines have been developed whereby plants grown in modules can be placed in the soil directly (see Fig. 3.14).

This retains the process of module production in compost blocks and reduces very substantially the requirements for field staff. Transplanting can also be automated by changing the manner by which plants are raised and the machinery for field insertion. Plants are raised in bandoliers consisting of continuous belts of paper tubes filled with compost, as opposed to modular trays. When the plants reach the required growth stage the bandoliers are transferred to the field. They are then fed through transplanting equipment which automatically places the plants at required distances and in the desired cropping geometry. This requires considerable re-tooling of the transplant raising and transplanting systems which is expensive.

MONITORING GROWTH

Crop managers require real-time information on growth stage status for precise decision making of when applications of nutrients, water, crop protection agents and harvesting dates are scheduled (van der Heijden *et al.*, 2007).

Fig. 3.14. Automated brassica transplanting from glasshouse to field placement. (Courtesy of Richard Parish, Standen/Ferrari).

Remote sensing allows the estimating and mapping of individual plant growth stage variations within crops. Calibrating remote imagery, using fast and non-destructive close-range (below 1.3 m height) sensing equipment, allows this to take place over large areas of crops at relatively low costs. Reflected radiation can be recorded with an active close-range sensing device, consisting of visible light and NIR imaging spectrographs, and a three-CCD camera, equipped with special band filters (central wavelengths are at 600 nm, 710 nm and 800 nm) using four-band UltraCam digital CCD cameras (Jongschaap, 2007). Crop characteristics, such as leaf area index (LAI) and canopy nitrogen contents (LeafNWt), control light interception and photosynthetic capacity for growth development, crop maturity and yields. Optical measurements provide non-destructive estimates of LAI and leaf chlorophyll content. Remote sensing also assesses crop health, collecting information using unmanned aerial vehicles (UAVs), manned aircraft and satellite platforms (Ferguson and Rundquist, 2018). Monitoring methodologies using the relationship analysis of a normalized difference vegetation index (rNDVI) are advocated by Wang *et al.* (2018), with an algorithm accuracy of 90.6% when tested in winter wheat (*Triticum aestivum*) crops.

REFERENCES

Abud, H.F., Cicero, S.M. and Gomes Junior, F.G. (2017) Computerized image analysis of seedlings to evaluate broccoli seed vigor. *Journal of Seed Science* 39(3), 303–310. DOI: 10.1590/2317-1545v39n3174582.

Ahuja, K.L., Badwal, S.S. and Labana, K.S. (1981) Qualitative and quantitative changes in the seed and oil content of *Brassica juncea* mutants at different times of harvesting. *Foods for Human Nutrition* 31(1), 61–66. DOI: 10.1007/BF01093888.

Alessio, P., Mosti, S., Tani, G., di Falco, P., Lazzeri, L. *et al.* (2010) Ultrastructural aspects of the embryo and different endosperm compartments, in *Eruca sativa* Hill cv. Nemat (Brassicaceae) during Heart and Torpedo stages. *Caryologia* 63(2), 197–210. DOI: 10.1080/00087114.2010.10589727.

Anon (1981) *Propagating and Transplanting Vegetables. Ministry of Agriculture, Fisheries and Food (MAFF) Reference Book 344*. Department for the Environment, Fisheries and Rural Affairs (Defra), London.

Anon (2018) *Growing Media Review*. Agricultural and Horticultural Development Board (Horticulture) (AHDB), Stoneleigh, UK. Available at: https://horticulture.ahdb.org.uk (accessed 26 May 2023).

Awan, S., Footitt, S. and Finch-Savage, W.E. (2018) Interaction of maternal environment and allelic differences in seed vigour genes determines seed performance in *Brassica oleracea*. *Plant Journal* 94(6), 1098–1108. DOI: 10.1111/tpj.13922.

Baenas, N., Villano, D., Garcia-Viguera, C. and Moreno, D.A. (2016) Optimizing elicitation and seed priming to enrich broccoli and radish sprouts in glucosinolates. *Food Chemistry* 204, 314–319. DOI: 10.1016/j.foodchem.2016.02.144.

Baker, K.F. (ed.) (1957) *The U.C. System for Producing Healthy Container–Grown Plants, Manual 23*. California Agricultural Experiment Station, University of California Division of Agricultural Sciences, Davis, CA.

Baranek, M. and Raddova, J. (2015) Evaluation of selected SSR markers for their capability to control the quality of cabbage F1 hybrids production. *Acta Horticulturae* 1100, 131–134. DOI: 10.17660/ActaHortic.2015.1100.19.

Basra, A.S. (ed.) (2006) *Handbook of Seed Science and Technology*. Food Products Press, New York.

Bettey, M. and Finch-Savage, W.E. (1996) Respiratory enzyme activities during germination in *Brassica* seed lots of differing vigour. *Seed Science Research* 6(4), 165–174. DOI: 10.1017/S0960258500003226.

Bettey, M. and Finch-Savage, W.E. (1998) Stress protein content of mature *Brassica* seeds and their germination performance. *Seed Science Research* 8(3), 347–355. DOI: 10.1017/S096025850000427X.

Bettey, M., Finch-savage, W.E., King, G.J. and Lynn, J.R. (2000) Quantitative genetic analysis of seed vigour and pre-emergence seedling growth traits in *Brassica oleracea*. *New Phytologist* 148(2), 277–286. DOI: 10.1046/j.1469-8137.2000.00760.x.

Bewley, J.D. and Black, M. (1994) Development regulation and maturation. In: Bewley, J.D. and Black, M. (eds) *Seeds: Physiology of Development and Germination*. Plenum, Press, London, pp. 117–145. DOI: 10.1007/978-1-4899-1002-8.

Bhat, R., Rashid, Z., Dar, S.B. and Mufti, S. (2017) Seed yield and quality parameters of cabbage (*Brassica oleracea* var. *capitata*) in relation to different sources and levels of sulphur. *Current Agriculture Research Journal* 5(2), 177–183. DOI: 10.12944/CARJ.5.2.04.

Black, M., Bewley, J.D. and Halmer, P. (eds) (2006) *The Encyclopedia of Seeds: Science, Technology and Uses*. CABI Publishing, Wallingford, UK. DOI: 10.1079/9780851997230.0000.

Bowden, L. and Landais, L. (2018) The impact of light and high light on seed germination and the radicle emergence test. *Seed Science and Technology* 46(3), 465–471. DOI: 10.15258/sst.2018.46.3.03.

Bray, C.M., Davison, M., Ashraf, M. and Taylor, R.M. (1989) Biochemical changes during osmopriming of leek seeds. *Annals of Botany* 63(1), 185–193. DOI: 10.1093/oxfordjournals.aob.a087722.

Burris, J.S., Edge, O.T. and Wahab, A.H. (1969) Evaluation of various indices of seed and seedling vigour in soybean (*Glycine max* (L.) Merr). *Proceedings of the Association of Official Seed Analysts* 59, 73–81.

Cardoso, A.I.I. and da Silva, N. (2009) Influence of cultivar and seed size on cauliflower production. [Portuguese]. *Revista Ceres* 56(6), 777–782.

Cardoso, A.I.I., Nomura, E.S. and Silveira, V.N. (2002) Influence of cabbage seed size on seedlings production. [Portuguese]. *Cientifica (Jaboticabal)* 30(1/2), 53–61.

Carvajal, M., Martinez-Ballesta, M.C., Moreno, D.A., Bernabeu, J. and Garcia-Viguera, C. (2015) Seed coating increase broccoli nutrient content and availability after cooking. *Journal of Agricultural Science* 7(1), 182–191. DOI: 10.5539/jas. v7n1p182.

Chamling, N., Devhade, P.G. and Basu, A.K. (2017) Assessment of pattern in seed deterioration during ambient storage in some *Brassica* genotypes. *Journal of Crop and Weed* 13(2), 79–83.

Costa, M.A. and Taranto, O.P. (2005) The effect of initial size on the germination of pelleted broccoli seeds. *Transactions of the ASAE* 48(5), 1677–1680.

Dasgupta, S. and Mandal, R.K. (1993) Compositional changes and storage protein synthesis in developing seeds of *Brassica campestris*. *Seed Science and Technology* 21, 291–299.

Deleuran, L.C., Olesen, M.H., Shetty, N., Gislum, R. and Boelt, B. (2018) Importance of seed quality for the fresh cut chain. *Acta Horticulturae* 1209(5), 35–40. DOI: 10.17660/ActaHortic.2018.1209.5.

Dell'Aquila, A. (2003) Image analysis as a tool to study deteriorated cabbage (*Brassica oleracea* L.) seed imbibition under salt stress conditions. *Seed Science and Technology* 31(3), 619–628. DOI: 10.15258/sst.2003.31.3.11.

Dell'Aquila, A. (2004) Cabbage, lentil, pepper and tomato seed germination monitored by an image analysis system. *Seed Science and Technology* 32(1), 225–229. DOI: 10.15258/sst.2004.32.1.24.

Dell'Aquila, A., van Eck, J.W. and van der Heijden, G.W.A.M. (2000) The application of image analysis in monitoring the imbibition process of white cabbage (*Brassica oleracea* L.) seeds. *Seed Science Research* 10(2), 163–169. DOI: 10.1017/S0960258500000179.

Dickson, M.H. and Wallace, D.H. (1987) Cabbage breeding. In: Bassett, M.J. (ed.) *Breeding Vegetable Crops*. AVI Publishing, Westport, CT, pp. 395–432.

Dogra, H. and Dani, K.G.S. (2019) Defining features of age-specific fertility and seed quality in senescing indeterminate annuals. *American Journal of Botany* 106(4), 604–610. DOI: 10.1002/ajb2.1265.

Ellis, R.H. and Roberts, E.H. (1981) The quantification of ageing and survival in orthodox seeds. *Seed Science and Technology* 9(2), 373–409.

Ellis, R.H., Hong, T.D. and Roberts, E.H. (1987) The development of desiccation-tolerance and maximum seed quality during seed maturation in six grain legumes. *Annals of Botany* 59(1), 23–29. DOI: 10.1093/oxfordjournals.aob. a087280.

Ferguson, R. and Rundquist, D. (2018) Remote sensing for site-specific crop management. In: Shannon, D.K., Clay, D.E. and Kitchen, N.R. (eds) *Precision Agriculture Basics*. American Society of Agronomy, Crop Science Society of America, Soil Science Society of America, Madison, WI, pp. 103–118. DOI: 10.2134/precisionagbasics.

Finch-Savage, W.E., Clay, H.A., Lynn, J.R. and Morris, K. (2010) Towards a genetic understanding of seed vigour in small-seeded crops using natural variation in *Brassica oleracea*. *Plant Science* 179(6), 582–589. DOI: 10.1016/j. plantsci.2010.06.005.

Fischer, W., Bergfield, R., Plachy, C., Schäfer, R. and Schopfer, P. (1988) Accumulation of storage materials, precocious germination and development of desiccation tolerance during seed maturation in mustard (*Sinapis alba* L.). *Botanica Acta* 101(4), 344–354. DOI: 10.1111/j.1438-8677.1988.tb00055.x.

Fitzgerald, D.M., Barry, D., Dawson, P.R. and Cassells, A.C. (1997) The application of image analysis in determining sib proportion and aberrant characterization in F1 hybrid *Brassica* populations. *Seed Science and Technology* 25(3), 503–509.

Goldberg, R.B., Barker, S.J. and Perez-Grau, L. (1989) Regulation of gene expression during plant embryogenesis. *Cell* 56(2), 149–160. DOI: 10.1016/0092-8674(89)90888-x.

Gols, R., Wagenaar, R., Poelman, E.H., Kruidhof, H.M., van Loon, J.J.A. *et al.* (2015) Fitness consequences of indirect plant defence in the annual weed, *Sinapis arvensis*. *Functional Ecology* 29(8), 1019–1025. DOI: 10.1111/1365-2435.12415.

Górnik, K., de Castro, R.D., Liu, Y.Q., Bino, R.J. and Groot, S.P.C. (1997) Inhibition of cell division during cabbage (*Brassica oleracea* L.) seed germination. *Seed Science Research* 7(4), 333–340. DOI: 10.1017/S0960258500003731.

Gowtham, H.G., Duraivadivel, P., Ayusman, S., Sayani, D., Gholap, S.L. *et al.* (2021) ABA analogue produced by *Bacillus marisflavi* modulates the physiological response of *Brassica juncea* L. under drought stress. *Applied Soil Ecology* 159, 103845. DOI: 10.1016/j.apsoil.2020.103845.

Groot, S.P.C., Birnbaum, Y., Rop, N., Jalink, H., Forsberg, G. *et al.* (2006) Effect of seed maturity on sensitivity of seeds towards physical sanitation treatments. *Seed Science and Technology* 34(2), 403–413. DOI: 10.15258/sst.2006.34.2.16.

Gurusamy, C. and Thiagarajan, C.P. (1998) The pattern of seed development and maturation in cauliflower (*Brassica oleracea* L. *var botrytis*). *Phyton* 38(2), 259–268.

Haas, J., Lozano, E.R., Haida, K.S., Mazaro, S.M., Vismara, E. de S. *et al.* (2018) Getting ready for battle: do cabbage seeds treated with jasmonic acid and chitosan affect chewing and sap-feeding insects? *Entomologia Experimentalis et Applicata* 166(5), 412–419. DOI: 10.1111/eea.12678.

Halmer, P. (2008) Seed technology and seed enhancement. *Acta Horticulturae* 771, 17–26. DOI: 10.17660/ActaHortic.2008.771.1.

Hampton, J.G., Leeks, C.R.F. and McKenzie, B.A. (2009) Conductivity as a vigour test for Brassica species. *Seed Science and Technology* 37(1), 214–221. DOI: 10.15258/ sst.2009.37.1.24.

Harrington, J.F. (1972) Seed storage and longevity. In: Kozlowski, T.T. (ed.) *Seed Biology Volume 3*. Academic Press, New York, pp. 145–243.

Hassini, I., Martinez-Ballesta, M.C., Boughanmi, N., Moreno, D.A. and Carvajal, M. (2017) Improvement of broccoli sprouts (*Brassica oleracea* L. var. *italica*) growth and quality by KCl seed priming and methyl jasmonate under salinity stress. *Scientia Horticulturae* 226, 141–151. DOI: 10.1016/j.scienta.2017.08.030.

Heather, D.W. and Sieczka, J.B. (1991) Effect of seed size and cultivar on emergence and stand establishment of broccoli in crusted soil. *Journal of the American Society for Horticultural Science* 116(6), 946–949. DOI: 10.21273/JASHS.116.6.946.

Huang, K.-Y. and Cheng, J.-F. (2017) A novel auto-sorting system for Chinese cabbage seeds. *Sensors* 17(4), 886. DOI: 10.3390/s17040886.

Huh, M.K. and Choi, J.S. (2009) Seed purity test and genetic diversity evaluation using RAPD markers in radish (*Raphanus sativus* L.). *Korean Journal of Crop Science* 54(4), 346–350.

Ibeas, M.A., Grant-Grant, S., Navarro, N., Perez, M.F. and Roschzttardtz, H. (2017) Dynamic subcellular localization of iron during embryo development in Brassicaceae seeds. *Frontiers in Plant Science* 8, 2186. DOI: 10.3389/fpls.2017.02186.

Jett, W.L. and Welbaum, G.E. (1996) Changes in broccoli (*Brassica oleracea* L.) seed weight, viability, and vigour during development and following drying and priming. *Seed Science and Technology* 24(1), 127–137.

Jalink, H., van der Schoor, R., Birnbaum, Y.E. and Bino, R.J. (1999) Seed chlorophyll content as an indicator for seed maturity and seed quality. *Acta Horticulturae* 504, 219–228. DOI: 10.17660/ActaHortic.1999.504.23.

Jett, L.W., Welbaum, G.E. and Morse, R.D. (1996) Effects of matric and osmotic priming treatments on broccoli seed germination. *Journal of the American Society for Horticultural Science* 121(3), 423–429. DOI: 10.21273/JASHS.121.3.423.

Jongschaap, R.E.E. (2007) Sensitivity of a crop growth simulation model to variation in LAI and canopy nitrogen used for run-time calibration. *Ecological Modelling* 200(1/2), 89–98. DOI: 10.1016/j.ecolmodel.2006.07.015.

Joosen, R.V.L., Kodde, J., Willems, L.A.J., Ligterink, W., van der Plas, L.H.W. *et al.* (2010) GERMINATOR: a software package for high-throughput scoring and curve fitting of *Arabidopsis* seed germination. *Plant Journal* 62(1), 148–159. DOI: 10.1111/j.1365-313X.2009.04116.x.

Kaminski, P., Marasek-Ciolakowska, A., Podwyszyńska, M., Starzycki, M., Starzycka-Korbas, E. *et al.* (2020) Development and characteristics of interspecific hybrids between *Brassica oleracea* L. and *B. napus* L. *Agronomy* 10(9), 1339. DOI: 10.3390/agronomy10091339.

Kataki, P.K. and Taylor, A.G. (1997) Ethanol, a respiratory by-product: an indicator of seed quality. In: Ellis, R.H., Black, M., Murdoch, A.J. and Hong, T.D. (eds) *Basic and Applied Aspects of Seed Biology. Proceedings of the Fifth International Workshop on Seeds.* Springer, Dordrecht, Netherlands, pp. 421–427. DOI: 10.1007/978-94-011-5716-2.

Kelly, A.F. (1988) *Seed Production of Agricultural Crops.* Longman Scientific & Technical, Harlow, UK.

Khattra, S., Sharma, K. and Singh, G. (1993) The physiology of extremely desiccated *Brassica juncea* L. seeds. *Acta Agrobotanica* 46(2), 5–13. DOI: 10.5586/aa.1993.011.

Khulbe, H., Prabha, S.S., Khulbe, D. and Prasad, S. (2010) Effect of curd size on seed yield and seed quality parameters of cauliflower (*Brassica oleracea* var *botrytis* L.). *Trends in Biosciences* 3(2), 130–132.

Kikuti, P.A.L. and Marcos-Filho, J. (2008) Drying and storage of cauliflower (*Brassica oleracea* var. *botrytis*) hydroprimed seeds. *Seed Science and Technology* 36(2), 396–406. DOI: 10.15258/sst.2008.36.2.13.

Komba, C.G., Brunton, B.J. and Hampton, J.G. (2007) Effect of seed size within seed lots on seed quality in kale. *Seed Science and Technology* 35(1), 244–248. DOI: 10.15258/sst.2007.35.1.23.

Kottler, E.J., VanWallendael, A. and Franks, S.J. (2018) Experimental treatment with a hypomethylating agent alters life history traits and fitness in *Brassica rapa*. *Journal of Botany* 2018, 1–10. DOI: 10.1155/2018/7836845.

Lee, Y.-H., Kim, K.-S., Lee, J.-E., Cha, Y.-L., Moon, Y.-H. *et al.* (2018) Comprehensive transcriptome profiling in relation to seed storage compounds in tetralocular *Brassica rapa*. *Journal of Plant Growth Regulation* 37(3), 867–882. DOI: 10.1007/s00344-018-9784-0.

Lema, M., Ali, M.Y. and Retuerto, R. (2019) Domestication influences morphological and physiological responses to salinity in *Brassica oleracea* seedlings. *AoB Plants* 11(5), plz046. DOI: 10.1093/aobpla/plz046.

Leskovar, D.I., Crosby, K.M., Palma, M.A. and Edelstein, M. (2014) Vegetable crops: linking production, breeding and marketing. In: Dixon, G.R. and Aldous, D.E. (eds) *Horticulture Plants for People and Places. Volume 1 Production Horticulture*. Springer, Dordrecht, Netherlands, pp. 75–96. DOI: 10.1007/978-94-017-8578-5.

Liou, T.D. (1987) Studies on germination and vigour of cabbage seeds. PhD. Dissertation, The Agricultural University, Wageningen, Netherlands.

Liu, G., Liu, L., Gong, Y., Wang, Y., Yu, F. *et al.* (2007) Seed genetic purity testing of F1 hybrid cabbage (*Brassica oleracea* var. *capitata*) with molecular marker analysis. *Seed Science and Technology* 35(2), 477–486. DOI: 10.15258/sst.2007.35.2.21.

Martinez-Laborde, J., Ibanez, M., Rey-Mazon, E., Pita-Torres, N., Draper, D. *et al.* (2016) Assessment of genetic drift in *ex situ* conserved seeds of rocket (*Eruca vesicaria*) after regeneration. *Seed Science and Technology* 44(2), 342–356. DOI: 10.15258/sst.2016.44.2.15.

Matthews, S. and Powell, A.A. (1987) Controlled deterioration test. In: Fiala, F. (ed.) *The Handbook of Vigour Test Methods*. International Seed Testing Association (ISTA), Zürich, Switzerland, pp. 49–56.

McDonald, M.B. (1975) A review and evaluation of seed vigour tests. *Proceedings of the Association of Official Seed Analysts* 65, 109–139.

McIlrath, W.J., Abrol, Y.P. and Heiligman, F. (1963) Dehydration of seeds in intact tomato fruits. *Science* 142(3600), 1681–1682. DOI: 10.1126/science.142.3600.1681.

Min, T.G., Choi, B.S. and Hong, B.R. (2013) Predicting germination probability of radish (*Raphanus sativus* L.), Chinese cabbage (*Brassicarapa* ssp. *pekinensis*), and cabbage (*B. oleracea* var. *capitata* L.) seeds via amino acid leakage parameters. *Horticulture, Environment, and Biotechnology* 54(5), 388–398. DOI: 10.1007/s13580-013-0071-5.

Morris, K., Barker, G.C., Walley, P.G., Lynn, J.R. and Finch-Savage, W.E. (2016) Trait to gene analysis reveals that allelic variation in three genes determines seed vigour. *The New Phytologist* 212(4), 964–976. DOI: 10.1111/nph.14102.

Norton, G. and Harris, J.F. (1975) Compositional changes in developing rape seed (*Brassica napus* L.). *Planta* 123(2), 163–174. DOI: 10.1007/BF00383865.

Opeña, R.T., Kuo, G.C. and Yoon, J.Y. (1988) *Breeding and Seed Production of Chinese Cabbage in the Tropics and Subtropics. Bulletin 17*. Asian Vegetable Research and Development Centre, Shanhua, Taiwan.

Osborne, D.J. (1983) Biochemical control systems operating in the early hours of germination. *Canadian Journal of Botany* 61(12), 3568–3577. DOI: 10.1139/b83-406.

Perez-Balibrea, S., Moreno, D.A. and García-Viguera, C. (2011) Genotypic effects on the phytochemical quality of seeds and sprouts from commercial broccoli cultivars. *Food Chemistry* 125(2), 348–354. DOI: 10.1016/j.foodchem.2010.09.004.

Perry, D.A. (1978) Report of the vigour test committee 1974–1977. *Seed Science and Technology* 6, 159–181.

Powell, A.A. and Matthews, S. (1984) Application of the controlled deterioration vigour test to detect seed lots of Brussels sprouts with low potential for storage under commercial conditions. *Seed Science Technology* 12, 649–657.

Powell, A.A. and Matthews, S. (2005) Towards the validation of the controlled deterioration vigour test for small seeded vegetables. *Seed Testing International* 129, 21–24.

Powell, A.A., Yule, L.J., Jing, H.C., Groot, S.P.C., Bino, R.J. *et al.* (2000) The influence of aerated hydration seed treatment on seed longevity as assessed by the viability equations. *Journal of Experimental Botany* 51(353), 2031–2043. DOI: 10.1093/jexbot/51.353.2031.

Ren, Y., He, Q., Ma, X. and Zhang, L. (2017) Characteristics of color development in seeds of brown- and yellow-seeded heading chinese cabbage and molecular analysis of *Brsc*, the candidate gene controlling seed coat color. *Frontiers in Plant Science* 8, 1410. DOI: 10.3389/fpls.2017.01410.

Rutzke, C.F.J., Taylor, A.G. and Obendorf, R.L. (2008) Influence of aging, oxygen, and moisture on ethanol production from cabbage seeds. *Journal of the American Society for Horticultural Science* 133(1), 158–164. DOI: 10.21273/JASHS.133.1.158.

Sakamoto, K., Kusaba, M. and Nishio, T. (2000) Single-seed PCR-RFLP analysis for the identification of S haplotypes in commercial F_1 hybrid cultivars of broccoli and cabbage. *Plant Cell Reports* 19(4), 400–406. DOI: 10.1007/s002990050747.

Sakata, T. and Tagawa, A. (2009) Characteristics of water absorption and volume change in primed cabbage (*Brassica oleracea* L. var. *capitata*) seeds. *Transactions of the ASABE* 52(4), 1231–1238. DOI: 10.13031/2013.27765.

Samec, D., Kruk, V. and Ivanisevic, P. (2019) Influence of seed origin on morphological characteristics and phytochemicals levels in *Brassica oleracea* var. *acephala*. *Agronomy* 9(9), 502. DOI: 10.3390/agronomy9090502.

Schausberger, C., Roach, T., Stoggl, W., Arc, E., Finch-Savage, W.E. *et al.* (2019) Abscisic acid-determined seed vigour differences do not influence redox regulation during ageing. *Biochemical Journal* 476(6), 965–974. DOI: 10.1042/BCJ20180903.

Shaw, R.H. and Loomis, W.E. (1950) Bases for the prediction of corn yields. *Plant Physiology* 25(2), 225–244. DOI: 10.1104/pp.25.2.225.

Silva, P.P., Freitas, R.A., Lima, G.P. and Nascimento, W.M. (2018) Seed size, physiological quality and brassica transplant development. *Acta Horticulturae* 1204, 223–228. DOI: 10.17660/ActaHortic.2018.1204.29.

Snapp, S., Price, R. and Morton, M. (2008) Seed priming of winter annual cover crops improves germination and emergence. *Agronomy Journal* 100(5), 1506–1510. DOI: 10.2134/agronj2008.0045N.

Soriano, F., Claudio, M.T.R. and Cardoso, A.I.I. (2019) Germination of broccoli organic seeds treated with thermotherapy. *Acta Horticulturae* 1249, 73–77. DOI: 10.17660/ActaHortic.2019.1249.14.

Sreeramulu, N., Tesha, A.J. and Kapuya, J.A. (1992) Some biochemical changes in developing seeds of bambarra groundnut (*Voandzeia subterranea* Thouars). *Indian Journal of Plant Physiology* 35, 191–194.

Still, D.W. and Bradford, K.J. (1994) Development of seed quality in red cabbage. *HortScience* 29(3), 552 (Abstract). DOI: 10.21273/HORTSCI.29.5.552f.

Still, D.W. and Bradford, K.J. (1998) Using hydrotime and ABA-time models to quantify seed quality of brassicas during development. *Journal of the American Society for Horticultural Science* 123(4), 692–699. DOI: 10.21273/JASHS.123.4.692.

Stoute, A.I., Varenko, V., King, G.J., Scott, R.J. and Kurup, S. (2012) Parental genome imbalance in *Brassica oleracea* causes asymmetric triploid block. *Plant Journal* 71(3), 503–516. DOI: 10.1111/j.1365-313X.2012.05015.x.

Sukthong, M. (2008) Influence of inflorescence thinning on seed quality and seed yield of broccoli-Chinese kale hybrid in Loei province, Thailand. *Acta Horticulturae* 771, 83–88. DOI: 10.17660/ActaHortic.2008.771.11.

Taylor, A.G., Allen, P.S., Bennett, M.A., Bradford, K.J., Burris, J.S. *et al.* (1998) Seed enhancements. *Seed Science Research* 8(2), 245–256. DOI: 10.1017/ S0960258500004141.

Taylor, A.G., Min, T.G. and Mallaber, C.A. (1991) Seed coating system to upgrade Brassicaceae seed quality by exploiting sinapine leakage. *Seed Science and Technology* 19(2), 423–433.

Tokumasu, S., Kato, M. and Yano, F. (1975) The dormancy of seed as affected by different humidities during storage in *Brassica*. *Japanese Journal of Breeding* 25(4), 197–202. DOI: 10.1270/jsbbs1951.25.197.

Tyc, O., Putra, R., Gols, R., Harvey, J.A. and Garbeva, P. (2020) The ecological role of bacterial seed endophytes associated with wild cabbage in the United Kingdom. *MicrobiologyOpen* 9, e00954. DOI: 10.1002/mbo3.954.

van der Heijden, G.W.A.M., Clevers, J.G.P.W. and Schut, A.G.T. (2007) Combining close-range and remote sensing for local assessment of biophysical characteristics of arable land. *International Journal of Remote Sensing* 28(23/24), 5485–5502. DOI: 10.1080/01431160601105892.

Varier, A., Vari, A.K. and Dadlani, M. (2010) The subcellular basis of seed priming. *Current Science* 99(4), 450–456.

Vieira, J.F., Abreu Juinior, J.S., Castanho, F.R., Almeida, T.L., Villela, F.A. *et al.* (2019) Physiological and sanitary quality of kale seeds. *Acta Horticulturae* 1249, 209– 213. DOI: 10.17660/ActaHortic.2019.1249.40.

Wang, L.M., Liu, J., Yao, B.M., Ji, F.H. and Yang, F.G. (2018) Area change monitoring of winter wheat based on relationship analysis of GF-1 NDVI among different years. *Transactions of the Chinese Society of Agricultural Engineering* 34(8), 184–191.

Watanabe, S. (1953) Studies on the dormancy of seed in cruciferous vegetables. *Journal of the Japanese Society for Horticultural Science* 10, Abstract.

Wellington, P.S. (1970) Handbook for seedling evaluation. *Proceedings of the International Seed Testing Association* 35(2), 449–597.

Willis, C.G., Hall, J.C., Rubio de Casas, R., Wang, T.Y. and Donohue, K. (2014) Diversification and the evolution of dispersal ability in the tribe Brassiceae (Brassicaceae). *Annals of Botany* 114(8), 1675–1686. DOI: 10.1093/aob/ mcu196.

Wills, A.B., Fyfe, K. and Wiseman, E.M. (1979) Testing F1 hybrids of *Brassica oleracea* for sibs by seed isoenzyme analysis. *Annals of Applied Biology* 91(2), 263–270. DOI: 10.1111/j.1744-7348.1979.tb06498.x.

Yadav, S.K. and Singh, R. (2005) Developmental changes in activities of enzymes of glycolytic and pentose phosphate pathways in cytosolic and leucoplastic fractions of developing seeds of *Brassica campestris* L. *Physiology and Molecular Biology of Plants* 11(1), 71–80.

Yadav, S.K., Jalink, H., Groot, S.P.C., van der Schoor, R., Yadav, S. *et al.* (2015) Quality improvement of aged cabbage (*Brassica oleracea* var. *capitata*) seeds using chlorophyll fluorescence sensor. *Scientia Horticulturae* 189, 81–85. DOI: 10.1016/j. scienta.2015.03.043.

Yarra, R. and Xue, Y. (2020) Ectopic expression of nucleolar DEAD-Box RNA helicase *OsTOGR1* confers improved heat stress tolerance in transgenic Chinese cabbage. *Plant Cell Reports* 39(12), 1803–1814. DOI: 10.1007/s00299-020-02608-x.

Zeng, C.-L., Wang, J.-B., Liu, A.-H. and Wu, X.-M. (2004) Seed coat microsculpturing changes during seed development in diploid and amphidiploid *Brassica* species. *Annals of Botany* 93(5), 555–566. DOI: 10.1093/aob/mch080.

Developmental Physiology

4

Geoffrey R. Dixon*

School of Agriculture, Policy and Development, Earley Gate, Whiteknights Road, PO Box 237, University of Reading, Reading, Berkshire RG6 6EU and GreenGene International, Hill Rising, Horsecastles Lane, Sherborne, Dorset DT9 6BH, UK

Abstract

Understanding the subtle physiological attributes which operate during the development and maturation growth stages of brassicas provides the research platforms for crop scheduling and uniform harvesting.

Growth stages are described in this chapter for: cauliflower, green broccoli (calabrese), cabbage, Chinese cabbage, mibuna, mizuna, garden rocket, wall rocket, turnip, swede, radish, watercress and wasabi. For many crops, accumulated temperatures (thermal time) above a critical value provided a basis for growth stage and eventual maturity predictions. This model, however, requires adjustment for the complex flowering forms such as cauliflower and green broccoli where specific stages of juvenility, vernalization and curd or spear maturity are significantly affected by nutrition, light-induced *in planta* manufacture of carbohydrates, genotype (especially F_1 hybrids) and the increasingly potent influence of climatic changes. Growth stages and maturity in other brassicas, especially Chinese cabbage, have been similarly dissected in molecular terms identifying phases of gene up- and down-regulation.

*geoffrdixon@gmail.com

© Geoffrey R. Dixon and Rachel Wells 2024. *Vegetable Brassicas and Related Crucifers,* 2nd edition. (G.R. Dixon and R. Wells)
DOI: 10.1079/9781789249170.0004

INTRODUCTION

Physiology describes and measures the genetically driven processes of growth and reproduction, relating them to environmental factors such as temperature, radiation, photoperiod, water and nutrient availability. Crop physiology has immediate practical significance by permitting the prediction of growth and maturity rates as affected by changing weather and other events. Managing brassica crops is notoriously difficult because of the impact of periods of high or low temperatures accelerating or retarding maturity. Increasingly, detailed knowledge of the phases of plant growth and their alteration by temperature is permitting horticulturists to plan and predict planting and harvest dates, as described by Wien and Stutzel (2020). This means that harvesting schedules are made more reliable, supermarket buyers can formulate the promotion campaigns of particular commodities and profits along the food supply chain can be maintained.

For many crops, the use of accumulated temperature values above a specified base (thermal time) provides a unifying timescale by which progress to maturity can be monitored. This method is applicable only when the rate of maturation is a linear function of temperature (Monteith, 1981). In cereals and grasses, for example, the rate of leaf appearance and the reciprocal of the time to anthesis (flowering) are linearly related to temperature (Gallagher, 1979). Leaf appearance in some dicotyledonous crops, such as sugarbeet (Milford *et al.*, 1985), is also related to temperature. Regrettably, defining the relationship of *Brassica* genotypes to their environment and predicting maturity has proved less straightforward.

The genetic constitution of most of the important brassica crops has changed significantly over the last 20 years. Plant breeders have produced F_1 hybrid cultivars largely replacing open-pollinated forms (Singh *et al.*, 2018). These offer much improved consumer desirability and crop uniformity linked with consistency of vigour, yield and quality. Because commercial seed production of these cultivars faces technical difficulties, their cost has become a major element in crop budgeting. These difficulties are overcome by developing a robust system of self-incompatibility (SI), Ogura cytoplasmic male sterility (CMS) and doubled haploid (DH) parental lines and careful management of pollinators.

BRUSSELS SPROUTS (*BRASSICA OLERACEA* VAR. *GEMMIFERA*)

The imperative for improved understanding of growth and maturity in brassica crops began with Brussels sprouts (*B. oleracea* var. *gemmifera*) in the early 1960s as production for quick-freeze processing took an increasing share of the market, especially in the early and mid-season periods. The processing companies demanded that regular, high-quality and predictable supplies entered their factories so that the production lines operated at maximum

Fig. 4.1. Brussels sprout Irene F_1. (Geoff Dixon).

efficiency as had already been achieved for vining peas and green beans. Plant breeders contributed towards increased uniformity of growth through the early production of F_1 hybrid cultivars. Brussels sprouts were one of the first brassica crops where F_1 hybrids became widely available (see Fig. 4.1). This advance was supported by the achievements of physiologists in understanding the manner in which crop efficiency could be increased.

Brussels sprouts, like other *Brassica* spp., are photoperiodically day-neutral and flowering is induced at low temperatures. In heated conditions, the plants remain in a vegetative stage as do cuttings taken from them. Plants sown or transplanted into the field in the spring first pass through a juvenile stage when the plant cannot be induced to flower (see Fig. 4.2). An adult stage is reached in summer when the axillary buds (sprouts), which form the ware or commercial crop, develop. If the plants are exposed to low temperatures in

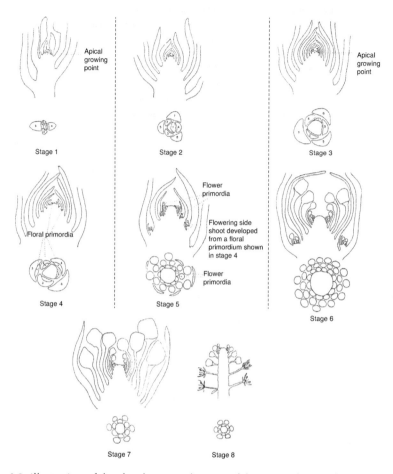

Fig. 4.2. Illustration of the developmental stages of the apex of Brussels sprouts (*Brassica oleracea* var. *gemmifera*). (After Stokes and Verkerk, 1951).

the autumn and winter, normal flowering begins in the following spring. The degree of flowering and rapidity of bolting are determined by plant age and the length of cold exposure (Stokes and Verkerk, 1951). Older plants form larger numbers of axillary buds and hence produce more flowering shoots. Plants given longer periods of cold break bud more quickly and produce more flowers, compared with plants given only the minimum period of cold necessary for flowering.

During the juvenile phase an increasingly large proportion of dry matter is transferred into the stem. This swells as storage products accumulate. The ratio of leaf and root to stem dry weights reaches a maximum value in the juvenile stage and thereafter remains approximately constant as the plants age (see Fig. 4.3).

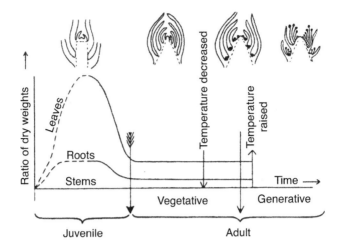

Fig. 4.3. Illustration of the changes in the apical growing points and distribution of growth during the juvenile, vegetative adult and generative stages of Brussels sprouts (*Brassica oleracea* var. *gemmifera*). Dry weights of leaves and roots are represented as multiples of stem weight. (After Stokes and Verkerk, 1951).

The formation of axillary buds (sprouts) and changing morphology at the apical growing point accompany the maximum accumulation of storage materials in the stem.

Growth stages

The growth stages of Brussels sprouts (*B. oleracea* var. *gemmifera*) are (also see Fig. 4.2):

Juvenile

1. Stem apex flat and very small; perhaps four rudimentary leaves and three primordia present.
2. At puberty the stem apex becomes pointed, the growing point enlarges and the apical bud becomes swollen by the increasing numbers of leaves and primordia. Once the growing point becomes dome-shaped Brussels sprout plants will flower following exposure to a period of cold.

Adult

3. Rapid enlargement of the growing point and top bud; the growing point becomes a globular structure on the apex of the stem and the leaf primordia enlarge.

4. A. This stage is reached once the plants have been subjected to cold. The first floral primordia become visible in the axils of the leaves. Primordia on the actively growing buds, usually only those in the apical buds, continue to full development. Buds lower down the plant will develop much later and, if given sufficient cold, these buds can develop into flowering shoots. At this point, the triangular arrangement of the apex is lost and the shape rounded as progressively larger numbers of initials segregate round the growing point.

Although this is the first indication of potentially generative structures, the change is reversible since if the plant is returned to warmer temperatures the buds will form only leafy shoots.

Plants resulting from the reversal of environment from cold to warm retain the phyllotaxis of a generative plant. They bolt but form no flowers and the apices of the leafy shoots remain in Stage 3.

B. As more buds segregate from the apex, the development of flowering ones takes precedence over that of leaf primordia, until the latter show solely as rudimentary lobes and eventually completely disappear.

5. At this stage, only flower primordia segregate from the apex.

6. The apical meristem reaches maximum size containing large numbers of flowers at varying stages of development and 30 or more leaves. The bud is visible to the naked eye when the leaves are removed.

7. This is the flower bud stage where the buds open out displaying individual flowers prior to bolting. The apical meristem decreases in size.

8. This is the bolting stage where the apical meristem is very small and only a few initials are cut off together. The growing point is permanently retained but no flowers are formed. Some primordia become arrested in growth so that the plant eventually dies with a terminal growing point still present.

The change to the flowering state is accompanied by alterations to the general morphology of the plant. This follows a reduction in petiole length and the production of long, strap-shaped leaves towards the top of the plant. These leaves have an increasingly wider angle with the stem (Stokes and Verkerk, 1951).

In Brussels sprouts, the number of leaves determines the number of potential bud sites for the formation of a crop of harvestable sprouts. During crop growth there are two distinct phases of leaf formation. In early crop growth, the rate of leaf formation is faster than in later periods. Leaf initial formation increases with later planting, but the lower rate of leaf formation is reached earlier with later planting.

In the first phase of crop growth, planting density has no influence on the number of leaves formed. Subsequently, the number of leaves formed at high cropping densities lags behind that formed at wider spacing. Late and higher-density planting restricts leaf formation and ultimately ware sprout yield but these factors appear to operate independently of each other.

Leaf area index (LAI), the leaf area per unit area of land (Watson, 1947), reaches maximum values (5–6) at between 80 and 100 days after transplanting and then decreases to values of 1–2.5 in the second half of the crop-growing period and through to harvesting. Development of LAI is faster with crops that are planted later but is heavily influenced by planting density. Specific values of LAI are attained more quickly in earlier planted crops.

Stem length is greatest in the early crop growth phases with later plantings. The rate of increase of stem length decreases most quickly with later plantings compared with early and second planting dates. Planting density has little effect on the ultimate crop height. Dry matter production at harvest was highest with early plantings although crops planted very late in the season (late June or early July) showed the fastest initial accumulation of dry matter.

Sprout bud initiation is greatly influenced by planting date. Buds start appearing between 21 and 32 days earlier with later plantings (i.e. further into the calendar year) and are unaffected by plant density. Bud dry weight was always lower with later planting dates. Buds finally formed in approximately 80% of the leaf axils. Increasing plant density reduces the final fresh weight per bud.

Brussels sprout crops intercept as much as 90% of available radiation at LAI 3.5–3.8; this stage is reached 50–70 days after transplanting. LAI values of 3.5 are attained most quickly by later transplanted crops. But this has little practical value since radiation use efficiency declines within these crops linked with the seasonal reduction in incoming radiation. Crops transplanted early in the season make the most efficient use of the environmental resources available to them. The grower's most effective strategy is developing high LAI sufficiently quickly to exploit the maximum available incoming radiation. Where planting dates are similar, bud initiation commences more quickly in early maturing cultivars compared with mid- and late-season maturing ones (Fisher and Milbourn, 1974). Nitrogen fertilizers encourage the marketable bud yield by increasing leaf area, and hence the interception of incoming radiation, raising total biomass (Booij, 2000).

Bud initiation tends to start earlier with low crop planting densities commencing when the total amount of assimilates produced can no longer be utilized completely by the growing apex. Such assimilate 'saturation' may trigger an hormonally regulated start to bud growth. Bud initiation continues up to harvesting, leading to a spread of bud ages up the plant. This resultant spread of bud maturity affects postharvest quality and shelf life. Traditionally, the sprout buds were selectively harvested by hand labour 'picking-over' the crop several times per season. The advent of machine harvesting did not allow grading in the field. Produce which is mechanically harvested in the field and destined for the fresh market may be graded for size and maturity in the packing shed. The yields of quality buttons in specific size ranges are described by a normal distribution where the mean represents the mean diameter of buttons and the standard deviation the spread of button sizes. The model enables the calculation of marketable yield (Hamer, 1993). Additionally, yield and timing of development may be estimated from the harvests of early maturing cultivars (Hamer, 1991). Higher yields are achieved most efficiently by planting as early as field conditions permit. But this is adversely affected by low temperatures which encourage vernalization and flowering, in preference to sprout bud formation (Everaarts *et al.*, 1998).

A common husbandry practice in Brussels sprouts is the removal of apical plant growth a few weeks before harvest. This promotes the development of axillary buds (sprouts) and ensures higher yields. Decapitation encourages the accumulation of glucosinolates and hydroxycinnamic acids (Jakopic *et al.*, 2016). Glucosinolate and flavonoid content tends to be highest in the sprouts from the upper stem.

CAULIFLOWER (*BRASSICA OLERACEA* VAR. *BOTRYTIS*)

Summer (early and late) and autumn maturing cultivars

Brassica oleracea is highly polymorphic and includes varieties which exhibit a headed phenotype (a large pre-inflorescence): the curd of cauliflower and 'Romanesco' (var. *botrytis*), and the spear of green broccoli (var. *italica*). This headed phenotype results from highly iterative patterns of activity at the primary meristems. Differences in the morphology of curds and spears are accounted for by three quantitative variables: the rate of production of branch primordia on the flanks of the apical meristems, the number of branch primordia produced before the first-formed begin producing their own branch primordia (the iteration interval) and the duration of the pre-inflorescence stage (before production of flower primordia). A relatively long pre-inflorescence stage in cauliflower and 'Romanesco' results in the curd surface being composed largely of branch primordia, whereas in green broccoli this stage is short and the spear surface is made up of flower buds (Kieffer *et al.*, 1998). Complementary lines of molecular evidence suggest that the cauliflower curd arose in southern Italy from a heading green broccoli via an intermediate Sicilian crop type (Smith and King, 2000). Consequently, both crops are closely related.

Cauliflower (see Fig. 4.4) is the brassica crop which has probably changed most significantly resulting from the introduction of F_1 hybrid cultivars. These cultivars have brought much improved precision for crop scheduling, substantially increased uniformity at maturity and, consequently, enabled more reliable harvest volumes. All of these improvements bring opportunities for the application of automated, robotic harvesting, reducing the need for manual field staff at a time when availability is becoming restricted. In addition, plant breeders have diversified curd colour and shape, thereby increasing consumer choice. For example, there is a *Purple (Pr)* gene mutation in cauliflower which confers an abnormal pattern of anthocyanin accumulation. Resultant phenotypes have an intense purple curd (Chiu *et al.*, 2010).

Fig. 4.4. Cauliflower (*Brassica oleracea* var. *botrytis*) high-quality F$_1$ hybrid. (Geoff Dixon).

Cauliflowers are short caulescent plants with shoot tips composed of young leaves and leaf primordia situated around an apical dome, separated by expanding internodes. By comparison with most brassicas, the shoot tip components of the cauliflower are large and easy to detach, measure and analyse following environmental changes. The growers' cauliflower consists of a large immature inflorescence (the curd), formed at the stem tip after a period of vegetative growth. The curd is a specialized organ and the most important product of cauliflower. Inflorescence development in cauliflower is particularly complex, presenting unique challenges for those seeking to predict and manage flowering time. An integrated physiological and molecular approach has been used to clarify the environmental control of cauliflower reproductive development at the molecular level. A functional allele of *BoFLC2* was identified for the first time in an anual brassica, along with an allele disrupted by a frameshift mutation (*boflc2*). The correlations observed between gene expression and flowering time in controlled environment experiments were validated with gene expression analyses of cauliflowers grown outdoors under 'natural' vernalizing conditions. These correlations indicated potential for transcript levels of flowering genes to form the basis of predictive assays for curd initiation and flowering time (Ridge *et al.*, 2015). The mechanism underlying the regulation of curd formation and development remains largely unknown, but a novel homologous gene containing the OSR (organ size-related) domain, namely *CURD DEVELOPMENT ASSOCIATED GENE 1 (CDAG1)*, was identified in cauliflower (Li *et al.*, 2017).

The regulation of reproductive development in both cauliflower and green broccoli is botanically unusual as most enlargement occurs while development is arrested at a distinct stage. Cauliflower and green broccoli curds are composed of inflorescence meristems and flower buds, respectively. Initiation of floral primordia and enlargement of floral buds in green broccoli and cauliflower are not controlled solely by homologues of the genes that do so in the simpler model brassica *Arabidopsis thaliana* (Duclos and Björkman, 2008).

The growth phases of both cauliflower and green broccoli can be divided as follows (see Fig. 4.5):

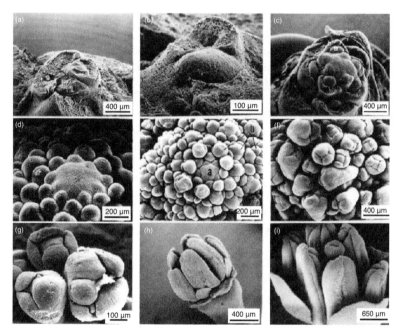

Fig. 4.5. Illustration of scanning electron microscope observations of the development of cauliflower (*Brassica oleracea* var. *botrytis*) and green broccoli (*Brassica oleracea* var. *italica*). (a–e) cv. Snow Queen and (f–i) cv. Wase-midori. (a) The vegetative stage; shoot apex initiating leaf primordia is pointed and narrow. (b) The dome-forming stage; shoot apex is flat and wide. (c) The early curd-forming stage; curd formation is evident by the initiation of inflorescence primordia 'a' and bract primordia 'b'. (d) The intermediate curd-forming stage; flat dome is covered with inflorescence primordia. (e) The late curd-forming stage; inflorescence primordia are newly initiated around the primordium of first-order inflorescence ('a' = apex of the main axis; 'b' = inflorescence primordia). (f) The sepal-forming stage; four sepals are initiated in every flower bud. (g) The stamen and pistil-forming stage; six stamens and a pistil are initiated inside the sepals. (h) The petal-forming stage; four petals are initiated at the outer base of the stamens. (i) The petal-elongation stage; petals are notably elongated. (After Fujime and Okuda, 1996).

a = vegetative stage;

b = dome-forming stage;

c = early curd-forming stage (initiation of inflorescence primordia);

d = intermediate curd-forming stage (increase of inflorescence primordia or flower buds);

e = late curd-forming stage (newly initiation of inflorescence primordia around the first-order inflorescence primordium);

f = sepal-forming stage (initiation of flower organs in the floret);

g = stamen and pistil-forming stage;

h = petal-forming stage;

i = petal-elongation stage.

Normally, the inflorescence primordia of cauliflower do not develop beyond the state of primary protuberance (stages d and e) and increase only in number during the curd-forming and thickening stages. Only after the curds have matured and the peduncles of the inflorescences have elongated is flower bud development resumed, leading to the formation of floral organs in the floret such as sepals, stamens and pistils.

The mature curd of cauliflower is composed of a single flower stalk and numerous first-order inflorescences which branch several times and whose tips are shortened considerably. There are numerous inflorescence primordia in a state of primary protuberance on the surface of a cauliflower curd.

In the field, development of cauliflower plants from transplanting to harvesting may be divided into three physiologically distinct phases which respond differently to ambient temperatures: juvenile phase, curd induction or initiation phase and curd growth phase.

Juvenile phase

Juvenility is recognized where plants grow and form leaves at a temperature- and size-dependent rate but cannot be induced to reproduce. The end of juvenility is identified in the cauliflower by the initiation of curds at the main stem apices. This morphological change may relate to the plant having produced a critical number of leaves and this varies with different genotypes (Hand and Atherton, 1987; Booij and Struik, 1990). The presence of leaf numbers ranging from 8 to 19 marks the end of juvenility in different cultivars. The initiation of new leaves continues until the stem apex changes from vegetative to generative growth at curd initiation. The rate at which leaves are initiated is not uniform and increases in the latter stages of juvenility.

Leaf initiation rate is related to temperature (Wiebe, 1972a) and the rapidity with which growth restarts after transplanting. But this shift is not linked to the end of juvenility according to Booij and Struik (1990). Booij (1987) identifies the juvenile phase as being similar in both direct-seeded and transplanted crops. The cauliflower reaches a 'physiological age' before curds are initiated following a period of lower temperatures. Some research suggests that the rate at which physiological age is attained may be manipulated by seed treatment since Fujima and Hirose (1979) enhanced curd induction by applying cold treatment to germinating seed.

Curd induction or initiation phase

After juvenility there follows a phase of curd induction for which relatively low temperatures are required. Factors controlling initiation have been investigated by Wiebe (1972a, b, c), Wurr *et al.* (1981a), Hand and Atherton (1987), and Booij and Struik (1990). The term vernalization is sometimes used to describe the response of cauliflower initiation to temperature. Vernalization implies that cool temperatures hasten curd initiation. Wurr *et al.* (1990a), in studies over three seasons, showed this may not always be the case (see Fig. 4.5).

Cool temperatures in spring are well recognized as delaying curd initiation more than warmer ones in summer. Historically, this effect was well recognized in older cultural systems since warmth applied to pot-raised, glasshouse-grown, early spring cauliflowers resulted in 'buttoning', that is curd initiation was accelerated but resulted in small, immature, low-quality curds. At that time, the growers' practical objective was to move these plants into the field as early as soil and weather conditions permitted, preventing premature curd initiation.

The stimulus of curd initiation in early summer maturing genotypes is thought to be quantitative. The early summer group form marketable curds in the period June to August following transplanting in March. Transplants must be retained in the vegetative juvenile state without curd initiation during propagation. If curds form before transplanting, they will fail to reach marketable size and quality in the field.

The juvenile state is maintained by retaining the transplants in modules at 0–10°C and limiting the availability of inorganic nitrogen. Curd maturity could then be extended over a period of 2–6 weeks (Atherton *et al.*, 1987). Salter (1969) correlated the period of maturity to the time over which curds were initiated. With the older open-pollinated cultivars, closely synchronizing curd initiation had immense implications for the improvement of harvesting and land-use efficiency. Much research was invested in understanding how the precision of maturity in these genotypes could be synchronized. For examples, see Wiebe (1972b), Fujima (1983), Atherton *et al.* (1987), Roberts and Summerfield (1987), and Wurr *et al.* (1993).

In practice, because of the effects of weather conditions on crop development, counts of leaf number will still provide the grower with information concerning the time of curd induction and temperature conditions during that phase. The relationship between mean temperature and the final number of leaves is dependent on the specific genotype.

Curd growth phase

This phase follows on from curd initiation. The diameter of the curd increases with temperatures up to a maximum (Wiebe, 1975; Wurr *et al.*, 1990b). The contrasting effects of temperature on development in different growth phases are the most likely causes of the major problems encountered by producers in maintaining the continuity of maturation of cauliflower curds in the field. Ultimately, the duration of curd growth is fairly constant and independent of eventual weight at harvest. Lower temperatures increase the duration of curd growth. The relationship between radiation and curd growth is connected with the time of transplanting or direct seeding, since with plantings made later in the season, day length (in Europe) is declining and radiation quantity is reduced. Consequently, the duration of curd growth and its interaction with incident radiation could be an expression of declining day length which determines maturity as related to elongation of the inflorescence.

Use of a 'degree day' system (a product of temperature and time above a specified critical value) failed to improve the precision of estimating cauliflower maturity. A positive relationship exists between temperature and time by measuring weight changes during curd growth (Salter, 1969; Salter and Fradgley, 1969). But temperature and time products failed to predict harvest dates with sufficient precision for practical use. Competition for light during plant raising and the increased age of the transplant also lengthened the maturity period once they reached the field. More recent studies indicate that total above-ground dry matter increases linearly with accumulated incident radiation integral after curd initiation. However, under lower radiation conditions, the rate of increase per unit incident radiation integral was greater than under higher radiation conditions. Curd growth also increased linearly with increasing accumulated incident radiation integral and a greater mean relative curd dry matter increase per megajoule (MJ) under lower incident radiation conditions than higher incident radiation levels (Rahman *et al.*, 2007). Greater rates of curd growth (curd length, diameter, fresh and dry weights) were achieved at warmer night temperatures than day temperatures. Although greater leaf and stem growth (leaf area, stem length, fresh and dry weights) were achieved when day temperatures were warmer than night temperatures, even with the same overall mean temperatures (Rahman *et al.*, 2013).

Some authors have suggested that a model of the early development of cauliflower could be based on meteorological observations and then used to predict curd initiation. After this point, there is a simpler relationship between crop development and temperature and therefore the date of harvest can be predicted with greater accuracy (Wurr *et al.*, 1990b).

Modelling the early period of development and combining it with a harvest prediction models would give growers an early warning of disturbances in their production schedules (Grevsen and Olesen, 1994). These Danish authors describe the duration of the juvenile phase by a temperature sum starting at transplanting and using the base of 0°C. The temperature sum requirements (thermal time in Kelvin days (Kd)) for six genotypes ranged between 26 and 83 Kd compared with 250 Kd for two genotypes in similar Dutch experiments. Leaf initiation is well accounted for at temperatures above 0°C. Counting the number of leaves that have been initiated provides an indication of the end of juvenility. A strong description of leaf initiation (R^2 values of 0.9) was obtained when a gradual acceleration in the leaf initiation rate with increasing leaf number and rising temperature values were used. The end of the juvenile phase occurred when the plants had 12 leaves initiated in these Danish experiments compared with 17–19 leaves in the Dutch experiments. The duration of the curd induction phase in the model is described by using linear responses to temperature that are symmetrical below and above an optimum temperature. A common optimum temperature for curd initiation was estimated to be 12.8°C for two cultivars in the Dutch experiment. The base temperature was estimated to be 0°C and the maximum temperature, therefore, is taken to be 25.6°C. The best fits of data in the combined model of juvenile and curd induction phases show R^2 values between 0.4 and 0.6.

This model needs validation that takes account of the season of production (the genotypes used were late summer and early autumn maturing types). There will be differences for each model dependent on the location of production areas and most probably micro-geographical variations within such regions. Ultimately, validation of such models must be done at the farm and field level to provide the standard of confidence which is required for crop scheduling in order to meet market demands. The market itself will impose constraints on the producer since it will demand greater levels of supply at particular times within the week. This requires the integration of short-term crop storage with the scheduling of crop maturity and harvesting.

Forms of curd abnormality resulting from the effects of temperature are shown in Fig. 4.6.

Influence of nutrition

Manipulation of the supply of inorganic nitrogen to cauliflower (*B. oleracea* var. *botrytis*) plants may substitute for chilling and restrict growth in terms of leaf area and shoot size. In warm, non-promotive temperatures nitrogen starvation delayed curd initiation both in terms of time and the increase in the number of leaves formed before curds began to develop. Nitrogen restriction did not affect the vernalization rate in cool inductive temperatures. Nitrogen limitation reduces the dry matter contents of the apical dome and young leaves. At the husbandry level, intercropping cauliflower with annual clover (*Trifolium resupinatum*) provides a sustainable means for optimizing nitrogen input and reducing synthetic fertilizer demands in crop rotations (Tempesta *et al.*, 2019).

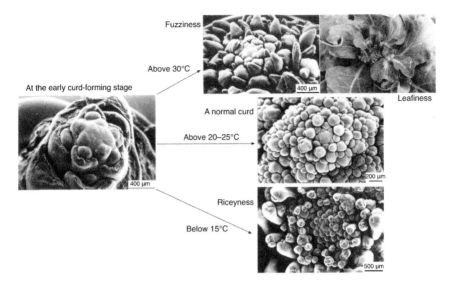

Fig. 4.6. Illustration of abnormal curds where cauliflower (*Brassica oleracea* var. *botrytis*) cv. Snow Queen was grown at various temperatures after curd formation. (After Fujime and Okuda, 1996).

Carbohydrates

Carbohydrates have long been associated with vernalization processes. They increase in the shoot tip as the process advances. Further vernalization may only proceed in excised buds and embryos when an external supply of sugars is available. Atherton *et al.* (1987) postulate that feeding sucrose to the shoot tip of intact cauliflower (*B. oleracea* var. *botrytis*) may partially substitute for the low-temperature stimulus for curd initiation. They suggest that the events leading to curd initiation in cauliflower include preferential redistribution of carbohydrates to the apical dome at the expense of young leaves, leaf primordia and adjacent stem tissues. This phenomenon is also reported for white mustard (*Sinapis alba*) (Bodson and Remacle, 1987).

Development of curds (cauliflower) or spears (green broccoli) is a developmental event as distinct from growth processes which involve the accumulation of dry matter and increasing leaf area. For both crops, models which aid in the prediction of maturity dates exist based on the use of accumulated thermal time values and logarithmic values of curd diameter (Salter, 1969; Hand, 1988; Wurr *et al.*, 1990a, b, 1991) (see Fig. 4.7). Where curd diameter is known then the thermal time required for maturity may be estimated using mean temperatures obtained from meteorological records for a particular locality.

Curd solidity is a critical component of cauliflower quality; a solidity index, curd voidage and analysis of texture involving deformation allowed Zhao *et al.* (2013) to quantify this property in 20 cultivars for breeding and production purposes.

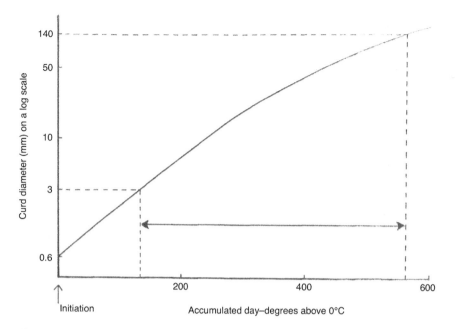

Fig. 4.7. Illustration of the prediction of curd size in cauliflower (*Brassica oleracea* var. *botrytis*). (After Wurr *et al.*, 1990c).

Genotype effects

Pearson *et al.* (1994) attempted to formulate models which predicted the time required for curd initiation and the duration or rate of curd growth of summer and autumn maturing cauliflower (*B. oleracea* var. *botrytis*). Optimum temperature for curd growth varied with genotype as supported by Wurr *et al.* (1990b) who derived a regression model based on constant thermal time between curd initiation and the attainment of a specific curd diameter. Variations from this model were apparent with different genotypes. The later maturing types required more thermal time to reach maturity.

The time taken to reach maturity is controlled by a combination of genotypic and environmental factors and major variations are due to differences in the time taken to satisfy the requirements for curd induction. Identifying the number of genes involved, and the manner by which they exert their action, would help plant breeders to produce cultivars with more controlled induction as suggested by Wurr (personal communication). Such genetic studies begin offering alternative approaches since the development of cauliflower is highly dependent on temperature due to its vernalization requirement, which often causes delay and unevenness in maturity during months with warm temperatures. A quantitative trait loci (QTL) model resulted in a moderately accurate prediction of time to curd induction ($R^2 = 0.42–0.51$), while a growth stage model generated slightly better results ($R^2 = 0.52–0.61$). Predictions of time to curd induction of test hybrids from independent DH lines were less precise with $R^2 = 0.40$ for the QTL and $R^2 = 0.48$ using the growth stage model (Rosen *et al.*, 2018).

Temperatures above 20–22°C inhibit development towards curding even in many summer cultivars. Significant QTL × environment interactions for final leaf number and days to curd initiation (DCI) on chromosomes C06 and C09 suggest that these hotspot regions have major influences on temperature-mediated curd induction. Quantitative trait loci for leaf area ratio (LAR) were detected repeatedly in several environments on C01, C04 and C06. Negative correlations between LAR and DCI, and QTL co-localizations on C04 and C06, suggest that LAR also has effects on development towards curd induction (Hasan *et al.*, 2016).

Climate change

An increase of +1°C at temperatures below 15.5°C will reduce the time taken to cauliflower (*B. oleracea* var. *botrytis*) curd initiation. Where temperatures are above 15.5°C, time to curd initiation will be likely to increase since curd initiation is reduced at elevated temperatures. Hence, crop duration from plantings later in the season is likely to be extended. Increases of carbon dioxide (CO_2) content raised curd weight (biomass), especially dry matter content, but has a much lower effect on curd diameter which can be slightly reduced in size (Wheeler *et al.*, 1995).

Cauliflower production in India demonstrates that over time hybrids suitable for warmer conditions can be developed. During the 19th century, cauliflower production was initiated in India. As a result, cultivars suitable for higher temperatures have been selected. In India, a wide diversity exists in tropical cauliflowers in terms of their adaptation to different temperatures and maturity duration (Bhatia *et al.*, 2017). These authors suggest that the development of triploid F_1 hybrids using tetraploid lines could be an alternative to the conventional hybrid breeding of cauliflower because of limited heterosis. The number of leaves required to induce curd initiation is less than nine in tropical cauliflower at temperatures of 18–30°C (Lin *et al.*, 2019). Studies showed that under non-vernalized high temperatures, cultivars of tropical cauliflower can initiate curd development but with a different pattern from those cultivars grown in temperate zones. The switch from vegetative development to curd formation in cauliflower, referred to as the generative switch, is strongly temperature responsive in the majority of cultivars. But high temperatures do not delay the timing of expression by the majority of genes in meristems and leaves of sensitive cultivars that were delayed in the switch to the generative stage. Expression of a few potential flowering-time genes was affected by the high-temperature treatment in a sensitive cultivar, making them potential candidates for causing the observed delay in generative switching (Sun *et al.*, 2018).

Winter-hardy cauliflower (Roscoff types)

This overwintered crop is transplanted into the field in mid to late summer and forms a large framework of leaves in the juvenile phase. Limited, and in cases rudimentary, information is available concerning curd initiation by winter cauliflower (*B. oleracea* var. *botrytis*). Information available for this crop in general compares poorly with summer and autumn types where recent research has made significant advances in defining the parameters controlling plant juvenility and maturation and the subsequent initiation and development of curds and spears in green broccoli (*B. oleracea* var. *italica*). Historic evidence by Wellington (1954) identified an empirical correlation between the time of curd initiation and temperature during the vegetative juvenile phase. This indicated variations of up to 2 months in the maturity of overwintered cauliflower cultivars, depending upon seasonal weather conditions. Later quantification showed that plants maintained above 15.6°C remained vegetative and time of exposure to cool conditions was cumulative (Haine, 1959). The most likely reasons for such variations are differences in the time of curd initiation. Subsequently, Wurr *et al.* (1981b) showed that the temperature threshold below which curd induction proceeds is genotypically controlled. Wurr and Fellows (1998), using two Roscoff cauliflower selections, 'December/January' and 'March' (ex CODEBRIC Seed Group), showed that both light and temperature are important regulators of leaf production. Raising the cultural temperature by growing under polythene covers increased the time to curd initiation in 'December/January' by up to 93 days and in 'March' by up to 57 days. For overwintered cauliflower types, for instance, low light intensity and the periods of cold treatment affect the extent of juvenility. Thus overwintered cauliflower can form in excess of 100 leaves before curd initiation commences. Whereas with summer/autumn cauliflower types, juvenility may end once 17–22 leaves have formed (Wurr *et al.*, 1994). Once this is achieved, then growth towards initiation is determined by temperature. The vernalization rate then increases with temperature to a maximum and thereafter declines (Atherton *et al.*, 1987; Wurr *et al.*, 1988) and, as suggested by Nieuwhof (1969), temperatures in excess of 23°C prevented curd initiation in annual cauliflower. Similar comparative studies are now required for the F_1 hybrid cultivars which currently dominate winter production.

GREEN BROCCOLI (*BRASSICA OLERACEA* VAR. *ITALICA*)

Agronomically, green broccoli (*B. oleracea* var. *italica*) (see Fig. 4.8) growth may be divided into two phases: first, from drilling or transplanting to spear initiation and second, from initiation to maturity.

Predicting the duration of both growth periods has vital relevance to crop scheduling, for land occupancy and hence rotational growing, and for adherence to marketing schedules (Lindemann-Zutz *et al.*, 2016). This is particularly

Fig. 4.8. Green broccoli cv. Beaumont. (Courtesy of Elsoms Seeds Ltd and Bejo Zaden).

important for green broccoli since three or more crops may be grown within a single season as multiple production, thereby substantially reducing the overhead costs attributed to each one. This is a crop which maintains a relatively high fraction of photosynthetically active tissue up to harvesting compared with other brassicas. Agronomically, intensive green broccoli production has a high demand for nitrogen (see Chapter 5 section, Crop Responses) which, without careful management, may cause ground and surface water pollution (Conversa *et al.*, 2019). Head quality is defined by colour, smoothness, size, shape and bead uniformity (Stansell *et al.*, 2017). Other qualities include a convex head shape, tight branching angle, small and uniform size and regular form of flower buds (Grabowska *et al.*, 2013).

Leaf number defines the crop growth stage. First, this is related to genotype and second, it increases with later planting dates partly controlled by rising temperatures that accelerate growth. There is some evidence that the time from transplanting to spear initiation varies more than the time from initiation to maturity with earlier season crops but the reverse is the case for later season cultivars (Wurr *et al.*, 1991).

Opinions vary as to whether green broccoli has a vernalization requirement for spear initiation. Possibly, this is a facultative cold requirement which means that cooler conditions, while not essential for spear initiation, do accelerate earlier head formation when applied. Temperature affects the number of nodes produced prior to spear initiation. Variation in the time from transplanting to maturity between crops emphasizes the practical difficulties of maintaining continuity schedules. Head growth is also affected by solar radiation, reflecting the high planting densities at which green broccoli is grown (four to fivefold higher than may be used with cauliflower (*B. oleracea* var. *botrytis*)). Consequently, interplant competition commences much earlier in the crop's life.

Morphologically, green broccoli follows similar developmental patterns to cauliflower with respect to juvenile, head induction (initiation) and head growth phases. Head initiation requires the development of 14 leaves, seven of which are visible to the naked eye. Periods of warm weather delay initiation leading to greater numbers of leaves being formed (Farnham and Bjorkman, 2011). There is a strong genotype effect here and some older cultivars, such as cv. Shogun, are unsuited to warm temperatures that result in head malformation. The minimum leaf number required before head initiation may also indicate the length of the juvenile phase in different green broccoli cultivars and is consistent with late maturing types, such as cv. Shogun, forming nine leaves in the juvenile phase compared with early maturing ones, such as cv. Emperor, forming seven.

The total number of leaves initiated at the stem apex is directly related to the number of visible ones (this also applies to the cauliflower). This provides a means of determining the growth stages of crops in the field by counting visible leaves. The quadratic relationship between the natural logarithm of head diameter (measured in millimetres) and a temperature sum from head initiation provided an accurate method for predicting head maturity. A two-dimensional search for the optimal combination of base and ceiling values for the temperature sum expression gave base at 0°C and ceiling at 17°C as limits to the minimum residual sum of squares. The fit of the quadratic expression is more sensitive to changes in the base temperature limits over 0°C.

Head diameter proved to be a better parameter for calculating time to harvest compared with the use of plant or head weight. Some workers, such as Pearson and Hadley (1988), used a linear relationship between head diameter and temperature sum as opposed to a curved one. The good fit they achieved with a linear relationship may result from the fact that they were experimenting with smaller green broccoli heads. Others, such as Wurr *et al.* (1992), have used logistic models but the difference in goodness to fit between logistic and quadratic relationships was very small. Indeed, Wurr *et al.* (1992) changed to using quadratic models when investigating interactions between plant density and harvest maturity.

Plant density has a high impact on head size in green broccoli. Grevsen (1998) reports that densities of 5–10 plants/m² are the normal Danish husbandry practice in order to achieve harvestable 150–250 g heads with diameters of approximately 90–120 mm, respectively. Some other variables which have been examined as components for these models include genotype, solar radiation and planting month, with only the latter enhancing the accuracy of the model sufficiently for use in practice. Wurr *et al.* (1991) suggested utilizing 'effective day-degrees' (Scaife *et al.*, 1987) as a means of predicting time to crop maturity. Effective day-degrees (EDD) (effective temperature sum) are a combination of temperature with radiation (R) in MJ/m²/day, DD is the temperature sum (day-degree) with a certain base and ceiling temperature and is a factor denoting the importance of radiation relative to temperature:

$$1/EDD = 1/DD + a/R$$

Some research has indicated a high degree of association between temperature and radiation using this equation, especially where high plant densities (22 plants/m²) are employed. High plant density results in competition for light early in the growth of the crop. Only relatively small proportions of the overall variance are accounted for by including radiation or planting month values in the predictive equation. Hence, these Danish authors advocated use of a simple temperature sum to determine harvest date for commercial purposes.

The Marshall and Thompson (1987a, b) model was applicable to direct-sown crops. But this form of husbandry is now largely discarded, in Europe at least, in favour of using module-grown transplants because of their reliability and the increasingly high cost of seed. With direct sowing the period to maturity is 90–140 days and their model could predict harvest date for 90% of crops to within ±7 days and for 100% to within ±12 days.

Later models (Wurr *et al.*, 1992) started when heads initiated, and at 10 mm, and predicted harvest date to within ±2–3 days using head diameters and weather data. The problem with the Wurr method is that it predicts small head diameters too late and large head diameters too early (comment taken from Grevsen, 1998). More recently, a strong correlation was identified between the diameter of stem, plant biomass and the diameters of stem and curd. The growth of stem and curd diameters was dependent on the days after transplantation in the field, but the dependence was even stronger if related to the sum of maximal daily temperature (Waltert and Theiler, 2003).

Improving horticultural quality in regionally adapted green broccoli and other *B. oleracea* crops is challenging due to complex genetic control of traits affecting morphology, development and yield. Mapping horticultural quality traits to genomic loci is an essential step in these improvement efforts. Three key genomic hotspots associated with pleiotropic control of the green broccoli heading phenotype were identified. One phenology hotspot reduces days to flowering by 7 days and includes an additional *FLOWERING LOCUS C* (*FLC*) homologue *Bo3g024250* that does not exhibit epistatic effects with the three horticultural quality hotspots. Strong candidates for other horticultural traits were identified: *BoLMI1* (*Bo3g002560*) associated with serrated leaf margins and leaf apex shape, *BoCCD4* (*Bo3g158650*) implicated in flower colour, and *BoAP2* (*Bo1g004960*) implicated in the hooked sepal horticultural trait. The BolTBDH reference mapping population provides a framework for *B. oleracea* improvement by targeting key genomic loci contributing to high horticultural quality green broccoli and enabling *de novo* mapping of currently unexplored traits (Stansell *et al.*, 2019).

In green broccoli inflorescence primordia form at the early curd stage (stage c, Fig. 4.5) and thickening stages. This increases primordia and flower bud development in inflorescences progress together. Thus, the flower buds of green broccoli reach the petal-forming stage prior to spear maturation. Cauliflower, by comparison, possesses a mechanism that stops the progression to flower bud development during curd forming and thickening. The flower head of green broccoli is composed of similar organs to that of the cauliflower but their development progresses differently. There are numerous flower buds, each initiating sepals, stamens and a single pistil on the green broccoli spear surface. This means that the quality of green broccoli spears, as defined by the retention of a wholly undeveloped green stage, is more difficult to achieve. Yellowing of the spear, resulting from the appearance of sepals, happens far more quickly compared with cauliflower curds, which complicates marketing-chain management.

Cauliflower and green broccoli curd or spear maturity

Predicting when a green broccoli (*B. oleracea* var. *italica*) or cauliflower (*B. oleracea* var. *botrytis*) crop will mature is vital to modern marketing practice because it determines when supplies of the product will be available and enables the grower or cooperative to adjust marketing strategy in anticipation of crop maturity.

The maturity of a green broccoli head (spear) is primarily determined by the developmental states of the florets. The head is harvested shortly before the flowers open. Since flowering, as in most species, follows the cessation of leaf production on the stem, the reciprocal of the time taken from sowing to maturity is essentially an estimate of the rate of development and would be expected to correlate closely with average temperature. Hollow stem is a significant defect found in mature green broccoli. Studies of incidence and severity of hollow stem, using digital image analysis, found it increased with wider plant spacing but was not attributable to boron deficiency (Boersma *et al.*, 2009).

Water stresses applied during different stages of head development significantly affects the shelf life of green broccoli heads. Final head size is affected by water stress applied immediately after head initiation and completed shortly after heads are 5 mm in diameter, with reduced head weight and diameter as well as a shortened shelf life. There are also significant deleterious effects of water stress, while heads are small, on stem turgor, head colour, bud elongation and floret looseness at maturity, indicating that water stress at least 3 weeks before maturity induces differences in each postharvest character (Wurr *et al.*, 2002).

Maturity is also judged by the size of spears. Hence, solar radiation which governs crop growth also affects maturity date. There are apparently no reports of day length (photoperiod) affecting the development of green broccoli. In green broccoli, Fujima and Hirose (1979) and Marshall and Thompson (1987a) showed that the concept of thermal time could be applied to the prediction of maturity using a multiple linear regression model. Maturity is defined as the point where half of the spears had been harvested. Using daily records of maximum and minimum air temperature and total solar radiation, maturity could be predicted to within ±7 days for 90% of crops tested over a 4-year period and within any 1 year the precision improved to ±5 days. The crops progressed to maturity when the air temperature exceeded defined base temperatures. This is termed the 'active duration'. When the air temperature was below this base temperature the crops ceased growing and this was defined as 'inactive duration'. For green broccoli in particular, growth ceased at a base temperature of 0°C.

Crop duration steadily declined from about 140 days down to about 90 days as the sowing date advanced from late February (calendar day 55) to early May (calendar day 130). The actual date of harvest changed by only 25 days compared with 75 days difference in sowing date. Crop duration remained at 85 days until late June (day 175). Thereafter, duration increased to around 120 days for mid-July sowings (day 202). In practice, the forecasted harvest date and intervals between harvests for successive cropping requires continual revision throughout the season using the current year's weather records combined with data for long-term averages. A dynamic approach to harvest forecasting can reduce the spread of intervals between harvests to 1.8 days (Marshall and Thompson, 1987b).

CABBAGE (*BRASSICA OLERACEA* VAR. *CAPITATA*)

European cabbage (*B. oleracea* var. *capitata*) (see Fig. 4.9) has become part of staple diets worldwide. Crop types vary significantly according to their maturity seasons.

These extend from early spring cabbage leaves, which lack a substantial head, through hearted 'Primo' types to the late summer and autumn hearted forms of Savoy and Christmas Drumheads. White cabbage, capable of being stored over winter, have long formed part of the basic diet, especially in regions with harsh winter climates which preclude a continuity of field cropping. Cabbages are also the major ingredients for preserves such as sauerkraut. This diversity is discussed in detail by Nieuwhof (1969). Over succeeding decades plant breeders have modified and improved cabbage types substantially by developing F_1 hybrids which deliver vastly increased uniformity, succulence, reliability and attractiveness for the consumer. Speedy growth is especially important for the production of succulent and attractive forms of cabbage. Harvestable individual weights of spring cabbage shoots (cv. Myatts

Fig. 4.9. Mature early spring cabbage (*Brassica oleracea* var. *capitata*). (Geoff Dixon).

Offenham Compacta) correlated with times of seedling emergence. Differences of 5–10 days accounted for 46% variation in weight 65 days after sowing (Shanmuganathan and Benjamin, 1993).

Predicting the maturity of cabbage heads using methods of heat-unit accumulation was suggested by Isenberg *et al.* (1975), but has not found acceptance in practice. Estimates of maturity still remain subjective and empirical, based on visual estimates and finger tests of firmness. None the less, quantitative measurements of yield become more necessary as crop producers require increasingly greater precision in the use of their resources and capacity for providing predictable supplies to their customers. Yield can be established accurately by use of the relationship between head number, size and density (Kleinhenz, 2003; Radovich and Kleinhenz, 2004). Head weight is also positively related to diameter and height; and negatively to the number of outer leaves and plant breadth (Du *et al.*, 2017). Similar parameters are suggested by Özer (2021), specifically for red-headed cabbage, and Adzić *et al.* (2012) more generally.

Growth stages of head cabbage can be categorized by a decimal code as: germination, emergence and establishment, leaf formation, head formation and maturity. Maturity is defined as heads reaching a marketable weight and density, and standing ability is the number of days from 90% of the crop being marketable, descending to 75% of the crop being marketable (Everaarts, 1994).

Cabbage heads around 2 kg are preferred for the fresh market since this is suitable for consumers. Weight can be regulated by choice of cultivar and cultural practices, including rates of fertilizer application (Kolota and Chohura, 2015) and planting density. Their studies suggested that close spacing increased dry matter content, vitamin C and total sugars and decreased nitrate accumulation. Use of ammonium sulfate and urea also increased soluble sugars (Sady *et al.*, 1999). Using nitrogen applications of 300 kg N/ha, the yield of heads greater than 1 kg increased. Excessive use of nitrogen increased nitrate accumulation and decreased calcium content. Nitrogen, especially when applied as ammoniacal formulations, leaches easily particularly when crops are grown on light sandy soils with low water-holding capacity (da Silva *et al.*, 2020). Use of calcium cyanamide formulations avoids these problems and encourages beneficial soil microbial populations which enhance crop health (Dixon, 2017).

CHINESE CABBAGE AND PAK CHOI (*BRASSICA RAPA* SSPP. *PEKINENSIS* AND *CHINENSIS*, RESPECTIVELY)

A very detailed description of the origins, physiology, breeding and these crops is contained in Talekar and Griggs (1981). Chinese cabbage (*B. rapa* ssp. *pekinensis*) (see Fig. 4.10) responds to inductive temperatures from germination, forming flower initials when vernalization is completed.

Minimum plant raising temperatures of 18°C will reduce bolting in the field. The growing point height is determined at or before transplanting.

Fig. 4.10. Head-forming Chinese cabbage (*Brassica rapa* ssp. *pekinensis*). (Geoff Dixon).

A linear model incorporating temperature and solar radiation describes subsequent extension of the growing point. Chinese cabbage plants go through seedling and rosette stages before forming their leafy head. Chinese cabbage plants resemble pak choi plants at their seedling stage, but in their rosette stage the leaves of Chinese cabbage differentiate as they increase in size with shorter petioles. Overall, comparisons of the transcriptomes between leaves of two very different plants, Chinese cabbages with pak choi, during plant development allowed the identification of specific gene categories associated with leafy head formation (Sun *et al.*, 2019). In Chinese cabbage, leaf adaxial–abaxial (ad–ab) polarity is tightly related to leaf incurvature, an essential factor for the formation of leafy heads (Gao *et al.*, 2020). Midrib bending is induced autonomously during leaf growth in Chinese cabbage. Bending occurs in the basal midrib during early head formation. The main developmental factor for erect leaves during early head formation is likely to be autonomously induced midrib bending during growth (Nishijima and Fukino, 2006). The processes of vernalization and growing point extension are controlled separately. It is known that stem elongation can occur without flower formation. The influence of solar radiation varies with older well-established genotypes, e.g. being greater in the bolting-susceptible cv. Jade Pagoda and less in cv. Kasumi, although temperature has similar effects on both cultivars. Bolting is encouraged by long days. Head development and flower stalk extension appear to be separate processes. The former is driven predominantly by temperature and the latter by temperature, subject to an upper maximum, and solar radiation (Wurr *et al.*, 1996). Recently, Kawanabe *et al.* (2016) noted that epigenetic regulation is crucial for the development of plants and for adaptation to a changing environment. Results suggest that the epigenetic state during vernalization is an important aspect in breeding for high bolting resistance in *B. rapa*.

In Chinese cabbage, leafy head-related traits are directly related to yield and marketability; but there is scant information on the mechanisms involved in the head formation under the influence of low temperatures. Results indicated that the leafy head formation of Chinese cabbage has involved a complex regulatory pattern (Zhang *et al.*, 2016), but seedlings exposed to low temperatures produced lower fresh and dry matter content (by 3.35 g/plant and 0.98%, respectively) and accumulated less soluble sugar (by 0.44%). No significant effects of temperature were identified for chlorophyll *a*, chlorophyll *b* and carotenoid content of Chinese cabbage seedlings (Kalisz and Cebula, 2006).

As emphasized by Choi *et al.* (2020), traditional breeding methods for all brassicas usually involve field tests carried out by experienced breeders which are costly and time-consuming. Recently, with the development of next-generation sequencing technology, molecular markers can be utilized for selection processes in breeding. For Chinese cabbage breeding in particular, these single nucleotide polymorphism marker sets are not only efficient for selecting of early fixed lines as background selection but are also useful for testing the purity of F_1 hybrids.

MIBUNA AND MIZUNA (*BRASSICA RAPA* SSP. *NIPPOSINICA*)

Brassica rapa shows a wide range of morphological variations, in addition to Chinese cabbage and pak choi as discussed in the preceding section. For example, the leaf morphologies of the Japanese traditional leafy vegetables mibuna and mizuna (*B. rapa* ssp. *nipposinica*) (see Fig. 4.11) are distinctly different, even though they are closely related cultivars that are easy to cross.

As well as the differences in the gross morphology of leaves, some cultivars of mibuna (Kyo-nishiki) have many trichomes on their leaves, whereas mizuna (Kyo-mizore) do not. Results indicate that *BrGL1* on linkage group A9 is one of the candidate genes responsible for the difference in trichome number between mibuna and mizuna (Kawakatsu *et al.*, 2017).

Fig. 4.11. Mizuna (*Brassica rapa* ssp. *nipposinica*). (Geoff Dixon).

Evaluations of different growing dates (transplants were planted out to the field in the middle of, and at the end of, August – first and second production terms, respectively) and cultivars (mibuna, mizuna) on morphological parameters, yielding and chemical composition of the plants are reported by Kalisz *et al.* (2012). No statistical differences in yields among cultivars were detected but the growing date did have an effect. In the second term, compared with the first, the total and commercial yield was higher by 3.34 and 3.77 t/ha, respectively. During harvests the content of chlorophylls, carotenoids and L-ascorbic acid was higher in the rosettes of the mibuna cultivar, while mizuna had more dry matter and soluble sugars.

SALAD ROCKET (*ERUCA SATIVA*) AND WALL ROCKET (*DIPLOTAXIS TENUIFOLIA*)

These crops are frequently associated with mibuna and mizuna leaves in bags of salad leaves and are examples of genera of crucifers which have become widely popular recently. There are differences between the annual garden rocket and perennial wall rocket when cultivated in production conditions (Ivanova *et al.*, 2018) in morphology, chromosome number and glucosinolate content. Some of the morphological features used included: the size, shape and characteristics of the spout of the pod; sheet form and attachment to the stem; and shape, venation and colour of the corolla. Perennial wall rocket is more resistant to stress factors, more suitable for mechanized harvesting and has a longer shelf life. There was a decrease in the length of the distal portion, leaf number and fresh and dry weights with increasing seedling density in garden rocket (Reghin *et al.*, 2004), but total yield rose.

TURNIP (*BRASSICA RAPA* SSP. *RAPIFERA*)

The crop species *B. rapa* has worldwide economic importance, as discussed in detail by McNaughton (1976) and Goldman (2020). Crop domestication produced diversity resulting in the turnip (*B. rapa* ssp. *rapifera*) (Bird *et al.*, 2017), which can be subdivided into European and Asian forms. Subspecies of *B. rapa* are a morphologically and genetically very diverse group of vegetables cultivated worldwide (Park *et al.*, 2019). The presence of subspecies-specific divergence in transposable elements may have influenced this phenotypic diversity within the one species. The turnip root or tuber develops by forming supernumerary cambia (Goldman, 2020) with associated storage parenchyma cells produced as secondary tissues arising from the vascular cambium. Growth in girth results from the expansion of vascular cambia producing inner xylem and outer phloem cells. The purpose of this growth is to act as a storage organ, providing energy for growth and reproduction in the following season.

Turnip is a biennial plant grown in temperate regions from spring through to autumn and forming fleshy tubers (Zheng *et al.*, 2018). Depending on climate, they may be overwintered *in situ* or harvested and replanted in the following spring for flowering and seed production. *FLOWERING LOCUS C (FLC)* is a MADS-box transcription factor that acts as a major repressor of floral transition by suppressing the flowering promoters *FT* and *SOC1*. Vernalization represses further tuber growth and promotes flowering. Sucrose transporters, or sucrose carriers, are found in turnip which move photosynthates from sources to sinks in the root (Liu *et al.*, 2017).

Many Japanese turnip cultivars are heirloom vegetables, such as 'Omikabu', an ancestral white cultivar; red- or purple-skinned cultivars may produce coloured flesh and petioles, which may be used for producing pickled vegetables (Kubo *et al.*, 2019). Some turnip lines formed genetic clusters with 'Jonansensuji mizuna' – this suggests past hybridizations.

Turnip has potentially important characteristics for improving human health (Stepanova *et al.*, 2020). They report that turnips grown in extremely harsh climates can be stored over winter at 0–1°C and 90–95% relative humidity. Characteristics of odour, taste and flavour were enhanced by storage at 5°C or 10°C compared with 0°C. Warmer conditions increased a muddy odour and decreased juiciness in turnip tissues. Low-temperature storage preserved the sucrose content in turnip (Helland *et al.*, 2016). Fresh leafy vegetables are grown from Italian and Spanish cultivars yielding greens and turnip tops (Cartea *et al.*, 2020) which also deliver bioactive compounds favourable for human nutrition. Early season white-rooted Milan fresh turnips are produced in the open in milder Mediterranean regions or under protection further north.

SWEDE (*BRASSICA NAPUS* SSP. *RAPIFERA*)

Swede (*Brassica napus* ssp. *rapifera*) is a widely grown root crop which is noted as a healthy vegetable with high vitamin C content (Thomsen *et al.*, 2017). Vitamin C content rose and nitrate levels fell where only modest amounts of nitrogen fertilizer were used. This practice also increased total, aliphatic, indole and individual glucosinolates, on a dry matter basis, when grown on high organic content soils. Cropping on sandy soils was associated with an increased sweet taste (Thomsen *et al.*, 2017).

Swede tends to be most popular as a vegetable in northern latitudes such as Scotland, Scandinavia and parts of Canada. Production in higher latitudes with long photosynthetic periods and photoperiods increases sweet, less bitter tastes and reduces glucosinolate content (Mølmann *et al.*, 2018). Crop production at 21°C increased the pungent odour, bitterness, astringency and fibrousness of roots, while lower temperatures (9°C) were associated with acidic odour, sweet taste, crispiness and juiciness (Johansen *et al.*, 2016). Low temperatures and the light qualities of northern Scandinavia improved the eating quality of swede roots (Mølmann *et al.*, 2020). Far-red light-emitting diode (LED) (740 nm) light and 15°C resulted in longer leaves, purple skin and rounder roots compared with other treatments. At 9°C, increased concentrations of D-fructose and D-glucose and the glucosinolate, progoitrin, were detected. The latter compound is also encouraged by potassium fertilizers.

RADISH (EUROPEAN RADISH – *RAPHANUS SATIVUS* AND DAIKON – *RAPHANUS SATIVUS* VAR. *LONGIPINNATUS*)

Differences in growth of the upper part of the taproot shapes the long, tapering roots of radish cv. Long White Icicle and the smaller, round cv. Cherry Belle (Ting and Wren, 1979a), and resulted in thickening from increased cambial activity, cell number and size. Cultivar Icicle contained more cells in the secondary xylem and phloem and secondary cortex than cv. Cherry Belle, and the non-lignified xylem parenchyma cells were also larger in cv. Icicle. In addition, the upper part of the taproot elongated in cv. Icicle but not in cv. Cherry Belle, due to increasing cell numbers in the longitudinal plane. Auxins, cytokinins and inositol contents appear not to affect storage organ development (Ting and Wren, 1979b). Several functional proteins, including cell division cycle 5-like protein (CDC5), expansin B1 (EXPB1) and xyloglucan endotransglucosylase/hydrolase protein 24 (XTH24), were responsible for cell division and expansion during the radish taproot thickening process (Xie *et al.*, 2018). Asian radish or daikon (*R. sativus.* var. *longipinnatus*) (see Fig. 4.12) produces much larger and longer roots compared with the much smaller European radish (*R. sativus*).

Fig. 4.12. Asian radish or daikon (*Raphanus sativus* var. *longipinnatus*). (Geoff Dixon).

In summer-grown crops, root splitting can be a major defect (Lockley *et al.*, 2021). Radish diameter was positively associated with compression failure force, suggesting that larger roots are more resistant to compressive splitting. Increased hypocotyl water content may increase splitting susceptibility by raising the water potential of the radish tissue. Results suggested that splitting is reduced after harvest by processing at a higher, ambient, temperature followed by storing at a low temperature and high humidity. Signal transductions and metabolic, biosynthetic and abiotic stress-responsive pathways have important roles in reducing stress-induced damage and enhancing heat tolerance in radish (Wang *et al.*, 2018a). Heat-stress-responsive genes are activated which repair damaged proteins and enhance thermotolerance in radish. Proteomic analysis and omics association analysis revealed a heat-stress-responsive regulatory network in radish. The initial sensing of heat stress occurred at the plasma membrane, and then key components of stress signal transduction triggered heat-responsive genes which establish homeostasis and enhance thermotolerance (Wang *et al.*, 2018b).

Premature bolting is a major factor affecting radish cultivation (Pei *et al.*, 2019). Abscisic acid and indole acetic acid have roles in delaying bolting and flowering in tetraploid radish. Gene *FLC1.1* was highly expressed in tetraploid plants at the flowering stage. Expression levels of genes positively associated with flowering and bolting were much lower in tetraploid compared to diploid radishes at flowering and bolting stages. An analysis found 144 MADS-box family members in the radish identifying gene-mediated molecular mechanisms underlying flowering and floral organogenesis in this crop (Li *et al.*, 2016). *CONSTANS-like* (*CO-like, COL*) genes help control the circadian clock in radish ensuring regular development (Hu *et al.*, 2018). The expression levels of *RsaCOL-02* and *RsaCOL-04* were significantly increased during vernalization treatment, while the expression of *RsaCOL-10* was significantly decreased. These results offer opportunities for genetically controlling bolting and flowering in radish.

Radish is a crop which might be produced by factory methods using red plus blue LEDs. Cherry radish formed commercially acceptable roots when light intensity was equal to or over 240 μmol/m²/s (Zha and Liu, 2018). Cherry radish production is primarily dependent on light intensity, followed by light quality and photoperiod.

AQUATIC CRUCIFERS

Watercress (*Nasturtium officinale*)

Production of two aquatic crucifers has increased considerably over the past 20 years and consequently they are now included in this book. Grown internationally, watercress (*Nasturtium officinale*) (see Fig. 4.13) is highly rated as promoting health and well-being. It is a nutrient-intense, leafy crop and is consumed raw as a salad vegetable or in soups (Payne *et al.*, 2015).

Fig. 4.13. Watercress (*Nasturtium officinale*) crop. (Courtesy Tom Amery, The Watercress Company).

Stem length, stem diameter and antioxidant potential varied across the accessions and genes coding for glucosinolate production were identified in a dwarf phenotype. Quality is assessed by leaf colour, sheen, aroma, flavour, texture and the overall acceptability and appearance (Spence, 1982). Axillary stem growth produces a primary axillary bud and numerous roots from primordia adaxial to the bud (Selker and Lyndon, 1996).

Research compared the growing cycle (spring vs winter) and nutrient-solution aeration (no aeration, low aeration or high aeration) on yield, quality and shelf life as a fresh-cut product of watercress grown in a floating system. Growing cycles lasted 25 days in spring and 39 days in winter. In the spring cycle, the plants had significantly higher yield and antioxidant capacity and lower specific leaf area, total root length and diameter; and oxalate content than in the winter cycle. The absence of aeration increased the antioxidant capacity and vitamin C content in both cycles. Several adventitious roots developed exogenously from the watercress stem at the nodes as a morphological adaptation to oxygen depletion, particularly in the absence of aeration. The mild dehydration problems observed in the winter cycle led to a slightly lower overall product quality that could be the result of the development of thinner leaves and also the differences in the respiration rates compared with the spring cycle. In general, the spring cycle led to higher productivity, antioxidant capacity, and calcium and potassium contents, and lower oxalate content (Niñirola *et al.*, 2014).

Wasabi (*Wasabia japonica*)

Wasabi (see Fig. 4.14) is an aquatic perennial crucifer native to Japan where it is found growing adjacent to mountain streams and valued for its astringent flavour, similar to European horseradish (*Armoracia lapathifolia*).

Fig. 4.14. Wasabi (*Wasabia japonica*) crop. (Courtesy Tom Amery, The Wasabi Company).

This crop is now produced commercially and regularly in Japan, Australia and the UK. Growing 30–40 cm tall with characteristic coarse, heart-shaped leaves and producing finger-thick rhizomes up to 20 cm long, it can be cultivated in fast-flowing, clean spring water. Cultivation in nutrient film systems (Hassan *et al.*, 2001) demonstrated the yield advantages obtained from warm water. However, seed germination is favoured by cool (5°C), moist conditions (Palmer, 1990) and inhibited at temperatures greater than 10°C. More recently, Hoang *et al.* (2019) reported successful propagation using *in vitro* hydroponic culture for micropropagation, rapidly producing high-quality wasabi plantlets.

REFERENCES

Adzić, S., Pavlović, S., Jokanovic, M.B., Cvikić, D., Pavlović, N, *et al.* (2012) Correlation of important agronomic characteristics and yield of medium late genotypes of head cabbage. *Acta Horticulturae* 960, 159–164. DOI: 10.17660/ActaHortic.2012.960.22.

Atherton, J.G., Hand, D.J. and Williams, C.A. (1987) Curd initiation in the cauliflower (*Brassica oleracea* var. *botrytis* L.). In: Atherton, J.G. (ed.) *Manipulation of Flowering: Proceedings of the Sutton Bonnington Easter School in Agricultural Science.* Butterworth, London, pp. 133–145.

Bhatia, R., Dey, S.S., Sood, S., Sharma, K., Parkash, C. *et al.* (2017) Efficient microspore embryogenesis in cauliflower (*Brassica oleracea* var. *botrytis* L.) for development of plants with different ploidy level and their use in breeding programme. *Scientia Horticulturae* 216, 83–92. DOI: 10.1016/j.scienta.2016.12.020.

Bird, K.A., An, H., Gazave, E., Gore, M.A., Pires, J.C, *et al.* (2017) Population structure and phylogenetic relationships in a diverse panel of *Brassica rapa* L. *Frontiers in Plant Science* 8, 321. DOI: 10.3389/fpls.2017.00321.

Bodson, M. and Remacle, B. (1987) Distribution of assimilates from various sources-leaves during floral transition of *Sinapis alba* L. In: Atherton, J.G. (ed.) *Manipulation of Flowering: Proceedings of the Sutton Bonnington Easter School in Agricultural Science.* Butterworth, London, pp. 341–350.

Boersma, M., Gracie, A.J. and Brown, P.H. (2009) Relationship between growth rate and the development of hollow stem in broccoli. *Crop and Pasture Science* 60(10), 995–1001. DOI: 10.1071/CP09391.

Booij, R. (1987) Environmental factors in cauliflower curd initiation and growth. *Netherlands Journal of Agricultural Science* 35(4), 435–445. DOI: 10.18174/njas. v35i4.16703.

Booij, R. (2000) Yield formation in Brussels sprouts: effects of nitrogen. *Acta Horticulturae* 533, 377–384. DOI: 10.17660/ActaHortic.2000.533.46.

Booij, R. and Struik, P.C. (1990) Effects of temperature on leaf and curd initiation in relation to juvenility of cauliflower. *Scientia Horticulturae* 44(3–4), 201–214. DOI: 10.1016/0304-4238(90)90120-4.

Cartea, M.E., Di Bella, M.C., Velasco, P., Soengas, P., Toscano, S. *et al.* (2020) Evaluation of Italian and Spanish accessions of *Brassica rapa* L.: effect of flowering earliness on fresh yield and biological value. *Agronomy* 11(1), 29. DOI: 10.3390/agronomy11010029.

Chiu, L.W., Zhou, X.J., Burke, S., Wu, X.L., Prior, R.L. *et al.* (2010) The purple cauliflower arises from activation of a MYB transcription factor. *Plant Physiology* 154(3), 1470–1480. DOI: 10.1104/pp.110.164160.

Choi, S.R., Oh, S.H., Dhandapani, V., Jang, C.S., Ahn, C.-H. *et al.* (2020) Development of SNP markers for marker-assisted breeding in Chinese cabbage using Fluidigm genotyping assays. *Horticulture, Environment, and Biotechnology* 61(2), 327–338. DOI: 10.1007/s13580-019-00211-y.

Conversa, G., Lazzizera, C., Bonasia, A. and Elia, A. (2019) Growth, N uptake and N critical dilution curve in broccoli cultivars grown under Mediterranean conditions. *Scientia Horticulturae* 244, 109–121. DOI: 10.1016/j.scienta.2018.09.034.

da Silva, A.L.B.R., Candian, J.S., Zotarelli, L., Coolong, T. and Christensen, C. (2020) Nitrogen fertilizer management and cultivar selection for cabbage production in the southeastern United States. *HortTechnology* 30(6), 685–691. DOI: 10.21273/HORTTECH04690-20.

Dixon, G.R. (2017) Managing clubroot disease (caused by *Plasmodiophora brassicae* Wor.) by exploiting the interactions between calcium cyanamide fertilizer and soil microorganisms. *The Journal of Agricultural Science* 155(4), 527–543. DOI: 10.1017/S0021859616000800.

Duclos, D.V. and Björkman, T. (2008) Meristem identity gene expression during curd proliferation and flower initiation in *Brassica oleracea*. *Journal of Experimental Botany* 59(2), 421–433. DOI: 10.1093/jxb/erm327.

Du, H.-P., Cao, M.-L., Yan, S.-J. and Liu, J. (2017) Studies on main phenotypic traits affecting single head weight of spring cabbage. [Chinese] *Journal of Henan Agricultural Sciences* 46(5), 106–111.

Everaarts, A.P. (1994) A decimal code describing the developmental stages of head cabbage (*Brassica oleracea* var. *capitata*). *Annals of Applied Biology* 125(1), 207–214. DOI: 10.1111/j.1744-7348.1994.tb04962.x.

Everaarts, A.P., Booij, R. and de Moel, C.P. (1998) Yield formation in Brussels sprouts. *The Journal of Horticultural Science and Biotechnology* 73(5), 711–721. DOI: 10.1080/14620316.1998.11511038.

Farnham, M.W. and Bjorkman, T. (2011) Breeding vegetables adapted to high temperatures: a case study with broccoli. *HortScience* 46(8), 1093–1097. DOI: 10.21273/HORTSCI.46.8.1093.

Fisher, N.M. and Milbourn, G.M. (1974) The effect of plant density, date of apical bud removal and leaf removal on the growth and yield of single-harvest Brussels sprouts (*Brassica oleracea* var. *gemmifera* D.C.): I. Whole plant and axillary bud growth. *The Journal of Agricultural Science* 83(3), 479–487. DOI: 10.1017/S0021859600026964.

Fujima, Y. (1983) Studies on thermal conditions of curd formation and development in cauliflower and broccoli, with especial reference to abnormal curd development. *Memoirs of the Faculty of Agriculture, Kagawa University* 40, 1–123.

Fujima, Y. and Hirose, T. (1979) Studies on thermal conditions of curd formation and development in cauliflower and broccoli. *Journal of the Japanese Society of Horticultural Science* 114, 552–556.

Fujime, Y. and Okuda, N. (1996) The physiology of flowering in *Brassicas*, especially about cauliflower and broccoli. *Acta Horticulturae* 407, 247–254. DOI: 10.17660/ActaHortic.1996.407.30.

Gallagher, J.N. (1979) Field studies of cereal leaf growth: 1. Initiation and expansion in relation to temperature and ontogeny. *Journal of Experimental Botany* 30(117), 625–636. DOI: 10.1093/jxb/30.4.625.

Gao, Y., Lu, Y., Li, X., Li, N., Zhang, X. *et al.* (2020) Development and application of SSR markers related to genes involved in leaf adaxial-abaxial polarity establishment in Chinese cabbage (*Brassica rapa* L. ssp. *pekinensis*). *Frontiers in Genetics* 11, 773. DOI: 10.3389/fgene.2020.00773.

Goldman, I.L. (2020) Carrot, beet and turnip. In: Wien, H.C. and Stützel, H. (eds) *The Physiology of Vegetable Crops*, 2nd edn. CAB International, Wallingford, UK, pp. 399–420. DOI: 10.1079/9781786393777.0000.

Grabowska, A., Sękara, A., Bieniasz, M., Kunicki, E. and Kalisz, A. (2013) Dark-chilling of seedlings affects initiation and morphology of broccoli inflorescence. *Notulae Botanicae Horti Agrobotanici Cluj-Napoca* 41(1), 213–218. DOI: 10.15835/nbha4118272.

Grevsen, K. (1998) Effects of temperature on head growth of broccoli (*Brassica oleracea* L. var. *italica*): parameter estimates for a predictive model. *The Journal of Horticultural Science and Biotechnology* 73(2), 235–244. DOI: 10.1080/14620316.1998.11510970.

Grevsen, K. and Olesen, J.E. (1994) Modelling cauliflower development from transplanting to curd initiation. *Journal of Horticultural Science* 69(4), 755–766. DOI: 10.1080/14620316.1994.11516510.

Haine, K.E. (1959) Time of heading and quality of curd in winter cauliflower. *Journal of the National Institute of Agricultural Botany* 8, 667–674.

Hamer, P.J.C. (1991) *The IT vegetable farm, Part III. Analysis of the variation of the yield and timing of development of Brussels sprout buttons.* Divisional Note (DN 1616). Silsoe Research Institute, Silsoe, UK.

Hamer, P.J.C. (1993) Model parameters to describe the development of marketable yield of Brussels sprouts. *Journal of Horticultural Science* 68(6), 871–882. DOI: 10.1080/00221589.1993.11516426.

Hand, D.J. (1988) Regulation of curd initiation in the summer cauliflower. PhD. Thesis, University of Nottingham, UK.

Hand, D.J. and Atherton, J.G. (1987) Curd initiation in the cauliflower: 1. Juvenility. *Journal of Experimental Botany* 38(197), 2050–2058. DOI: 10.1093/jxb/38.12.2050.

Hasan, Y., Briggs, W., Matschegewski, C., Ordon, F., Stützel, H. *et al.* (2016) Quantitative trait loci controlling leaf appearance and curd initiation of cauliflower in relation to temperature. *Theoretical and Applied Genetics* 129(7), 1273–1288. DOI: 10.1007/s00122-016-2702-6.

Hassan, M., Fujime, Y., Okuda, N., Matsui, T. and Suzuki, H. (2001) Effects of solution temperature on the growth and development of Wasabi grown in NFT. *Technical Bulletin of the Faculty of Agriculture, Kagawa University* 53, 13–18.

Helland, H.S., Leufvén, A., Bengtsson, G.B., Pettersen, M.K., Lea, P. *et al.* (2016) Storage of fresh-cut swede and turnip: effect of temperature, including sub-zero temperature, and packaging material on sensory attributes, sugars and glucosinolates. *Postharvest Biology and Technology* 111, 370–379. DOI: 10.1016/j.postharvbio.2015.09.011.

Hoang, N.N., Kitaya, Y., Shibuya, T. and Endo, R. (2019) Development of an in vitro hydroponic culture system for wasabi nursery plant production—effects of nutrient concentration and supporting material on plantlet growth. *Scientia Horticulturae* 245, 237–243. DOI: 10.1016/j.scienta.2018.10.025.

Hu, T., Wei, Q., Wang, W., Hu, H., Mao, W. *et al.* (2018) Genome-wide identification and characterization of *CONSTANS-like* gene family in radish (*Raphanus sativus*). *PLoS ONE* 13(9), e0204137. DOI: 10.1371/journal.pone.0204137.

Isenberg, F.M.R., Pendergress, A., Carroll, J.E., Howell, L. and Oyer, E.B. (1975) The use of weight, density heat units and solar radiation to predict maturity of cabbage for storage. *Journal of the American Society for Horticultural Science* 100(3), 313–316. DOI: 10.21273/JASHS.100.3.313.

Ivanova, M.I., Bukharov, A.F., Litnetsky, A.V., Razin, A.F., Meshcheryakova, R.A. *et al.* (2018) Principal differences between perennial wall rocket (*Diplotaxis tenuifolia* (L.) DC.) and annual garden rocket (*Eruca sativa* Mill.) in cultivation in production conditions: overview. [Russian] *The Agrarian Scientific Journal* 1(1), 14–19. DOI: 10.28983/asj.v0i1.319.

Jakopic, J., Weber, N., Cunja, V., Veberic, R. and Slatnar, A. (2016) Brussels sprout decapitation yields larger sprouts of superior quality. *Journal of Agricultural and Food Chemistry* 64(40), 7459–7465. DOI: 10.1021/acs.jafc.6b03486.

Johansen, T.J., Hagen, S.F., Bengtsson, G.B. and Mølmann, J.A.B. (2016) Growth temperature affects sensory quality and contents of glucosinolates, vitamin C and sugars in swede roots (*Brassica napus* L. ssp. *rapifera* Metzg.). *Food Chemistry* 196, 228–235. DOI: 10.1016/j.foodchem.2015.09.049.

Kalisz, A. and Cebula, S. (2006) The effect of temperature on growth and chemical composition of Chinese cabbage seedlings in spring period. *Folia Horticulturae* 18(1), 3–15.

Kalisz, A., Sekara, A. and Kostrzewa, J. (2012) Effect of growing date and cultivar on the morphological parameters and yield of *Brassica rapa* var. *japonica*. *Acta Scientiarum Polonorum – Hortorum Cultus* 11(3), 131–143.

Kawakatsu, Y., Nakayama, H., Kaminoyama, K., Igarashi, K., Yasugi, M. *et al.* (2017) A *GLABRA1* ortholog on LG A9 controls trichome number in the Japanese leafy vegetables Mizuna and Mibuna (*Brassica rapa* L. subsp. *nipposinica* L. H. Bailey): evidence from QTL analysis. *Journal of Plant Research* 130(3), 539–550. DOI: 10.1007/s10265-017-0917-5.

Kawanabe, T., Osabe, K., Itabashi, E., Okazaki, K., Dennis, E.S. *et al.* (2016) Development of primer sets that can verify the enrichment of histone modifications, and their application to examining vernalization-mediated chromatin changes in *Brassica rapa* L. *Genes and Genetic Systems* 91(1), 1–10. DOI: 10.1266/ggs.15-00058.

Kieffer, M., Fuller, M.P. and Jellings, A.J. (1998) Explaining curd and spear geometry in broccoli, cauliflower and 'romanesco': quantitative variation in activity of primary meristems. *Planta* 206(1), 34–43. DOI: 10.1007/s004250050371.

Kleinhenz, M.D. (2003) A proposed tool for preharvest estimation of cabbage yield. *HortTechnology* 13(1), 182–185. DOI: 10.21273/HORTTECH.13.1.0182.

Kolota, E. and Chohura, P. (2015) Control of head size and nutritional value of cabbage by plant population and nitrogen fertilization. *Acta Scientiarum Polonorum – Hortorum Cultus* 14(2), 75–85.

Kubo, N., Ueoka, H. and Satoh, S. (2019) Genetic relationships of heirloom turnip (*Brassica rapa*) cultivars in Shiga Prefecture and other regions of Japan. *The Horticulture Journal* 88(4), 471–480. DOI: 10.2503/hortj.UTD-071.

Li, C., Wang, Y., Xu, L., Nie, S., Chen, Y. *et al.* (2016) Genome-wide characterization of the MADS-box gene family in radish (*Raphanus sativus* L.) and assessment of its roles in flowering and floral organogenesis. *Frontiers in Plant Science* 7, 1390. DOI: 10.3389/fpls.2016.01390.

Li, H., Liu, Q., Zhang, Q., Qin, E., Jin, C. *et al.* (2017) Curd development associated gene (*CDAG1*) in cauliflower (*Brassica oleracea* L. var. *botrytis*) could result in enlarged organ size and increased biomass. *Plant Science* 254, 82–94. DOI: 10.1016/j.plantsci.2016.10.009.

Lin, C.-Y., Chen, K.-S., Chen, H.-P., Lee, H.-I. and Hsieh, C.-H. (2019) Curd initiation and transformation in tropical cauliflower cultivars under different temperature treatments. *HortScience* 54(8), 1351–1356. DOI: 10.21273/HORTSCI13881-19.

Lindemann-Zutz, K., Fricke, A. and Stützel, H. (2016) Prediction of time to harvest and its variability of broccoli (*Brassica oleracea* var. *italica*) part II. Growth model description, parameterisation and field evaluation. *Scientia Horticulturae* 200, 151–160. DOI: 10.1016/j.scienta.2016.01.009.

Liu, Y., Yin, X., Yang, Y., Wang, C. and Yang, Y. (2017) Molecular cloning and expression analysis of turnip (*Brassica rapa* var. *rapa*) sucrose transporter gene family. *Plant Diversity* 39(3), 123–129. DOI: 10.1016/j.pld.2017.05.006.

Lockley, R.A., Beacham, A.M., Grove, I.G. and Monaghan, J.M. (2021) Postharvest temperature and water status influence postharvest splitting susceptibility in summer radish (*Raphanus sativus* L.). *Journal of the Science of Food and Agriculture* 101(2), 536–541. DOI: 10.1002/jsfa.10662.

Marshall, B. and Thompson, R. (1987a) A model of the influence of air temperature and solar radiation on the time to maturity of calabrese (*Brassica oleracea* var. *italica*). *Annals of Botany* 60(5), 513–519. DOI: 10.1093/oxfordjournals.aob. a087474.

Marshall, B. and Thompson, R. (1987b) Application of a model to predict the time to maturity of calabrese (*Brassica oleracea* var. *italica*). *Annals of Botany* 60(5), 521–529. DOI: 10.1093/oxfordjournals.aob.a087475.

McNaughton, I.H. (1976) Turnip and relatives. In: Simmonds, N.W. (ed.) *Evolution of Crop Plants*. Longman, London, pp. 45–48.

Milford, G.F.J., Pocock, T.O. and Riley, J. (1985) An analysis of leaf growth in sugar beet. 1. Leaf appearance and expansion in relation to temperature under controlled conditions. *Annals of Applied Biology* 106(3), 163–172. DOI: 10.1111/j.1744-7348.1985.tb03106.x.

Mølmann, J.A., Hagen, S.F., Bengtsson, G.B. and Johansen, T.J. (2018) Influence of high latitude light conditions on sensory quality and contents of health and sensory-related compounds in swede roots (*Brassica napus* L. ssp. *rapifera* Metzg.). *Journal of the Science of Food and Agriculture* 98(3), 1117–1123. DOI: 10.1002/jsfa.8562.

Mølmann, J.A.B., Hansen, E. and Johansen, T.J. (2020) Effects of supplemental LED light quality and reduced growth temperature on swede (*Brassica napus* L. ssp. *rapifera* Metzg.) root vegetable development and contents of glucosinolates and sugars. *Journal of the Science of Food and Agriculture* 101(6), 2422–2427. DOI: 10.1002/jsfa.10866.

Monteith, J.L. (1981) Climatic variation and the growth of crops. *Quarterly Journal of the Royal Meteorological Society* 107(454), 749–774. DOI: 10.1002/qj.49710745402.

Niñirola, D., Fernández, J.A., Conesa, E., Martínez, J.A. and Egea-Gilabert, C. (2014) Combined effects of growth cycle and different levels of aeration in nutrient solution on productivity, quality, and shelf life of watercress (*Nasturtium officinale* R. Br.) plants. *HortScience* 49(5), 567–573. DOI: 10.21273/HORTSCI.49.5.567.

Nieuwhof, M. (1969) *Cole Crops: Botany, Cultivation and Utilization*. Leonard Hill, London.

Nishijima, T. and Fukino, N. (2006) Autonomous development of erect leaves independent of light irradiation during the early stage of head formation in Chinese cabbage (*Brassica rapa* L. var. *pekinensis* Rupr.). *Journal of the Japanese Society for Horticultural Science* 75(1), 59–65. DOI: 10.2503/jjshs.75.59.

Özer, M.Ö. (2021) Morphological variability of red head cabbage (*Brassica oleracea* L. var. *capitata* L. subvar. *rubra*) populations. *Genetic Resources and Crop Evolution* 68(3), 1033–1043. DOI: 10.1007/s10722-020-01046-8.

Palmer, J. (1990) Germination and growth of wasabi (*Wasabia japonica* (Miq.) Matsumara). *New Zealand Journal of Crop and Horticultural Science* 18(2–3), 161–164. DOI: 10.1080/01140671.1990.10428089.

Park, H.R., Kang, T., Yi, G., Yu, S.H., Shin, H. *et al.* (2019) Genome divergence in *Brassica rapa* subspecies revealed by whole genome analysis on a doubled-haploid line of turnip. *Plant Biotechnology Reports* 13(6), 677–687. DOI: 10.1007/s11816-019-00565-w.

Payne, A.C., Clarkson, G.J.J., Rothwell, S. and Taylor, G. (2015) Diversity in global gene expression and morphology across a watercress (*Nasturtium officinale* R. Br.) germplasm collection: first steps to breeding. *Horticulture Research* 2, 15029. DOI: 10.1038/hortres.2015.29.

Pearson, S. and Hadley, P. (1988) Planning calabrese production. *The Grower* 100, 21–22.

Pearson, S., Hadley, P. and Wheldon, A.E. (1994) A model of the effects of temperature on the growth and development of cauliflower (*Brassica oleracea* L. botrytis). *Scientia Horticulturae* 59(2), 91–106. DOI: 10.1016/0304-4238(94)90076-0.

Pei, Y., Yao, N., He, L., Deng, D., Li, W. *et al.* (2019) Comparative study of the morphological, physiological and molecular characteristics between diploid and tetraploid radish (*Raphunas sativus* L.). *Scientia Horticulturae* 257, 108739. DOI: 10.1016/j. scienta.2019.108739.

Radovich, T.J.K. and Kleinhenz, M.D. (2004) Rapid estimation of cabbage head volume across a population varying in head shape: a test of two geometric formulae. *HortTechnology* 14(3), 388–391. DOI: 10.21273/HORTTECH.14.3.0388.

Rahman, H.U., Hadley, P., Pearson, S. and Dennett, M.D. (2007) Effect of incident radiation integral on cauliflower growth and development after curd initiation. *Plant Growth Regulation* 51(1), 41–52. DOI: 10.1007/s10725-006-9146-y.

Rahman, H.U., Hadley, P., Pearson, S. and Khan, M.J. (2013) Response of cauliflower (*Brassica oleracea* L. var. *botrytis*) growth and development after curd initiation to different day and night temperatures. *Pakistan Journal of Botany* 45(2), 411–420.

Reghin, M.Y., Otto, R.F. and Vinne, J.V.D. (2004) Efeito da densidade de mudas por célula e do volume da célula na produção de mudas e cultivo da rúcula. *Ciência e Agrotecnologia* 28(2), 287–295. DOI: 10.1590/S1413-70542004000200006.

Ridge, S., Brown, P.H., Hecht, V., Driessen, R.G. and Weller, J.L. (2015) The role of *BoFLC2* in cauliflower (*Brassica oleracea* var. *botrytis* L.) reproductive development. *Journal of Experimental Botany* 66(1), 125–135. DOI: 10.1093/jxb/eru408.

Roberts, E.H. and Summerfield, R.J. (1987) Measurement and prediction of flowering in annual crops. In: Atherton, J.G. (ed.) *Manipulation of Flowering: Proceedings of the Sutton Bonnington Easter School in Agricultural Science.* Butterworth, London, pp. 17–50.

Rosen, A., Hasan, Y., Briggs, W. and Uptmoor, R. (2018) Genome-based prediction of time to curd induction in cauliflower. *Frontiers in Plant Science* 9, 78. DOI: 10.3389/fpls.2018.00078.

Sady, W., Rozek, S., Leja, M. and Mareczek, A. (1999) Spring cabbage yield and quality as related to nitrogen fertilizer type and method of fertilizer application. *Acta Horticulturae* 506, 77–80. DOI: 10.17660/ActaHortic.1999.506.8.

Salter, P.J. (1969) Studies on crop maturity in cauliflower: I. Relationship between the times of curd initiation and curd maturity of plants within a cauliflower crop. *Journal of Horticultural Science and Biotechnology* 44(2), 129–140. DOI: 10.1080/00221589.1969.11514301.

Salter, P.J. and Fradgley, J.R. (1969) Studies on crop maturity in cauliflower: II. Effects of cultural factors on the maturity characteristics of a cauliflower crop. *Journal of Horticultural Science and Biotechnology* 44(2), 141–154. DOI: 10.1080/00221589.1969.11514302.

Scaife, A., Cox, E.F. and Morris, G.E.L. (1987) The relationship between shoot weight, plant density and time during the propagation of four vegetable species. *Annals of Botany* 59(3), 325–334. DOI: 10.1093/oxfordjournals.aob.a087321.

Selker, J.M.L. and Lyndon, R.F. (1996) Leaf initiation and de novo pattern formation in the absence of an apical meristem and pre-existing patterned leaves in watercress (*Nasturtium officinale*) axillary explants. *Canadian Journal of Botany* 74(4), 625–641. DOI: 10.1139/b96-079.

Shanmuganathan, V. and Benjamin, L.R. (1993) The effect of time of seedling emergence and density on the marketable yield of spring cabbage. *Journal of Horticultural Science* 68(6), 947–954. DOI: 10.1080/00221589.1993.11516435.

Singh, B.K., Singh, B. and Singh, P.M. (2018) Breeding cauliflower: a review. *International Journal of Vegetable Science* 24(1), 58–84. DOI: 10.1080/19315260.2017.1354242.

Smith, L.B. and King, G.J. (2000) The distribution of *BoCAL-a* alleles in *Brassica oleracea* is consistent with a genetic model for curd development and domestication of the cauliflower. *Molecular Breeding* 6(6), 603–613. DOI: 10.1023/A:1011370525688.

Spence, R.-M.M. (1982) The development of a vocabulary and sensory profile for the assessment of watercress quality. *International Journal of Food Science & Technology* 17(5), 633–648. DOI: 10.1111/j.1365-2621.1982.tb00222.x.

Stansell, Z., Björkman, T., Branham, S., Couillard, D. and Farnham, M.W. (2017) Use of a quality trait index to increase the reliability of phenotypic evaluations in broccoli. *HortScience* 52(11), 1490–1495. DOI: 10.21273/HORTSCI12202-17.

Stansell, Z., Farnham, M. and Björkman, T. (2019) Complex horticultural quality traits in broccoli are illuminated by evaluation of the immortal BolTBDH mapping population. *Frontiers in Plant Science* 10, 1104. DOI: 10.3389/fpls.2019.01104.

Stepanova, A.G., Davydenko, N.I., Golub, O.V. and Stepanova, E.N. (2020) Effect of storage methods on various sorts of Siberian turnip (*Brassica rapa* L.). [Russian] *Food Processing* 50(3), 470–479. DOI: 10.21603/2074-9414-2020-3-470-479.

Stokes, P. and Verkerk, K. (1951) Flower formation in Brussels sprouts. *Mededelingen van de Landbouw Hogeschool te Wageningen* 50, 143–160.

Sun, X., Bucher, J., Ji, Y., van Dijk, A.D.J., Immink, R.G.H. *et al.* (2018) Effect of ambient temperature fluctuation on the timing of the transition to the generative stage in cauliflower. *Environmental and Experimental Botany* 155, 742–750. DOI: 10.1016/j.envexpbot.2018.06.013.

Sun, X.-X., Basnet, R.K., Yan, Z., Bucher, J., Cai, C. *et al.* (2019) Genome-wide transcriptome analysis reveals molecular pathways involved in leafy head formation of Chinese cabbage (*Brassica rapa*). *Horticulture Research* 6, 130. DOI: 10.1038/s41438-019-0212-9.

Talekar, N.S. and Griggs, T.D. (eds) (1981) *Chinese Cabbage: Proceedings of the First International Symposium.* Asian Vegetable Research and Development Centre, Shanhau, Taiwan.

Tempesta, M., Gianquinto, G., Hauser, M. and Tagliavini, M. (2019) Optimization of nitrogen nutrition of cauliflower intercropped with clover and in rotation with lettuce. *Scientia Horticulturae* 246, 734–740. DOI: 10.1016/j.scienta.2018.11.020.

Thomsen, M.G., Riley, H., Borge, G.I.A., Lea, P., Rødbotten, M. *et al.* (2017) Effects of soil type and fertilization on yield, chemical parameters, sensory quality and consumer preference of swede (*Brassica napus* L. ssp. *rapifera*). *European Journal of Horticultural Science* 82(6), 294–305. DOI: 10.17660/eJHS.2017/82.6.4.

Ting, F.S.-T. and Wren, M.J. (1979a) Storage organ development in radish (*Raphanus sativus* L.). 1. A comparision of development in seedlings and rooted cuttings of two contrasting varieties. *Annals of Botany* 46(3), 267–276. DOI: 10.1093/oxfordjournals.aob.a085917.

Ting, F.S.-T. and Wren, M.J. (1979b) Storage organ development in radish (*Raphanus sativus* L.). 2. Effects of growth promoters on cambial activity in cultured roots, decapitated seedlings and intact plants. *Annals of Botany* 46(3), 277–284. DOI: 10.1093/oxfordjournals.aob.a085918.

Waltert, B. and Theiler, R. (2003) Cauliflower and broccoli: growth pattern and potential yield. [German] *Agrarforschung* 10(7), 276–281.

Wang, R., Mei, Y., Xu, L., Zhu, X., Wang, Y. *et al.* (2018a) Differential proteomic analysis reveals sequential heat stress-responsive regulatory network in radish (*Raphanus sativus* L.) taproot. *Planta* 247(5), 1109–1122. DOI: 10.1007/s00425-018-2846-5.

Wang, R., Mei, Y., Xu, L., Zhu, X., Wang, Y. *et al.* (2018b) Genome-wide characterization of differentially expressed genes provides insights into regulatory network of heat stress response in radish (*Raphanus sativus* L.). *Functional & Integrative Genomics* 18(2), 225–239. DOI: 10.1007/s10142-017-0587-3.

Watson, D.J. (1947) Comparative physiological studies on the growth of field crops: I. Variation in net assimilation rate and leaf area between species and varieties, and within and between years. *Annals of Botany* 11(1), 41–76. DOI: 10.1093/oxfordjournals.aob.a083148.

Wellington, P.S. (1954) The heading of broccoli: factors affecting quality and time. *Agriculture* 61, 431–434.

Wheeler, T.R., Ellis, R.H., Hadley, P. and Morison, J.I.L. (1995a) Effects of CO_2, temperature and their interaction on the growth, development and yield of cauliflower (*Brassica oleracea* L. *botrytis*). *Scientia Horticulturae* 60(3–4), 181–197. DOI: 10.1016/0304-4238(94)00725-U.

Wiebe, H.J. (1972a) Wirkung von Temperatur und Licht auf Wachsum und Entwicklung von Blumenkohl: I. Dauer der Jungendphase fur Vernalisation. *Gartenbauwissenschaft* 37, 165–178.

Wiebe, H.J. (1972b) Wirkung von Temperatur und Licht auf Wachstum und Entwicklung von Blumenkohl: II. Optimale Vernalisationstemperatur UNF Vernalisationdauer. *Gartenbauwissenschaft* 37, 293–303.

Wiebe, H.J. (1972c) Wirkung von Temperatur und Licht auf Wachtsum und Entwiklung von Blumenkohl: III. Vegitative phase. *Gartenbauwissenschaft* 37, 455–469.

Wiebe, H.J. (1975) Effect of temperature on the variability and maturity date of cauliflower. *Acta Horticulturae* 52, 69–76. DOI: 10.17660/ActaHortic.1975.52.7.

Wien, H.C. and Stutzel, H. (2020) Cauliflower, broccoli, cabbage and Brussels sprouts. In: Wien, H.C. and Stützel, H. (eds) *The Physiology of Vegetable Crops*, 2nd edn. CAB International, Wallingford, UK, pp. 357–388. DOI: 10.1079/9781786393777.0000.

Wurr, D.C.E., Kay, R.H., Allen, E.J. and Patel, J.C. (1981a) Studies of the growth and development of winter-heading cauliflowers. *The Journal of Agricultural Science* 97(2), 409–419. DOI: 10.1017/S0021859600040855.

Wurr, D.C.E., Kay, R.H. and Allen, E.J. (1981b) The effect of cold treatments on the curd maturity of winter-heading cauliflowers. *The Journal of Agricultural Science* 97(2), 421–425. DOI: 10.1017/S0021859600040867.

Wurr, D.C.E., Elphinstone, E.D. and Fellows, J.R. (1988) The effect of plant raising and cultural factors on the curd initiation and maturity characteristics of summer/autumn cauliflower crops. *The Journal of Agricultural Science* 111(3), 427–434. DOI: 10.1017/S0021859600083593.

Wurr, D.C.E., Fellows, J.R. and Hiron, R.W.P. (1990a) The influence of field environmental conditions on the growth and development of four cauliflower cultivars. *Journal of Horticultural Science and Biotechnology* 65(5), 565–572. DOI: 10.1080/00221589.1990.11516094.

Wurr, D.C.E., Fellows, J.R. and Hiron, R.W.P. (1990b) Relationships between the times of transplanting, curd initiation and maturity in cauliflower. *The Journal of Agricultural Science* 114(2), 193–199. DOI: 10.1017/S0021859600072191.

Wurr, D.C.E., Fellows, J.R., Sutherland, R.A. and Elphinstone, E.D. (1990c) A model of cauliflower curd growth to predict when curds reach a specified size. *Journal of Horticultural Science and Biotechnology* 65(5), 555–564. DOI: 10.1080/00221589.1990.11516093.

Wurr, D.C.E., Fellows, J.R. and Hambidge, A.J. (1991) The influence of field environmental conditions on calabrese growth and development. *Journal of Horticultural Science and Biotechnology* 66(4), 495–504. DOI: 10.1080/00221589.1991.11516179.

Wurr, D.C.E., Fellows, J.R. and Hambidge, A.J. (1992) The effect of plant density on calabrese head growth and its use in a predictive model. *Journal of Horticultural Science and Biotechnology* 67(1), 77–85. DOI: 10.1080/00221589.1992.11516223.

Wurr, D.C.E., Fellows, J.R., Phelps, K. and Reader, R.J. (1993) Vernalisation in summer/autumn cauliflower (*Brassica oleracea* var *botrytis* L.). *Journal of Experimental Botany* 44(266), 1507–1514. DOI: 10.1093/jxb/44.9.1507.

Wurr, D.C.E., Fellows, J.R., Phelps, K. and Reader, R.J. (1994) Testing a vernalization model on field-grown crops of four cauliflower cultivars. *Journal of Horticultural Science and Biotechnology* 69(2), 251–255. DOI: 10.1080/14620316.1994.11516452.

Wurr, D.C.E., Fellows, J.R. and Phelps, K. (1996) Growth and development of heads and flowering stalk extension in field-grown Chinese cabbage in the UK. *Journal of Horticultural Science and Biotechnology* 71(2), 273–286. DOI: 10.1080/14620316.1996.11515406.

Wurr, D.C.E. and Fellows, J.R. (1998) Leaf production and curd initiation of winter cauliflower in response to temperature. *The Journal of Horticultural Science and Biotechnology* 73(5), 691–697. DOI: 10.1080/14620316.1998.11511035.

Wurr, D.C.E., Hambidge, A.J., Fellows, J.R., Lynn, J.R. and Pink, D.A.C. (2002) The influence of water stress during crop growth on the postharvest quality of broccoli. *Postharvest Biology and Technology* 25(2), 193–198. DOI: 10.1016/S0925-5214(01)00171-5.

Xie, Y., Xu, L., Wang, Y., Fan, L., Chen, Y. *et al.* (2018) Comparative proteomic analysis provides insight into a complex regulatory network of taproot formation in radish (*Raphanus sativus* L.). *Horticulture Research* 5, 51. DOI: 10.1038/s41438-018-0057-7.

Zha, L.-Y. and Liu, W.-K. (2018) Effects of light quality, light intensity, and photoperiod on growth and yield of cherry radish grown under red plus blue LEDs. *Horticulture, Environment, and Biotechnology* 59(4), 511–518. DOI: 10.1007/s13580-018-0048-5.

Zhang, C., Wei, Y., Xiao, D., Gao, L., Lyu, S. *et al.* (2016) Transcriptomic and proteomic analyses provide new insights into the regulation mechanism of low-temperature-induced leafy head formation in Chinese cabbage. *Journal of Proteomics* 144, 1–10. DOI: 10.1016/j.jprot.2016.05.022.

Zhao, Z., Gu, H., Wang, J., Sheng, X. and Yu, H. (2013) Development and comparison of quantitative methods to evaluate the curd solidity of cauliflower. *Journal of Food Engineering* 119(3), 477–482. DOI: 10.1016/j.jfoodeng.2013.06.025.

Zheng, Y., Luo, L., Liu, Y., Yang, Y., Wang, C. *et al.* (2018) Effect of vernalization on tuberization and flowering in the Tibetan turnip is associated with changes in the expression of *FLC* homologues. *Plant Diversity* 40(2), 50–56. DOI: 10.1016/j.pld.2018.01.002.

CROP AGRONOMY

5

GEOFFREY R. DIXON*

School of Agriculture, Policy and Development, Earley Gate, Whiteknights Road, PO Box 237, University of Reading, Reading, Berkshire RG6 6EU and GreenGene International, Hill Rising, Horsecastles Lane, Sherborne, Dorset DT9 6BH, UK

Abstract

Knowledge of the responses of brassica genotypes cultivated as crops in a productive environment is the fabric of agronomic science as delivered in this chapter. First, this contains detailed information covering nutrient requirements, resultant crop responses, means for improving and sustaining soil health and quality, soil analyses achieved by chemical and physical processes, the impact of pH on yields, minimizing artificial fertilizer use and the increasing value of biostimulants and biofertilizers. Responses by particular crops, such as Brussels sprouts, cauliflower, green broccoli (calabrese), cabbage and bulbous brassicas, are evaluated. Thereafter, irrigation, water conservation and quality are discussed in terms of smart scheduling, sensing soil moisture stocks, tracking and electrically powered systems.

Changing systems of motive power and evolving science and technologies used for primary and secondary cultivations are assessed. These are reducing the greenhouse gas emissions resulting from brassica cultivation, which helps mitigate climatic change.

INTRODUCTION

Agronomy provides the scientific rationale which describes why and how crops respond efficiently in growth, yield and ultimate financial profitability with the husbandry resources provided by farmers and growers. This is the

*geoffrdixon@gmail.com

© Geoffrey R. Dixon and Rachel Wells 2024. *Vegetable Brassicas and Related Crucifers,* 2nd edition. (G.R. Dixon and R. Wells) DOI: 10.1079/9781789249170.0005

most direct and strongest expression of the relationship between horticultural research and commercial production. It now leads the drive for increasing crop yields worldwide concomitantly with rising human populations and mitigating the effects of climate change. Since the publication of the first edition of *Vegetable Brassicas and Related Crucifers* in 2007, attitudes towards the science of agronomy have changed significantly, becoming more positive and realistic. Previous antagonistic attitudes, especially in developed nations, have been replaced by a realisation that reliable, safe food production is a national and international priority. Simultaneously, however, there is also a realisation that increased production must go hand-in-hand with protection of the environment and conservation of biodiversity. Balancing these factors can only be achieved in ways which are compatible with limiting climate change. Current global warming is noticeably influencing the practices involved in crop production. The range of crops is increasing in some previously temperate areas but being severely limited in more tropical regions (Dixon, 2009, 2012b; Dixon *et al.*, 2014). Crucially, agronomy will have great importance in coping with these challenges, both in feeding people and designing landscapes capable of reducing the volumes of damaging greenhouse gases entering the atmosphere. Agronomists occupy the lead in responding to increasing public demands for conservation of the environment and the avoidance of soil and aerial pollution while still increasing global food supply, quality and security.

Public, and by inference political, demands for a convergence between food production and environmental sustainability will be met only by careful and measured husbandry systems validated by reliably sound research and development. As a result, agronomists are developing ever more sophisticated methods which deliver a clearer understanding of the interactions between crops, other organisms and the aerial and edaphic environments. Demands for higher yields, food quality and security achieved with reduced environmental impact converge under the umbrella title of 'sustainability' (see Chapter 6). This term means many things to many people, not least its urban protagonists who live at a distance from the realities of crop production. As a result, the roles of agronomists also include providing reliable, factually accurate information and educating consumers and politicians as, for example, given by Aitken *et al.* (2012). By inference, this responsibility also includes refuting assertions based on pseudoscience which are so often promulgated by those with hidden and biased agendas.

Agronomy embraces both organic growing, which avoids using soluble mineral fertilizers or synthetic pesticides, and highly productive, integrated resource management that delivers food in sufficient volumes and quality to satisfy increasing consumer requirements. For both these forms of production, agronomists interpret the efficiency of returns from resources used. But introducing the term 'sustainability' means that efficiency is redefined. Factors relating the interaction of particular crops with the entire production system, the environment and onwards through the complete supply chain to the ultimate consumer, become parts of the sustainability equation. Increasingly, agronomists are in demand to integrate the economic, environmental and social factors affecting crop production. None the less, the producer must know that the price paid for the products will adequately and profitably meet variable resource costs and the fixed overheads of land, labour and finance. Against this changing background, the second edition of *Vegetable Brassicas and Related Crucifers* reviews aspects of crop agronomy.

The response of field-grown vegetables to adverse weather conditions is strongly coupled to the timing of those events and crop growth stage. Agronomists understand the sensitivity of crop growth stages and the management actions needed. Historical records have come to light in the Czech Republic, which include details, for example, of yields of vulnerable vegetables, such as early kohlrabi, summer Savoy cabbage, late cauliflower and late cabbage, and a high-resolution historical climate data set (seven daily meteorological variables, recorded at a 10×10 km resolution) over a 54-year period. In these records, years with yield gains were substantially more common than years with crop losses for brassicas, in all of the cultivated regions (Potopova *et al.*, 2017). But subsequently, Europe is experiencing the impact of global warming, bringing hotter summers and very severe weather events which might now provide evidence indicating the rapidly shifting environment with which agronomists and producers must cope.

Cruciferous vegetables are prized for their nutritive value and have been selected and cultivated for thousands of years. There are numerous wild western Mediterranean species in the family Brassicaceae, and it is therefore assumed this centre of diversity is also the region of origin. Within the tribe, the *Brassica nigra* and *Brassica oleracea* clades contain three diploid *Brassica* crops: *B. oleracea*, *Brassica rapa* and *B. nigra* (see Chapter 1). These three species hybridized forming the tetraploid crop species *Brassica juncea*, *Brassica carinata* and *Brassica napus*. Collectively, these crop brassicas have been thought to be closely related because they can still hybridize. There is evidence that the family originated around the intersection forming between the Arabian Peninsula and Saharan Africa, approximately 24 million years ago (Mya). Data also suggest that the maternal genomes of the three diploid crop brassicas are not closely related and that the *B. nigra* and *B. oleracea* clades diverged 20 Mya. Analyses indicate that the core *Oleracea* lineage gave rise to *B. oleracea* and *B. rapa* originating ~3 Mya in the north-eastern Mediterranean, from where ancestors of *B. oleracea* spread through Europe and *B. rapa* to Asia (Arias *et al.*, 2014). Subsequently, selection in cultivation resulted in the multitude of European and Asian brassica crops currently grown commercially and domestically. This process is continued by plant breeders producing imaginative hybrids within and between the species suitable for the increasingly sophisticated consumer demands (see Chapter 1).

Brassicas are the predominant group of arable field vegetables worldwide, harvested either for immediate fresh consumption or with the minimum of postharvest preparation, or used in making preserves, such as sauerkraut, suan cai (suan tsai) and kimchi. Increasingly, agronomic research is required which improves the effectiveness of husbandry and responds to changing market and social demands in the fresh produce food chain. Available global production and value statistics focus largely on cabbage crops (*B. oleracea* var. *capitata*). Estimates of world trade totals suggest annual world production of 70 million tonnes of cabbage valued at about US$2 trillion. The United Nations Food and Agriculture Organization (FAO) estimates cabbage and allied brassica occupy about 2.6 million hectares of land worldwide (compared with 772 million tonnes of wheat grown on 215 million hectares of land). Consequently, these crops, which are essential ingredients of diets supporting human health and welfare, come from intensive production, possibly with several harvests in 1 year from the same land area which utilizes resources very effectively.

GENETIC AND ENVIRONMENTAL INTERACTIONS

In general, plants originating from low-resource environments have an inherently high root–shoot ratio (Grime, 1979) and reduced nutrient supply leads to phenotypes with active fine roots, high root–shoot ratio, many root hairs and a slow maximum rate of net photosynthesis which maintains balanced growth at low nitrogen availability (Robinson, 1991). Raising nutrient supply increases the amounts of resources allocated to aerial organs but not necessarily to leaf blades. In the specific case of Asian 'green cabbage', growth increased in the leaf petioles not the leaf blades. When there is competition for light, the plants increased resource allocation for stem growth and expanded internodal length. Species with different morphologies invest more resources in those organs that provide greatest advantage in capturing light, thus 'green cabbage' extended the petioles. The ability to compete for resources is not solely a function of biomass allocation patterns but also depends on morphological characters such as leaf area ratio (LAR) and specific leaf area (SLA), indicating the importance of morphological plasticity.

Brassicas are an example of these responses since the original wild progenitors of cultivated brassicas evolved in inhospitable arid conditions and soils with minimal nutrient availability (see Chapter 1). But in contrast, cultivated brassicas are very responsive to increasing supplies of nutrients, particularly nitrogen and water. In practice, therefore, they are used as the leading crop in rotations planted following the application of animal manure or composts or growing legumes, emphasizing their responsiveness to nitrogen fertilizers in particular.

Improving the efficiency by which fertilizers are used in brassica crops requires ecological knowledge of how biomass allocation, morphological plasticity and competitive abilities interact. This may have especial relevance where crops are subjected to shading, either from surrounding plants or when, for example, they are grown under plastic mulches or pest-protecting netting. The maintenance of balanced growth requires the adjustment of morphological characters and physiological behaviour that maximizes whole plant growth and hence the eventual yield, especially where this is composed of either leaves or swollen roots and hypocotyls. Plants respond to changes in the external supply of nutrient resources by adjusting the relative size and distribution of organs. Ecologists have suggested that reduced nutrient supply leads to a higher root–shoot ratio, thereby compensating for loss in root-foraging capabilities. Additionally, lower light availability due to shading results in greater shoot growth.

In research studies, a photosynthesis–respiration-based dry matter production model was used in order to derive functional relationships between light use efficiency (LUE) (about 25 μg/J) and irradiance, leaf area index (LAI) and temperature. The light-saturated photosynthesis rate (P_{max}) showed an optimum response to temperature and increased with increasing nitrogen content of leaves. The model analysis demonstrated that LUE is independent of the light integral over a range of 5–10 MJ/m^2/day for photosynthetically active radiation (PAR), assuming an adaptation of P_{max} within the canopy and over time according to the incident irradiance. Acclimatization within the canopy and higher LAIs reduce the decrease of LUE with irradiance but a substantial decline remains, even for LAI values of 4 (Kage et al., 2001). Consequently, the amount of resource captured by plants is at least partially related to the age, area or volume of the organ system responsible for obtaining the resource. Hence, species with a higher biomass allocation to the root system should perform better at low soil nutrient levels compared with those allocating more to stems and leaves (Tilman, 1988). Further studies of these factors with the brassicas are reported by Li et al. (1999) who used Asian 'green cabbage' B. rapa cv. Natsurakuten, a leafy variety, and a turnip B. rapa var. rapifera cultivar, the disease-tolerant Hikari.

The innate responsiveness of brassicas towards nutrients, especially nitrogen, causes practical problems where excessive quantities of fertilizers are applied in efforts to boost yields. This is because physiologically the brassicas are not very efficient in their use of such resources. Nutrient imbalances in the soil and plant tissues cause toxicity and deficiency syndromes that impair growth, resulting in stress disorders and the development of off-flavours in harvested produce. For these reasons alone it is important that fertilizers are applied with due regard to the nutrient reserves in the soil and the demands that each growing brassica crop imposes. This requires managerial knowledge of the structure and texture of the soil type and previous cropping history of each field, its current nutrient status as determined by soil analysis and the likely response to added resources. Increasingly, the nutrient status of a field is being monitored and recorded in great detail resulting in the identification of areas where additional fertilizers are needed and those places where there is already ample supply. Differences in brassica yields may not be wholly governed by the amount of nitrogen applied, but its availability has been demonstrated by Scandinavian studies of the usefulness of green manures (Sorensen and Thorup-Kristensen, 2011).

Molecular studies are now leading towards a better understanding of the genetic controls of nitrogen uptake and utilization which is an important step towards its more efficient use and concomitantly reduced pollution. Ammonium transporter (AMT) proteins mediate its transport across plasma membranes. Investigating whether AMTs are regulated at the post-transcriptional level required a gene construct consisting of the cauliflower mosaic virus 35S promoter driving the *Arabidopsis thaliana AMT1;1* gene which was introduced into tobacco (*Nicotiana tabacum*). Ectopic expression of *AtAMT1;1* in transgenic tobacco lines led to higher transcription and protein levels at the plasma membrane and translated into an approximately 30% increase in root uptake capacity for N^{15}-labeled ammonium in hydroponically grown transgenic plants (Yuan *et al.*, 2007). Results indicated that the accumulation of *AtAMT1;1* transcripts is regulated in a nitrogen- and organ-dependent manner and suggest that mRNA turnover is an additional mechanism for the regulation of *AtAMT1;1* in response to the nitrogen status of plants.

Environmental factors may also be involved in the uptake and use of other nutrients. Brassicas are generally considered to grow well in soils with low phosphate availability. Assessments of the role of the rhizosphere in growth and phosphate uptake by three *Brassica* genotypes (mustard, *B. juncea* cv. Chinese greens and canola, *B. napus* cvs Drum and Outback) revealed that at maturity, each genotype had a distinct microbial community. The production of high biomass at low shoot phosphate concentrations, as well as the capacity for maintaining high phosphate availability and mobilization in the rhizosphere, explained the differences in plant growth and nutrient uptake (Marschner *et al.*, 2007).

NUTRITION

On some soils, large amounts of fertilizers are required to satisfy the prolonged growth and maturity stages of brassica crops. Demands are satisfied either by single fertilizer applications being made before drilling or transplanting, or by dividing the requirements between basic applications during secondary cultivation and the addition of 'top dressings' at appropriate crop growth stages. Applying the entire crop requirement during secondary cultivations is wasteful of increasingly expensive fertilizer resources because the plants cannot immediately use them. Consequently, much nutrient, especially nitrogenous forms, is leached into field drains resulting in water catchment area pollution; or in luxury root uptake and subsequent lush foliage growth which attracts pest and pathogen invasion. Good environmental and financial incentives, therefore, encourage the precision application and sparing use of all brassica fertilizers. Increasing fertilizer use efficiency by brassica crops improves the predictability of growth and maturity for the current crop and soil health, thereby benefitting subsequent crops. Achieving these objectives requires the planned use of soil nutrient status data (Anon, 2020b).

Variation in the nutrient application rates from centrifugal fertilizer spreaders previously inhibited automation (Lawrence and Yule, 2007). The Global Positioning System (GPS) and geographic information systems (GIS) have removed much of this variation, identifying driver accuracy and driving method as important factors in the effectiveness of automation. In Italian studies, tractors used for variably controlled fertilizer distribution were equipped with: first, GPS and sensors controlling forward speed, fertilizer flow and the measurement of distributor performance; second, devices which read zone maps linked with the GPS and other computer systems; third, computer-controlled fertilizer release; and fourth, mechanical, hydraulic or electrical devices regulating output (Sartori and Bertocco, 2005). These authors concluded that careful calibration of the machine and the use of fertilizers with uniform granule sizes are essential requirements for automated distribution. Fertilizer spreaders capable of variable-rate application are an increasingly important means for enhancing nutrient management in horticultural crops, such as brassicas, because they improve the accuracy of placement and increase nutrient use efficiency (Schumann, 2010). Matching fertilizer application with crop requirements reduces growers' costs, pollution, waste and accurately provides for nutritional needs, increasing productivity. Variable-rate fertilization is a precision agriculture technology made possible by embedded high-speed computers, accurate GPS receivers, GIS, remote sensing, yield and soil maps, actuators, and electronic sensors capable of measuring and even forecasting crop properties in real time. Similar results are reported by Amiama-Ares *et al.* (2011) who compared the use of GPS-tracked and non-guided applications. Satellite and airborne images form part of a French decision support tool (Farmstar) used in managing fertilizer application for oilseed rape (*B. napus*) crops (Desbourdes *et al.*, 2008). Detection of variations in nitrogen deficiency in overwintered cauliflower, cabbage and Brussels sprouts is achieved by measuring changes in foliar chlorophyll fluorescence, identifying variations in crop colour and correlating these with fertilizer need. Tractor-mounted monitors pass though crops and automatically vary fertilizer application rates (see Fig. 5.1).

Soil nutrient indexing

Reliable and repeatable analyses that measure nutrient reserves present in soil allow reasonable predictions of the effects of fertilizer application and are available for most elements with the general exception of nitrogen. Soil mineral nitrogen levels vary continuously due to the effects of mineralization, fertilizer dressings, leaching, denitrification and plant uptake. Consequently, the timing of soil sampling is critical and measurements are only of use for crop management decisions immediately after sampling.

Laboratory analyses of soil samples are used which determine available macronutrients such as phosphorus, potassium and magnesium expressed

Fig. 5.1. Tractor-mounted nitrogen fertilizer distributor and sensor. (Courtesy Tom Decamp, YARA).

in mg/l. For practical simplicity, these values are converted to indices (see Table 5.1) which indicate the relative quantities of nutrient available to the crop ranging from 0 (deficiency) to 9 (excess). Soils intended for brassica production should be maintained at index = 3 for phosphorus and index = 2 for potassium and magnesium. Soils with these indices require only maintenance quantities of additional fertilizer; below these values larger amounts are necessary to ensure economic returns from the crop and to restore the nutrient reserves of the soil.

Table 5.1. Relationship between available phosphorus, potassium and magnesium determined by laboratory analysis and the soil index system. (After Anon, 1985).

Index	Phosphorus (mg/l)	Potassium (mg/l)	Magnesium (mg/l)
0	0–9	0–60	0–25
1	10–15	61–120	26–50
2	16–25	121–240	51–100
3	26–45	241–400	101–175
4	46–70	401–600	176–250
5	71–100	601–900	251–350
6	101–140	901–1500	351–600
7	141–200	1501–2400	601–1000
8	201–280	2401–3600	1001–1500
9	>280	>3600	>1500

CROP RESPONSES

Brussels sprouts

Brussels sprout (*B. oleracea* var. *gemmifera*) crops require up to 300 kg N/ha for optimal yield. Total aerial biomass production and nitrogen uptake rates increase strongly with nitrogen applications, mainly because of their pronounced effect on leaf area expansion (Booij *et al.*, 1996). A high nitrogen uptake rate rapidly depletes nitrate reserves in the soil (Booij *et al.*, 1993). Crop growth of Brussels sprouts has two phases (see Chapter 4, Figs 4.2 and 4.3) (Abuzeid and Wilcocksen, 1989). In the first phase, biomass increase is mainly by leaf and stem growth, and in the second phase, it is concentrated towards axillary and terminal bud growth. The period of bud growth coincides with increasing leaf senescence. When nitrate reserves in the soil are depleted before the onset of bud growth, its rate is then dependent on nitrogen mineralization from soil organic matter. Hence, the well-established practice of applying large quantities of organic fertilizer (such as farmyard manure) in advance of growing Brussels sprout crops.

The low availability of soil nitrogen and high rate of leaf senescence during bud growth implies that nitrogen remobilization takes place to support bud growth. Efficient use of nitrogen fertilizers and the maintenance of the environment require an understanding of how nitrogen is removed from the leaves and distributed to the buds. Nitrogen in the non-marketable portions of this crop (roots, stems and leaves) may remain in the field, depending on the harvesting method used. Some harvesting techniques require that the stems bearing sprouts are removed to a static bud stripper located at some distance from the crop, whereas other systems may employ mobile strippers which remove the buds in the field leaving residues behind (see Fig. 5.2).

Where residues remain in the field, they are a potential source of further soil nitrogen losses through mineralization (Whitmore and Groot, 1994) and subsequent leaching by winter rains into adjacent waterways (Greenwood, 1990). To diminish this risk the nitrogen harvest index should be high. Remobilization of nitrogen results in a higher nitrogen harvest index than the harvest index for total biomass. From an environmental perspective, the rapid depletion of the soil and the later relocation of nitrogen during bud growth lowers the risk of leaching from the soil as nitrogen is conserved in the crop. Booij *et al.* (1997) found that with Brussels sprouts at the onset of bud (sprout) growth 60–80% of the total biomass has been produced and an equivalent amount of nitrogen taken up. Eventual harvest bud weight and bud nitrogen content were both positively correlated with total biomass and nitrogen content at the onset of bud growth.

Partitioning of biomass and nitrogen among the different aerial plant organs was hardly affected by the availability of nitrogen and the timing of fertilizer applications. During bud growth, the leaves senesced rapidly. Biomass and nitrogen, in particular, were remobilized from the leaves to the buds before

Fig. 5.2. Mobile Brussels sprout (*Brassica oleracea* var. *gemmifera*) harvesting. (Courtesy of Phillip Effingham).

abscission. Fisher and Milbourn (1974) and Wilcockson and Abuzeid (1991) concluded that where leaves were removed before they became senescent (commonly a cultural practice of defoliation, particularly for early maturing cultivars), remobilization of dry matter from leaves did not contribute to bud growth. During bud growth, loss of nitrogen from leaves was up to 50% of the nitrogen increase in buds. Where nitrogen was applied at the onset of bud growth, it was accelerated and increased greening, with a delay in leaf shedding. When nitrogen was applied as a split application, half at transplanting and the rest at the onset of bud growth, the nitrogen content of buds was increased. The partitioning of biomass and final bud yield, however, was unchanged compared with treatments in which the entire fertilizer application was made at transplanting.

Booij *et al.* (1997) further concluded that to reach an acceptable bud yield, the nitrogen application rate should aim at optimal biomass at the onset of bud growth. Biemond *et al.* (1995) showed that for Brussels sprouts, sufficient nitrogen should be available for unrestricted growth at the beginning of the season. Where this is not the case, bud growth is delayed. Apparently, under these conditions most of the nitrogen is removed from the field as crop (buds) at harvest with relatively small amounts of nitrogen remaining in the field as residues. This has environmental benefits, since less nitrogen is leached into groundwater following the mineralization of crop residues. A larger total green leaf area, attained with more nitrogen, resulted mainly from larger leaves; the number of leaves increased only slightly. Larger leaves were the result of higher rates of leaf expansion. Neuvel (1990) showed that when Brussels sprouts were either direct-drilled in April or transplanted in May, with an application of 300 kg/ha of fertilizer nitrogen, the total dry matter production in October was on average 14.5 t/ha and nitrogen uptake was 335 kg/ha (Everaarts and van Beusichem, 1998). Maximum sprout yield of about 4.5 t/ha of dry matter was usually attained with the largest amounts of nitrogen applied, i.e. 300 or 375 kg/ha. Scaife (1988) showed that in Brussels sprouts pre-planting nitrogen fertilizer applications resulted in nitrate-nitrogen increasing for 43 days from the date of application. Thereafter, it declined to near zero in November at the time of single harvesting.

Green broccoli

Green broccoli (*B. oleracea* var. *italica*) crops maintain a relatively high fraction of photosynthetically active tissue approaching the harvest compared to other brassicas (Conversa *et al.*, 2019). This is another nitrogen-demanding brassica and its availability determines crop productivity and quality. But the nitrogen not taken up may cause ground and surface water pollution. Uptake of nitrogen increases with availability and application rate as shown, for example, by Everaarts and de Willigen (1999). Recommendations vary between 220 and 270 kg N/ha (Greenwood *et al.*, 1980; Vågen, 2003). In relating these rates with crop growth for green broccoli, Vågen *et al.* (2004) showed that crop biomass and LAI accumulated, intercepted PAR and radiation use efficiency (RUE) increased with rising nitrogen rate, but RUE was not changed significantly with rates between 120 and 240 kg N/ha. Hence, applying higher rates of nitrogen increased crop biomass by increasing intercepted radiation rather than by additional RUE. The saturation level of RUE was reached when the sum of applied nitrogen and the mineral nitrogen of the soil before planting was approximately 200 kg/ha.

There was a strong effect of nitrogen application on early plantings where climatic factors were suboptimal. Low temperature, and possibly oversaturation of light radiation, were the most likely reasons for low LAI, RUE and relative growth rate (RGR) following low or nil nitrogen applications. The nitrogen concentration decreased as total biomass increased. Two alternative relationships of critical nitrogen concentration (N_c) for green broccoli were suggested by Greenwood *et al.* (1986, 1996); of these, the linear equation fitted best. It gave higher estimates, and both equations were evaluated on the correspondence between relative nitrogen concentration $N/N_c = 1$ and maximum relative values for biomass, LAI, accumulated intercepted PAR and RUE. A curvilinear function might, however, produce a higher correlation over the whole range of plant nitrogen concentrations. RUE and accumulated intercepted PAR approached saturation level at lower relative nitrogen concentration than did biomass and LAI, suggesting that nitrogen was more limiting for biomass production than for PAR interception and RUE at nitrogen rates below the highest application rate. Incident radiation was fully absorbed at quite a low LAI, resulting in low RUE values.

Cauliflower

Cauliflower (*B. oleracea* var. *botrytis*) curds are large pre-inflorescences with a complex morphology characterized by a high degree of ramification with little internode extension and an accumulation of meristematic domes. Biometric analysis of curd ramification in cv. Plana revealed a lack of dominance between branches of different order (primary hierarchical sequence) and position (secondary hierarchical sequence), and a constancy of both organogenic and plastochronic apical activity. It is shown that $>10 \times 10^6$ apical meristems are carried by a curd of marketable size (150–200 mm) (Kieffer *et al.*, 1996). The rate of nitrogen uptake by the cauliflower crops increases rapidly from about 4 weeks after planting; concurrently, the amount of mineral nitrogen in the soil starts to decrease. Most of the nitrogen was taken up from the 0–30 cm soil layer. The amount of nitrogen in the crop at harvest, with the recommended amount of nitrogen applied, ranged between 170 and 250 kg/ha, while 7–100 kg/ha of mineral nitrogen remained in the soil (0–60 cm layer). Crop residues contained about 95–140 kg N/ha. In this study, no evidence was found for leaching of fertilizer nitrogen during crop growth. It is concluded that the potential for loss of nitrogen to the environment is greater after crop harvest, when nitrogen may be lost from crop residues and soil, than during growth (Everaarts, 2000). Promoting temporary storage of nitrogen in cauliflower leaves and stems before curd initiation could stimulate head translocation, which could result in reduction of post-transplanting inputs (Li, 2012). During breeding, both general and specific combining abilities of cauliflower cultivars were important components for heterosis of mineral content in Snowball cauliflower (Ram *et al.*, 2018).

Leaching losses are substantially reduced by using nitrification inhibitors, such as 3,4-dimethylpyrazole phosphate (DMPP) or dicyandiamide (DCD). There was an accumulation of ammonium and nitrate-nitrogen after incorporation of cauliflower residues in untreated soil incubation experiments. In treated soil samples, DCD inhibited the nitrification from crop residues for 50 days and DMPP for at least 95 days. Hence, DMPP in particular shows a potential to reduce nitrate leaching after incorporation of crop residues (Chaves *et al.*, 2006). Alternative studies suggest that the intercropping system, cauliflower–clover, can be a sustainable tool to optimize nitrogen input and reduce artificial fertilizer requirements for the succeeding crop (Tempesta *et al.*, 2019).

Solarization is a valuable technique for cauliflower cropping where there are consistent high temperatures during fallow periods. In preparation for sowing cauliflower seed, beds were covered with transparent polyethylene sheets and subsequent solarization significantly altered soil properties compared with untreated areas. The mean pH, electrical conductivity, calcium, magnesium, nitrogen, phosphate, potassium and carbon contents recorded in solarized soil were higher than in non-solarized. Heating reduced the populations of fungi from 25.68×10^4 to 4.8×10^4, bacteria from 20.28×10^6 to 5.66×10^6 and actinomycetes from 31.60×10^5 to 4.40×10^5. Cauliflower seedlings subsequently grown in solarized soil had higher vigour indices compared with those in non-solarized soil (Sofi *et al.*, 2014).

Cabbage

Strong heterosis in F_1 hybrid cabbage (*B. oleracea* var. *capitata*), while resulting in higher yields, better plant stands, earlier maturity, larger and more uniform heads, uniformity in head compactness, and disease tolerance, has not so far produced root systems with improved nutrient uptake efficiencies (Singh *et al.*, 2009).

Summer and autumn maturing cabbage used 3.8 kg N/t of edible cabbage yield and produced crop residues contained 300 kg N/ha. Almost 50% of the nitrogen absorbed by cabbage ends up in the crop residues. Cabbage was found to make efficient use of available nitrogen, absorbing more than 100 kg/ha from unfertilized fields. Uptake was slow at the beginning of the spring growing phase but increased thereafter, continuing into September. Broadcasting nitrogen, as opposed to band placement, apparently encouraged cabbage growth. But as the crops matured their demand for nitrogen declined. Husbandry systems for cabbage require careful nitrogen management, which increases their sustainability. This can be achieved by improving nitrogen use efficiency through increasing nitrogen uptake, nitrogen utilization efficiency and nitrogen harvest index (van Bueren and Struik, 2017). Head-forming crops, such as cabbage, depend on the prolonged photosynthesis of outer leaves which provides the carbon sources for continued nitrogen supply and the growth of the photosynthetically less active, younger inner leaves. Improving root performance is important for all brassica types, but especially short-cycle forms which benefit from early below-ground vigour such as spring and early summer season greens and cabbage or green broccoli (*B. oleracea* var. *italica*). There is sufficient genetic variation available among modern brassica cultivars for further improvement in nitrogen use efficiency but this requires integration of agronomy and crop physiology knowledge. The benefits of increased nitrogen use efficiency through breeding are potentially large but realising these benefits is challenged by the huge genotype-by-environment interaction and the complex behaviour of nitrogen in brassica husbandry.

Spring maturing cabbage and overwintered cauliflower (*B. oleracea* var. *botrytis*) benefit greatly in terms of yield and quality from top dressings of nitrogen. Green broccoli and cauliflower are frequently grown intensively as several successive crops occupying a single area of land; there can be sequential double or even triple cropping with these crops. In this form of intensive husbandry, there may be macronutrients besides nitrogen carried over between crops. Reductions in the quantities of phosphorus applied to succeeding crops may be appropriate. By comparison, cabbage crops absorb particularly large quantities of potassium and it may be necessary to increase applications for succeeding crops.

Bench studies of five potted ornamental cabbage cultivars (*B. oleracea* ssp. *acephala*) identified that while fertilizing with 10 ml of nitrate per pot produced high-quality plants, it resulted in inadequate tissue nitrogen, phosphorus, potassium, calcium, magnesium and iron (Cardarelli *et al.*, 2015).

Bulbous brassicas

There is little quantitative data which defines the growth and development of 'bulbous' brassicas such as turnips (*B. rapa* ssp. *rapifera*), swedes (*Brassica napus* ssp. *rapifera*) or kohlrabi (*B. oleracea* var. *gongylodes*) and consequently limited prediction models and decision support systems are available. Andreucci *et al.* (2014) measured the growth and development of cultivars 'Barkant' and 'Green Globe' turnips showing that temperature was a principal factor involved.

NITROGEN USE

Internationally, there is concern regarding the use of nitrogen fertilizers and their potential for causing groundwater pollution, as typified by Chinese research (Jin, 2012). None the less, nitrogen has very pronounced effects promoting the growth, yield and quality of all vegetable brassicas within the limits of crop need, as discussed under the section 'Genetic and Environmental Interactions' earlier in this chapter. Unlike other nutrients, nitrogen requirements are not usually based on soil analyses but on the specific demands of particular crops and the status of field soil, making an allowance for residues from the previous crop(s) and applications of organic manure.

Consideration is mainly given to the last crop grown when determining the nitrogen index, but for brassica crops following lucerne (alfalfa) or other legumes, long leys (grass pastures of several years duration), permanent pasture or intensive brassica production, longer cropping histories should be taken into account. Three levels of soil nitrogen index are used (see Table 5.2). Fields in index = 0 have low nitrogen reserves and require more nitrogen compared with those with index = 1. Index = 2 soils have the highest soil nitrogen reserves. This approach may be satisfactory for most agricultural crops but the increasing complexity of horticultural brassica production has stimulated research seeking greater precision and efficiency in the use of nitrogen. Consequently, management strategies in reducing nitrogen losses in brassica crops (such as green broccoli (*B. oleracea* var. *italica*) or cauliflower (*B. oleracea*

Table 5.2. Nitrogen index system based on the last crop grown. (After Anon, 1985).

Nitrogen index = 0	Nitrogen index = 1	Nitrogen index = 2
Cereals	Beans	Any crop in field receiving
Forage crops removed	Forage crops grazed	large frequent dressings of farmyard manure or slurry
Leys[a] (1–2 years) cut	Leys (1–2 years) grazed, high N[c]	Long leys, high N[c]
Leys (1–2 years) grazed, low N[b]	Long leys, low N[b]	Lucerne (alfalfa)
Maize	Oilseed rape	Permanent pasture – average
Permanent pasture, poor quality, matted	Peas	Permanent pasture – high N[c]
	Potatoes	
Sugarbeet, tops removed	Sugarbeet, tops ploughed in	
Vegetables receiving < 200 kg N/ha	Vegetables receiving > 200 kg N/ha	

[a]Ley is a European term for extended cropping with a forage crop, especially grass which may occupy the land for several seasons. Land occupied for 1–3 years would in a 'short ley' and land occupied for 3 years or more would be a 'long ley'.
[b]Low N – <250 kg N/ha per year and low clover content.
[c]High N – >250 kg N/ha per year or high clover content.

var. *botrytis*)) require an understanding of factors such as: total crop yield and the nitrogen content required to provide marketable produce; nitrogen supply at critical periods of crop growth; major sources of available nitrogen in the soil; methods of application (broadcast or placement); and the form of nitrogen applied (ammonium, nitrate or calcium cyanamide in solid or liquid form). Computer simulation models now offer means for identifying strategies which reduce losses through leaching (Rahn, 2002).

Efficient use of nitrogen involves accurate estimation of crop nitrogen demand, the choice of application method and the timing of application. The effects of band placement, rate of fertilization, dry matter accumulation, yield and uptake have been studied by Salo (1999). The concentration of a particular nutrient in brassicas may increase and decrease during the life cycle of the plant. A decrease is due to greater production of carbon-rich compounds relative to accumulation of nutrient ions, such as nitrogen, phosphate, potassium, magnesium and calcium. Starch and cell walls are the principal carbon-rich compounds. Early in a brassica growth cycle, nutrients increase due to high rates of uptake in young roots and the high RGR of young foliage.

As plants mature, the LAI, and hence the degree of mutual shading, rises resulting in decreased net photosynthesis per unit leaf area, reduced RGR and declining nutrient concentration. The onset of this decline can be altered by the timing of fertilizer application, possibly avoiding shortages of nutrient in the root zone. Raising nitrogen status increases the content in plants. There may also be parallel increases of phosphorus and cations as nitrogen is raised. Key factors in these processes are the particular nutrients in question, plant growth stage, crop type and environmental variables including soil fertility and temperature, light interception and water availability.

All fertilizer recommendations are given in kg/ha. The number of kg of nutrient in the standard 50 kg bag of fertilizer is obtained by dividing the percentage of the nutrient by 2 ($100/2 = 50$). For example, one 50 kg bag of 20-10-10 NPK compound fertilizer will contain 10 kg of nitrogen, 5 kg of phosphate (P_2O_5) and 5 kg of potash (K_2O) (see Table 5.3).

Using the indexing system, standard tables relating to fertilizer application are available for brassica crops. Those shown in Tables 5.4 and 5.5 are applicable to the UK and northern temperate Europe and are based on the UK Ministry of Agriculture, Fisheries and Food (MAFF) (now known as the Department for Environment, Food and Rural Affairs (Defra)) recommendations. The interpretation of these tables requires knowledge of the particular soil type involved and its previous cropping history. This is especially true with regard to nitrogen. In general, highly organic soils supply more available nitrogen than mineral soils. Hence, the recommended rates should be reduced by about 10% for peat-rich soils.

The data in Tables 5.4 and 5.5 are drawn from the *Reference Book 209, Fertiliser Recommendations*, generally referred to as 'RB 209' 1985 version (Anon, 1985; reprinted in Anon, 2000) published by MAFF. RB 209 is now owned by the UK Agriculture and Horticulture Development Board (AHDB).

More recently, RB 209 (Anon, 2017b) has been extensively revised by the AHDB. A more sustainable approach is taken in this revision where the recommendations for artificial nitrogen fertilizers applied to brassica crops are calculated relative to soil nitrogen status (see Tables 5.6–5.8) and supplemented by Table 5.9 which identifies how recommendations should be interpreted in the light of individual crop and contract requirements relative to cultivar, weather, season and commercial requirements. These data are applicable for the UK, much of northern Europe and worldwide in regions with comparable

Table 5.3. Conversion factors between common nutrient elements and their oxides. (After Anon, 1985).

Nutrient element	Oxide	Conversion factor
Calcium (Ca)	Calcium oxide (CaO)	1:1.40
Phosphorus (P)	Phosphorus pentoxide (P_2O_5)	1:2.291
Potassium (K)	Potassium oxide (K_2O)	1:1.205
Magnesium (Mg)	Magnesium oxide (MgO)	1:1.67
Sulfur (S)	Sulfur trioxide (SO_3)	1:2.50

Table 5.4. Specific fertilizer requirements for Brussels sprouts, cabbage (including Chinese cabbage), swede and turnip. (After RB 209: Anon, 1985).

Crop and soil type	N, P, K or Mg index						Top dressing
	0	1	2	3	4	>4	
Brussels sprouts: market picking or single harvest							
Silt and brickearth soils							
Nitrogen (N)[a]	200	150	100	–	–	–	
Other soils							
Nitrogen (N)[a]	300	250	200	–	–	–	
All soils							
Phosphate (P_2O_5)	175	125	75	50	25	Nil	
Potash (K_2O)	200	175	125	60	Nil	Nil	
Cabbage							
Summer, autumn and Chinese							
Nitrogen (N)[a]	300	250	200	–	–	–	
Winter and Savoy							
Pre-Christmas cutting							
Nitrogen (N)[a]	300	250	200	–	–	–	
Post-Christmas cutting							
Nitrogen (N)[a]	150	125	100	–	–	–	0–75
Winter white for storage							
Nitrogen (N)[a]	250	200	150	–	–	–	
Spring							
Nitrogen (N)[b]	75	50	25	–	–	–	200–400
Early frame-raised							
Nitrogen (N)[a]	250	200	125	–	–	–	60–120
Cabbage all types[c]							
Phosphate (P_2O_5)	200	125	75	50	25	Nil	
Potash (K_2O)	300	250	175	75	Nil	Nil	
Brussels sprouts and cabbage							
Sands and light loams							
Magnesium (Mg)	90	60	Nil	Nil	Nil	Nil	
Other soils							
Magnesium (Mg)	60	30	Nil	Nil	Nil	Nil	
Swedes							
Nitrogen (N)	100	50	Nil	–	–	–	
Phosphate (P_2O_5)	150	100	50	50	25	Nil	
Potash (K_2O)	250	200	150	75	Nil	Nil	
Turnips							
Early bunching							

Continued

Table 5.4. Continued

Crop and soil type	N, P, K or Mg index						Top dressing
	0	1	2	3	4	>4	
Nitrogen (N)	150	100	50	–	–	–	
Maincrop							
Nitrogen (N)	100	50	Nil	–	–	–	
All crops							
Phosphate (P$_2$O$_5$)	150	100	50	50	25	Nil	
Potash (K$_2$O)	250	200	150	75	Nil	Nil	

All values are given as kg/ha.
[a]For direct-drilled crops or transplants on sands and light loams, nitrogen in excess of 100 kg/ha
should be top-dressed to reduce the risk of damage to seedlings or young plants and applied at singling or
within 1 month of transplanting. Extra top dressing may be required, especially on shallow or sandy soils,
when rainfall greatly exceeds transpiration within 2 months of applying basal nitrogen. Top dressings are
unnecessary if nitrogen is injected either before drilling or between the rows up to 1 month after emergence
or transplanting.
[b]A fully grown crop can use up to 400 kg N/ha. Smaller crops of greens, harvested in early spring, may
need less than half of this amount. Applications should be in dressings of 100–200 kg N/ha, and related
mainly to growth, but the potential marketing period
and weather conditions should also be considered.
[c]When spring cabbage follows a crop leaving substantial residues, reduce the phosphate application by
half and the potash application by 60 kg/ha.

soils and climatic conditions. Additionally, they provide guidance for other regions where local nutrient and fertilizer data are lacking. Providing the RB 209 data from both 1985 and 2017 editions in this book gives readers and users opportunities for understanding the changes in approach now possible with more refined knowledge of soil and plant performance. The earlier edition is, however, still a valuable guide for the overall nutrient requirements of vegetable brassicas.

These changes found in RB 209 (Anon, 2017b) resulted at least in part from the research work of Rahn *et al.* (2001) who compared and eventually rationalized nitrogen fertilizer recommendation systems from 15 European countries as part of ENVEG, a European Union-funded project investigating environmental problems associated with the nitrogen fertilization of field-grown vegetable crops generally. Previously, recommendation systems worldwide ranged in complexity from those based on experience, or on simple 'look-up' tables, to those relying on measurements of soil mineral nitrogen (SMN) and computer models.

From the perspective of brassica crops, returning of high nitrogen content crop residues to soil, particularly in autumn, can result in considerable environmental pollution and substantial loss of soil biological health. Pollution arises both from nitrate leaching into watercourses, and the generation of nitrous oxides, which are implicated in global warming. Improved management of such residues can reduce losses following harvest and increase cycling of nitrogen between crops, particularly in nitrate-sensitive areas (the

Table 5.5. Specific fertilizer requirements for green broccoli and cauliflower. (After RB 209: Anon, 1985).

Crop and soil types	N, P, K or Mg index						Top dressing
	0	1	2	3	4	>4	
Green broccoli							
Nitrogen (N)[a]	250	200	160	–	–	–	
Phosphate (P_2O_5)	150	75	60	60	30	Nil	
Potash (K_2O)	150	100	75	50	Nil	Nil	
Sands and light loams							
Magnesium (Mg)	90	60	Nil	Nil	Nil	Nil	
Other soils							
Magnesium (Mg)	60	30	Nil	Nil	Nil	Nil	
Cauliflower							
Early summer, late summer and autumn, e.g. Flora Blanca and Australian types							
Nitrogen (N)[a]	250	200	125	–	–	–	
Winter, Roscoff types					–	–	
Nitrogen (N)[b]	75	40	Nil	–	–	–	60–125
Winter, hardy types[c]					–	–	
Nitrogen (N)[b]	75	40	Nil	–	–	–	125–200
All types							
Phosphate (P_2O_5)[d]	175	125	75	50	25	Nil	
Potash (K_2O)[d]	300	200	125	60	Nil	Nil	
All types							
Sands and light loams							
Magnesium (Mg)	90	60	Nil	Nil	Nil	Nil	
Other soils							
Magnesium (Mg)	60	30	Nil	Nil	Nil	Nil	

All values are given as kg/ha.
[a]For direct-drilled crops or transplants on sands and light loams, nitrogen in excess of 100 kg/ha should be top-dressed to reduce the risk of damage to seedlings and should be applied at singling or within 1 month of transplanting.
[b]For February cuttings in frost-free areas, apply the top dressing in the late autumn. For crops to be harvested later than mid-February, apply top dressings in mid-February.
[c]Top dressings should be applied in the January to March period.
[d]When winter cauliflower follows a crop leaving substantial residues, reduce the phosphate by one half and the potash applications by 60 kg/ha.

UK term is nitrate vulnerable zones). Studies have shown that an effective material for reducing nitrogen losses to the environment is compactor waste, which is derived from the recycling of cardboard. Co-incorporating compactor waste into field soils at a rate equivalent to 3.75 t/ha of carbon, together

Table 5.6. Nitrogen, phosphate, potash and magnesium for Brussels sprouts and cabbage. (After Anon, 2017b; courtesy of the Agriculture and Horticulture Development Board, AHDB).

Crop	Soil nitrogen supply, P, K or Mg index						
	0	1	2	3	4	5	6
Nitrogen (N)[a] – all soil types							
Brussels sprouts	330	300	270	230	180	80	0[b]
Storage cabbage	340	310	280	240	190	90	0[b]
Head cabbage, pre-31 December	325	290	260	220	170	70	0[b]
Head cabbage, post-31 December	240	210	180	140	90	0[b]	0[b]
Collards, pre-31 December	210	190	180	160	140	90	0[b]
Collards, post-31 December	310	290	270	240	210	140	90
Phosphate (P_2O_5)[c], all crops	200	150	100	50	0	0	0
Potash (K_2O)[c]	300	250	200 (2–) 150 (2+)	60	0	0	0
Magnesium (MgO), all crops	150	100	0	0	0	0	0

All values are given as kg/ha.

[a]On light soils, where leaching may occur or where crops are established by direct seeding, no more than 100 kg N/ha should be applied prior to seeding or transplanting. On retentive soils in drier parts of the country, where leaching risk is low and spring-planted brassicas are established from modules, more nitrogen can be applied prior to planting. The remainder of the nitrogen requirement should be applied after establishment.

[b]A small amount of nitrogen may be needed if soil nitrogen levels are low in the top 30 cm of soil.

[c]Phosphate and potash requirements are for average crops and it is important to calculate specific phosphate and potash removals based on yields, especially for the larger yielding cabbage crops. As a general rule for cabbage crops, increase potash application by 40 kg/ha K_2O for every 10 t/ha fresh weight yield over 40 t/ha.

with sugarbeet residues, had a significant effect in reducing both autumn SMN levels and subsequent losses by leaching and denitrification (Rahn *et al.*, 2003).

In Belgium, for example, large amounts of inorganic fertilizers are used to obtain rapid growth, high brassica yields and quality. This practice results in an exceedance of the legally imposed residual soil nitrate threshold value of 90 kg N-NO_3/ha at the end of the growing season (Heuts *et al.*, 2016). A soil transport model focusing on mineralization and nitrate leaching was linked to a generic crop model, including transpiration and water-demand, and nitrogen demand and uptake. A life cycle assessment focused for each fertilization treatment on the differences in input loads and their corresponding environmental impacts in terms of different categories: global warming, cumulative energy demand, human toxicity, acidification and eutrophication. A higher nitrogen

Table 5.7. Nitrogen, phosphate, potash and magnesium for cauliflowers and green broccoli. (After Anon, 2017b; courtesy of the Agriculture and Horticulture Development Board, AHDB).

Crop	Soil nitrogen supply, P, K or Mg index						
	0	1	2	3	4	5	6
Nitrogen (N) – all soil types							
Cauliflower, summer/autumn	290	260	235	210	170	80	0[b]
Cauliflower, winter hardy/ Roscoff[a]							
Seed bed	100	100	100	100	60	0[a]	0[a]
Top dressing	190	160	135	110	100	80	0[b]
Green broccoli[a]	235	200	165	135	80	0[b]	0[b]
Phosphate (P_2O_5), all crops	200	150	100	50	0	0	0
Potash (K_2O), all crops	275	225	179 (2–) 125 (2+)	35	0	0	0
Magnesium (MgO), all crops	150	100	0	0	0	0	0

All values are given as kg/ha.
[a]The recommendations assume overall application, band spreading of nitrogen may be beneficial.
[b]A small amount of nitrogen may be needed if soil nitrogen levels are low in the top 30 cm of soil.

application rate resulted in an overall higher impact in all categories. Fertilizer production has a high impact on human toxicity, global warming potential, acidification and cumulative energy demand. While fertilizer application increases eutrophication, acidification and global warming potential, fertigation dramatically increases the cumulative energy impact.

Tillage apparently increased nitrogen leaching from green broccoli crops. Between 20% and 60% of the nitrogen content of residues of green broccoli was leached. Consequently, removing crop residues after harvest will reduce this source of pollution (de Ruijter *et al.*, 2010). The soluble nitrate anion has a high potential for contaminating groundwater due to its mobility in soil. German studies showed that mean nitrate leaching over years decreased when ploughing was deferred until later in the season. But the disadvantages of this strategy were that as temperatures increased with the advancing season, leaching was encouraged and later ploughing decreased yields (Schwarz *et al.*, 2010).

This invaluable source of information has been revised and brought up to date as the *Nutrient Management Guide* (RB 209) (Anon, 2017b). The revision emphasizes the importance of Integrated Plant Nutrient Management. The changed approach identifies that plants obtain their nutrients from several sources including:

- mineralization of soil organic matter;
- deposition from the atmosphere, mainly referring to nitrogen and sulfur;

Table 5.8. Nitrogen, phosphate, potash and magnesium for radish, swede and turnips. (After Anon, 2017b; courtesy of the Agriculture and Horticulture Development Board, AHDB).

	Soil nitrogen supply, P, K or Mg index						
	0	1	2	3	4	5	6
Radish							
Nitrogen (N) – all soil types	100	90	80	65	50	20	0[a]
Phosphate (P_2O_5)	175	125	75	25	0	0	0
Potash (K_2O)	250	150 (2−) 100 (2+)	0	0	0	0	0
Magnesium (MgO)	150	100	0	0	0	0	0
Swede							
Nitrogen (N) – all soil types	135	100	70	30	0[a]	0[a]	0[a]
Phosphate (P_2O_5)	200	150	100	50	0	0	0
Potash (K_2O)	300	250	200 (2−) 150 (2+)	60	0	0	0
Turnip							
Nitrogen (N) – all soil types	170	130	100	70	20	0[a]	0[a]
Phosphate (P_2O_5)	200	150	100	50	0	0	0
Potash (K_2O)	300	250	200 (2−) 150 (2+)	60	0	0	0

All values are given as kg/ha.
[a]A small amount of nitrogen may be needed if nitrogen levels are low in the top 30 cm of soil.

- weathering of soil minerals, especially supplying potash;
- biological nitrogen fixation by legumes;
- applications of organic materials such as farmyard manure and composts; and
- applications of manufactured fertilizers.

Recommendations contained in the 2017 edition of RB 209 (Anon, 2017b) for nitrogen applications are now based on estimates of the soil nitrogen supply (SNS) which is defined as 'the amount of nitrogen (kg N/ha) available for uptake from the soil by the crop throughout its entire life'. This accounts for nitrogen losses but excludes nitrogen applied to the crop in manufactured fertilizers or manures. This is a more accurate and environmentally sustainable approach which is particularly important for brassicas because of the

Table 5.9. Data required for modification of brassica crop nitrogen requirements. (After Anon, 2017b; courtesy of the Agriculture and Horticulture Development Board, AHDB).

Crop	Fresh market yield (t/ha)	% dry matter marketable	Dry weight harvest index	Total dry matter (t/ha)	Relation N% and dry matter yield[a]			Total N uptake (kg/ha)	Mineralized (kg/ha)	Period dates	Root depth (cm)
					a	b	% N				
Brussels sprouts	20.3	17.0	0.26	13.3	2.5	3.5	2.8	368	121	20/05–17/12	90
White cabbage storage	110	8.6	0.65	14.6	2.55	0.8	2.6	378	122	01/05–12/11	90
Head cabbage, pre-31 December	60	8.6	0.48	10.8	2.55	0.8	2.7	270	44	18/05–19/07	90
Head cabbage, post-31 December	53	8.6	0.46	10.0	2.55	0.80	2.7	203	74	31/07–15/01	90
Collards, pre-31 December	20	8.6	0.34	5.1	3.45	0.60	4.0	260	51	16/07–24/09	45
Collards, post-31 December	30	8.6	0.38	6.8	3.45	0.6	3.8	300	41	15/09–15/01	60
Cauliflower over winter	–	–	–	8.1	3.45	0.60	3.7	300	85	30/07–10/03	75
Cauliflower summer	30.6	8.2	0.37	6.8	3.45	0.60	3.8	259	44	21/05–21/07	75
Green broccoli	16.3	10.4	0.17	10.0	1.80	3.50	2.3	226	36	27/04–25/06	90
Radish	50	–	–	–	–	–	–	100	24	21/05–11/06	30
Turnip	48	–	–	–	–	–	–	241[b]	92	30/03–27/08	90
Swede	84.4	11.7	0.62	16.0	1.35	1.87	1.4	222	92	30/03–27/08	90

[a]See original publication for details of a and b.
[b]N uptake taken from German KNS system, 2007

large demands for nitrogen required in achieving high-quality, predictable crop maturity and high-quality yields.

> SNS = soil mineral + estimates of nitrogen + estimates of mineralizable nitogen (SMN) already in the crop soil nitrogen

Where:

- SMN (kg N/ha) is the nitrate-nitrogen and ammonium-nitrogen content of the soil within the maximum rooting depth of the crop;
- nitrogen already in the crop (kg N/ha) is the total content of nitrogen in the crop when soil is sampled for SMN; and
- mineralizable soil nitrogen (kg N/ha) is the estimated amount of nitrogen which becomes available for crop uptake from mineralization of soil organic matter and crop debris during the growing season after sampling for SMN.

Soil nitrogen supply varies across a field, from field to field and season to season and is influenced by:

- nitrogen residues left in the soil from fertilizer applied for the previous crop;
- nitrogen residues from any organic manure applied for the previous crop and in previous seasons;
- soil type and soil organic matter content;
- losses of nitrogen by leaching and other processes such as quantities of winter rainfall; and
- nitrogen made available for crop uptake from mineralization of soil organic matter and crop debris during the growing season.

Phosphate, potash and magnesium supplies are important nutrients for producing brassica crops. Recommendations are usually based on the index system. A phosphate index of 3 (26–45 mg/l) and a potash index of 2+ (121–240 mg/l) are recommended where crops are grown as part of an arable rotation. Continuous brassica cropping may require increased supplies of these nutrients. Magnesium is an important micronutrient for brassica crops. Deficiency may be corrected by applying 5 t/ha of magnesian limestone which adds approximately 750 kg MgO/ha (magnesium oxide).

Details of the recommendations as contained in RB 209 (Anon, 2017b), which highlight the use of soil indexing are given in Tables 5.6–5.8.

Table 5.9 indicates parameters which allow for the customization of nitrogen applications because brassica crops are planted at many different times of year, have a range of expected yields and each individual crop is unique depending on weather, season and previous land uses. Variations in nitrogen applications can be made relative to:

1. Size of the crop – the size, frame or weight of the crop needed to produce economic yields or fulfil the requirements of contracts negotiated with customers such as the supermarkets.

2. Nitrogen uptake – the optimum nitrogen uptake associated with a crop of the size capable of fulfilling contractual commitments; or varying relative to prevailing weather events.

3. Supply of nitrogen – based on the nitrogen supply from the soil within rooting depth, including any nitrogen mineralized from organic matter during the growing season. This can vary for brassica crops depending on the previous crops present in a rotation and used for cover cropping or intercropping.

FERTILIZER APPLICATIONS

Fertilizer requirements for a range of brassica vegetables are given in Tables 5.3–5.9. These bear out the more specific requirements indicating that these crops benefit from substantial applications of major nutrients. To avoid damage to the root systems by increasing soil conductivity to dangerous concentrations, it is advisable to apply nitrogen especially as split dressings, with half applied to the seed bed or at transplanting and the residue 2 weeks later.

Nitrogen and potassium, when available in excess, can reduce germination and damage seedling root systems, particularly on dry, sandy soils. Thorough incorporation of fertilizers into the soil before drilling or transplanting is essential. Both potassium and phosphorus may be applied some weeks in advance of seeding or transplanting. Pre-transplanting application of phosphates enhanced photosynthesis and root elongation, and increased water and nutrient absorption from soil for cabbage (Watanabe *et al.*, 2011). There is a danger that nitrogen applied too early in the growing season may be lost by leaching. Noticeably, potassium use efficiencies varied among six Chinese cabbage (*B. rapa* ssp. *pekinensis*) cultivars and appropriate rates resulted in greater development of root systems and ion uptake (Li *et al.*, 2015) (see Table 5.10).

Soil health and quality

Most soils suffer from degradation, such as compaction, declining soil organic matter content, nutrient leaching and erosion. The most commonly agreed and used soil indicators can be grouped in the three categories of (i) biological, (ii) chemical and (iii) physical parameters indicating soil health (Anon, 2016). A comprehensive review of literature relating to soil health and quality is contained in Vieweger *et al.* (2016). Soil health is defined as 'the capability of soil to support productivity and ecosystem services' (Kibblewhite *et al.*, 2008) and soil quality is viewed as the 'fitness of soil for use' (Karlen *et al.*, 2003). There are obviously contrasting opinions on differentiating between these terms and the value of keeping them distinct.

Organic husbandry systems are advocated as a means for improving soil health. The main challenges of organic forms of cropping are improving nutrient management, increasing yields and reducing the risk of nitrogen losses (Pinto *et al.*, 2020). Combinations of farmyard manure and compost with

Table 5.10. Nutrient requirements for other brassica crops. (After Siemonsma and Piluek, 1993).

Crop	Nutrient (kg/ha)			Comments
	N	P_2O_5	K_2O	
Brassica juncea (brown mustard)	90–100	90	90	Compost 10 t/ha; nitrogen applied as split application; half as basal dressing and half as side dressing 2 weeks later
Brassica oleracea (cauliflower and green broccoli)	NPK depends on soil type, soil reserves and expected yields; top dressings are applied to stimulate head formation			Compost 20 t/ha
Brassica oleracea (Chinese kale)	250 kg/ha NPK (15-15-15)			10–20 t/ha organic manure
Brassica oleracea (head cabbage)	Cabbage crop of 25 t/ha absorbs 140 kg N, 40 kg P, 180 kg K			20–50 t/ha compost or organic manure
Brassica rapa (Asian greens)	Very responsive to N			Yield 30–50 t/ha
Brassica rapa (caisin)	60–110	40–60	80–100	10–15 t/ha compost, nitrogen applied as split dressing; half as basal fertilizer and half 2 weeks later
Brassica rapa (Chinese cabbage)	120–200	40–60	70–150	Soluble fertilizer applied as split dressing; half at planting and the rest 10–14 days later
Brassica rapa (pak choi)	55–75 kg N, 40–80 kg P, 80–110 kg K at planting			55–75 kg/ha applied 14 days after planting
Raphanus sativus (radish)				Apply compost and adequate NPK pre-sowing and N at regular intervals thereafter

green manure built up soil fertility and reduced nitrogen leaching. Brassica crops need significant amounts of calcium, magnesium and sulfur. The organic farming approach, however, restricts the use of chemical fertilizers considerably, challenging balanced mineral nutrition of brassica crops such as cabbage (*B. oleracea* var. *capitata*). This may be alleviated by using polyhalite, a natural mineral that is authorized for use with organic crops in several countries. It contains 14% potash (K_2O), 48% sulfate (SO_3), 6% magnesium oxide (MgO) and 17% calcium oxide (CaO).

Application of organic nitrogen amendments and reduction of chemical fertilizer are considered as an effective approach in sustainable agriculture (Xu *et al.*, 2019a). For pak choi (*B. rapa* ssp. *chinensis*) and Chinese cabbage (*B. rapa* ssp. *pekinensis*), applications of appropriate organic fertilizers at sowing resulted in increased yields and product quality, meanwhile reducing the environmental risk of nitrate-nitrogen.

Green manure obtained from incorporating cover crops stands out in organic vegetable production as a complementary fertilizer alternative to organic compound incorporation, contributing to reduce production costs and improve the physical, chemical and biological features of soil

(Bento *et al.*, 2020). The use of jack beans (*Canavalia ensiformis*) as a cover crop associated with biomass incorporation was shown to be promising for the production of cabbage.

Soil analyses

The results of soil analyses reflect the quality of the sampling methods used. Samples must be representative of the area and taken to a standardized depth, usually 15 cm. Field areas may differ significantly in soil type, previous cropping and applications of manure, fertilizer or lime, so they should be considered as separate samples. Small areas that are known to differ significantly from the rest of the field should be excluded from the main samples and tested separately. Traditionally, a minimum of 25 individual subsamples (auger cores) will be adequate for a uniform area. Subsampling points are selected systematically and evenly distributed across the area. This is usually achieved by following a 'W' pattern and taking subsamples along the legs of this pattern at regular intervals. Samples should not be collected from the vicinity of gateways, headlands or close to trees and hedges. Fields used for brassica production should be subjected to soil analysis at a minimum of once every 3 years and more frequently where the land is either cropped several times in one season or rented from a third-party owner.

Soil sampling and analysis for brassica crops is now mechanized and automated (see Fig. 5.3).

Sample collections are recorded by GPS mapping identifying areas of low nutrient status within fields with capabilities for precise, accurately placed return visits and testing after treatment. Soil samples are sent for laboratory analysis by automated wet chemistry. Accumulated over years, data for soil nutrient reserves, fertilizer use and resultant yields are now well-established growers' decision making tools, which maximize returns on the resources employed, minimize waste and limit pollution.

Mapping nutrient status

Previously, relatively few samples could be collected from randomized field locations and bulked before submission for laboratory analysis. Resultant recommendations were guidance on a 'whole-field' or 'whole-crop' basis. Now, precise, mechanized soil sampling using GPS location and recording is available at increased intensity and regularity. Soil mapping by GPS, combined with chemical analyses, provides recommendations for variable-rate fertilizer applications by growers and their contractors. Mobile *in situ* soil-analysis laboratories carried in the back of vehicles (Lobsey *et al.*, 2010) result in recommendations being even more quickly available. This is achieved using a multi-ion measuring system which rapidly quantifies nutrient status. Samples

Fig. 5.3. Mechanized soil-sampling quad bikes. (Courtesy of David Norman, Fresh Produce Consultancy).

can be collected by staff using low ground pressure vehicles and sampling large field areas in relatively limited times.

Physical soil analyses

As alternatives to chemical testing, physical analysis systems now offer large-scale, rapid, precision and automated *in situ* field analysis (see Fig. 5.4).

The measurement of soil properties using indirect spectral responses from an online sensor is reported by Marìn-Gonzâlez *et al.* (2013). Equipment consists of a tractor-trailed subsoiler tine (pan-buster) coupled with an 'AgroSpec' mobile, fibre-type, visible and near-infrared (vis–NIR) spectrophotometer (tec5 Technology for Spectroscopy, Oberursel, Germany). This has a measurement range of 305–2200 nm which defines soil spectra using a diffuse reflectance mode. Extending the usefulness of this equipment for all soil types, including heavy clay soils, has been achieved by adding a single-ended, shear-beam load cell which measures draught, a vis–NIR sensor measuring moisture content and a wheel gauge equipped with a draw-wire linear sensor estimating depth, permitting the calculation of bulk density which identifies the degree of soil compaction (Quraishi and Mouazen, 2013). Potentially, this machinery quantifies a range of soil properties including: pH, cation-exchange capacity, exchangeable calcium, organic carbon, total nitrogen, phosphorus content,

Fig. 5.4. Tru-Nject spectrographic mobile soil analyser. (Courtesy Andrew Manfield).

bulk density, moisture content and exchangeable magnesium. Analyses of soil properties can be collected from more than 1000 points/ha offering a level of nutritional guidance of previously unachievable speed, accuracy and precision in commercial practice. Growers can, therefore, accumulate libraries of data from soil scans and by adding yield maps identify resource need correlated with soil properties, crop type and resultant yields. Commercial equipment using these principles is being developed by engineering companies and should become readily available in the coming years.

Soil type

Soil type significantly affects shoot growth and concentrations of phosphorus, zinc, potassium, calcium, magnesium and manganese, but not the iron concentration of red-headed cabbage (*B. oleracea* var. *capitata*) (Pongrac *et al.*, 2019). In the shoots and roots of Chinese cabbage (*B. rapa* ssp. *pekinensis*), Sung *et al.* (2018) found a quantitative increase in amino acid levels in response to magnesium deficiency. This possibly resulted from increased protein production and also a marked increase in the levels of quinate, a precursor of the shikimate pathway, following cation (potassium, calcium and magnesium) deficiency. Dicyandiamide inhibitors slow down nitrogen release speed, increasing utilization of nitrogen and the yield of Chinese cabbage (Li *et al.*, 2018), which is especially valuable for organic cropping. Calcium nitrate limits the occurrence of tipburn and bacterial rotting of Chinese cabbage, but

Table 5.11. Guide to pH values below which crop productivity is reduced. (After MAFF).

Crop	Soil pH
Brussels sprouts	5.7
Cabbage	5.4
Cauliflower	5.6
Mustard	5.4
Swede	5.4
Turnip	5.4

the weather conditions during cultivation have the greatest impact on the severity of tipburn (Borkowski *et al.*, 2016).

SOIL pH AND CALCIUM CONTENT

In studies of genetic variation in wild brassicas in England and Wales, Watson-Jones *et al.* (2005) looked at 15 populations of *B. nigra*, eight populations of *B. oleracea* and nine populations of *B. rapa*, using amplified fragment length polymorphism and relating this to soil pH in the sites of origin. There was a significant correlation between pH and coarse sandy soils with high percentage polymorphism. Within populations at higher alkaline pH values there was less polymorphism. Commercial brassica crops are most productive when grown on land with an approximately neutral pH. The ideal is pH 6.5 for mineral soils and pH 5.8 for organic soils. This rule should be altered where soil-borne pathogens are present such as *Plasmodiophora brassicae*, the causal agent of clubroot disease (see Chapter 7 section, *Plasmodiophora brassicae* – clubroot). Land where even very low levels of infection are present should be raised to a pH in excess of 7.0. Brassica crops vary in their sensitivity to acidic pH (see Tables 5.11 and 5.12) and the point at which crop productivity begins to diminish is shown in Table 5.11.

The lime requirement (t/ha ground limestone or chalk) is obtained by multiplying the liming factor for each soil type by the difference between the initial (measured) and target pH. As a rule of thumb, applications of ground limestone or chalk of 1 t/ha will change the soil pH by 0.5 units. Ground limestone and chalk applications take 3–6 months, depending on the time of year, before altering pH values.

Lime requirements of acidic soils are expressed in t/ha of ground limestone or ground chalk. The amount of lime recommended for soils of similar pH may vary with soil texture and soil organic matter content. Usually, the recommendations aim to maintain the top 20 cm depth of mineral soil to a pH of 6.5 and an organic soil to a pH of 5.8. Where there are variations in soil acidity across the profile, then larger applications of lime may be needed. Applications of lime in excess of 12 t/ha should be made as several separate

Table 5.12. Lime requirements for arable land. (Courtesy Agriculture and Horticulture Development Board (ADHB)).

Initial pH	Sands and sandy loams	Sandy loams and silts	Clay loams and clays	Organic soils[a]	Peaty soils[b]
Liming factor					
	6	7	8	8	16
Lime applied (t/ha)					
6.2	3	4	4	4	0
6.0	4	5	6	6	0
5.5	7	8	10	10	8
5.0	10	12	14	14	16

Ground limestone or ground chalk, neutralizing value = application of 50 t/ha (modified from Anon, 2017b, Section 1, page 14).
[a]For mineral and organic soils the target pH is 6.7 for continuous arable cropping.
[b]For peaty soils the target pH is 6.0 for continuous arable cropping.

dressings. Lime should be applied well before sowing or transplanting. Several months are required for changes in soil acidity to take place. There is an increasing tendency for growers of brassicas using highly intensive systems to apply calcium oxide (CaO) as a liming agent (also known as hotlime or quick-lime). This has the advantage of acting very quickly to alter pH and is applied at about one-third the rate of carbonate forms. Also, the liming effect on pH is lost by the end of the season and this permits growers to plant potatoes in the following year with lower risks of infection by *Spongospora subterranea* and *Streptomyces scabies*, the causes of powdery and common scab diseases, respectively. Normally, brassica vegetables should not be grown immediately after liming a very acid soil (pH below 5.0). But where a crop is failing because of slight acidity then some improvement may be achieved by top dressing across a standing crop. This is most likely to be successful where calcium is applied in readily accessible forms such as calcium cyanamide or calcium nitrate. Over liming, bringing the pH beyond 7.5, should be avoided, especially on sands, light loam and organic soils since this can result in induced deficiencies of trace elements such as boron and manganese.

When soils become increasingly acidic, then microbial activity declines and mineralization slows or stops. This has a big impact on nutrient availability and the waste of resources. For example, where soils are at pH 5.5, 33% of NPK fertilizer could be wasted. The financial loss is substantial. For example, a 28.2 t articulated lorry load of 20-10-10 fertilizer, priced at UK£240/t (price valid on 30 November 2020 and tripled by autumn 2022), wastes UK£2233 by applying a fertilizer that is unavailable to the crop due to reduced microbial activity slowing or stopping mineralization. At a neutral soil of pH 7, NPK nutrients are fully available for root uptake and crop growth (Trapnell, 2021), which illustrates the importance of soil testing and subsequent application of lime.

TRACE ELEMENTS

Depending on soil type, soil pH and crop sensitivity, trace element deficiencies can develop and cause significant crop losses. Deficiencies of trace elements have substantial effects on the yield, quality and storability of brassica crops.

Cauliflower (*B. oleracea* var. *botrytis*) and swede (*B. napus* ssp. *rapifera*) are susceptible to boron deficiency, especially when grown on light soils with pH values above 6.5. Boronated fertilizers should be used as a matter of routine, or applications made to the seed bed, or prior to transplanting, at 20 kg/ha of borax (sodium tetraborate) or 10 kg/ha of Solubor™.

Copper deficiency is encountered less frequently but may develop on soils with a high organic matter content, sands (especially reclaimed heathland or moorland) and humus soils which overlay chalk. This deficiency is corrected by applying copper oxychloride or cuprous oxide at 2 kg/ha plus a wetting agent at either high- or low-volume sprays, or treating the soil with 60 kg/ha of copper sulfate.

Manganese deficiency develops on soils similar to those prone to copper deficiency and is corrected by high- or low-volume sprays containing 9 kg/ha of manganese sulfate plus a wetting agent.

Cauliflower crops are especially prone to molybdenum deficiency, causing typical whip-tailing symptoms of the foliage with reduced lamina and a prominent main vein. This condition is associated with acidic soils. Hence, the soil should be maintained at pH 6.5–7.5. Where treatment is necessary, fields should receive 300 g/ha of sodium or ammonium molybdate, or transplants are drenched with 0.25 g/l at the plant propagation stage.

Sulfur applications of up to 100 kg/ha significantly increased the yield of seed cabbage crops (*B. oleracea* var. *capitata*). This resulted in less days to first flower, 75% seed maturity and the highest number per plant of branches, siliquae, seed count/siliqua, 1000-seed weight, seed yield and seed yield/ha.

MINIMIZING FERTILIZER RESOURCE USE

Brassica crops seldom utilize all of the nutrients applied, leading to the excess remaining in the soil and potentially available for leaching into groundwater and ultimately causing pollution hazards. Public concern regarding the use of fertilizers is leading to a search for more effective means of application which diminish the quantities used and focuses them into the crop root zone enabling more efficient utilization without compromising yield or product quality.

Nitrogen is derived from several sources, both natural and artificial. Once nitrogen is in the nitrate form it becomes subject to movement in the groundwater and contributes to contamination. USA environmental standards have a maximum of 10 mg of nitrate-nitrogen in drinking water and the European acceptable daily intake of nitrate-nitrogen is 3.65 mg/kg of body weight (Anon, 1997). All vegetables, and brassica crops in particular, have high financial values and are intensively managed, requiring substantial inputs of fertilizer, especially nitrogen, and irrigation water. The problem of groundwater contamination may be exacerbated where growers set unrealistic yield, as opposed to quality, targets and attempt to realise these by excessive use of fertilizers. Overfertilization contributes to groundwater contamination through leaching and is also a wasteful use of resources by the producer. In California, USA, studies of nitrogen balances in vegetable crops as early as 1984 (Pratt, 1984) showed that nitrogen leaching losses ranged between 90 and 260 kg/ha, demonstrating a wide, and wasteful, range of application rates of fertilizers. Nitrogen applications tend to be made in response to known crop requirements, as opposed to deficits established by soil analyses, as used for other nutrient elements. Consequently, this level of disparity between growers' use of nitrogen and its successful uptake by crops is not surprising. Environmental and financial considerations are now forcing growers to look for more efficient use of nitrogen to avoid losses by leaching into groundwater. and statutory penalties are beginning to be imposed in parts of Western Europe and North America.

Nitrogen forms and sources

Ammoniacal forms of nitrogen tend to be recommended due to their lower cost and higher soil retention compared with nitrate-nitrogen sources. However, in sandy soils, that are frequently used for brassica crops, with limited cation-exchange capacity, such retention is less likely. There may be opportunities for the use of controlled-release nitrogen sources that have relatively rapid breakdown characteristics. Alternatively, the use of organic fertilizers may become more attractive, especially as municipal organic wastes are increasingly disposed of through green waste composting procedures.

Split applications

Traditionally, brassica crops have received part of their nitrogen requirement by means of top dressing applications, particularly with the longer term crops such as overwintered cauliflower (*B. oleracea* var. *botrytis*) (Hochmuth, 1992). Applications of granular fertilizer, that are broadcast and then incorporated in the traditional manner, initially only enrich a very small volume of soil surrounding the fertilizer granule. With direct-drilled crops in particular, there is therefore, a delay before the seedling roots penetrate into a zone of nutrient-enriched soil. This delay can lead to short-term nutrient deficiencies resulting in poor early growth that may diminish the final yield and increase the time taken to reach maturity.

Traditionally, this problem has been minimized by retaining high residual levels of potassium and phosphate in the soils and applying excess nitrogen during cropping that exceeded the requirement for optimal yields. Such strategies aggravate the environmental problems associated with fertilizer use and impose an additional financial penalty on the grower. Split applications result in a less steep decline in soil nitrogen content. They also result in soil nitrogen content being sustained more effectively through to harvest (Biemond *et al.*, 1995). In white cabbage (*B. oleracea* var. *capitata*), Sorensen (1988) showed that split applications of nitrogen resulted in increases of total nitrogen, nitrate-nitrogen, phosphate, potassium, calcium and magnesium that then decreased as plant growth increased. Gardner and Roth (1989) applied nitrogen to white cabbage and cauliflower 2–3 times weekly and even here nitrogen concentrations declined as the plants aged. It decreased from 1.4% at week 7 to 0.5% at week 15 after transplanting. There is a reasonable presumption that the initial increase in nutrient concentration is caused by the high RGR of young roots and shoots and the high uptake by young roots. During growth, constituents such as lignin and polysaccharides (cellulose, hemicellulose and pectin) increase. Hara and Sonoda (1981) showed that in young cabbage concentrations of total soluble carbohydrates and starch increased during early plant growth.

Nutrient conditioning

Stand establishments of direct-seeded brassica crops are adversely affected by high or low temperatures, and high soil and water salt levels, leading to erratic germination, emergence and variable stands. Transplanting allows more efficient use of expensive F_1 hybrid seed, ensures uniform populations at desired spacing, improves adult plant uniformity and hence results in earlier harvesting. Increasing uniform maturity raises the percentage of the crop that is cut and minimizes the number of harvests. Successful transplant establishment relies on plants surviving the stresses associated with replanting and rapidly resuming growth, often in an hostile environment. Successful establishment is increased by: traditional hardening techniques, use of antitranspirant chemicals, application of growth regulators, pruning or defoliation of leaves, mechanical brushing prior to transplanting and transplant nutrient conditioning.

In the USA, McGrady (1996) showed that cauliflower cv. Snow Crown seedlings could be germinated in a mixture of equal parts peat and vermiculite with a limited nutrient supply. The macronutrients were then applied as transplant conditioning nutrient treatments. Seedling fresh weight, leaf area and stem diameter increased linearly in response to the main effects of nitrogen and phosphate. Transplanting shock, as measured by the number of yellowing leaves, increased with the highest levels of applied nitrogen, but it also encouraged maximum subsequent growth. Effects were also genotype dependent, since with the cauliflower cv. Olympus the greatest percentage of heads were cut in a single harvest from the lower nitrogen treatments.

Transplanting shock can be defined as severe necrosis appearing on true leaves and affecting more than 50% of the area of one leaf. Recovery from transplanting shock is defined as the resumption of growth as indicated by an increase in leaf number and/or stem diameter. Experiments showed that nutrient conditioning of cauliflower transplants was beneficial to plants grown in arid climates such as the Yuma area of Arizona, USA. The technique can also be applied with benefit to crops grown in cool temperate conditions.

Band placement

Band placement of fertilizer for brassica crops significantly increases their efficiency compared with broadcasting (Everaarts, 1993). Placement provides better availability of fertilizer nitrogen with either a higher yield or similar yield for lower resource input compared with broadcasting. The degree of response is dependent on crop type, seasonal conditions and husbandry systems (Everaarts and de Moel, 1995).

In the Netherlands, the recommended nitrogen rate is 300 kg/ha, minus the mineral nitrogen (N_{min}) in the soil layer 0–60 cm at planting. A minimum of 50 kg and maximum of 250 (minus N_{min}) kg/ha should be broadcast at planting and 50 kg/ha at 6 weeks after planting. There may be opportunities to reduce this recommendation to 225 kg/ha, at least for sandy clay soils. An additional 25 kg/ha might be applied in situations where additional nitrogen could be beneficial, such as with early spring crops where soil nitrogen mineralization is slow. In the UK, the nitrogen applications for cauliflower (*B. oleracea* var. *botrytis*) range between 100 and 300 kg/ha (Greenwood *et al.*, 1980) and would be applied in a similar split manner to those in the Netherlands. In direct-seeded North American green broccoli (*B. oleracea* var. *italica*) crops (Bracy *et al.*, 1995), the effects of side-dressing and pre-planting applications of nitrogen were compared. Comparisons of nitrogen applied at rates between 134 and 258 kg/ha and total nitrogen rates of 179 and 348 kg/ha as side dressings detected little effect on overall yield. Similarly, no differences between broadcasting and banding fertilizer applications were found in terms of total yield but substantial economies in nutrient use were achieved.

Starter fertilizers

An alternative to solid compound fertilizers is to use small volumes of liquid 'starter' fertilizer applied close to the transplant, making it readily available to the roots emerging from the propagation module. Several studies have demonstrated that this technique may increase the rate of the early growth phases of crops and is ultimately expressed in additional yield. Such benefits have been achieved even where the soil has a high residual nutrient status or where ample fertilizer has been applied as broadcast granules (Costigan, 1998). Starter fertilizers usually contain phosphates of ammonia since both ions become strongly adsorbed to soil particles resulting in little change to the soil pC. Use of ammonium ion encourages roots to excrete hydrogen ions (H^+) in preference to bicarbonate (HCO_3^-) ions, with resultant acidification of the rhizosphere and increased phosphate uptake (Marschner, 1995); application of calcium nitrate would have the reverse effect, which would help build up soil health. The presence of excess ammonium can be deleterious in reducing potassium absorption that will adversely affect the early growth of seedlings. Hence, inclusion of potassium in starter fertilizers is beneficial. But there can be seedling damage where the form of potassium also releases chloride ions, thereby raising the pC. Use of liquid potassium phosphate may alleviate the problem (Stone, 1998). Evidence suggests that starter fertilizers offer little advantage where the soil residual phosphate and potassium are high, but provide opportunities for maintaining yields where these values are diminished by lowering applications of granular fertilizers. This is in accordance with the aim of reducing fertilizer use to minimize environmental hazards.

More recently, one approach that has been examined for minimizing the nitrogen inputs to brassica crops is the use of cultivars which are more efficient in the use of nitrogen. Field trials in the Netherlands and Germany with cauliflower (*B. oleracea* var. *botrytis*) used an optimum nitrogen supply of 250 kg/ha, composed of the inorganic nitrogen content of the soil (N_{min}) and applied fertilizer nitrogen, and a limited nitrogen supply which was solely the N_{min} value. Yield was measured in terms of total dry matter and quality (percentage class 1 curds). The cv. Marine produced both the highest yield and quality and could be regarded as nitrogen efficient, whereas other cultivars were either nitrogen inefficient or behaved inconsistently across sites and seasons. Reducing the supply of nitrogen increased the number of loose curds and is suggested to encourage bolting as well. Some of the nitrogen inefficient cultivars exhibited buttoning of the curd. Rather *et al.* (1999) concluded that nitrogen-efficient cultivars achieved a higher uptake capacity through greater root activity and/or made more effective utilization of nitrogen.

Biostimulants and biofertilizers

Biostimulants (including biofertilizers or crop strengtheners) are 'everything that can be added to plants or the soil which stimulate natural processes benefiting them, beyond fertilization or pesticidal action alone' (Anon, 2017a). This definition is extended by the European Biostimulants Industry Council (EBIC) as 'substances and/or microorganisms ... stimulating natural processes which enhance or benefit nutrient uptake, nutrient efficiency, tolerance to environmental stress and crop quality'. Many commercial products contain, either singly or in various formulations, seaweed extracts, micronutrients, calcium and potassium phosphite for use with brassica crops. Biofertilizers may be considered as a subcategory of biostimulants (du Jardin, 2015) which increase nutrient use efficiency, open new routes for nutrient acquisition by plants and also promote aspects of biocontrol of pests, pathogens and abiotic stress. The effects of microbial additions to biofertilizers which increase these properties is discussed by Raimi *et al.* (2020).

Seaweed extracts have long histories of successful use in the raw state and as formulated products. Research identified four basic properties. They encourage populations of benign soil microbes which counter pests and pathogens producing soil suppressiveness (Dixon and Walsh, 1998); they increase cell multiplication resulting in additional root growth with greater nutrient uptake; they enhance chlorophyll content accelerating photosynthesis; and they improve tolerance against stresses such as cold, heat or drought. Currently research is identifying how genes are activated by their products, explaining how stress tolerance, greater carbohydrate manufacture and cell multiplication are achieved. The effects of seaweed extracts have been ascribed to their growth regulator content, micronutrient value and that they contain specific polysaccharides, betaines, polyamines and phenolic compounds. Transcriptomics and metabolomics are now providing an understanding of modes of action ascribed to seaweed extracts (De Saeger *et al.*, 2019).

Seaweed extracts are often marketed with added nutrients, such as boron, zinc, nitrogen and potassium, which enhance the effectiveness of sugar alcohols in improving micronutrient mobility. The amino acid content aids plant vigour and resilience during times of environmental stress, while added organic acids also improve micronutrient uptake. Stress is further reduced by adding polysaccharides and vitamins. These products are particularly targeted towards encouraging root growth and stress tolerance and as such contribute to integrated crop management strategies (Giordano *et al.*, 2021). The specific usefulness of biofertilizers in the sustainable production of commercial crops of wall rocket (*Diplotaxis tenuifolia*) is reported by Giordano *et al.* (2020). Abiotic stresses trigger responses which include the production of primary and secondary plant metabolites. These may influence the nutritional value of crops for human consumption; an aspect which requires further studies (Teklic *et al.*, 2020).

Other biostimulants/biofertilizers exploit the well-established properties of calcium, boron, iron, sulfur and silicon, plus amino acids and chelating agents such as mannitol, fucoidans and alginic acids. These encourage nutrient use efficiency through vigorous root development. Potassium phosphite is a well-established compound which promotes root growth and crop health. It can be particularly valuable in aiding recovery from stresses associated with transplanting brassicas. Transplanting removes plants from a highly protected glasshouse environment, exposing them to the rigours of field soils and variable weather when swift root establishment is vitally important. Applications of biostimulants via overhead spray lines on to modules are recommended 7 days ahead of transplanting. Biostimulants prepared using standardized formulations of ingredients are now becoming available and some have positive stimulatory effects, particularly of root growth (see Fig. 5.5).

Stabilized liquid silicon in a readily available form is capable of entering plants via foliage or roots, where it stimulates growth and especially root hair extension. Mixtures of silicon and salicylic acid aid wound repair processes, sealing off injured tissues and preventing damage from opportunist pathogens such as grey mould (*Botrytis cinerea*). Root-stimulating properties are particularly valuable, encouraging increases in fine, fibrous root development as discussed further by Shahrajabian *et al.* (2021).

Readily accessible forms of calcium may also act as biostimulants, encouraging cell wall development, acting as signalling molecules and encouraging forms of pathogen and environmental stress tolerance against, for example, drought, salt, nutrient imbalances and temperature fluctuations. Used on cauliflower (*B. oleracea* var. *botrytis*) and green broccoli (*B. oleracea* var. *italica*) crops they improve quality and shelf life, and reduce opportunities for damage caused by bacteria and fungi. Rapidly available calcium released by calcium cyanamide fertilizer is similarly valuable for increasing vigorous root growth, encouraging head quality and reducing pathogen impact (Dixon, 2012a, 2016). Similarly, combinations of calcium nitrate and boron aid vigorous root growth. Polyhalites may now be included in the biostimulant group of fertilizers and used on organic crops (Terrones *et al.*, 2020).

MANIPULATING DRILLING OR PLANTING DATES AND CROP GEOMETRY

Direct drilling was originally initiated for agricultural kale fodder crops in the 1960s. It was unsuitable, however, for field vegetables for the following reasons: the effects of soil type and soil conditions; the machinery used; the effects on pests, diseases and soil nutrients; and the leaving behind of unwanted live weeds, straw and trash (Elliott, 1975). That said, increasing plant population

Fig. 5.5. The biostimulant Actisil® encouraging vigorous root growth. (Courtesy of Tom Decamp, YARA).

Table 5.13. Seeding, transplanting and harvesting dates for pak choi cv. Troy F_1. (After Siomos, 1999).

Growing period	Sowing date	Transplanting date	Harvesting date	Sowing to transplanting	Transplanting to harvesting	Sowing to harvesting
					Days from	
1	25 Oct	3 Dec	27 Jan	39	55	94
2	23 Dec	28 Jan	28 Mar	36	59	95
3	25 Feb	30 Mar	14 May	33	45	78

Results for Greece were similar to those from the Netherlands and Australia for glasshouse/polythene-grown pak choi; after May, the crops tend to bolt under protection.

densities is a useful technique for raising yield and potential profits in brassicas. But for successful high-density, cole crop production, nitrogen applications should increase to accommodate increased nutrient demands. The use of high-density populations will have certain disadvantages. Green broccoli (*B. oleracea* var. *italica*) and cauliflower (*B. oleracea* var. *botrytis*) yields per unit area normally increase with closer planting densities but are associated smaller head size. While this may increase the numbers of heads and the total yield, maturity is often delayed and quality reduced (Salter and James, 1975). A smaller row spacing (0.6×0.4 m) and a boron dose of 6.01 kg/ha provided the highest cabbage (*B. oleracea* var. *capitata*) yield (Eberhardt Neto *et al.*, 2016).

Manipulating resource use by altering drilling or planting dates is exemplified by the work of Siomos (1999) who studied pak choi (*B. rapa* var. *chinensis* group) grown under unheated plastic in Greece, in three periods during 1993–1994 December–January, January–March and March–May (see Tables 5.13–5.15). Only when the temperatures and light incidence were high (period 3) did increased plant density (10 plants increased to 16.7 plants/m^2) raise yield. Here, individual fresh weight fell but the total number of plants rose. Changing the planting date and within-row plant spacing had little effect on dry matter, total soluble solids and fibre content.

Table 5.14. Effect of the growing period and plant spacing on yield (kg/m^2) of pak choi cv. Troy F_1. (After Siomos, 1999).

Plant spacing (cm)	1	2	3	Mean
		Growing period		
15×40	5.14 cd	5.63 c	12.66 a	7.81A
25×40	4.02 d	5.71 c	9.29 b	6.30B
Mean	4.58C	5.67B	10.98A	

Means followed by differing letters are significantly different at the 0.05 level of probability (Duncan's multiple range test).
Comparisons between means are differentiated by capital letters.

Nutrient requirements alter with changes to crop density and spacing. In the south-eastern USA, vegetables are usually planted on raised beds (Parish, 2000), with either single or double rows to each bed. Double rows offer higher yields per unit area, but may be difficult to maintain because of the erosion of the sides of the bed caused by localized heavy rainfall. Beds also provide advantages of quicker and earlier soil warming and allow the use of remotely guided (GPS) cultivation, such as steerage hoeing. In some areas, beds promote the avoidance of soil-borne pathogens such as *P. brassicae*, the causal agent of clubroot disease, because the soils are drier and this inhibits the movement of primary zoospores towards the host root hairs. In areas of moderate rainfall, the bed structure is retained and the greatest yields come from multiple row plots. Beds are especially popular for rapidly maturing brassicas, such as the leafy greens, mustard (*B. juncea*), turnip (*B. rapa* ssp. *rapifera*) and collard (*B. oleracea* Acephala group), which have become popular for both processing and fresh markets. Growing six rows on 2 m wide beds proves very effective, producing higher yields compared with fewer rows on narrower beds.

In cauliflower and green broccoli crops, results show that with increasing rectangularity of spacing, that is between-row spacing divided by within-row spacing, crop yields are decreased (Chung, 1982). This indicates that it is more advantageous to grow crops in a square formation rather than in a rectangular pattern (Salter *et al.*, 1984; Sutherland *et al.*, 1989). Modification of plant population densities is used to control cauliflower curd weight. A number of studies demonstrate that curd weight decreases with increasing plant density (Dufault and Waters, 1985; Singh and Naik, 1991; Sorensen and Grevsen, 1994).

Commercially, spacing varies substantially depending on location, genotype and husbandry systems. For example, in Europe, summer cultivars require much smaller spacing compared with overwintered types. In Western Australia, wider spacing is normal, thus between-row spacing of 0.75–0.80 m and, for most cultivars, a within-row spacing of 0.40–0.50 m, and two rows of cauliflower per planting bed. Resultant curd size varies in the range between 0.5 and 2.0 kg. Field experiments in Western Australia (Stirling and Lancaster, 2005) demonstrated

Table 5.15. Effect of growing period and plant spacing on mean daily growth increments (g/m²/day) of pak choi cv. Troy F_1. (After Siomos, 1999).

Plant spacing (cm)	Growing period			Mean
	1	2	3	
15 × 40	93.4c	95.4c	281a	156.6A
25 × 40	73.0c	96.8c	206.5b	125.5B
Mean	89.2B	96.1B	243.8A	

Means followed by differing letters are significantly different at the 0.05 level of probability (Duncan's multiple range test).
Comparisons between means are differentiated by capital letters.

that plants grown in a four-row configuration produced significantly ($P = 0.007$) higher total yields than control plants grown in a two-row configuration (see Fig. 5.6). Within the four-row configuration, a significant ($P = 0.019$) linear trend was observed, with yield falling by 0.3 t/ha for every 0.01 m increase in within-row plant spacing.

Uniformity of curd maturation improved when the number of plant rows per bed was increased from two to four (see Table 5.16). The majority of curds from plants grown in four rows were removed in the first two harvests, with only a small proportion of curds remaining at the final harvest. An increase in the uniformity

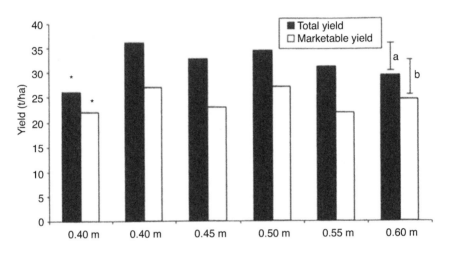

Fig. 5.6. Illustration of total and marketable yield of cauliflower (*Brassica oleracea* var. *botrytis*) cv. Summer Love produced by plants spaced at 0.40, 0.45, 0.50, 0.55 and 0.60 m. *Two-row treatment data. Bars indicate the least significant difference between all treatments (5%) = 5.2 (a) and 7.0 (b). (After Stirling and Lancaster, 2005).

Table 5.16. Percentage of cauliflower curds cut at each harvest time. (After Stirling and Lancaster, 2005).

Number of rows per 1.6 m bed	Spacing between plants within row (m)	Harvest 1 (%)	Harvest 2 (%)	Harvest 3 (%)	Total (%)
2	0.40	10.79	74.59	14.62	100.00
4	0.40	31.97	65.24	2.78	100.00
4	0.45	34.74	61.72	3.54	100.00
4	0.50	24.53	73.06	2.41	100.00
4	0.55	22.81	69.88	7.31	100.00
4	0.60	25.32	67.63	7.04	100.00
LSD between all treatments (5%)		10.87	15.16	14.46	
LSD between four-row treatments only		12.16	11.88	11.21	

LSD = least significant difference.

of mature curds was identified in plants grown in four rows, spaced at 0.40, 0.45 and 0.50 m. Curd weight decreased significantly ($P < 0.001$) when the planting configuration was altered from two to four rows (see Table 5.17). Within the four-row configuration, there was a significant ($P = 0.011$) linear effect of plant spacing on curd weight, which decreased with increasing plant density. Average curd weight decreased by 4.1 g for every 0.01 m decrease in plant spacing. There was a significant ($P = 0.003$) decrease in the average diameter of all curds harvested per treatment when plants were grown in four rows compared with two (see Table 5.17). Uniformity of curd maturity improved when the row number per bed was increased from two to four. This is an important consideration to cauliflower producers as it has a major influence on variable costs. Crops that mature in unison require fewer harvests and substantially reduce labour costs.

In Minnesota, USA, as cauliflower populations were increased from 24,000 to 72,000 plants/ha, with nitrogen rates held constant at either 112 or 224 kg N/ha, marketable curd weights decreased linearly, as did yields at the higher densities. Increasing the nitrogen rate to 112 kg/ha or higher reduced cull production at 24,000 plants/ha, but not at populations of 36,000 plants/ha or higher. Cauliflower yields were optimized at 24,000 plants/ha and 112 kg N/ha based on reduced cull production, satisfactory curd weights and transplant economy (Dufault and Waters, 1985).

MODELLING NUTRIENT NEED

A significant barrier to more efficient use of nitrogen fertilizer by brassica growers is a lack of information on the seasonal, soil-related and cultural variations in the supply of mineral nitrogen from the soil and the requirements for nutrients by the crop. Considerable information was accumulated for the WELL_N model in the UK in an attempt to remedy these deficiencies. WELL_N is a user-friendly computer program, conceived by Warwick

Table 5.17. Average curd weight (g) and diameter (cm) of marketable curds. (After Stirling and Lancaster, 2005).

Number of rows per 1.6 m bed	Spacing between plants within row (m)	Average curd weight (g)	Average curd diameter (cm)
2	0.40	887.71	15.64
4	0.40	723.63	14.92
4	0.45	733.44	14.78
4	0.50	784.76	15.17
4	0.55	750.66	15.00
4	0.60	816.35	15.21
LSD between all treatments (5%)		69.74	0.40
LSD between four-row treatments only		67.4	0.42

LSD = least significant difference.

University–Warwick Crop Centre at Wellesbourne, UK. It provides fertilizer recommendations and management advice on the use of nitrogen for a range of brassicas and tailors recommendations for different weather conditions, soil factors and cultural practices at each site. It was accessed via MORPH, which is a package of decision support tools designed for use by growers and their consultants in the fresh fruit and vegetable industries. Currently, the software used in this package is aged and attempts are being made to bring it up to date. The package has been compared with the paper-based system of recommendation for those testing midrib sap (Gardner and Roth, 1989). The electronic model allows users to manually select a particular model suited to their conditions. From that point, the programme has the potential to set up automatically the optimum model for the user under particular field conditions. In field trials, a substantial database of measurements was built up from farm-scale experiments that were run over two seasons on 37 sites across the UK, including a range of brassica crops. The database was constructed to store the descriptions and results of field trials and make them readily available for future research and development uses.

IRRIGATION AND WATER USE

Water and soil are considered two of the world's most precious and diminishing resources. All field vegetables, and brassicas in particular, need readily available soil moisture which delivers succulence, freshness and quality. Early, very practical advice (Secrett, 1935) from an authoritative grower working on very light sandy land in southern UK (Milford, Surrey) was 'never irrigate unless obliged to do so for the welfare of crops'. This advice was based on 'adverse effects in the soil' and 'heavy labour costs', which were incurred when moving equipment around the farm. A century later, water is becoming an increasingly scarce and costly resource, as highlighted by Knox et al. (2020a, b), emphasizing the need for water economy. Efficient water management is a prerequisite for effective nitrogen management. Nitrate-nitrogen leaching can be minimized by matching irrigation applications to evapotranspiration (ET) need. In cauliflower (B. oleracea var. botrytis), for example, water is needed throughout the crop's life but is most effective at the onset of curd formation (Salter, 1961; Wiebe, 1981). Improved product quality is greater on soils with higher water-holding capacity, but soil type has less effect than nitrogen fertilizer (Nilsson, 1980). Increasing nitrogen from 150 to 300 kg N/ha significantly increased yield. Yields increased up to 500 kg N/ha. Polish growers add nitrogen up to 500 kg N/ha with irrigation (probably accompanied by huge leaching losses into groundwaters) and this correlated with increasing nitrate-nitrogen in the curds. At these levels of application, there was a linear increase in nitrate-nitrogen in cauliflower leaves and curds (Kaniszewski and Rumpel, 1998). Water content in the root zone can now be monitored in real time using horizontally oriented soil moisture sensors linked to data logging and telemetry. The data obtained can automatically trigger drip irrigation for commercially grown field vegetables. Excessive use of fertigation can, however, contribute to nitrogen pollution and waste of resources (Monaghan *et al.*, 2010).

Cabbage is intermediately susceptible to water stress with the head formation stage more sensitive than the preceding growth stages (Smittle *et al.*, 1994). Critical periods for water stress are in the 3–4 weeks before harvest. Yields of vegetable crops, including cabbage, are reduced when soil water tension is >25 kPa. Crops irrigated when the soil moisture tension is <25 kPa at 10 cm produced the highest total and marketable yields. This regime requires more water to be applied but the water use efficiency rate is similar to that for cabbage irrigated at 50 and 75 kPa.

Several methods exist for measuring ET from climatic data. The modified Penman and Jensen–Haise methods use combinations of solar radiation, temperature, humidity, wind velocity and vapour pressure measurement to estimate ET for a reference crop. This then requires a crop coefficient to adjust the values obtained for the reference crop to estimate the ET of the crop to be irrigated. The crop coefficient values (ET of the irrigated crop/ET of the reference crop) are multiplied by the ET values estimated by the specific method to estimate the ET of the irrigated crop.

Pan evaporation (E_p) incorporates the climatic factors influencing ET into a single measurement and has been used to schedule irrigation for several crops. The single crop factor value (ET/E_p) usually results in applications of excessive irrigation during some growth stages and water deficits in others. A generalized curve was developed to describe crop factor value changes during growth but the generalized curve lacked precision. Smittle *et al.* (1994) have developed regression equations to calculate the daily crop factor values during the growth of several vegetables and have incorporated these equations to estimate ET from E_p data into irrigation scheduling models.

Precision irrigation

Field vegetables in general, and brassicas in particular, have considerable requirements for water which comes from either from rain-fed, or more frequently irrigation, sources (Dixon, 2015). Irrigation delivers water more precisely which is essential for the production of high-quality, succulent brassica vegetables. Continuing irrigation, which satisfies crop need, depends on water supplies which are monitored and rationed by water authorities or environment agencies and are becoming increasingly costly (Knox *et al.*, 2020a). The value of irrigation for increasing yields and quality in vegetable brassicas began being scientifically measured in the 1950s. Questions, such as 'how much and how often' should irrigation be applied, were answered by, for example, Salter (1961). He identified the irrigation requirements of early summer cauliflower (*B. oleracea* var. *botrytis*) in relation to stage of crop growth, plant spacing and available nitrogen supplies. Irrigation increased yield, especially when applied immediately prior to harvest. Maintaining adequate water availability for brassicas, particularly in summer months, ensures uninterrupted growth, succulence and maturation (see Fig. 5.7).

Uneven and inadequate irrigation has a devastating effect on brassica growth, as was demonstrated with the ornamental crucifer, Brompton stocks (*Matthiola incana*) by Goto and Yoshida (2003). The technology of precision irrigation is now driven by the ingenuity of specialist engineering companies that operate worldwide, developing equipment managed by computers and microwave communications (see Fig. 5.8).

Scheduling irrigation

Originally, irrigation schedules were either rotational sequences following crop maturation around a holding or applications timed more rationally by estimating water needs using the modified Penman and Jensen–Haise methods. Techniques for assessing soil water content, calculating crop need and controlling the volumes and format of water application have become increasingly more accurate, reliable and remotely controlled. 'Irrigation is no longer a marginal activity. It is now part of a sophisticated production system linked with measuring and monitoring soil moisture which should be

Fig. 5.7. Wide-boom irrigation of cabbage. (Courtesy of Briggs Irrigation Ltd).

Fig. 5.8. Irrigation control and data transmission connecting with mobile phones, laptops and farm offices. (Courtesy of Jones Engineering).

standard best practice' (Knox *et al.*, 2020a). Commercially, this is defined as 'precision irrigation (which) is placing the right amount of water at the right time as required by the crop growth stage and soil moisture status' (Simon Turner, Agri-tech, Shefford, UK, personal communication). Greater water, energy and labour savings are achieved by basing automatic irrigation on the use of volumetric soil water content as recently reviewed by Vera *et al.* (2021).

Smarter irrigation

Commercial monitors are now capable of measuring rainfall, irrigation and soil moisture by increments of 10 cm down to depths of 60 cm. These function on any mobile phone network, resulting in improved crop performance and quality. The user sees the effects of rainfall and irrigation applications in the soil profile on their own mobile platform. Consequently, water management becomes more efficient and economical. Growers can now receive crop irrigation strategies at daily or weekly intervals using smart sensors. Information and advice are cloud-based and sent to dashboards on growers' PCs, laptops or smartphones. The smart data includes weather monitoring, and assessments of soil moisture, rainfall and irrigation requirements. This should be preceded by on-site surveys of field topography, runoff, water retention and percolation factors, resulting in the design of effective and economical irrigation systems suitable for owner-occupied and rented land.

Aerial irrigation

Hose reels fitted with rain guns are probably still the most frequently used irrigators for long-season brassicas. Advocates claim rain guns are robust, versatile, labour efficient and fit in well with mechanized farming. They do, however, have a reputation for wasting water and energy. It appears that many fresh produce growers in particular are switching from rain guns and seeking more uniform irrigation systems which reduce soil splash, cause less crop damage and potentially lower disease incidence. 'Booms are more satisfactory for irrigating fresh produce crops' (Adrian Colwill, Briggs Irrigation, Corby, UK, personal communication). The booms are 50–76 m wide and fitted on to a gantry with a wheeled chassis. These will irrigate 50–96 m widths on each traverse depending on the nozzle or sprinkler fitments. Commercial equipment has capabilities for controlling droplet size which reduces soil capping, erosion and also protects delicate crops from damage. Solid-set systems using rotary sprinklers are also becoming more commonly used by growers of seasonal brassica vegetables and salads. They can be equipped with electronic control systems, using remotely activating or closing solenoid valves, providing flexibility and precision.

Ground-level irrigation

Various forms of drip and low-level irrigation have been used for orchards and other long-term woody crops for decades. Now, they are becoming used more frequently for vegetables. In personal communications with the author, Anthony Hopkins of Wroot Water, Doncaster, UK, sets out the benefits of ground-level irrigation. 'Low-level irrigation results in significant economies by reducing water losses whereas overhead systems experience losses caused by evaporation from aerial sprays and wind-driven drift.' Drip and lay-flat systems 'preserve soil structure, avoid erosion and retain equal volumes of air and water within the soil which promote healthier root systems and save water'. 'Drip irrigation is suitable for all soil types and particularly useful in areas where water availability is limited'. Key requirements for successful low-level systems are: a reliable water supply, an effective filtration system and efficient pumps. Water flow rates of between 50 and 70 m³/h are required depending on field topography and distances between the water sources and irrigation sites. The tapes can be buried to a depth of 3–5 cm as required. Radio frequency (RF) systems automatically control the pumps, filters and valves centrally. Irrigation pumps must be suitable for use with drip or tape irrigation and not pump more water than needed. Additionally, there are requirements for tractor-mounted machinery which initially lays out and then recoils the tapes after use.

Sensing soil moisture status

The objectives of precise, integrated irrigation systems are accurate and inexpensive sensing of within-field differences and then optimal control of variable-rate applications of water as required (Melian-Navarro *et al.*, 2010). Decision making systems using wireless sensor networks (WSN) (Evans *et al.*, 2012) can be displayed on GPS-identified maps at a computer base station. Low-cost Bluetooth wireless RF communications from distributed WSN monitors can control irrigators and then change water applications in real time. Using graphical user interfaces crop producers have remote, automated control of irrigation patterns, locations and output from home, office or else-where via wireless RF communications (see Fig. 5.9).

Satellite sensing

Satellite-based control systems can ensure that site-specific irrigation matches field needs, and reduces consumption, runoff and nutrient leaching into the groundwaters (Privette *et al.*, 2011). This is achieved by remote, real-time continuous measurements of soil moisture at specified depths. Sensors determine soil moisture by recording the GPS signal reflected from the earth's surface. A modified GPS delay mapping receiver tracks and measures the direct, line-of-sight, right-hand circularly polarized signal of a GPS satellite and simultaneously measures the delayed, earth-reflected, near-specular, left-hand circularly polarized GPS signal. The receiver estimates the surface scattering coefficient and path delays between the direct and reflected GPS signals. Scattering coefficients over land will then indicate changes in soil moisture content. Space-based technology has great potential for determining soil volumetric moisture values. The GPS reflectivity increases as the soil moisture content rises. Studies show that the sensitivity of the L-band signal (1.575 GHz) to soil moisture changed with soil type and sampling depth. The

Fig. 5.9. Soil moisture monitor. (Courtesy of James Martin, Agrovista UK Ltd).

sensitivity decreased with sampling depth in light soils and increased in heavy soils. L-band signals are used because they can penetrate clouds, fog, rain, storms and vegetation (see Fig. 5.10).

Tracking irrigation

Continually moving automated irrigation provides the variations in water application across fields as needed by crops and soil requirements (Chavez *et al.*, 2010). Originally, equipment used programmable logic controller technology which worked well with on-site management but was very expensive

Fig. 5.10. Satellite imaging and monitoring crops. (Courtesy of Environmental Systems Ltd).

when adding remote, real-time monitoring and control, as is now feasible by wireless sensor networks and the internet. An alternative approach used a single-board computer (SBC) with the Linux operating system installed which controls solenoids connected with individual or groups of nozzles and with application maps. The main control box contained an SBC connected with a sensor network radio, a GPS unit, and an ethernet radio creating a wireless connection with a remote server. A C Software control programme (C Software Ltd, Tewkesbury, UK) was installed in the SBC, controlling the on and off timing for each group of nozzles based on instructions derived from remote application maps. Integrating sensors identifying soil conditions and the performance of irrigation equipment provided precise real-time irrigation and data which can be stored as records for future farm planning and decision making.

Electrically powered irrigation

Policies which limit greenhouse gas emissions are leading towards the development and use of electrically powered pumps which draw energy supplies from on-farm solar panels (Mathur, 2019), replacing the use of petrochemical fuels. Electrical power generated from solar panels has the added advantage that excesses provide a new, sustainable revenue stream for farm enterprises. Combinations of solar panels, energy-efficient pumps and micro-irrigation equipment can minimize water use and possibly increase fertilizer use efficiency in crops. Potential side benefits of on-farm electricity generation include the use of more energy-efficient electric soil cultivators and harvesters. Avoiding greenhouse gas emissions at the user level contributes towards zero greenhouse gas agriculture and decarbonization of the electricity grid. Currently, initial investment costs are high but potentially result in significant lifetime benefits.

Water quality

Grieve *et al.* (2001) determined the effect of salinity and timing of water stress on leaf ion concentration in the field using pak choi (*B. rapa* var. *chinensis* group), tatsoi (*B. rapa* var. *narinosa* group), kale (*B. oleracea* ssp. *acephala* group), cooking greens (*B. rapa*) and brown mustard (*B. juncea*). The experiment used saline solutions which simulated the high-sodium and high-sulfate drainage waters typical of the San Joaquin Valley of California, USA. The mineral ion concentrations in leaves were significantly affected by the increasing salinity of irrigation water. But the stage at which salinity was applied had little effect. With increasing salinity, calcium ions and potassium ions decreased in the leaves of all species whereas sodium ions and total sulfur significantly increased. Magnesium also rose in the leaves of brassicas with increasing salinity; there was also an increase in chloride ion content. The use of moderately saline irrigation water did not adversely affect crop quality as rated by colour, texture and the mineral nutrient content available to consumers. In many areas of the world, water is now the most valuable and scarce resource and this shortage is set to become more acute. Large-scale areas of intensive vegetable production, such as parts of the USA (California, Florida), southern Europe (southern Spain, Portugal, Italy, Greece) and the Middle East (Israel, Egypt) are being forced to confront the problems of increasing salinity in their irrigation water and the implications of this for the quality of vegetables made available to the consumer. The latter is becoming ever more conscious of the health risks associated with nitrate-nitrogen loading, especially in leafy vegetables of which brassicas are a major portion. One of the on-farm management options is the reuse of agricultural drainage effluents. This strategy is especially attractive because significant amounts of good-quality water are preserved and because it substantially reduces the need to dispose of large volumes of drainage water (Sorensen, 2000).

In the suggested drainage water reuse system proposed for the San Joaquin Valley of California, USA, selected crops would be grown and irrigated in sequence, starting with the very salt-tolerant species (Johnston *et al.*, 2011). Drainage effluents from these crops would be used to irrigate crops of higher salt tolerance. At each step in the sequence, the drainage water becomes progressively more salinated. The composition of the drainage effluent waters in this region are typically a mixture of salts with sodium > sulfate > chloride > magnesium > calcium predominating in that order (see Table 5.18).

Table 5.18. Water content and mineral ion concentration (mg/100 g edible portion) of selected *Brassica* crops irrigated with saline water (11 dS/m) compared with non-saline conditions.[a] (After: 1 = Grieve *et al.*, 2001; 2 = Rubatzky and Yamaguchi, 1997; 3 = Ensminger *et al.*, 1995).

Vegetable	Water (%)	Ca	Mg	Na	K	P	S	Cl	Reference
Tatsoi	<u>93.7</u>	<u>123</u>	<u>26</u>	<u>236</u>	<u>323</u>	<u>24</u>	<u>63</u>	<u>208</u>	1
Mustard greens									
Vitamin	<u>95.0</u>	<u>106</u>	<u>20</u>	<u>220</u>	<u>262</u>	<u>23</u>	<u>48</u>	<u>490</u>	1
Red Giant	<u>94.1</u>	<u>116</u>	<u>24</u>	<u>61</u>	<u>419</u>	<u>23</u>	<u>52</u>	<u>118</u>	1
	89.3	181	29	33	374	46	–	–	2
	92.6	138	–	18	220	32	–	–	3
Kale	<u>89.0</u>	<u>306</u>	<u>91</u>	<u>222</u>	<u>476</u>	<u>53</u>	<u>175</u>	<u>246</u>	1
	89.7	202	45	43	420	55	–	–	2
	91.2	134	–	43	221	46	–	–	3
Pak choi	<u>94.0</u>	<u>118</u>	<u>24</u>	<u>248</u>	<u>426</u>	<u>24</u>	<u>53</u>	<u>192</u>	1
	93.8	118	19	71	234	39	–	–	2

[a]Data from crops grown with saline water underlined.

Members of the Brassicaceae family are relatively tolerant of salinity (see Chapter 1 section, Origins and Diversity of *Brassica* Crops). But even with these crops, abnormally high levels of salinity severely limit plant growth. Leafy vegetables are the primary source of mineral nutrients for human diets. Numerous tables of food composition list the major constituents of vegetables and give values for predominant mineral ions. These values are only estimates in so far as the data are based on a limited number of samples and vary due to biological and environmental factors such as maturity, analytical procedures and processing. In addition, the availability, uptake and partitioning of mineral ions within the plant are controlled by numerous environmental factors including the concentration and composition of solutes in the soil solution.

Under saline conditions, mineral ion interactions in the external media may affect the internal requirements of elements essential for plant growth and development. These imbalances often influence the growth and nutrition of the crop, which in turn may affect crop quality in terms of colour, texture and nutritive value. Calcium plays a vital nutritional and physiological role in plant metabolism. Under saline conditions, ion imbalances in the substrate or plant may adversely affect calcium nutrition. Substrate levels of calcium that are adequate for plant requirements under non-saline conditions may be nutritionally inadequate and growth-limiting when the plant is salt-stressed. The calcium status of the plant is strongly influenced by the ionic composition of the external medium in that the presence of other salinizing ions in the substrate may reduce calcium activity and limit the availability of calcium to the plant. Cations, such as sodium and magnesium, may disrupt calcium acquisition, uptake and transport. Leaf calcium concentrations in all vegetables tend to decrease as salinity increases despite added quantities being available in the soil. Salinity-induced calcium deficiency can result in physiological disorders in brassica vegetables, such as internal tipburn, browning and necrosis of the inner leaves (see Chapter 8 section, Physiological Disorders: Tipburn).

Under salinity stress, maintenance of adequate levels of potassium is essential for plant survival. High levels of external sodium not only interfere with potassium acquisition through the roots but may also disrupt the integrity of root membranes and alter the selectivity of the root system for potassium as compared with sodium. Total sulfur content in the leaves of all brassica vegetables increases as sulfate values rise. The brassicas are especially active sulfur accumulators. Members of the Brassicaceae are among the 15 plant families which biosynthesize significant quantities of sulfur-rich glucosinolates. Hydrolysis of these compounds yields 'mustard oils' that impart the characteristic spicy tastes to these vegetables. Increases in total sulfur accumulation in response to irrigation with moderately saline, sulfur-dominated waters may enhance the flavour, benefits to human health and consumer acceptability of brassica vegetables.

MOTIVE POWER AND CULTIVATIONS

Since the 1950s, fresh produce crops, including the brassicas, have moved progressively from employing manual labour towards mechanization. Tractors became widespread during and following World War II, triggered by their increased availability from assembly-line manufacture in the USA (Gunston, 1947) and by the imperative of demands for increased food production. Ploughing was one of the earliest operations to be mechanized, quickly followed by seed drilling, transplanting and spraying.

Mechanization has evolved over the subsequent 70 years into a worldwide drive towards complete automation using robotics and the application of artificial intelligence in crop husbandry. Market pressures played important roles in promoting first mechanization and then latterly automation. The arrival of supermarkets into the fresh produce supply chains (Dawson and Shaw, 1989; Shaw and Gibbs, 1996) significantly stimulated requirements for mechanisation and automation. This was led by large retail companies and their consumers demanding an increasing range of high-quality brassica foods delivered 'on spec and on time' from bigger, more innovative producer, packer and processor companies.

Additionally, in rapidly urbanizing societies, the manual husbandry and harvesting of brassica crops are increasingly unattractive physical occupations resulting in a reduction of easily available local labour supplies. Opportunities for employing rurally aware and skilled people emanating from lower-wage economies are now declining. Changes are being driven, therefore, by cultural, social and economic factors.

Mechanization and automation are facilitated agronomically by vastly increased understandings of brassica genetics and plant physiology which are applied to breeding new, more tailored cultivars. This requires the transfer of research and development findings into the production industry (Dixon, 2002). In practice, the implementation of change also depends on commercial engineering companies profitably designing and delivering compatible, reliable and cost-effective machinery and on retail markets providing equitable returns for growers, which in turn finances the uptake of new technologies. The need for change is now ever more urgent because of climate change which demands sustainable intensification of increased production based on smart technologies (Grieve *et al.*, 2019).

CHANGES IN PRODUCTION AND MARKETING

High-quality, health-promoting and attractive brassicas are required in a bigger range of value-for-money forms, available irrespective of seasonality (see Chapter 8 section, Introduction). Supermarket specifications require their production, processing and delivery effectively and efficiently at minimal prices. Fulfilling these criteria requires brassica cultivars which grow and mature uniformly as attractive and increasingly novel products. Plant breeders are providing these cultivars in a widening range, developed by F_1 hybrid, male sterile and marker-assisted breeding and now utilizing gene-transfer and gene-editing techniques (see Chapters 1 and 2). These cultivars require precise cultivation, nutrition, irrigation and plant protection supplied via increasingly sophisticated electronic monitoring and means of resource application (see also Chapter 4 section, Introduction).

NEW TECHNOLOGIES

Machine operations guided by artificially intelligent systems which distinguish crop plants from weeds, identify incipient pest and pathogen invasions (see Chapter 6 and 7 Introduction sections), select mature harvestable produce and transport it for preparation, storage, packaging and despatch, are either available or in development. Advantageously, these advances reduce energy consumption, and minimize pollution and wasteful losses ensuring delivery into retail markets in the volumes, maturity and quality required (Anon, 2018). For example, mechanical weed control using robotically controlled machinery (see Fig. 5.11 and Chapter 6 section, Weed Control in Practice).

Automation and robotics are equally applicable in organic forms of production as conventional husbandries. Indeed, organic production benefits from simpler husbandry and harvesting, thereby encouraging market expansion. As automated systems become more widely available their price will fall,

Fig. 5.11. Robotic autonomous weeding of cabbage crops. (Courtesy of Robotti).

resulting from 'economies of scale' (Smith, 1776, 1976). That encourages the transfer of new technologies into smaller enterprises supplying more localized markets. Developing new technologies is an expensive process requiring the assembly of multidisciplinary teams. Consequently, the large, broadacre crops are the initial target markets. The evolution, application and future prospects of these technologies are considered here in wider contexts, therefore demonstrating agricultural opportunities and highlighting wherever possible applications for brassica crops.

Primary cultivation

Soil inversion by ploughing is one of the most ancient of humankind's husbandry techniques, reaching back at least 4000 years to Ancient Egypt (Janick, 2002). It aims at burying weeds and crop residues and incorporating manures and municipally derived composts which help develop the well-structured, nutrient-rich soils required by all vegetable brassica crops. Increasing soil organic matter content is aided by the actions of beneficial soil microbes and macrobes such as earthworms. Weathering by frosting in cool-temperate regions suited for brassica production breaks down the soil structure into smaller clods which subsequent tillage converts into friable tilths suitable for seed drilling or transplanting (Dixon, 2019). Ploughing also exposes overwintering pests and weed seeds for avian predation and microbial degradation. The resultant fertile, healthy soils are well suited for brassica production.

Tractor power

Tractor-drawn ploughing was one of the first aspects of brassica husbandry to benefit from mechanization. This increased the scale and scope of brassica production as more powerful and efficient tractors and ploughs appeared from the 1920s onwards, culminating in very powerful, heavy tractors capable of hauling six row reversible ploughs (see Fig. 5.12).

Manufacturers are now seeking means for reducing the weight of tractors and hence limiting soil compaction. Newer machines are guided by GPS delivered from satellites which increases the accuracy of land preparation using precise receivers or local positioning systems, such as electronic compasses or inertial navigation systems (Gomez-Gil *et al.*, 2011). Cost and complexity, these authors suggest, are reduced by placing the GPS receiver ahead of the tractor, and by obtaining the midpoint position and orientation of the tractor rear axle by applying the principles of dynamic motion resulting in a straight trajectory suitable for bed building, subsequent transplanting and computerized recording of plant positions in the field. Tractor power for torque and engines is also be monitored by GPS. This results in information which can be utilized since it identifies developing mechanical faults, thereby avoiding unnecessary downtime, breakdown costs and pollution (Pexa *et al.*, 2011). 'Predictive maintenance' programmes identify potential malfunctioning and alert both growers and repair engineers, resulting in their automated call-outs being equipped with the required spare parts which have been identified and selected remotely, enabling considerable efficiency gains.

Reducing greenhouse gases

As part of the reduction in greenhouse gas emissions, diesel and petrol fuels are being phased out as energy sources, replaced by tractors and implements that are powered by electricity and hydrogen (Xu *et al.*, 2019b; Hoy and

Fig. 5.12. Traditional primary cultivation machinery. (Geoff Dixon).

Kocher, 2020). The suitability of these renewable power sources for organic crop production is emphasized by Dumitru *et al.* (2019). Electric-powered machinery can have lower running costs compared with those run on fossil fuels such as petrochemicals (Mathur, 2019), although the initial capital outlays are greater and the tractors need onboard storage of electricity in batteries. Hydraulic systems for tractors and other 'off-highway' vehicles now use environmentally acceptable fluids which are readily biodegradable, based on oilseed rape or other bioderived viscous hydrocarbons.

Limiting primary tillage

Non-tillage and direct-drilling husbandries involving land preparation without soil inversion are advocated as promoting environmental sustainability. Direct drilling of seed is more suitable for broadacre oilseed brassicas (*B. napus*) compared with vegetable brassicas. Studies of direct drilling identified problems associated with the residual crop residues in which pests and pathogens perennate, added weed competition (especially perennial deeply rooted species), soil compaction and impeded drainage (Cannell, 1985). Advantages of non-tillage include increased structural stability of the soil, erosion resistance and beneficial biological activity. Several seasons of no-till husbandry may result in reduced crop nitrogen requirements, retention of phosphate ions and improved self-structuring at the surface, especially with calcareous soils and greater water permeability (Anon, 2020a). Effects associated with no-till husbandries, such as weed persistence and greater risks from pest and pathogens, especially slugs, make them less attractive for vegetable brassica production where the final product quality and uniform maturation are paramount.

Secondary tillage

Secondary land preparation ahead of seed drilling or transplanting has similarly benefitted from increasing tractor power and machine design and manufacture. The range and power of secondary cultivation equipment designed for broadacre agricultural crops has increased enormously with the aim of reducing farmers' variable costs and increasing productivity. Many of these machines are not, however, suitable for fresh produce crops, especially brassicas, because they 'force' tilths and disrupt soil health. Consequently, specialist producers are increasingly using machinery specifically designed for vegetable crops, particularly brassicas. Reducing the number of soil movements with lighter machinery lowers costs and causes less structural damage to the soil. Machinery selection requires on-farm assessments of soil structure, compaction and consolidation risks, residue management from previous cropping, demands for weed and soil-borne pathogen control and suitability for vegetable brassica crops.

Soil compaction

Increasingly, soil compaction is recognized as a major husbandry hurdle preventing healthy brassica growth. Compaction reduces soil aeration and water infiltration, and increases the risks of waterlogging. Root growth is minimized in compacted soils, limiting water and mineral uptake. Measuring *in situ* soil compaction permits guided tillage and controlled traffic farming (CTF) using GPS guidance and computerized field layouts. In cropped areas, this encourages rapid surface drainage limiting waterlogging and erosion. As a result, soil health and quality are improved and structural damage of the soil is avoided. The result is easier seed sowing and transplanting on to raised-bed formations, which more adequately hold moisture needed for germination and growth and yet drain excesses efficiently. Over time, CTF encourages a slow natural increase in the soil organic matter content and is supplemented with added organic matter, such as farmyard manure or composts. The resultant organic matter store is preserved because CTF reduces oxidation through less tillage, which enhances carbon storage. Additionally, less powerful farm machinery is needed reducing fuel consumption and further lowering pollution and costs. In practice, once GPS has provided the coordinates for each field, layouts are established and recorded in the tractor's computer, then drivers will use the same wheel spacings for many years by simply selecting their location in the memory.

Although referring to broadacre brassica crops, the findings of Bulinski and Niemczyk (2011) are indicative of the damage incurred by soil compaction and resultant penalties. These findings, for oilseed rape (*B. napus*), include decreases of 23% in the number of plants per linear metre, 43% in siliquae and 54% in seed yield, all as a result of 43% increased compaction and soil density raised by 0.25 g/cm^3 compared with areas without wheel trafficking. There are concerns that tillage may adversely affect populations of beneficial insects which help limit brassica pests. Studies of spring broccoli (*B. oleracea* var. *italica*) crops, however, showed that tillage did not reduce the numbers of predators which provide natural biocontrol of cabbage root fly (*Delia radicum*) populations (Mesmin *et al.*, 2020). Soil tillage did, however, reduce the emergence of carabid species overwintering as larvae, possibly because of increased bird predation. All secondary cultivations require timely planning with regard to weather windows and advanced guidance from wholly automated delivery of meteorological information. Commercially available equipment provides soil moisture monitoring which measures rainfall amounts and/or applications of irrigation. Inputs and resultant soil moisture changes are measured in increments of 10 cm down to a depth of 60 cm in the profile and linked with weather stations in the field. Data are relayed on to mobile platforms as described by Agrovista (Nottingham, UK).

Frequently, land which is suitable for growing brassica vegetables is rented. This has the advantage of brassica crops being integrated into mixed rotations and minimizes risks from the perennating soil-borne pathogens such as *P. brassicae* (the cause of clubroot disease) (Dixon, 2020). For high-quality brassica production, especially root crops, such as swede (*B. napus* ssp. *rapifera*) and turnip (*B. rapa* ssp. *rapifera*), or baby leaf brassicas, such as salad rocket (*Eruca sativa*) and wall rocket (*D. tenuifolia*), there may be requirements for de-stoning ahead of further land preparations. Machines are readily available which crush stones *in situ*, returning the residual gravel into the soil. Trailed harrows break down ploughed soil into tilths suitable for cropping and the tractors providing motive power are guided by GPS. Similarly, Cambridge rollers, suitable for heavier clay soils, consist of sets of independently rotating rings which crush larger clods reducing them to a finer tilth. Flat-bed rollers very effectively seal soil surfaces after drilling, retaining soil moisture and aiding speedier germination. On light, highly fertile land, such as in Lincolnshire, UK, after ploughing further preparation is completed by using Dutch harrows, or less frequently power harrows, ready for seeding and transplanting. In wetter areas, such as Fife, UK, bed-formers are used which provide crops with increased drainage and improved root aeration (Andrew Richardson, Allium and Brassica Centre, Kirton, UK, personal communication).

Sowing and transplanting tilths

Frequently, where growers are producing several crops on a single area of land in one season, the soil will be reworked with a rotavator. These machines have a rapidly rotating shaft connected with a tractor's power take-off link and are fitted with a series of L-shaped tines. Rotavators are very effective, especially when weather windows for drilling or transplanting successional crops are

Fig. 5.13. Crop cover rewinder for conservation of sheeting. (Courtesy of Edwards Farm Machinery Ltd).

limited. Their disadvantage is that frequent use results in impermeable pans forming just below the working depth of the tines. Pans prevent effective drainage into the subsoil with consequent waterlogging and asphyxiation of crop roots. They are disrupted with a single 'pan-busting' tine trailed behind a powerful tractor. Where crops are protected with insect-excluding netting or early growth encouraged with polythene sheets, machinery is required which will lay and remove these crop covers (see Fig. 5.13).

REFERENCES

Abuzeid, A.E. and Wilcocksen, S.J. (1989) Effects of sowing date, plant density and year on growth and yield of Brussels sprouts (*Brassica oleracea* var. *bullata* subvar. *gemmifera*). *The Journal of Agricultural Science* 112(3), 359–375. DOI: 10.1017/S0021859600085816.

Aitken, A., Hewett, E., Warrington, I., Hale, C. and McCaffrey, D. (2012) *Harvesting the Sun: A Profile of World Horticulture. Scripta Horticulturae No. 14.* International Society for Horticultural Science (ISHS), Leuven, Belgium.

Amiama-Ares, C., Bueno-Lema, J., Alvarez-Lopez, C.J. and Riveiro-Valino, J.A. (2011) Manual GPS guidance system for agricultural vehicles. *Spanish Journal of Agricultural Research* 9(3), 702–712. DOI: 10.5424/sjar/20110903-353-10.

Andreucci, M.P., Moot, D.J. and Black, A.D. (2014) Quantifying growth and development of bulb turnips. *European Journal of Agronomy* 55, 1–11. DOI: 10.1016/j.eja.2013.12.007.

Anon (1985) *Fertiliser Recommendations 1985–1986. Reference Book No. 209.* Ministry of Agriculture, Fisheries and Food for England & Wales, London, pp. 15–17.

Anon (1997) European Commission regulation (EC) No.194/97 dated 31st January 1997. *Official Journal of the European Community No. L* (31), 48–50.

Anon (2000) *Fertiliser Recommendations for Agricultural and Horticultural Crops. Reference Book No. 209.* Department for Environment, Food and Rural Affairs (Defra) [previously, Ministry of Agriculture, Fisheries and Food], London.

Anon (2016) Developing an integrated approach to soil health assessment and improvement. *CP 107b: Growing Resilient Efficient And Thriving Soils (GREAT) Soils – Annual Report 2016.* Agriculture and Horticulture Development Board (AHDB), Stoneleigh, UK.

Anon (2017a) *Crop Biostimulants: Factsheet.* Agriculture and Horticulture Development Board (AHDB), Stoneleigh, UK.

Anon (2017b) *Nutrient Management Guide (RB 209). Section 1 Principles of Nutrient Management and Fertilizer Use; Section 6 Vegetables and Bulbs.* Agriculture and Horticulture Development Board (AHDB), Stoneleigh, UK.

Anon (2018) Crop module 6615: Brussels sprouts. Red Tractor Assurance for Farms – Fresh Produce. Available at: www.redtractorassurance.org.uk (accessed 12 June 2023).

Anon (2020a) *Great Soils: Arable Soil Management: Cultivations and Crop Establishment.* Agricultural and Horticultural Development Board (Cereals & Oil Seeds) (AHDB), Stoneleigh, UK, 36 pp. Available at: www.ahdb.org.uk

Anon (2020b) *Nutrient Management Guide (RB 209) Section 6 Field Vegetables.* Agricultural and Horticultural Development Board (AHDB), Stoneleigh, UK, 48 pp. Available at: www.ahdb.org.uk

Arias, T., Beilstein, M.A., Tang, M., McKain, M.R. and Pires, J.C. (2014) Diversification times among brassica (Brassicaceae) crops suggest hybrid formation after 20 million years of divergence. *American Journal of Botany* 101(1), 86–91. DOI: 10.3732/ajb.1300312.

Bento, T. da S., de Carvalho, M.A.C., Yamashita, O.M., Dallacort, R., da Silva, I.V. *et al.* (2020) Application of several green manures to produce organic cabbage (*Brassica oleracea* var. Capitata) and their influence on soil biological properties. *Australian Journal of Crop Science* 14(9), 1372–1378. DOI: 10.21475/ajcs.20.14.09.p2167.

Biemond, H., Vos, J. and Struik, P.C. (1995) Effects of nitrogen accumulation and partitioning of dry matter and nitrogen of vegetables. 1. Brussels sprouts. *Netherlands Journal of Agricultural Science* 43(4), 419–433. DOI: 10.18174/njas.v43i4.564.

Booij, R., Enserink, C.T., Smit, A.L. and van der Werf, A. (1993) Effects of nitrogen availability on crop growth and nitrogen uptake of Brussels sprouts and leek. *Acta Horticulturae* 339, 53–66. DOI: 10.17660/ActaHortic.1993.339.5.

Booij, R., Kreuzer, A.D.H., Smit, A.L. and van der Werf, A. (1996) Effect of nitrogen availability on dry matter production, nitrogen uptake and light interception of Brussels sprouts and leeks. *Netherlands Journal of Agricultural Science* 44(1), 3–19. DOI: 10.18174/njas.v44i1.554.

Booij, R., Kreuzer, A.D.H., Smit, A.L. and van der Werf, A. (1997) Effects of nitrogen availability on the biomass and nitrogen partitioning in Brussels sprouts (*Brassica oleracea* var. *gemmifera*). *Journal of Horticultural Science and Biotechnology* 72(2), 285–297. DOI: 10.1080/14620316.1997.11515515.

Borkowski, J., Dyki, B., Oskiera, M., Machlanska, A. and Felczynska, A. (2016) The prevention of tipburn on Chinese cabbage (*Brassica rapa* L. var. *pekinensis* (Lour.) Olson) with oliar fertilizers and biostimulators. *Journal of Horticultural Research* 24(1), 47–56. DOI: 10.1515/johr-2016-0006.

Bracy, R.P., Parish, R.L. and Bergeron, P.E. (1995) Side dress N application methods for broccoli production. *Journal of Vegetable Crop Production* 1(1), 63–71. DOI: 10.1300/J068v01n01_07.

Bulinski, J. and Niemczyk, H. (2011) Effect of tractor traffic in traditional cultivation technology on the yield of winter rape. *Annals of Warsaw University of Life Sciences – SGGW, Agriculture (Agricultural and Forest Engineering)* (57), 5–14.

Cannell, R.Q. (1985) Reduced tillage in north-west Europe: a review. *Soil and Tillage Research* 5(2), 129–177. DOI: 10.1016/0167-1987(85)90028-5.

Cardarelli, M., Rouphael, Y., Muntean, D. and Colla, G. (2015) Growth, quality index, and mineral composition of five ornamental cabbage cultivars grown under different nitrogen fertilization rates. *HortScience* 50(5), 688–693. DOI: 10.21273/HORTSCI.50.5.688.

Chaves, B., Opoku, A., de Neve, S., Boeckx, P., Van Cleemput, O. *et al.* (2006) Influence of DCD and DMPP on soil N dynamics after incorporation of vegetable crop residues. *Biology and Fertility of Soils* 43(1), 62–68. DOI: 10.1007/s00374-005-0061-6.

Chavez, J.L., Pierce, F.J., Elliott, T.V. and Evans, R.G. (2010) A remote irrigation monitoring and control system for continuous move systems. Part A: description and development. *Precision Agriculture* 11(1), 1–10. DOI: 10.1007/s11119-009-9109-1.

Chung, B. (1982) Effects of plant density on the maturity and once-over harvest yields of broccoli. *Journal of Horticultural Science and Biotechnology* 57(3), 365–372. DOI: 10.1080/00221589.1982.11515065.

Conversa, G., Lazzizera, C., Bonasia, A. and Elia, A. (2019) Growth, N uptake and N critical dilution curve in broccoli cultivars grown under Mediterranean conditions. *Scientia Horticulturae* 244, 109–121. DOI: 10.1016/j.scienta.2018.09.034.

Costigan, P.A. (1998) The placement of starter fertilisers to improve the early growth of drilled and transplanted vegetables. *Proceedings of the Fertiliser Society* 274, 1–24.

Dawson, J.A. and Shaw, S.A. (1989) The move to administered vertical marketing systems by British retailers. *European Journal of Marketing* 23(7), 42–52. DOI: 10.1108/EUM0000000000578.

de Ruijter, F.J., ten Berge, H.F.M. and Smit, A.L. (2010) The fate of nitrogen from crop residues of broccoli, leek and sugar beet. *Acta Horticulturae* 852, 157–162. DOI: 10.17660/ActaHortic.2010.852.18.

De Saeger, J., Van Praet, S., Vereecke, D., Park, J-H., Jacques, S. *et al.* (2019) Toward the molecular understanding of the action mechanism of *Ascophyllum nodosum* extracts on plants. *Journal of Applied Phycology* 32(1), 573–597. DOI: 10.1007/s10811-019-01903-9.

Desbourdes, C., Blondlot, A. and Douche, H. (2008) Variable nitrogen application with satellite view. In: *Proceedings of the 9th International Conference on Precision Agriculture, Denver, Colorado, USA, 20–23 July, 2008.* Precision Agriculture Center, University of Minnesota, Department of Soil, Water and Climate, St Paul, MN.

Dixon, G.R. (2002) HorTIPs (Horticultural Technology into Profits): the achievements. *The Horticulturist* 11(4), 11–13.

Dixon, G.R. (2009) The impact of climate and global change on crop production. In: Letcher, T.M. (ed.) *Climate Change: Observed Impacts on Planet Earth.* Elsevier, Oxford, pp. 307–324.

Dixon, G.R. (2012a) Calcium cyanamide – a synoptic review of an environmentally benign fertiliser which enhances soil health. *Acta Horticulturae* 938, 211–217. DOI: 10.17660/ActaHortic.2012.938.27.

Dixon, G.R. (2012b) Climate change: impact on crop growth and food production. *Canadian Journal of Plant Pathology* 34(3), 362–379. DOI: 10.1080/07060661.2012.701233.

Dixon, G.R. (2015) Water, irrigation and plant diseases. *CABI Reviews* 10(9), 1–18. DOI: 10.1079/PAVSNNR201510009.

Dixon, G.R. (2016) Managing clubroot disease (caused by *Plasmodiophora brassicae* Wor.) by exploiting the interactions between calcium cyanamide fertilizer and soil microorganisms. *The Journal of Agricultural Science* 155(4), 527–543. DOI: 10.1017/S0021859616000800.

Dixon, G.R. (2019) *Garden Practices and Their Science.* Routledge, Oxford. DOI: 10.4324/9781315457819.

Dixon, G.R. (2020) Controlling clubroot: a case study. *The Vegetable Farmer* July, 26–28.

Dixon, G.R. and Walsh, U.F. (1998) Suppression of plant pathogens by organic extracts a review. *Acta Horticulturae* 469, 383–390. DOI: 10.17660/ActaHortic.1998.469.41.

Dixon, G.R., Collier, R.H. and Bhattachrya, I. (2014) An assessment of the effects of climate change on horticulture. In: Dixon, G.R. and Aldous, D.E. (eds) *Horticulture: Plants for People and Places, Vol. 2. Environmental Horticulture.* Springer, Dordrecht, Netherlands, pp. 817–858.

Dufault, R.J. and Waters, L. (1985) Interaction of nitrogen fertility and plant populations on transplanted broccoli and cauliflower yields. *HortScience* 20(1), 127–128. DOI: 10.21273/HORTSCI.20.1.127.

du Jardin, P. (2015) Plant biostimulants: definition, concept, main categories and regulation. (Special issue: Biostimulants in horticulture.) *Scientia Horticulturae* 196, 3–14. DOI: 10.1016/j.scienta.2015.09.021.

Dumitru, I., Persu, D.I., Cristea, C., Iuga, M., Stroescu, D. *et al.* (2019) Study on modern propulsion solutions of automatic agricultural equipment. In: *Proceedings of the Sixth International Conference on 'Research People and Actual Tasks on Multidisciplinary Sciences', Lozenec, Bulgaria, 12–15 June 2019.* Bulgarian National Multidisciplinary Scientific Network of the Professional Society for Research Work, Lozenec, Bulgaria, pp. 173–178 (Vol. 1).

Eberhardt Neto, E., Benett, K.S.S., Benett, C.G.S., dos Santos, E. da C.M., Reboucas, T.N.H. *et al.* (2016) Plant spacing and boron (B) topdressing fertilisation for purple cabbage crop (*Brassica oleracea* var. *capitata*) variety purple giant. *Australian Journal of Crop Science* 10(11), 1529–1533. DOI: 10.21475/ajcs.2016.10.11.PNE67.

Elliott, J.G. (1975) Developments in direct drilling in the United Kingdom. In: *Proceedings of the 12th British Weed Control Conference, Brighton, 1974,* British Crop Protect Council, pp. 1041–1049.

Ensminger, A.H., Ensminger, M.E., Konlande, J.E. and Robson, J.R.K. (1995) *The Concise Encyclopedia of Food.* CRC Press, Boca Raton, FL. DOI: 10.1201/9781420048186.

Evans, R.G., Iversen, W.M. and Kim, Y. (2012) Integrated decision support, sensor networks, and adaptive control for wireless site-specific sprinkler irrigation. (Special issue: Advances in irrigation.) *Applied Engineering in Agriculture* 28(3), 377–387. DOI: 10.13031/2013.41480.

Everaarts, A.P. (1993) Strategies to improve the efficiency of nitrogen fertilizer use in the cultivation of Brassica vegetables. *Acta Horticulturae* 339, 161–173. DOI: 10.17660/ActaHortic.1993.339.14.

Everaarts, A.P. (2000) Nitrogen balance during growth of cauliflower. *Scientia Horticulturae* 83(3–4), 173–186. DOI: 10.1016/S0304-4238(99)00087-4.

Everaarts, A.P. and de Moel, C.P. (1995) The effect of nitrogen and the method of application on the yield of cauliflower. *Netherlands Journal of Agricultural Science* 43(4), 409–418. DOI: 10.18174/njas.v43i4.563.

Everaarts, A.P. and de Willigen, P. (1999) The effect of nitrogen and the method of application on yield and quality of broccoli. *Netherlands Journal of Agricultural Science* 47(2), 123–133. DOI: 10.18174/njas.v47i2.471.

Everaarts, A.P. and van Beusichem, M.L. (1998) The effect of planting date and plant density on nitrogen uptake and nitrogen harvest by Brussels sprouts. *The Journal of Horticultural Science and Biotechnology* 73(5), 704–710. DOI: 10.1080/14620316.1998.11511037.

Fisher, N.M. and Milbourn, G.M. (1974) The effect of plant density, date of apical bud removal and leaf removal on the growth and yield of single-harvest Brussels sprouts (*Brassica oleracea* var. *gemmifera* D.C.): I. Whole plant and axillary bud growth. *The Journal of Agricultural Science* 83(3), 479–487. DOI: 10.1017/S0021859600026964.

Gardner, B.R. and Roth, R.L. (1989) Midrib nitrate concentration as a means for determining nitrogen needs of cabbage. *Journal of Plant Nutrition* 12(9), 1073–1088. DOI: 10.1080/01904168909364023.

Giordano, M., El-Nakhel, C., Caruso, G., Cozzolino, E., de Pascale, S. *et al.* (2020) Stand-alone and combinatorial effects of plant-based biostimulants on the production and leaf quality of perennial wall rocket. *Plants* 9(7), 922. DOI: 10.3390/plants9070922.

Giordano, M., Petropoulos, S.A. and Rouphael, Y. (2021) Response and defence mechanisms of vegetable crops against drought, heat and salinity stress. *Agriculture* 11(5), 463. DOI: 10.3390/agriculture11050463.

Gomez-Gil, J., Alonso-Garcia, S., Gómez-Gil, F.J. and Stombaugh, T. (2011) A simple method to improve autonomous GPS positioning for tractors. *Sensors (Basel, Switzerland)* 11(6), 5630–5644. DOI: 10.3390/s110605630.

Goto, T. and Yoshida, Y. (2003) Relationships between plant size and amount of overhead irrigation required for uniform water supply to individual cell medium. *Scientific Reports of the Faculty of Agriculture, Okayama University* 92, 27–30.

Greenwood, D.J. (1990) Production or productivity: the nitrate problem? *Annals of Applied Biology* 117(1), 209–231. DOI: 10.1111/j.1744-7348.1990.tb04208.x.

Greenwood, D.J., Cleaver, T.J., Turner, M.K., Hunt, J., Niendorf, K.B. *et al.* (1980) Comparison of the effects of nitrogen fertilizer on the yield, nitrogen content and quality of 21 different vegetable and agricultural crops. *The Journal of Agricultural Science* 95(2), 471–485. DOI: 10.1017/S0021859600039514.

Greenwood, D.J., Neeteson, J.J. and Draycott, A. (1986) Quantitative relationships for the dependence of growth rate of arable crops on their nitrogen content, dry weight and aerial environment. *Plant and Soil* 91(3), 281–301. DOI: 10.1007/BF02198111.

Greenwood, D.J., Rahn, C., Draycott, A., Vaidyanathan, L.V. and Paterson, C. (1996) Modelling and measurement of the effects of fertilizer-N and crop residue incorporation on N-dynamics in vegetable cropping. *Soil Use and Management* 12(1), 13–24. DOI: 10.1111/j.1475-2743.1996.tb00525.x.

Grieve, C.M., Shannon, M.C. and Poss, J.A. (2001) Mineral nutrition of leafy vegetable crops irrigated with saline drainage water. *Journal of Vegetable Crop Production* 7(1), 37–47. DOI: 10.1300/J068v07n01_06.

Grieve, B.D., Duckett, T., Collison, M., Boyd, L., West, J. *et al.* (2019) The challenges posed by global broadacre crops in delivering smart agri-robotic solutions: a fundamental rethink is required. *Global Food Security* 23, 116–124. DOI: 10.1016/j.gfs.2019.04.011.

Grime, J.P. (1979) *Plant Strategies and Vegetative Processes.* John Wiley, Chichester, UK.

Gunston, J. (1947) *Farming To-day and To-morrow.* Methuen & Co., London.

Hara, T. and Sonoda, Y. (1981) The role of macronutrients in cabbage-head formation. IV. Effect of nitrogen supply or light intensity on the growth and $^{14}CO_2$ and $^{15}NO_3$-N assimilation of cabbage plants. *Soil Science and Plant Nutrition* 27(2), 185–194. DOI: 10.1080/00380768.1981.10431270.

Heuts, R.F., Schrevens, E., Vansteenkiste, J. and Diels, J. (2016) Model-based life cycle assessment of nitrogen fertilization in a cauliflower-leek rotation system. *Acta Horticulturae* 1112, 403–410. DOI: 10.17660/ActaHortic.2016.1112.54.

Hochmuth, G.J. (1992) Concepts and practices for improving nitrogen management for vegetables. *HortTechnology* 2(1), 121–125. DOI: 10.21273/HORTTECH.2.1.121.

Hoy, R.M. and Kocher, M.F. (2020) The Nebraska Tractor Test Laboratory: 100 years of service. *ASABE Distinguished Lecture Series* 41, 1–14.

Janick, J. (2002) Ancient Egyptian agriculture and the origins of horticulture. *Acta Horticulturae* 583, 23–39. DOI: 10.17660/ActaHortic.2002.582.1.

Jin, J. (2012) Changes in the efficiency of fertiliser use in China. *Journal of the Science of Food and Agriculture* 92(5), 1006–1009. DOI: 10.1002/jsfa.4700.

Johnston, W.R., Westcot, D.W. and Delmore, M. (2011) San Joaquin Valley, California: a case study. In: Wallender, W.W. and Tanji, K.K. (eds) *Agricultural Salinity*

Assessment and Management (ASCE Manual and Report on Engineering Practice No. 71), 2nd edn. American Society of Civil Engineers, Reston, VA, pp. 977–1031. DOI: 10.1061/9780784411698.

Kage, H., Alt, C. and Stutzel, H. (2001) Predicting dry matter production of cauliflower (*Brassica oleracea* L. *botrytis*) under unstressed conditions: I. Photosynthetic parameters of cauliflower leaves and their implications for calculations of dry matter production. *Scientia Horticulturae* 87(3), 155–170. DOI: 10.1016/S0304-4238(00)00177-1.

Kaniszewski, S. and Rumpel, J. (1998) Effects of irrigation, nitrogen fertilization and soil type on yield and quality of cauliflower. *Journal of Vegetable Crop Production* 4(1), 67–75. DOI: 10.1300/J068v04n01_08.

Karlen, D.L., Ditzler, C.A. and Andrews, S.S. (2003) Soil quality: why and how? *Geoderma* 114(3–4), 145–156. DOI: 10.1016/S0016-7061(03)00039-9.

Kibblewhite, M.G., Ritz, K. and Souft, M. (2008) Soil health in agricultural systems. *Proceedings of the Royal Society B: Biological Sciences* 363(1492), 685–701. DOI: 10.1098/rstb.2007.2178.

Kieffer, M., Fuller, M.P. and Jellings, A.J. (1996) Mathematical model of cauliflower curd architecture based on biometrical analysis. *Acta Horticulturae* 407, 361–368. DOI: 10.17660/ActaHortic.1996.407.46.

Knox, J., Kay, M. and Weathered, K. (2020a) *Switching Irrigation Technologies*. UK Irrigation Association, Rushden, UK. Available at: https://www.ukia.org/resources-booklets/ (accessed 10 July 2023).

Knox, J.W., Kay, M.G., Hess, T.M. and Holman, I.P. (2020b) The challenges of developing an irrigation strategy for UK agriculture and horticulture in 2020: industry and research priorities. *CABI Reviews: Perspectives in Agriculture, Veterinary Science, Nutrition and Natural Resources* 15, 050. DOI: 10.1079/PAVSNNR202015050.

Lawrence, H.G. and Yule, I.J. (2007) Estimation of the in-field variation in fertiliser application. *New Zealand Journal of Agricultural Research* 50(1), 25–32. DOI: 10.1080/00288230709510279.

Li, B., Suzuki, J.-I. and Hara, T. (1999) Competitive ability of two *Brassica* varieties in relation to biomass allocation and morphological plasticity under varying nutrient availability. *Ecological Research* 14(3), 255–266. DOI: 10.1046/j.1440-1703.1999.143298.x.

Li, H. (2012) Nitrogen leaf-stem source and head sink relations in three cauliflower hybrid cultivars. *Acta Horticulturae* 938, 219–226. DOI: 10.17660/ActaHortic.2012.938.28.

Li, H.Y., Si, D.X. and Lv, F.T. (2015) Differential responses of six Chinese cabbage (*Brassica rapa* L. ssp. *pekinensis*) cultivars to potassium ion deficiency. *The Journal of Horticultural Science and Biotechnology* 90(5), 483–488. DOI: 10.1080/14620316.2015.11668704.

Li, T., Chen, X.-G., Xu, W.-H., Chi, S.-L., Zhao, W.-Y. *et al.* (2018) Effects of coated slow-release fertilizer with urease and nitrification inhibitors on nitrogen release characteristic and uptake and utilization of nitrogen, phosphorus and potassium in cabbage. *International Journal of Agriculture and Biology* 20(2), 422–430. DOI: 10.17957/IJAB/15.0552.

Lobsey, C., Viscarra Rossel, R. and McBratney, A. (2010) An automated system for rapid in-field soil nutrient testing. In: *Proceedings of the 19th World Congress of Soil Science; Soil Solutions for a Changing World, 1–6 August 2010.* International Union of Soil Sciences. Published on DVD.

Marìn-Gonzãlez, O., Kuang, B., Quraishi, M.Z., Munóz-García, M.Á. and Mouazen, A.M. (2013) On-line measurement of soil properties without direct spectral response in near infrared spectral range. *Soil and Tillage Research* 132, 21–29. DOI: 10.1016/j. still.2013.04.004.

Marschner, H. (1995) *Mineral Nutrition in Higher Plants*. Academic Press, London.

Marschner, P., Solaiman, Z. and Rengel, Z. (2007) Brassica genotypes differ in growth, phosphorus uptake and rhizosphere properties under P-limiting conditions. *Soil Biology and Biochemistry* 39(1), 87–98. DOI: 10.1016/j.soilbio.2006.06.014.

Mathur, A. (2019) Crops, drops and climate challenge: using energy efficiency to configure the perfect sustainability storm. In: *Proceedings of the Crawford Fund 2019 Annual Conference, Weathering the Perfect Storm: Addressing the Agriculture, Energy, Water, Climate Change Nexus, 12–13 August 2019*. Canberra, Australia, pp. 55–61.

McGrady, J. (1996) Transplant nutrient conditioning improves cauliflower early growth. *Journal of Vegetable Crop Production* 2(2), 39–49. DOI: 10.1300/J068v02n02_05.

Melian-Navarro, A., Molina-Martinez, J.M. and Ruiz-Canales, A. (2010) Automation and remote irrigation, tools for saving water: application to the cultivation of broccoli in the southeastern Spain. *Agricola Vergel: Fruticultura, Horticultura, Floricultura, Citricultura* 29(341), 299–306.

Mesmin, X., Cortesero, A.-M., Daniel, L., Plantegenest, M., Faloya, V. *et al.* (2020) Influence of soil tillage on natural regulation of the cabbage root fly (*Delia radicum*) in brassicaceous crops. *Agriculture, Ecosystems & Environment* 293, 106834. DOI: 10.1016/j.agee.2020.106834.

Monaghan, J.M., Rahn, C.R., Hilton, H.W. and Wood, M. (2010) Improved efficiency of nutrient and water use for high quality field vegetable production using fertigation. *Acta Horticulturae* 852, 145–152. DOI: 10.17660/ActaHortic.2010.852.16.

Neuvel, J.J. (1990) *Nitrogen fertilisation of brussels sprouts*. Report 102. [Dutch] Research Station for Arable Farming and Field Production of Vegetables, Lelystad, Netherlands.

Nilsson, T. (1980) The influence of soil type and irrigation on yield, quality and chemical composition of cauliflower. *Swedish Journal of Agricultural Research* 10, 65–75.

Parish, R.L. (2000) Stand of cabbage and broccoli in single- and double-drill plantings on beds subject to erosion. *Journal of Vegetable Crop Production* 6(2), 87–96. DOI: 10.1300/J068v06n02_10.

Pexa, M., Cindr, M., Kubin, K. and Jurca, V. (2011) Measurements of tractor power parameters using GPS. *Research in Agricultural Engineering* 57(1), 1–7. DOI: 10.17221/18/2010-RAE.

Pinto, R., Brito, L.M., Mourao, I. and Coutinho, J. (2020) Nitrogen balance in organic horticultural rotations. *Acta Horticulturae* 1286, 127–134. DOI: 10.17660/ActaHortic.2020.1286.18.

Pongrac, P., McNicol, J.W., Lilly, A., Thompson, J.A., Wright, G. *et al.* (2019) Mineral element composition of cabbage as affected by soil type and phosphorus and zinc fertilisation. *Plant and Soil* 434(1–2), 151–165.

Potopova, V., Zahradnicek, P., Stepanek, P., Türkott, L., Farda, A. *et al.* (2017) The impacts of key adverse weather events on the field-grown vegetable yield variability in the Czech Republic from 1961 to 2014. *International Journal of Climatology* 37(3), 1648–1664. DOI: 10.1002/joc.4807.

Pratt, P.F. (1984) Nitrogen use and nitrate leaching in irrigated agriculture. In: Hauck, R.D. (ed.) *Nitrogen in Crop Production*. American Society of Agronomists, Madison, WI. DOI: 10.2134/1990.nitrogenincropproduction.c21.

Privette, C.V., Khalilian, A., Torres, O. and Katzberg, S. (2011) Utilizing space-based GPS technology to determine hydrological properties of soils. *Remote Sensing of Environment* 115(12), 3582–3586. DOI: 10.1016/j.rse.2011.08.019.

Quraishi, M.Z. and Mouazen, A.M. (2013) Calibration of an on-line sensor for measurement of topsoil bulk density in all soil textures. *Soil and Tillage Research* 126, 219–228. DOI: 10.1016/j.still.2012.08.005.

Rahn, C.R. (2002) Management strategies to reduce nutrient losses from vegetable crops. *Acta Horticulturae* 571, 19–29. DOI: 10.17660/ActaHortic.2002.571.1.

Rahn, C., De Neve, S., Bath, B., Bianco, V., Dachler, M. *et al.* (2001) A comparison of fertiliser recommendation systems for cauliflowers in Europe. *Acta Horticulturae* 563, 39–45. DOI: 10.17660/ActaHortic.2001.563.3.

Rahn, C.R., Bending, G.D., Lillywhite, R.D. and Turner, M.K. (2003) Novel techniques to reduce environmental N pollution from high nitrogen content crop residues. *Acta Horticulturae* 627, 105–111. DOI: 10.17660/ActaHortic.2003.627.12.

Raimi, A., Roopnarain, A., Chirima, G.J. and Adeleke, R. (2020) Insights into the microbial composition and potential efficiency of selected commercial biofertilisers. *Heliyon* 6(7), e04342. DOI: 10.1016/j.heliyon.2020.e04342.

Ram, H., Dey, S.S., Krishnan, S.G., Kar, A., Bhardwaj, R. *et al.* (2018) Heterosis and combining ability for mineral nutrients in snowball cauliflower (*Brassica oleracea* var. *botrytis* L.) using *Ogura* cytoplasmic male sterile lines. *Proceedings of the National Academy of Sciences India. Section B, Biological Science* 88(4), 1367–1376. DOI: 10.1007/s40011-017-0874-8.

Rather, K., Schenk, M.K., Everaarts, A.P. and Vethman, S. (1999) Response of yield and quality of cauliflower varieties (*Brassica oleracea* var. *botrytis*) to nitrogen supply. *The Journal of Horticultural Science and Biotechnology* 74(5), 658–664. DOI: 10.1080/14620316.1999.11511169.

Robinson, D. (1991) Strategies for optimising growth in response to nutrient supply. In: Porter, J.R. and Lawlor, D.W. (eds) *Plant Growth: Interactions with Nutrition and Environment. Society for Experimental Biology Seminar Series, Vol. 43*. Cambridge University Press, Cambridge, pp. 177–205.

Rubatzky, V.E. and Yamaguchi, M. (1997) *World Vegetables: Principles, Production, and Nutritive Values*. Chapman & Hall, New York. DOI: 10.1007/978-1-4615-6015-9.

Salo, T. (1999) Effects of band placement and nitrogen rate on dry matter accumulation, yield and nitrogen uptake of cabbage, carrot and onion. *Agricultural and Food Science* 8(2), 157–232. DOI: 10.23986/afsci.5624.

Salter, P.J. (1961) The irrigation of early summer cauliflower in relation to stage of growth, plant spacing and nitrogen level. *Journal of Horticultural Science and Biotechnology* 36(4), 241–253. DOI: 10.1080/00221589.1961.11514019.

Salter, P.J. and James, J.M. (1975) The effect of plant density on the initiation, growth and maturity of curds of two cauliflower varieties. *Journal of Horticultural Science and Biotechnology* 50(3), 239–248. DOI: 10.1080/00221589.1975.11514629.

Salter, P.J., Andrews, D.J. and Akehurst, J.M. (1984) The effects of plant density, spatial arrangement and sowing date on yield and head characteristics of a new form of broccoli. *Journal of Horticultural Science and Biotechnology* 59(1), 79–85. DOI: 10.1080/00221589.1984.11515172.

Sartori, L. and Bertocco, M. (2005) Criteria for choosing variable distribution spreaders of fertilizer. [Italian] *Informatore Agrario* 61(1), 33–38.

Scaife, A. (1988) Derivation of critical nutrient concentrations for growth rate from data from field experiments. *Plant and Soil* 109(2), 159–169. DOI: 10.1007/BF02202080.

Schumann, A.W. (2010) Precise placement and variable rate fertilizer application technologies for horticultural crops. *HortTechnology* 20(1), 34–40. DOI: 10.21273/HORTTECH.20.1.34.

Schwarz, A., Pfenning, J., Bischoff, W.-A. and Liebig, H.-P. (2010) Effects of N fertilization strategy and fixed ploughing date on nitrate leaching on field vegetable cultivation. *Acta Horticulturae* 852, 115–122. DOI: 10.17660/ActaHortic.2010.852.12.

Secrett, F.A. (1935) Irrigation of vegetable crops. *Scientific Horticulture* 3, 82–96.

Shahrajabian, M.H., Chaski, C., Polyzos, N. and Petropoulos, S.A. (2021) Biostimulants application: a low input cropping management tool for sustainable farming of vegetables. *Biomolecules* 11(5), 698. DOI: 10.3390/biom11050698.

Shaw, S.A. and Gibbs, J. (1996) The role of marketing channels in the determination of horizontal market structure: the case of fruit and vegetable marketing by British growers. *The International Review of Retail, Distribution and Consumer Research* 6(3), 281–300. DOI: 10.1080/09593969600000025.

Siemonsma, J.S. and Piluek, K. (eds) (1993) *Plant Resources of South-East Asia No. 8 Vegetables*. Pudoc, Wageningen, Netherlands.

Singh, R.V. and Naik, L.B. (1991) Response of cauliflower (cv. Early Kunwari) to plant density, nitrogen and phosphorus level. *Progressive Horticulture* 23(1–4), 51–54.

Singh, B.K., Sharma, S.R. and Singh, B. (2009) Heterosis for mineral elements in single cross-hybrids of cabbage (*Brassica oleracea* var. *capitata* L.). *Scientia Horticulturae* 122(1), 32–36. DOI: 10.1016/j.scienta.2009.04.007.

Siomos, A.S. (1999) Planting date and within-row plant spacing effects on pak choi yield and quality characteristics. *Journal of Vegetation Crop Production* 4(2), 65–73.

Smith, A. (1776, 1976) *An Inquiry into the Nature and Causes of the Wealth of Nations*. Clarendon Press, Oxford. DOI: 10.1093/actrade/9780199269563.book.1.

Smittle, D.A., Dickens, W.L. and Stansell, J.R. (1994) Irrigation regimes affect cabbage water use and yield. *Journal of the American Society for Horticultural Science* 119(1), 20–23. DOI: 10.21273/JASHS.119.1.20.

Sofi, T.A., Tewari, A.K., Razdan, V.K. and Koul, V.K. (2014) Long term effect of soil solarization on soil properties and cauliflower vigor. *Phytoparasitica* 42(1), 1–11. DOI: 10.1007/s12600-013-0331-z.

Sorensen, J.N. (1988) Assessment of the nutritional status of white cabbage. In: Hansen, H. (ed.) *Proceedings 7th International Colloquium for the Optimisation of Plant Nutrition*. Nyborg, Denmark, paper 480.

Sorensen, J.N. (2000) Ontogenetic changes in macro nutrient composition of leaf-vegetable crops in relation to plant nitrogen status: a review. *Journal of Vegetable Crop Production* 6(1), 75–96. DOI: 10.1300/J068v06n01_08.

Sorensen, L. and Grevsen, K. (1994) Effects of plant spacing on uniformity in broccoli for once-over harvesting. *HortScience* 59(3), 102–105.

Sorensen, J.N. and Thorup-Kristensen, K. (2011) Plant-based fertilizers for organic vegetable production. *Journal of Plant Nutrition and Soil Science* 174(2), 321–332. DOI: 10.1002/jpln.200900321.

Stirling, K. and Lancaster, R. (2005) Alternative planting configurations influence cauliflower development. *Acta Horticulturae* 694, 301–305. DOI: 10.17660/ActaHortic.2005.694.49.

Stone, D.A. (1998) The effects of "starter" fertilizer injection on the growth and yield of drilled vegetable crops in relation to soil nutrient status. *The Journal of Horticultural Science and Biotechnology* 73(4), 441–451. DOI: 10.1080/14620316.1998.11510997.

Sung, J.-K., Yun, H., Back, S., Fernie, A.R., Kim, Y.X. *et al.* (2018) Changes in mineral nutrient concentrations and C-N metabolism in cabbage shoots and roots following macronutrient deficiency. *Journal of Plant Nutrition and Soil Science* 181(5), 777–786. DOI: 10.1002/jpln.201800001.

Sutherland, R.A., Crisp, P. and Angell, S.M. (1989) The effect of spatial arrangement on the yield and quality of two cultivars of autumn cauliflower and their mixture. *Journal of Horticultural Science and Biotechnology* 64(1), 35–40. DOI: 10.1080/14620316.1989.11515924.

Teklic, T., Paradikovic, N., Spoljarevic, M., Zeljkovic, S., Loncaric, Z. *et al.* (2020) Linking abiotic stress, plant metabolites, biostimulants and functional food. *Annals of Applied Biology* 178(2), 169–191. DOI: 10.1111/aab.12651.

Tempesta, M., Gianquinto, G., Hauser, M. and Tagliavini, M. (2019) Optimization of nitrogen nutrition of cauliflower intercropped with clover and in rotation with lettuce. *Scientia Horticulturae* 246, 734–740. DOI: 10.1016/j.scienta.2018.11.020.

Terrones, C., Gelencser, T., Hauenstein, S., Sommer, L., Schwitter, P. *et al.* (2020) Testing polyhalite as a tool to overcome nutrient deficiencies in organic cabbage culture. *Electronic International Fertilizer Correspondent* 60, 23–30.

Tilman, D. (1988) *Plant Strategies and the Dynamics and Structure of Plant Communities. Monographs in Population Biology 26.* Princeton University Press, Princeton, NJ. DOI: 10.1515/9780691209593.

Trapnell, T. (2021) pH and nutrient availability of your nitrogen. *Mole Valley Newsletter* 680(January), 18.

Vågen, I.M. (2003) Nitrogen uptake in a broccoli crop. 1: nitrogen dynamics on a relative time scale. *Acta Horticulturae* 627, 195–202. DOI: 10.17660/ActaHortic.2003.627.25.

Vågen, I.M., Skjelvåg, A.O. and Bonesmo, H. (2004) Growth analysis of broccoli in relation to fertilizer nitrogen application. *The Journal of Horticultural Science and Biotechnology* 79(3), 484–492. DOI: 10.1080/14620316.2004.11511794.

van Bueren, E.T.L. and Struik, P.C. (2017) Diverse concepts of breeding for nitrogen use efficiency: a review. *Agronomy for Sustainable Development* 37(5), 50. DOI: 10.1007/s13593-017-0457-3.

Vera, J., Conejero, W., Mira-García, A.B., Conesa, M.R. and Ruiz-Sanchez, M.C. (2021) Towards irrigation automation based on dielectric soil sensors. *The Journal of Horticultural Science and Biotechnology* 96(6), 696–707. DOI: 10.1080/14620316.2021.1906761.

Vieweger, A., Capron, C. and Measures, M. (2016) Soil health assessment in UK Horticultural Systems – literature review (WP1). In: *AHDB CP 107b GREAT Soils Project.* Agriculture and Horticulture Development Board (AHDB), Stoneleigh, UK.

Watanabe, K., Niino, T., Murayama, T. and Nanzyo, M. (2011) Effects of the pre-transplanting phosphorus application on dry matter production, photosynthesis, root activities and nutrient absorption of cabbage and maize at initial growth stage. *Japanese Journal of Crop Science* 80(4), 391–402. DOI: 10.1626/jcs.80.391.

Watson-Jones, S., Maxted, N. and Ford-Lloyd, B. (2005) Genetic variation in wild Brassicas in England and Wales. *Crop Wild Relative* 3, 20–24.

Whitmore, A.P. and Groot, J.J.R. (1994) The mineralization of N from finely or coarsely chopped crop residues: measurements and modelling. *European Journal of Agronomy* 3(4), 367–373. DOI: 10.1016/S1161-0301(14)80168-8.

Wiebe, H.J. (1981) Influence of soil water potential during different growth periods on yield of cauliflower. *Acta Horticulturae* 119, 299–300. DOI: 10.17660/ActaHortic.1981.119.29.

Wilcockson, S.J. and Abuzeid, A.E. (1991) Growth of axillary buds of brussels sprouts (*Brassica oleracea* var. *bullata* sub var. *gemmifera*). *The Journal of Agricultural Science* 117(2), 207–212. DOI: 10.1017/S0021859600065291.

Xu, D.B., Si, G.H., Peng, C.L., Xu, X.Y., Xiong, Y.S. *et al.* (2019a) Effects of nitrogen in organic fertilizer instead of chemical fertilizer on the yield and quality of leaf vegetables and soil nitrogen residues. *Acta Horticulturae* 1253, 457–464. DOI: 10.17660/ActaHortic.2019.1253.60.

Xu, M., Zhou, J. and Yang, S.C. (2019b) Control characteristics of EPS system of electric tractor. *International Agricultural Engineering Journal* 28(4).

Yuan, L., Loque, D., Ye, F., Frommer, W.B. and von Wiren, N. (2007) Nitrogen-dependent posttranscriptional regulation of the ammonium transporter *AtAMT1;1*. *Plant Physiology* 143(2), 732–744. DOI: 10.1104/pp.106.093237.

Competitive Ecology and Sustainable Production

6

Geoffrey R. Dixon*

School of Agriculture, Policy and Development, Earley Gate, Whiteknights Road, PO Box 237, University of Reading, Reading, Berkshire RG6 6EU and GreenGene International, Hill Rising, Horsecastles Lane, Sherborne, Dorset DT9 6BH, UK

Abstract

Weed competition is a critical factor determining the yield and quality of brassicas. The ecology of weeds and their control is examined in this chapter. This involves defining weeds, their competitive abilities, growth and critical periods when crop productivity is reduced. Important factors are competition for light, space, soil nutrients and harbouring pests and pathogens which then invade into the crop canopies.

Integrated crop management is the sustainable strategy for weed management. The science and technology underpinning the use of tillage, non-tillage, intercropping and cover cropping, overseeding, crop covers, biostimulants and biofumigation for weed control are discussed. Fewer synthetic herbicides are available or permissible for brassica vegetables. Their precision application involving weed sensing and discrimination from crop plants is considered. Physical weeding using smart tillage guided by the Global Positioning System is examined. Weed contamination of seed crops poses critical problems for the delivery of pure samples of cultivars.

*geoffrdixon@gmail.com

© Geoffrey R. Dixon and Rachel Wells 2024. *Vegetable Brassicas and Related Crucifers,* 2nd edition. (G.R. Dixon and R. Wells)
DOI: 10.1079/9781789249170.0006

240

INTRODUCTION

Control of crop competitors and parasites prior to the 1940s mainly involved using inorganic active ingredients that were particularly targeted against fungal pathogens that caused severe damage in vegetable brassica crops (Russell, 2005) and applied by manually operated compression sprayers. Weed control relied on combinations of very basic, often household, chemicals and physical methods such as hand-hoeing and soil inversion. From the 1960s to early 1990s pre- and post-emergence and selective herbicides eliminated much manual weed control and synthetic organic pesticides controlled previously devasting invasions by pests and pathogens. One positive result of this agrochemical revolution was that consumers were supplied with copious, blemish-free produce at low prices. As a matter of routine pesticides were applied as 'insurance' measures, at short intervals, on the basis of 'spray and kill' schedules even where there was no evidence of pest or pathogen presence. That strategy is no longer acceptable – it wasted money, chemicals, damaged the environment, accumulated spray residues in wildlife and encouraged the emergence of chemically tolerant strains of pests and pathogens.

In the mid-20th century, weed control agronomists sought the most effective herbicides which would either eliminate or reduce weed competition with their crops. This strategy has changed very radically over the past 30 years, influenced by environmental, social and commercial considerations. Increasingly, stringent regulatory systems now control the discovery, testing and marketing of new herbicides and the retention or discarding of products in current use. Weed control or elimination is, however, vital for profitable brassica production. As a result, growers are using mechanical means of control and now these may also be robotically controlled. Additionally, combinations of robotics and artificially intelligent discrimination between weeds and crop plants are permitting weed elimination either mechanically or by very precisely targeted use of herbicides in minute volumes.

Strategies of 'integrated crop management' (ICM) linking all aspects of crop husbandry aim at producing uniformly maturing plants unaffected by weed competition. Specifically, ICM determines the costs and benefits of whether or not weed competition is impairing crop productivity before eradicating them. If the answer to that question is 'Yes' then consideration moves to the opportunities for cultural and husbandry weed-controlling methods in preference to using chemical herbicides. The availability, or otherwise, of herbicides suitable for brassica crops is still, however, a factor in this decision making process. Discovery, testing, registering, manufacturing and marketing of new products is very expensive. There are fewer and larger manufacturers of agrochemicals and these companies recognize that the amounts of active-ingredient chemicals required for vegetable crops are insufficiently profitable. Interest in the registration of effective active ingredients and their availability is, therefore, reliant in many instances on forms of 'off-label' schemes. These are operated under collaborative arrangements between manufacturers, grower–producer organizations and registration authorities. Useability is very crop, season and application specific and changes frequently. Consequently, discussion of herbicide availability for particular brassicas and other crucifer vegetables is, therefore, now limited in this chapter. Weed competition in brassicas is treated from an ecological (or 'sustainable') perspective which promotes environmental health and the mitigation of global warming.

WHAT ARE 'WEEDS'?

The ecologist defines weeds as: 'plants growing entirely or predominantly in situations (that are) disturbed by man without being deliberately cultivated' (Baker and Stebbins, 1965). This definition includes all plant types, not solely the flower-forming angiosperms, although these compose the majority of weed taxa. The world weed flora mainly contains plants from relatively few, highly advanced families (see Table 6.1) (Hill, 1977) which thrive in the fertile and nutrient-rich soils developed for crop production. Such plants are opportunistic and have evolved to exploit the crop environments created by humans and differing substantially from natural ecosystems; in other words, they are 'weeds of cultivation'.

THE EFFECTS OF WEEDS COMPETITION

The agronomist defines weeds as 'plants which interfere adversely with the production aims of the grower' (Spitters, 1990). Weeds affect crops by:

- reducing crop growth and yield, mainly due to competition that limits resources such as: light, water and nutrients;
- reducing the financial value of the harvested product, either by contaminating the produce or by diminishing its quality; and
- hampering husbandry practices, especially harvesting operations thereby increasing costs (see Fig. 6.1).

Table 6.1. List of angiosperm families whose species make up 60% of the 700 weeds introduced into eastern North America. (After Hill, 1977).

Family	Number of species
Compositae	122
Gramineae	65
Cruciferae	62
Labiatae	60
Leguminosae	54
Caryophyllaceae	37
Scrophulariaceae	30

Fig. 6.1. Radish (*Raphanus sativus*) crop infested with couch grass (*Elytrigia repens*). (Geoff Dixon).

Weeds reduce productivity either by directly lowering output (kg yield per ha × price per kg), or indirectly by either reducing the market price achieved through the diminution of quality attributes or by increasing husbandry and harvesting costs. Each of these penalties is imposed on brassica crops by weeds. Increasingly, their effects on crop quality are the primary consideration. These effects and their interactions are summarized by Table 6.2.

Factors that interact influencing the degree of mutual interference (competition) between weeds and crops are:

- *Crop factors:* type of crop, time of sowing and/or transplanting, time of thinning, and population density which is are a function of: seed rate, germination and thinning.
- *Weed factors:* species of weed(s), time of germination, start of growth and population density controlled by: seed population in soil, percentage germination and seed importation from other areas.

The level of competition resulting from these crop and weeds factors will be modified further by: weather conditions, soil type, soil fertility and cultivation practices. These effects have been quantified by Qasem (2009) in relation to the growth and yield of cauliflower (*Brassica oleracea* var. *botrytis* cv. 'White Cloud'). Periods of weed and cauliflower competition greatly reduced crop growth and head yield. Average reductions in shoot dry weight and head yield were 81% and 89%, respectively. Weed competition for 14 days after transplanting reduced cauliflower average head yield by 41% and the critical

Table 6.2. The effects of competition by an aggressor (A) species on a suppressed (B) species. (AfterDonald, 1958).

Effects	Competition for light only	Competition for nutrients only	Competition for both light and nutrients
Direct	(a) Intrusion of A into the light environment of B: reduced light supply for B	(c) Intrusion of A into the nutrient supply of B: reduced nutrient supply for B	B suffers: (a) reduced light supply (b) reduced nutrient supply
Indirect	(b) As a result of reduced light supply: B has reduced capacity to exploit its own nutrient supply	(d) As a result of reduced nutrient supply: B has reduced capacity to exploit its own light supply	(c) reduced capacity to exploit the nutrient supply (d) reduced capacity to exploit the light supply
Interactions	Interaction of (a) and (b)	Interactions of (c) and (d)	Interactions of ab, ac, bc, bd and cd plus any higher-order interactions

period of weed competition occurred at 0–38 days after transplanting which corresponded with the rapid increase in weed biomass. Consequently, early weed removal is necessary if yield loss is to be prevented.

IDEAL CHARACTERISTICS FOR A SUCCESSFUL WEED

The biological attributes favouring success as a weed are summarized in Table 6.3.

An open growth habit combined with the early induction of flowering means that small plants reproduce rapidly by seed and continue growing as further flowering takes place. This ensures continuing outputs of weed seeds. Plants growing in disturbed (cultivated) habitats tend to allocate greater proportions of their photosynthetic output to seed production than the more stable types as demonstrated in Table 6.4.

The competitive ability of weeds results from a combination of high reproductive potential and rapid vegetative spread (Fogg, 1975). This embraces physiological (efficient nutrient-ion uptake, rapid root growth) and morphological factors (ability to climb or scramble over competitors for light; production of rosettes which spread out close to the ground and smother competitors; large and vigorous habit). Brassica crops, especially where they are transplanted, may even improve conditions for the establishment of weeds after their germination by offering protection from extremes of temperature and the drying effects of wind. This provides a competitive advantage for some weed species by promoting an early start to their growth.

Table 6.3. Characteristics of a successful weed. (Composed from the text of Salisbury, 1961).

1.	Germination requirements fulfilled by many environments
2.	Discontinuous germination (internally controlled) and great longevity of seed
3.	Rapid growth through the vegetative to the flowering phase
4.	Continuous seed production during the entire growing season
5.	Self-compatible breeding system but not completely autogamous or apomictic
6.	Cross-pollinated by unspecialized visitors or wind
7.	Very high seed output in favourable environmental circumstances
8.	Produces at least modest seed yield in a wide range of adverse environments
9.	Adapted for short- and long-distance seed dispersal
10.	Where the life cycle is perennial, the plant has vigorous vegetative reproduction or regeneration from stem or root fragments
11.	Perennials have brittle stems or roots that are not easily drawn out of the ground
12.	Abilities to compete with other organisms by specialized means such as rosette habit, choking growth or the production of allelochemicals

Table 6.4. Average output of seeds or fruits for nine common weeds that compete with crops. (After Salisbury, 1961).

Weed species	Average output per plant	Seed (S) or fruit (F)
Groundsel (*Senecio vulgaris*)	1000–1200	F
Chickweed (*Stellaria media*)	2200–2700	S
Shepherd's purse (*Capsella bursa-pastoris*)	3500–4000	S
Greater plantain (*Plantago major*)[a]	13,000–15,000	S
Common poppy (*Papaver rhoeas*)	14,000–19,500	S
Prickly sowthistle (*Sonchus asper*)	21,500–25,000	F
Perforate St John's Wort (*Hypericum perforatum*)[a]	26,000–34,000	S
Canadian fleabane (*Conyza canadensis*)[a]	38,000–60,000	F
Hard rush (*Juncus inflexus*)[a]	200,000–234,000	S

[a]Perennial.

CRITICAL PERIODS OF COMPETITION

There are critical periods of growth when weeds provide competition with crops and other times when they are not exerting depressing effects on yield (Nieto *et al.*, 1968). Soils used for brassica production usually contain more than 2000 viable seeds of annual weeds/m² (20 million/ha). Only a small proportion of these germinate into seedlings at any one time but this can still lead to populations of 100–300 weed plants/m² competing with the crop. Yield losses vary according to the habit of the predominant weed species and the crop growth stage at which they begin competing with each other for resources. In drilled summer cabbage (*B. oleracea* var. *capitata*), for example, losses varied between 25% and 100% with a weed density of 100 plants/m². The scale of losses depended on the weed species present and the degree to which environmental conditions favoured either the crop or its competitors (Roberts *et al.*, 1976; Röhrig and Stützel, 2001).

In vegetable crops the concept 'critical periods' is highly developed, indicating times in which any competition from weeds will cause irreversible yield loss. A 'critical period' of weed competition is defined as: 'the minimum time over which weeds must be suppressed to prevent yield loss'. There are two separate components to such critical periods:

1. the time a crop must be weed-free after drilling or planting so that emerging weeds do not reduce yields; and
2. the subsequent time when weeds which emerge within the crop can remain before they begin to interfere with crop growth.

These factors define the optimum timing for weed removal to prevent yield loss rather than stopping intense competition. Use of this concept is limited to the timing of cultivation and other measures and is dictated by crop tolerance and the susceptible stages to weed growth. The critical period varies with crop type and cultivar, location, spectrum of weed species and their density in any particular season (Weaver, 1984). In cabbage the length of the weed-free period, weed density and ambient light reduction caused by competing weeds were important factors governing yield loss (Miller and Hopen, 1991). Maintaining the crop plants free from weeds in the first 4 weeks after sowing was crucial for the preservation of yield (see Table 6.5).

Table 6.5. Critical periods of weed competition in cabbage related to yield. (After Miller and Hopen, 1991).

Weeks weed-free after emergence[a]	Cabbage yield (kg/m²)
14	10.39
6	10.19
4	10.24
2	3.16
1	3.30
0	0.86
LSD (0.05)	1.52

[a]Weed = velvetleaf (*Abutilon theophrasti*).
LSD = least significant difference.

The concept of critical periods can be criticised for failing to take into account the influence of the full range of site-specific ecological and agronomic factors. These simulation models of competition tend to be generally empirical regression models based on one or more parameters such as weed density, relative leaf area or relative time of weed emergence. Such parameters vary widely between sites and over seasons. Such models also tend to lack physiological values dependent on the interaction between the crop and its surrounding weeds competing for resources. This competition is process-orientated and defined as the distribution of growth-limiting factors between species in a vegetation canopy and the efficiency with which each species uses resources for growth (Spitters, 1990).

The resources for which the crop and weed compete are: radiation (light), soil space, nutrients and water. The rate of use of these resources depends on the structure and growth pattern of crop plants and interactions with the weeds. Cauliflower (*B. oleracea* var. *botrytis*), for instance, has a relatively constant resource allocation pattern throughout the vegetative and reproductive growth stages and does not respond to competition for radiation by significantly changing dry matter distribution. Since it is a relatively tall plant, susceptible to damage from wind and heavy rainfall, it requires support from an appropriately sized stem which is responsible for the stability of the plant. The lack of side shoots and branches in cauliflower ensures that the leaves primarily determine the lateral expansion of the plant.

Radiation is the most crucial resource for which there is competition between brassica crops and weeds. Crops are usually planted in widely spaced rectangular patterns thus developing ecologically as single plants rather than 'closed' or row canopies. The influence of weeds on light availability in direct-sown cabbage is shown in Table 6.6.

MONITORING WEEDS

Integrated pest management (IPM) requires increasingly sophisticated monitoring of incipient invasions by pests and pathogens, as well as weeds (Zijlstra *et al.*, 2011). Crop monitoring and subsequent treatment is increasingly exploiting remotely operating systems, online and with radio transmission and the commissioning of responses resulting in more efficient and environmentally beneficial treatments. These need to be supported by libraries of information detailing previous cropping and the incidence of weeds, pests and pathogens linked with decision support systems which recommend precision spraying or alternative measures. The application of geographic information systems (Calegari *et al.*, 2013), linked with computer-aided design, increased the precision of chemical use by identifying crop location. In addition, land management systems integrating a range of farming practices are able, coincidentally, to trace final marketed product. Precision guidance and sprayer control reduces overlap, with associated cost savings in chemicals, fuel, and

Table 6.6. Effect of weed growth on light availability in direct-sown cabbage (*Brassica oleracea* var. *capitata*). (After Miller and Hopen, 1991).

Treatment[a]	Cabbage yield (kg/m^2)	Light availability to cabbage as percent full sunlight
Weed-free control	9.84	100.0
Weed-free 6 weeks	9.50	97.5
Weed-free 4 weeks	9.35	100.0
Weed-free 2 weeks	3.24	57.1
Weed-free 1 week	0.72	15.8
Unweeded 2 weeks	9.70	100.0
Unweeded 4 weeks	7.06	97.5
Unweeded 6 weeks	1.78	93.8
Unweeded 14 weeks	0.0	5.8
LSD (0.05)	1.35	14.1

[a]Weed = velvet leaf (*Abutilon theophrasti*).
LSD = least significant difference.

operator and machine time (Batte and Ehsani, 2006). The value of precision spraying systems also increases in proportion with the rising costs of the chemicals, numbers of applications and farm size.

POPULATION SIZE

For weed control, estimating the population size and its morphological characteristics using wide-view images is integral for real-time precision spraying (Gée *et al.*, 2008). In developing this approach, French researchers used colour-capturing cameras stationed above crop fields and constructed an algorithm which optimized simulated images and estimated weed masses. This algorithm was based on a Double Hough Transform with a region segmentation method using a blob colouring analysis. Although this proved to be a reliable means of crop row detection, differentiation between crop and weed plants still required further research. Subsequent progress has followed with site-specific weed management, drones for monitoring large areas, wider application of 'omics' and simulation models being developed (Shaner and Beckie, 2014). Weed control by site-specific weed management is also advocated by Peteinatos *et al.* (2014) as being more economical and having greater environmental benefits. They achieved this with mobile near-range sensor technologies which identified plants and their growth stages from spectral and morphological characteristics. Cameras, spectrometers, fluorometers and distance sensors are the usual tools suited for weed detection and measurement. Robotic weeders will be particularly useful for high-value speciality crops such as brassicas (Fennimore and Cutulle, 2019) because of increasing shortages of labour and the high costs of manual weeding. Such equipment would also be well suited for use in organic crops. Similar arguments also apply for the control of pests and pathogens in brassicas which were previously reliant on total-canopy applications of pesticides as a means of safeguarding product quality and yield (Warneke *et al.*, 2021). Consequently, sensor-controlled precision sprayers will be of major significance for pest and pathogen control. Studies in Florida, USA, have developed a smart sprayer equipped with machine vision and artificial intelligence (AI) which distinguishes target weeds from non-target vegetables and precisely applies spray on the desired foliage (Partel *et al.*, 2019). Two embedded graphics processing units (GPUs) were evaluated in the smart sprayer processing unit for image processing and target detection. The more powerful GPU (NVIDIA GTX 1070 Ti) had an overall precision of 71% and recall of 78%, for plant detection and target spraying accuracy on real plants, and 91% accuracy and recall with artificial plants. Finally, a real-time kinematic global positioning system (GPS) was connected to the smart sprayer and an algorithm was developed which automatically generated weed maps and visualized the collected data after every application. This smart technology integrates an AI based weed detection system and novel, fast, precision spraying and weed mapping systems.

Good husbandry ensures ample irrigation and fertilizer are provided so that water and nutrients are not limiting either for the crop or competing weeds. Hence, the spatial development of competing plants and their morphological adaptation to unfavourable growing conditions strongly influences the distribution of radiation within the canopy. The effect of the weed fat hen (*Chenopodium album*) (see Fig. 6.2) on the growth of spring- and summer-planted cauliflower (*B. oleracea* var. *botrytis*) was studied by Röhrig and Stützel (2001).

It is evident that this weed had much greater impact on crops planted later in the season. This is due to higher temperatures and the accelerating growth rate of the weed in summer. This encourages more vigorous growth of the weed and an increasing capacity to shade the cauliflower, thereby reducing radiation use efficiency in the crop. When this weed was growing at high density it substantially reduced cauliflower total dry weight, leaf area index (LAI) and curd diameter; crop height was reduced more in the late summer crops compared with earlier maturing types.

With spring and summer brassica crops grown from direct seeding or using transplants, provided the weeds are removed effectively in a single weeding, then there are few adverse effects of competition. The crucial period for weeding brassica crops is in the earliest stages of their establishment and as subsequent growth develops. The crop canopy closes soon after weeding and prevents further competition. But brassica crops show significant effects of row width and weed removal treatment on yield and quality. Yields tend to rise, as row spacing is reduced, up to critical values and thereafter they diminish. Weed removal becomes essential once competition between the individual crop plants and weed species commences after planting or drilling. Provided weeds are removed before this critical time, generally by a single weeding

Fig. 6.2. Fat hen (*Chenopodium album*). (Geoff Dixon).

operation (using either mechanical or herbicidal methods), then yield and quality losses are prevented. This requires the use of herbicides applied prior to or post-planting, or mechanical methods that remove the developing competition during crop growth. The period of brassica crop establishment is especially crucial in preventing weed competition developing since there are few, if any, environmentally acceptable herbicides that can be applied to established standing brassica crops. Single, thorough, mechanical weed removal must normally occur between 2 and 3 weeks after transplanting, depending on the crop row spacing employed. Weed competition commences more quickly at narrow row spacings compared with wider ones.

Weeds have limited effects in brassica crops grown during the autumn and winter provided they have been removed before active crop growth recommences in spring. Those weeds remaining into the spring compete severely with the crop resulting in smaller marketable cabbage (*B. oleracea* var. *capitata*) heads, reducing internal head quality, decreasing the number of plants forming heads and, in very severe cases, weeds can cause the death of crop plants (Lawson, 1972). Particularly successful weed species in Lawson's northern European (Scottish) studies included: chickweed (*S. media*), annual meadow grass (*Poa annua*), shepherd's purse (*C. bursa-pastoris*) and knotgrass (*Polygonum* spp.). A major factor in the competitive fitness of these weeds was an ability to grow rapidly in spring, outpacing and overtopping the crop and causing shading effects. Weed competition was analogous to increasing the density of a crop planting up to a point where cabbage population densities became so high that normal development was prevented. Conversely, comparison of cropped and uncropped plots indicated that the crop could exert considerable competitive pressure on the growth and development of weeds, particularly where their capacity for expansion had been retarded by treatment with herbicides.

The management decision of whether weed control is economically worthwhile depends on a costs and benefits analysis. This identifies that at a given level of weed infestation crop yield and/or quality are likely to be damaged if the weeds remain uncontrolled. Models of weed and crop competition are an essential part of both short- and long-term crop management planning.

The most favoured model that describes the relationship between weed infestation and crop yield loss is a rectangular hyperbolic curve (Cousens, 1985). Many discussions of the relationship between crop yield and weed densities have used either sigmoidal or quasi-sigmoidal models meaning that there is a threshold weed density below which there is no yield loss. In this relationship, there is virtually no competition between weed and crop at low weed density despite the plant size being at a maximum. This ignores the fact that the competitiveness of each individual weed plant is greatest when it has achieved maximum size and is exploiting nutrient and water resources from the soil. Support for the use of a sigmoidal relationship is based on the observation that at very low weed densities statistically significant reductions of yield are difficult to demonstrate in crops such as cereals. In vegetables such as the brassicas, however, losses may result from the damage to product quality that is difficult to quantify by the use of field experiments. Also, experimental designs are often incapable of detecting differences in yield of the magnitude expected at low weed densities. This does not mean that such yield differences do not exist and, especially with brassica vegetable crops, also exerting an indirect influence, by reducing quality, on their profitability. A hyperbolic model of the relationships between weed density and yield loss indicates, however, that wherever weeds are present in a crop there will be competition for resources such as nutrients and water. This occurs even when there is very limited or no shading effect reducing the light available to the crop plant. The hyperbolic model assumes that weeds are distributed at random in relation to crop plants. Consequently, as weed density increases the mean distance from the weed plant to crop plant remains constant (see Fig. 6.3).

Growers' practical experience emphasizes the influence of cultivar selection, row width and planting date on weed competition for brassica crops. These vary substantially in husbandry systems using a wide range of crop spacings, population densities and methods of establishment (direct seeding vs transplanting). Efficient weed management is essential because of the high intrinsic values of the products, limited availability of herbicides and high labour costs.

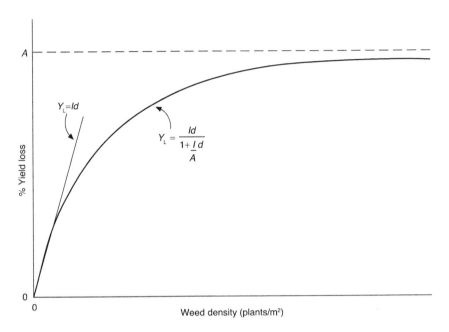

Fig. 6.3. The rectangular hyperbolic model for relating yield loss to weed density, illustrating the parameters *A* and *I*.

Y_L = percentage of yield lost because of weed competition

d = weed density

I = percentage of yield loss per unit weed density as $d \to 0$

A = percentage of yield loss as $d \to \infty$.

(After Cousens, 1985).

EFFECTS OF WEED TYPE

The morphological habit of a weed will substantially affect its impact on the growth and yield of the crop. For tall growing weeds, such as fat hen (*C. album*), there is a linear relationship between the extent of yield reduction and weed density; as few as three plants/m² were sufficient to cause statistically significant reductions to the yield of cabbage (*B. oleracea* var. *capitata*) crops (see Fig. 6.4). Low-growing weeds, such as chickweed (*S. media*), annual meadow grass (*P. annua*) and nettle (*Urtica urens*), caused smaller reductions in cabbage yield.

The presence or absence of brassicas such as cabbages does not normally affect the relative proportions of weed species. An exception to this seems to be shepherd's purse (*C. bursa-pastoris*) which, when treated with the herbicide trifluralin (α,α,α-trifluro-2,6-dinitro-N,N-dipropyl-*p*-toluidine), was then suppressed by the crop. Freyman *et al.* (1992) studied the competitive effect of shepherd's purse. This is one of the commonest and most difficult weeds to control in

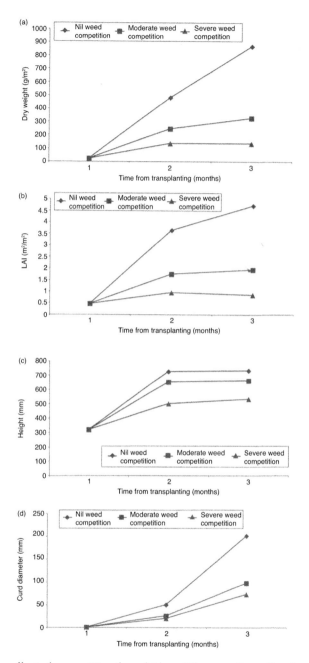

Fig. 6.4. The effect of competition from fat hen (*Chenopodium album*) on the growth and curd size of spring- and summer-transplanted cauliflower (*Brassica oleracea* var. *botrytis*) cv. Fremont. Simulated (lines) and observed value (symbols) for (a) aerial dry weight, (b) leaf area index (LAI), (c) height and (d) curd diameter of cauliflower. (After Röhrig and Stützel, 2001).

vegetable cole crops because of their botanical similarities. Consequently, this is where the use of herbicides would be most likely to be a cause damage to the crop plant. Results indicated that intra-row competition from the weed could be reduced by use of closer spacing within the rows. Closer intra-row spacing is normally associated with the use of wider inter-row spacing. Reducing the intra-row spacing diminished competition and made cultivation easier because of the increased distance between rows.

Chickweed was the main weed affecting winter cabbage in Lawson's (1972) studies. This weed was capable of surviving winter frost and then accelerating into rapid growth in the spring, becoming the dominant weed species and eventually shading the crop. Treatment with the herbicide propa-chlor (2-chloro-N-isopropylacetanilide) delayed the capacity of chickweed to cause competition. Combining this with the earlier use of trifluralin controlled this weed and allowed the crop to dominate the competitive relationship.

The dominance of weed species will change as a result of altering crop husbandry systems, including changing the spectrum of herbicides and other control techniques used and also resulting from several biological factors. This is what Lawson termed 'the ever moving target for weed control' (H. Lawson, personal communication). In California, USA, for example, cultivated radish (*Raphanus sativus*) and the weed (*Raphanus raphanistrum*, i.e. wild radish) are both introductions from Europe. Continual interspecific hybridization, since their arrival in America, has converted cultivated radish into a weed and the climatic range of charlock has been enlarged continuously (Panetsos and Baker, 1986).

The outcome of competition is determined by the timing of emergence of weed seedlings relative to those of the crop brassica and the extent to which environmental conditions during the early stages of growth favour either of the competitors. Difficulties arise where herbicides fail to be effective against the entire spectrum of weed species present in the crop. Under these circumstances those weeds that are uncontrolled are given a selective advantage and become predominant, producing populations with strong competitive advantages in comparison with the crop. These will provide major problems during crop harvesting and can act as reservoirs for pests and pathogens. They will also return substantial numbers of viable weed seeds into the soil causing an escalating problem for future years, especially where the opportunities for crop rotation are limited (Roberts *et al.*, 1976).

The use of transplanted brassicas provides the crop with an advantage since it is capable of competing with weed seedlings more quickly. The crop canopy closes rapidly, inhibiting the further growth of weed seedlings. This is usually linked to the use of a pre-planting, soil-incorporated selective herbicide capable of destroying germinating weed seedlings. There is a defect in this strategy caused by the presence of cruciferous weeds within the soil flora that are unaffected by such selective herbicides. This leads to an upsurge in weeds such as shepherd's purse and of volunteer crop brassica plants such as the increasingly problematical oilseed rape (*Brassica napus*) seedlings residual from preceding crops. There may also be other volunteer groundkeeping crops, such as potatoes, which by virtue of their presence as resilient tubers have a growth advantage even when in competition with transplanted brassicas.

COMPETITION FOR NUTRIENTS

Differences between crop species, and even between crop cultivars, in their competitive abilities against weeds are beginning to be explored. So also are the capacities of weeds to compete with each other for resources. These interactions potentially offer means by which the crop and weed relationship can be manipulated to the advantage of the former through their respective nutritional demands. This is because crops and weeds may compete for nutrient resources at different levels of effectiveness. Brassica crops, for example, are better competitors for soil nutrients than many other field vegetables (see Chapter 5 section, Introduction). In turn, it is known that cabbage plants have a considerable effect on the growth of various aggressive weed species such as fat hen (*C. album*) and to a lesser extent with groundsel (*S. vulgaris*) (see Fig. 6.5).

The presence of the crop plant markedly reduced the seed weight produced by fat hen (Qasem and Hill, 1993). The high responsiveness of many weed species to nitrogen may be a weakness to be exploited through development of fertilizer management methods that enhance crop competitiveness with weeds. An aim of plant breeders may, therefore, also be to develop cultivars that are sufficiently vigorous to be capable of out-competing weeds because of their increased efficiency in utilizing nutrients and water from the soil.

Fig. 6.5. Groundsel (*Senecio vulgaris*). (Geoff Dixon).

CROP INSPECTION

Integrated crop management identifies weed problems through regular crop inspections (so-called 'crop walking'). Increasingly, 'walking' is becoming a misnomer since growers and their advisors are using drones and other remote means of inspecting crops. None the less, inspection, combined with preventative, cultural, mechanical, biological and chemical control methods in a compatible manner, provides solutions for the management of weed competition. Integrated methods avoid relying solely on one management tool and help to reduce or eliminate the need for chemical weed control by combining a series of approaches to profitability, producing a marketable product while minimizing harm to the environment.

Weed control is an important aspect of integrated husbandry and illustrates many of the characteristics and problems also associated with pest and pathogen control (see Chapter 7 section, Invasion and Infection Forecasting and Management). Essentially, crop walking is no more than an application of the old Chinese adage that 'the best manure is the farmer's boot'. In other words, close and continuous attention to crops ensures the best results. Crop walking used to demand visiting the fields regularly, taking quadrat samples in which the numbers of weeds, their differing species and their locations within the crop are identified and recorded. Drone and satellite monitoring are becoming increasingly capable of identifying developing weed competition and relaying data to smartphone or PC platforms. This still permits particular attention being given to identifying the dominant species and also the occurrence of uncommon and perennial weeds that may pose particular problems for both the current and succeeding crops. It is an important aspect of integrated management that problems that might afflict future crops in the rotation are identified as early as possible, preferably before that crop is grown. Differentiating weed species by their life cycle is an essential aspect and is a key element that increases the success of integrated control.

Ephemeral and annual weeds reproduce primarily by seed and hence husbandry controls aim to prevent them from seeding. Many annual weeds of brassica crops have a periodicity of emergence and germinate in response to specific environmental cues. Cultivation strategies can be modified to provide 'stale seed beds', for example, where weeds are encouraged to germinate and are then destroyed by subsequent tillage prior to sowing or transplanting with the brassica crop.

Perennial plants, such as ground elder (*Aegopodium podegraria*), reproduce both sexually by seed and asexually by vegetative organs such as rhizomes and tillers which can be broken and distributed by ill-timed cultivation encouraging the spread and further propagation of the weed population (see Fig. 6.6).

Most perennial weeds tend to occur sporadically and patchily within a field. Mapping the location of these concentrations allows control measures to be targeted into the problem areas. This is a more efficient system of control and economizes on the resources used. Maintaining records of weed populations

Fig. 6.6. Ground elder (*Aegopodium podegraria*). (Geoff Dixon).

over a period of years provides an understanding of how weed species spread within fields and their response to control measures.

An essential aspect of ICM is the prevention of the spread of weeds into cropped areas. Weeds may be introduced on cultivation equipment into new areas. Cleaning equipment before moving between fields is an essential first step to preventing the entry of new species and for pest and pathogen control (see Fig. 6.7).

Animals can spread weeds by attachment to their coats and in their manure. They are also spread in propagation media, such as peat moss or shredded bark, used to produce transplants and in irrigation water. New infestations of weed species should be removed by chemical spot treatment while the colonies are small rather than waiting for a larger scale problem to develop before remedial action is taken. The effect of global climate change is encouraging plants to move into previously uncolonized regions and, because of the lack of predatory herbivores, they become very successful weeds that are exceedingly difficult to eradicate.

SUSTAINABLE CULTURAL PRACTICES

The principles of sustainable soil management and crop production are discussed by Anon (2019). In this context, cultural weed control aims to optimize sowing or planting dates, seed rates or transplant densities, spacing layouts, soil fertility, irrigation practices and cultivar selection to achieve a rapidity of crop growth which may outcompete weeds for resources. The aim should be

Fig. 6.7. Washing machinery. (Geoff Dixon).

to ensure that the crop plants either emerge first or transplants can establish ahead of weed development and close their canopy over the weeds and smother them.

Manipulating crop geometry can radically influence the grower's weed control. Thus, seeds or transplants should be placed at uniform depths to produce even and regular crop growth. Use of high cropping densities and narrow row spacing ensures that the crop canopy closes as quickly as possible. This may be aided by using artificial crop covers (see Chapter 7 section, Strategies for Pest and Pathogen Control) which increase the accumulation of units of soil heat and, therefore, encourage more rapid root and shoot growth. Integrated management requires that inputs, such as nutrients and water, are applied uniformly ensuring even growth without the development of stress within the crop. Similarly, biotic stresses caused by pests and pathogens should be minimized. Vigorous healthy crops have greater competitive advantages against weeds compared with those that are stressed.

Increasing organic carbon stocks in agricultural soils is currently of special interest because it can help mitigate global warming through atmospheric carbon sequestration (Camarotto *et al.*, 2020). Recommended management practices, such as conservation agriculture and conventional tillage with cover crops, could have significant implications for carbon sequestration potential. A long study period would provide detailed understanding of the potential benefits of conservation agriculture and cover crops on organic carbon sequestration from soil. Climate change may alter the seasonal environment for brassica crops. Increasing temperatures and abnormal cloud cover could change ambient ultraviolet B radiation which has the potential for altering crop–weed competitive interactions (Furness *et al.*, 2005).

NON-TILLAGE AND REDUCED TILLAGE

Reducing the number, scale and depth of cultivations has, over the past 20 years, become a major facet of sustainable crop management particularly aimed at increasing soil health and quality. Sound biological, environmental and financial reasons underpin this trend. Reduced frequency of cultivations limits the disruption of microbial and macrobial life in soil. It lessens the energy expended and consequently both moves production further towards net zero carbon and cuts the costs of growing brassicas. Tillage does, however, permit the incorporation of organic matter, farmyard manure or well decomposed compost, increases aeration and water percolation, breaks pans encouraging the deeper rooting of crop plants, and permits frosting (see Fig. 6.8) which increases soil friability and buries weeds.

For highly intensive vegetable brassica production these advantages still make tillage an important component of ICM. Specific studies offer varying evidence, for example, with spring cabbage (*B. oleracea* var. *capitata*) by Knavel and Herron (1981) detected little or no economic advantages with non-tillage methods. But by contrast, tilling soil at three different depths significantly decreased weed emergence (Egley and Williams, 1990).

These variations contrast with the benefits found for relatively low-intensity farm crops. Reduced tillage was discussed by Unger and McCalla (1980) for use on broadacre farm crops in the USA as means for preserving

Fig. 6.8. Well-frosted loam soil. (Geoff Dixon).

degraded soils. They associated it with similar methods such as: conservation tillage, direct drilling, eco-fallowing, limited tillage, minimum tillage, no-tillage and reduced tillage, and stubble mulching (Barnes and Putnam, 1983). The general aims of reduced tillage are: leaving sufficient plant residues on the soil surface at all times to reduce wind and water erosion; reducing energy use and to move increasingly towards net-zero carbon emissions in crop production and distribution; conserving soil health and quality; and minimizing water use (Brandsæter *et al.*, 1998).

Fallowing land, in the absence of crops, allows the use of intensive non-selective weed control and can be used to eliminate tenacious plant species that are otherwise difficult to manage. Land selection techniques can be linked with fallowing such that those parts of the holding infested with weeds which are difficult to control in brassica crops are avoided. This technique is a form of crop rotation and can be used to exploit forms of husbandry or herbicidal control which totally eradicate weeds in advance of specific brassica crops. Crop rotation itself has long been considered one of the simplest and most effective tools for managing weeds (Garrison *et al.*, 2014). Strategically arranging rotations involves 'crop stacking', which is increasing the number of consecutive plantings of the same crop within a rotation. This decreases the size of the weed seed bank, by forcing weeds to compete with each other in similar environments for longer periods of time, while still reaping the traditional benefits of crop rotation. This is a novel application of existing eco-logical theory, which improves weed management strategies. Fallowing and rotation produce significant benefits for vegetable brassica cropping by reduc-ing the inoculum potential of particularly tenacious soil-borne pathogens such as *Plasmodiophora brassicae* (the cause of clubroot disease) and *Sclerotinia sclerotiorum* (the cause of white mould) (see Chapter 7 section, Rotation and Tillage).

EFFECTS OF TILLAGE IMPLEMENTS

Vertical gradients in soil microclimate, such as water availability, temperature and light, occur in the field (Heydecker, 1972; Fenner, 1985), so that one of the major factors influencing the success of weed germination and emergence is the position of a weed seed within the soil profile. Chancellor (1982) identi-fied that weed seeds have mechanisms that respond to these gradients, thereby preventing germination at depths from which the seedling cannot emerge. Mechanisms, such as the need for light to stimulate germination in scented mayweed (*Matricaria recutita*), ensure that germination only takes place close to the soil surface. Each weed species has a characteristic emergence response to depth of burial. The distributions of weed seeds in soil banks are spatially het-erogeneous, both horizontally and vertically. The major means of seed move-ment in soils are the implements used in cultivation themselves. Consequently, cultivation practices can give rise to distinctive vertical distributions of seeds in the soil profile. While this distribution can be sampled and the products tested in controlled laboratory conditions to establish the spectrum of weed seed within the soil profile there is little evidence to determine how these seeds are placed in specific sectors of the profile in the first instance. Experiments have shown that spading and rotavating tends to spread weed seeds throughout the soil profile; whereas spring tine and power harrows concentrate weed seeds in the top 6 cm of soil (see Fig. 6.9).

Investigations of the effects of implements on weed seed dispersal in soil compared spader, rotavator, spring tine harrow and power harrow implements (Grundy *et al.*, 1999). The rotavator caused a backward movement of seeds (as represented by beads) but neither the spring tine nor spader had any significant effect on the horizontal displacement of seeds while the power harrow had the greatest capacity to move seeds forward beyond 0.5m in the soil.

Harrowing weeds before and after brassica crop emergence considerably reduces the reliance on herbicides. Encouraging the use of harrows requires that the number of cultivations should be minimized – currently, up to eight passes may be needed in a single weeding operation – and dependence on favourable weather conditions for soil working and weed killing should be reduced. Harrowing predominantly kills or suppresses small weeds in their early growth stages. A relatively small proportion of weeds (up to 25%) are uprooted. Most are simply covered by a loose layer of soil. As weed plants mature, they become taller and less flexible and hence more resistant to being covered by a layer of soil from which they can regrow after the cultivation operation is over.

Understanding and improving the effectiveness of harrowing demands knowledge of the relationships between: weed control and crop damage; weed density and crop yield; and crop damage and crop yield. Each of these relationships will be moderated by the prevailing weather and soil conditions at the time of cultivation. Harrowing depth, working speed and soil moisture content will have varying influences on the covering of different weeds, affecting the efficiency of weed control.

The prime factor affecting the efficiency of weed control appears to be the depth to which weeds are buried by the harrowing operation (see Fig. 6.10) (Kurstjens and Perdok, 2000).

The density of weed seedlings emerging in a crop following cultivation is related to:

- the ability of seedlings to emerge from various depths in the soil;
- the survival of seeds at different depths; and
- the depth of seed burial in no tillage, rotary tillage and plough tillage.

Where a great abundance of weed seeds is thoroughly mixed into the soil, probably the best approach is to use nil or minimum tillage and attempt to deplete the surface fraction of the weed seed pool. If this approach is to be successful then reseeding must be prevented, e.g. through the use of herbicides or by further shallow cultivation. Repeated surface cultivation has the

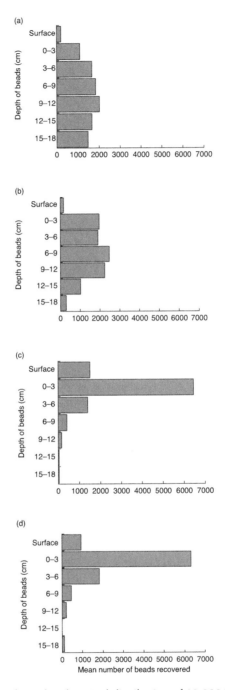

Fig. 6.9. Illustration of simulated vertical distribution of 10,000 beads initially sown on the soil surface following one pass of (a) spader, (b) rotavator, (c) spring tine and (d) power harrow. (After Grundy *et al.*, 1999).

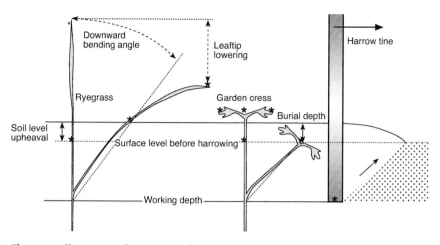

Fig. 6.10. Illustration of ryegrass and garden cress seedlings before and after harrowing. Digitized coordinates (marked *) show a harrow line, a ryegrass seedling and a garden cress seedling before and after harrowing, with an illustration of derived parameters (side view). (After Kurstjens and Perdok, 2000).

added benefit of promoting the germination of seeds already in the pool and thereby speeding the decline in the number of seeds that are close enough to the surface for emergence. When confronted by a year of weed control failure, in which many seeds are shed on to the surface of an otherwise clean soil, the best strategy will be to plough as deeply as possible and use minimal soil disturbance methods thereafter (Moss, 1985; Mohler, 1993).

EFFICACY OF TILLAGE WEED CONTROL FOR VEGETABLE BRASSICAS

Weed control is a major reason normally advocated for soil tillage. The use of primary cultivation (ploughing) and secondary cultivation (discing and harrowing) eliminate emerging annual weeds and suppress perennial weeds. Ploughing buries about 80% of surface weed seeds and only returns about 40% to the surface. It also brings perennial rhizomes and tillers to the surface allowing them to be killed by desiccation or freezing.

Conservation (limited) tillage or non-tillage systems have been advocated for oilseed brassicas (predominantly oilseed rape, *B. napus*) but are less frequently used for high-quality vegetable cole crops. The presence of crop residues on the surface tends to slow down the processes of soil warming, provides habitats for pests and pathogens, and may encourage the development of perennial and groundkeeper weed populations. Flame weeding has been used for selective control and the removal of crop residues and is applied prior to crop emergence in countries such as the USA. These methods are particularly suited to crops that emerge slowly. Hand weeding has largely been discarded, except as an expensive supplement to other strategies and when applied only as an emergency measure.

The side effects of mechanical weed control in brassica production are:

- In addition to the intended effects on weeds, mechanical weed control causes some negative side effects, such as physical damage to crops, and a positive effect – improved soil structure.
- Studies in cauliflower (*B. oleracea* var. *botrytis*) showed that where there was an overall treatment, this resulted in at least 4% loss of crop plants with ridging and tine harrowing. Yields of mechanically weed-controlled crops ranged between 82% and 111% with an average of 97% for untreated control crops (Laber and Stützel, 2000).

In specialized conditions, such as the cultivation of row crops on mountainous farmland, severe soil erosion may be generated due to low ground cover, especially in the early crop growth stages. Here, it was found (Arnhold *et al.*, 2014) that organic farming, due to the absence of herbicides, supported the development of weeds and increased the ground cover compared to conventional farming. But it was concluded that organic farming alone does not effectively control erosion, and that both organic and conventional farming methods require additional conservation measures, such as winter cover crops and residue mulching, for sufficient prevention of soil loss resulting from row crop cultivation.

INTERCROPPING AND COVER CROPPING

Increasing business specialization and intensification of brassica production has virtually eliminated the use of rotations, resulting in undesirable side effects, such as soil compaction, loss of soil structure and decreased organic matter content (Nicholson and Wien, 1983). Recognition of these problems has focused attention on the need for soil conservation and revived attention to the need for preserving and developing soil health and quality (Bronick and Lal, 2005; Walker *et al.*, 2011). Research indicates that intercropping with rows of plants, whose purpose is soil improvement, placed between the economic crop helps restore soil structure with minimal deleterious effects on the efficiency of the cash crop. Intercropping is still a common cultural practice in tropical areas but is much less frequently used in temperate regions (Theunissen and Ouden, 1980; Theunissen *et al.*, 1995). Intercropping can be part of an ICM strategy, contributing to ecologically and economically acceptable forms of sustainable horticulture. Favourable effects of intercropping include suppressing or delaying pest population expansion and crop colonization.

Criteria for intercropping species:

- they must offer 'intercropping effects' in terms of pest population suppression;
- competition with the main or cash crop must be minimal;

- they must not create weed problems for the following seasons;
- they must be predicable and manageable within the normal cropping pattern;
- the seed of the intercropping plant(s) must be commercially available in sufficient quantities;
- they must not encourage or support pests or pathogens; and
- leaching of nitrogen must be prevented or reduced.

The requirement that intercropping reduces pest insect populations is not fulfilled in all cases. Some species of pest seem to be unaffected by either the presence or absence of intercrops, or only react in specialized conditions as, for instance, with the small cabbage white butterfly (imported or European cabbage worm) (*Pieris rapae*) (Risch, 1981).

Intercropping may affect the morphology and development of brassicas. For example, cabbage heads taken from intercropped plots were smaller but more compact and this in turn can affect their sensitivity to pest attack. It is also suggested that stresses induced by polyculture may alter the physiology of the crop plant, rendering it less attractive or nutritious or even more toxic to its pests. Reports which describe the effects of monocropping and polycropping on the yield and value are infrequent. Comparisons of white cabbage (*B. oleracea* var. *capitata*) grown in the Netherlands (Theunissen *et al.*, 1995), with and without white and subterranean clover (*Trifolium repens* and *Trifolium subterraneum*), demonstrated the clear yield and financial benefits of polycropping, possibly because of the additional valuable nitrogen gained from these companion plants (see Fig. 6.11).

The successfulness of intercropping is increased where chemical suppression of the 'live mulch' prevents excessive competition. The most promising living mulches were the shorter and less vigorous turf grasses and clover. Chewing's fescue (*Festuca rubra* var. *commutata*), Kentucky bluegrass (smooth meadow grass) (*Poa pratensis*) and white clover did not affect the yield of cabbage. Some compromise is required between improving soil structure and limiting the competition with the brassica crop when growth of the live mulch is not controlled.

Conventional forms of tillage husbandry may deplete soil organic matter and are associated with the use of substantial inputs of resources, such as soluble fertilizers, pesticides and water. Alternative methods of weed, pathogen, pest and fertility management are becoming more regular techniques which reduce reliance on agrochemicals. They also reduce the carbon footprint of crops helping achieve net-zero carbon production.

The cover cropping approach is not new. Studies of aphid populations on Brussels sprouts (*B. oleracea* var. *gemmifera*) in the 1970s, for example, indicated that the presence of weed cover provided a useful means of reducing the size of aphid populations through increased natural predation (Smith, 1976a, b). Brussels sprouts grown in a weed-free culture were more attractive for colonizing aphids compared with those where weeds remained. Clean weeding

Fig. 6.11. Legume root showing nitrogen-fixing bacterial nodules. (Geoff Dixon).

of sprouts provided ideal conditions for colonization by aphids, whitefly and caterpillars of certain *Lepidoptera* spp. in the absence of their predators. These pests were attracted to the crop plants since they stood out against the bare background provided by a weed-free soil. In turn, colonization by predators of the pests of Brussels sprouts was related to plant density, height and the contrast or colour of the background that affected the optical attractiveness of the crop and it was encouraged by the presence of an understorey of non-crop plants. Variations in the species composition of the weed flora affected the differential attractiveness of crop backgrounds to a range of predators of insect pests as discussed by Tukahirwa and Coaker (1982).

Cover crops are grown during periods when the soil might otherwise be fallow (Dabney *et al.*, 2001). While actively growing, cover crops increase solar energy harvest and carbon flux into the soil, providing nutrients for soil macro- and microorganisms. Cover crops reduce sediment production from cropland by intercepting the kinetic energy of rainfall and by reducing the amount and velocity of runoff. Cover crops increase soil quality by improving biological, chemical and physical properties including: organic carbon content, cation-exchange capacity, aggregate stability and water infiltrability. Legume cover crops contribute nitrogen to subsequent crops. Other cover crops, especially grasses and brassicas, are better at scavenging residual nitrogen before it can be leached. Growth of these scavenging cover crops is usually nitrogen limited. Growing grass/legume mixtures often increases total carbon inputs without sacrificing nitrogen scavenging efficiency. Cover crops are best adapted to warm areas with abundant precipitation. Water use by cover crops can adversely affect yields of subsequent dryland crops in semi-arid areas. Similarly, cooler soil temperatures under cover crop residues can retard early growth of subsequent crops grown near the cold end of their range of adaptation. Development of systems that reduce the costs of cover crop establishment and overcome subsequent establishment problems will increase cover crop utilization and improve soil and water quality. Cover crop mixtures can provide both nitrogen retention and supply services. Grass–legume combinations maintain the soil nitrate-nitrogen concentrations (White *et al.*, 2017).

Comparisons of cover crop-based no tillage, strip tillage and conventional tillage in transitional organic green broccoli (calabrese) (*B. oleracea* var. *italica*) production produced data describing yield and quality, plant health, weed suppression, soil temperature, and cost of production (Jokela and Nair, 2016). A cover crop mixture of rye (*Secale cereale*) and hairy vetch (*Vicia villosa*) was seeded in all plots in September, and was ended by rolling and crimping for non-tillage and strip tillage plots or soil incorporation in late spring the following year. Results suggested that cover crop-based organic non-tillage and strip tillage systems may be useable options for organic green broccoli growers.

Mechanical methods (flail mowing and rolling) were tested for killing cover crops such as *Glycine max*, *Fagopyrum saggitatum* (*F. esculentum*), *Secale cereale* and *Setaria italica* and distributing and retaining their residues uniformly over the soil (Morse, 1999). Results showed that no-tillage green broccoli can be successfully produced without using herbicides where high-residue cover crops are effectively killed by flail mowing or rolling. In Uganda, banana (*Musa* spp.) residues have been tested successfully as organic inputs for cabbage production (Lekasi *et al.*, 2001).

Conservation agriculture includes the use of cover crops as a means for conserving soil quality by limiting erosion, adding organic matter and enhancing soil nutrient content, as well as water availability (Jaffuel *et al.*, 2017). All these factors can greatly influence the presence of soil organisms including, potentially, the persistence of entomopathogenic nematodes over the winter season. Unfortunately, higher numbers of entomopathogenic nematodes in agricultural soils do not necessarily translate into high infectivity, the key factor determining their effectiveness in controlling soil pests.

Some brassica cover crops have properties of biofumigation against plant-parasitic nematodes. Conversely, some species or cultivars are highly susceptible to the nematodes, posing an important challenge for pest management (Waisen *et al.*, 2019). Current research aims to exploit the ability of nematode-susceptible brassicas as trap crops by terminating these targeted pests before they undergo multiple life cycles; these are referred as 'open-end trap crops'. *Meloidogyne incognita*, *Meloidogyne javanica* and *Rotylenchulus reniformis* were targeted in this research. Brown mustard (*Brassica juncea* cv. Caliente 199) and oil radish (*R. sativus* cv. Sodbuster) were also used in these studies. Brown mustard suppressed soil populations of *Meloidogyne* spp. in two trials and *R.*

reniformis in one trial, but oil radish did not suppress the target nematodes in all trials. These results suggested that the trap crop effect may not only depend on brown mustard or oil radish susceptibility but also depends on the specific nematode targeted.

Cover cropping and intercropping may also be accompanied by the provision of flowering margins for fields which are beneficial for both the predators of injurious insects and the vital crop pollinators. ECOBEE is a colony monitoring scheme specifically intended for providing beekeepers and researchers with basic ecological data on honeybees in intensive agrosystems, as well as colony population dynamics (Odoux *et al.*, 2014). ECOBEE was launched in 2008 as a long-term ecological project with three specific aims: first, monitoring seasonal and interannual population dynamic parameters of honeybee colonies in a heterogeneous farming system; second, providing relevant and robust data sets to test specific hypotheses about bees such as the influence of landscape planning, agricultural inputs or human pressure; and third, offering opportunities for assessing the effectiveness of agroenvironmental schemes or the effects of changes in agricultural policies on honeybee well-being. For example, a sharp food shortage period in late spring between the flowering of oilseed rape and sunflowers, possibly temporarily constraining colony sustainability.

Crop management is known to influence biodiversity. Conservation tillage is often found to have a positive effect on biodiversity compared to conventional tillage. Conservation tillage is any tillage and planting system that covers 30% or more of the soil surface with crop residue, after planting, to reduce soil erosion. Studies of soil inversion methods found a greater abundance of birds with conservation tillage (Barré *et al.*, 2018). Results also highlighted that there is an even larger negative effect of herbicides compared with tillage on bird populations. This also emphasized the advantages of flowering field margins as these encourage increases in predators of crop pests (Stoate *et al.*, 2017) (see Fig. 6.12).

Non-tillage cover crops are sown in the autumn or very early spring and mown during late spring to form a mulch on the soil surface just before planting the main cash crop. Cover crop acreage has substantially increased over the last few years in the USA due to growers capitalizing on federal conservation payments and incorporating sustainable practices into agricultural systems (Palhano *et al.*, 2018). But despite all the known benefits, widespread adoption of cover crops remains limited due to potential cost and management requirements. Effective cover crop termination is crucial, because a poorly controlled cover crop can become a weed and lessen the yield potential of the current cash crop. Several herbicides are used for the control and elimination of cover crops; but none have adequately controlled oilseed rape groundkeepers which is an increasingly difficult problem for vegetable brassica growers, especially as they are reservoirs for foliar pathogens.

A primary challenge of managing vegetable production on a small land area is the maintenance and building of soil quality (Pfeiffer *et al.*, 2016).

Fig. 6.12. Biodiverse flowering field margins. (Geoff Dixon).

Studies have demonstrated the benefits of cover crops for improved soil quality. However, small North American growers struggle to fit cover crops into rotations. Cover crops have contributed to weed suppression for autumn crops of transplanted green broccoli (*B. oleracea* cv. Imperial). The East Coast green broccoli industry in the USA is growing, and optimizing weed control strategies in south-eastern US environments is important for the expansion of commercial green broccoli in the region (Cutulle *et al.*, 2019). Studies showed that cultivation is important for maximizing green broccoli yield characteristics in coastal South Carolina while herbicide use may be less critical.

These methods also include living mulches (Fracchiolla *et al.*, 2020) which provide many benefits to agroecosystems, such as erosion control, nitrogen fixation and nutrient recycling, increases in organic matter, weed and pest control, and increases in the number of soil organisms. For transplanted broccoli raab, also known as rapini (*Brassica rapa* ssp. *ruvo* cv. Grossa fasanese), results showed the positive effects of living mulch and organic fertilization in the sustainable production of this cultivar.

Cover crop mulch systems modify the micro-environment of the crop with subsequent effects on pest populations and crop yields. Plants in non-tillage systems tend to be smaller and paler than those from conventional tillage. This results from lower soil temperatures observed in non-tillage plots which decreases the rate of mineralization of soil nitrogen, reducing its availability to brassica crops. If a legume is included in the mulch then nitrogen is added into the soil. Suitable plant species for use as cover crops are given in Table 6.7.

Indirect benefits from cover crops

Some cover crops release chemicals that inhibit weed germination – a phenomenon termed allelopathy. Rye (*S. cereale*) residues release chemicals inhibitory to annual weeds and suppression is greater where the rye residues are left on the soil surface. Hairy vetch (*V. villosa*) has weed suppressive properties but to a lesser extent than rye. Cover crops may also affect the chemical constituents of soils. Cover crop residues may make nitrogen unavailable to vegetable crops due to assimilation or denitrification. Conversely, cover crops can also increase nitrogen available for brassicas, either by the decomposition of legume residues or by preventing nitrogen losses during winter. Rye also increases the concentration of exchangeable potassium near the soil surface. A detailed study of cover cropping in Norway was reported by Brandsæter (1996).

Vegetable growers in north-eastern USA use cereal rye as a winter cover crop, but rye does not fix nitrogen. Schonbeck *et al.* (1993) demonstrated the value of using hairy vetch with rye as cover cropping for green broccoli (*B. oleracea* var. *italica*) and cabbage crops. Cover crops such as red clover (*Trifolium pratense*), biennial sweet clover (*Melilotus* spp.) and lucerne (alfalfa) (*Medicago sativa*) will add nitrogen to the soil but take the land out of brassica production for 12 months. Winter annual legumes such as hairy vetch, crimson clover (*T. incarnatum*) and Austrian winter pea (*P. sativum* ssp. *arvense*) can add nitrogen and organic matter without interrupting production for an entire season. Legumes, when grown with a cereal grain or other grass (e.g. hairy vetch with rye), may produce more organic matter, protect the soil and suppress weeds better than either when grown alone.

Alternatively, there are disadvantages such as reduced yields of spring cabbage (*B. oleracea* var. *capitata*) in non-tillage crops compared with conventional tillage (Morse and Seward, 1986). By contrast, yields of autumn cabbage (*B. oleracea* var. *capitata*) increased with non-tillage systems. Differences in planting date influenced this response. Using legumes in non-tillage systems is a logical means of adding nitrogen. Hairy vetch is the most efficient legume in this respect, adding 90–100 kg N/ha annually. Many experiments fail to mitigate the effects of competition between the cash and cover crops and

Table 6.7. Plant species suitable as cover crops.

Grasses and cereals	Legumes
Rye – *Secale cereale*	Hairy vetch – *Vicia villosa*
Barley – *Hordeum vulgare*	Austrian winter pea – *Pisum sativum* ssp. *arvense*
Bread wheat – *Triticum aestivum*	Crimson clover – *Trifolium incarnatum*
Ryegrass – *Lolium multiflorum*	

hence find the crop growth components are less in living mulch treatments compared with bare soil control treatments. Hence, the biomass of cauliflower (*B. oleracea* var. *botrytis*) was reduced by 62% when grown in a vetch living mulch compared with clean cultivation (Altieri *et al.*, 1985). The yields of cauliflower and Brussels sprouts (*B. oleracea* var. *gemmifera*) were reduced by 42% and 61%, respectively, when overseeded with white clover (Dempster and Coaker, 1974; O'Donnell and Coaker, 1975). These authors planted Brussels sprouts into an established stand of white clover (*T. repens*) at differing levels of soil cover (25%, 50%, 75% and 100%) and yields were reduced by 38%, 40%, 56% and 81%, respectively.

Andow *et al.* (1986) found that, compared with clean cultivation, cabbage (*B. oleracea* var. *capitata*) head size was smaller in live mulch combinations consisting of cv. Idaho clover as opposed to the dwarf cv. Kent. Further, Ryan *et al.* (1980) found that when cabbage was interplanted with clover in every third row (33% cover) average head weight was 15% higher compared with the no cover treatment. Elmstrom *et al.* (1988) found that when a white clover cover crop was kept mown to 15 cm high the components of yield of green broccoli were statistically equivalent between no cover and living mulch treatments. The combined effects of cropping, tillage and herbicide treatment were described by Mangan *et al.* (1995). There is also evidence that water stress reduces the main (cash) crop yield in a living mulch system, hence there is need for added irrigation to adjust this deficiency. The amount of water required to produce a given cabbage yield increased when perennial ryegrass was added into the system as live mulch (Graham and Crabtree, 1987).

Cover cropping is seen as a particularly appropriate technique for the integrated management of brassicas. This is because the fields are finely tilled before sowing or planting and the soil is frequently wet during harvesting, which may involve the use of mechanical or mechanized harvesting, such as gantry systems, causing serious damage to the soil structure (Stivers-Young, 1998). Many brassicas are relatively inefficient users of nutrients; the rates of fertilizer applied frequently exceed crop demand and excess nutrients, especially nitrogen, are lost through leaching into the soil water and eventually into free-flowing water courses, such as streams and rivers. Nitrogen loss may also take place through denitrification and volatilization. Brassica crops tend to return relatively little organic matter to the soil and leave little surface residue that would protect the soil from wind and water erosion.

ECOLOGY OF CROP–PEST INTERACTIONS IN INTERCROPPING AND COVER CROPPING

The parasitism of brassica-weed insect herbivores that feed on reproductive plant tissues may have positive fitness consequences for charlock (*Sinapis arvensis*). The extent to which plant fitness may change depends on parasitoid lifestyle (solitary or gregarious), which is correlated with the amount of damage inflicted on these tissues by the parasitized host as explained by Gols *et al.* (2015). The interaction of weed species, crop type and the presence of insect pests has been summarized by Schellhorn and Sork (1997). Where the weeds are closely related botanically to the crop type, e.g. cruciferous weeds in a brassica crop, specialist feeding insects, such as flea beetles (*Phyllotreta* spp.), will be encouraged. Where the flora is a mixture of botanical types the numbers of specialist feeding insects will be depressed and more generalist feeders will be encouraged. Botanically mixed flora also encourage natural insect predators, such as coccinellids (see Fig. 6.13), carabids and staphylinids.

In turn, these predators could be responsible for the reduction in the presence of specialist feeding pests, such as imported cabbage worm (small cabbage white butterfly) (*P. rapae*) and diamondback moth larvae (*Plutella xylostella*). Polycultures of cash, inter-row and cover crops often support fewer insect pests at lower densities compared with those found in monocultures (Risch, 1981). Various biotic, structural and microclimatic factors in multispecies

Fig. 6.13. Seven-spot ladybird (*Coccinella septempunctata*). (Geoff Dixon).

plant communities work synergistically in producing pest control. Root (1973) suggested two ways by which this could be achieved. The first is the 'Enemies Hypothesis' which predicts that the increased abundance of insect predators and parasites in species-rich associations can increase control of herbivore populations. Richer plant associations supposedly supply a more favourable environment for predators and parasites, reducing the probability that they will leave or go locally extinct.

These conditions include:

- greater temporal and spatial distribution of nectar and pollen sources, both of which attract natural enemies and increase their reproductive potential;
- increased ground cover provided by a more diverse environment, particularly valuable to nocturnal predators; and
- improved herbivore richness, providing alternative hosts and prey when other hosts and prey are scarce or at inappropriate stages of their life cycle.

The second hypothesis is the 'Resources Concentration Hypothesis' (Root, 1973) and involves changes in the behaviour of the herbivorous insects themselves. Visual and chemical stimuli from the host and non-host plant affect both the rate at which herbivores colonize habitats and their behaviour in those habitats. The total strength of attractive stimuli for any particular herbivore species determines what Root (1973) called 'resource concentration' and it is the result of the following interacting effects:

- the number of host species present and the relative preference of the herbivores for each;
- the absolute density and spatial arrangement of each host species; and
- interference effects from the non-host plants.

An herbivorous insect will have greater difficulty locating a host plant when the relative resource concentration is low (Coaker, 1980, Coaker, 1988). Relative resource concentration may also influence the probability of the herbivore staying in a habitat once it has arrived. For instance, a herbivore may tend to fly sooner, farther or straighter after landing on a non-host plant rather than a host plant, resulting in a more rapid exit from those habitats with lower resource concentrations. Finally, reproductive behaviour can be affected, for example, when a herbivore tends to lay fewer eggs on host plants in an environment of lower resource concentration. The results of Risch's experiments suggest that the 'Resource Concentration Hypothesis' is correct as opposed to the 'Enemies Hypothesis'. The factors of importance were differences in levels of beetle colonization and residency time.

The impact of ground cover mulches on yield and quality is reported by Roberts and Cartwright (1991) and by Costello (1994). Brassica cover crop mulches may affect insects by interfering visually or olfactorily with host plant selection, thus reducing pest dispersal, reproduction and colonization of brassica crops. Beneficial insects, such as ground beetles, are favoured by reduced tillage. It was concluded that rye mulch, for example, offers significant levels of weed suppression for the most critical stages of cabbage and diminishes the populations of several important insect pests. But these improvements are at the expense of yield losses due to difficulties in crop management. The lower populations of diamondback moth, imported cabbage worm (small cabbage white butterfly) and aphids in rye mulch may have been related to the much smaller size and lower head weights of the crop plants. Cabbage planted in rye mulch and treated with Bt insecticide (*Bacillus thuringiensis*) had the lowest insect damage ratings of any of the treatments but yields were still less than those obtained by conventional tillage. A major yield constraint in the rye residue treatments was probably initial soil compaction and later competition with rye and red clover. Soil compaction was caused by equipment movement on wet soil necessary to mow the cover crop prior to planting of the cash crop.

Cabbage aphid (*Brevicoryne brassicae*) is a major pest of green broccoli (*B. oleracea* var. *italica*) (see Chapter 7 section, Important Pests and Pathogens of *Brassica* Crops). It colonizes the developing florets rendering them unmarketable. The effects of living mulch on aphid abundance are directly proportional to the amount of inter-row vegetation present; the aphids colonize more heavily plants surrounded by bare soil compared with those planted in vegetation (A'Brook, 1964, 1968; Gonzales and Rawlins, 1968).

Flea beetle populations are generally lower on brassicas in weedy habitats compared with bare ground monocultures. This is possibly related to their movement and host-finding behaviour. Flea beetles are extremely mobile and their host-finding ability is impeded by non-host odours (Tahvanainen and Root, 1972). Non-host foliage may inhibit movement resulting in faster leaving rates and lower colonization rates in living mulch plots. The response of the cabbage aphid *B. brassicae*, to mixtures of host and non-host species has been even more consistent than that of the crucifer flea beetle (*Phyllotreta cruciferae*). Compared with monocultures, aphid populations were lower in cole crops with weeds in experiments in both the USA and UK.

The response of the small white cabbage butterfly to mixtures of host and non-host plants has been variable both in the USA and UK experiments. The specific relationship between the physical and chemical structure of the cropping system and the precise host-searching behaviour of the small cabbage white butterfly may be critical. Oviposition behaviour of the small cabbage white butterfly is sensitive to plant size and development, plant water content and plant dispersion. Clover inhibited oviposition in the late summer generation of the small cabbage white butterfly but had no effect on oviposition in the midsummer generation.

Cabbage grown in polycropped systems showed less infection by onion thrips (*Thrips tabaci*) in terms of pest incidence and reduced population size and, in consequence, lower levels of physical damage were sustained. Limitation of damage by cabbage root fly (*Delia radicum*) and caterpillars (mainly cabbage moth (*Mamestra brassicae*) with smaller populations of small cabbage white butterfly and diamondback moth and the occurrence of the silver Y moth (*Plusia gamma*) in one experiment only) were quite substantial. The rates of infestation of heads by cabbage gall midges (*Confarinia nasturtii*) were low and evenly distributed across treatments. Feeding damage from flea beetles was low and concentrated in the monocropped plots. The causes of non-marketability in white cabbage (*B. oleracea* var. *capitata*) grown in monocropping and polycropping husbandry systems are reported by Theunissen *et al.* (1995). Markets may demand smaller heads and the increased quality noted where Kent wild white clover was used could be an added incentive to using these husbandry systems (Andow *et al.*, 1986). Intercropping cabbage and beans reduced oviposition by brassica root flies (*D. radicum* and *Delia floralis*) by 29% compared with monocultures (Hofsvang, 1991). There was also a reduction in oviposition when the crops were mixed with 'weeds' where reductions of 63% and 40% in eggs per cabbage plant were recorded in two seasons. Such effects are described by Hofsvang (1991).

Reduced tillage in combination with cover crop mulch systems can conserve beneficial insects. For example, predatory wasps nest in the ground and tillage interferes with their reproduction. Cover crop mulches may also reduce pest dispersal, reproduction and colonization of host plants. Plant compounds released by cover crop residues may influence host plant selection for oviposition and larval feeding. Cover crop mulches can confuse pests visually or olfactorily, reducing colonization of brassica crops. Important visual cues for insects are: leaf colour, area and visual prominence of the hosts. Twice as many cabbage root fly eggs were found on green and yellow models as compared with red or blue ones. Cabbage root fly landings increased linearly with host leaf area. The diamondback moth has a strong preference for egg laying on dark green hosts.

Tillage had a significant effect on cabbage maggot and diamondback moth incidence. Larger numbers of both pests associated with the tillage treatments compared with non-tillage treatments. Rye or hairy vetch (*V. villosa*) can, however, reduce cabbage yields. Rye residues have a high carbon-to-nitrogen ratio and their decomposition could immobilize soil nitrogen thereby reducing cabbage yields.

In several studies particular soil properties were improved by cover cropping and weed and insect pest populations were reduced but yields also fell. Yield reductions could have resulted from the immobilization of soil nitrogen, lower soil temperatures or allelopathy. Strip tillage, which cultivates the row where the brassicas are planted leaving residues between crop rows intact, may overcome reduced vegetable yields while combining with the advantages of both conventional tillage and cover crop mulch systems as identified in Mangan and colleagues' research (Mangan *et al.*, 1995).

This information can be used to predict how mixtures of plants can be used to reduce pest infestation and damage. So far, it has been difficult to make such predictions in relation to the manipulation of brassica habitats. But if animal behaviour is key to such a prediction, then it may be possible to identify herbivore abundance in novel horticultural systems. This requires a basic understanding of the pest's and predator's natural history (habitat preference, diet and general behaviour). That information can be used to suggest the taxonomic and structural plant diversity needed to achieve a degree of resistance developed by placing brassica crops in association with other plants.

OVERSEEDING

Overseeding is a technique used after the crop plants have been established to obtain a uniform stand of living mulch and suppress emerging weeds. It is particularly effective in a non-tillage system where the soil surface is left undisturbed before and after transplanting. Transplanted crops are suitable for overseeding because the transplants have a growth advantage over the post-transplant emergence and growth of weeds and overseeded species. In addition to controlling weeds, overseeding used to form living mulches can minimize erosion, decrease soil temperature, improve water infiltration rate, improve soil structure, favour microbial activity and increase crop yield.

In most non-tillage systems all live vegetation is desiccated with contact herbicide before planting to achieve a weed-free, stale seed bed environment. The stale seed bed is a form of limited tillage normally applied to plough and disc-harrowing systems in which a flush of new weed seedlings germinating after tillage is killed by mechanical or chemical means before planting the cash crop. Often, the desiccated residues rapidly decompose after planting, exposing the soil to potential erosion and soil moisture deficits. Overseeding can maintain the integrity of non-tillage systems by replacing the dead, decaying mulch with a non-competitive living mulch. After harvesting the cash crop and chopping the remaining crop residues to encourage decomposition, the overseeded living mulch becomes a valuable overwintering cover crop. It is claimed to provide a non-tillage system in which there is no competition with the cash crop for resources such as water, nutrients, space and light. The results of overseeding were found to neither increase nor decrease the yield of green broccoli (*B. oleracea* var. *italica*).

ARTIFICIAL CROP COVERS

In a field experiment conducted near Krakow, Poland, from 1991 to 1993, seedlings of green broccoli (*B. oleracea* var. *italica* cv. Corvet F_1) were planted on two dates (in the first and second halves of April) at a spacing of 45×45 cm (Kunicki *et al.*, 1996). Row cover treatments consisted of 0.04-mm-thick perforated polyethylene plastic (100 holes of 10 mm diameter/m²), the polypropylene non-woven fabric Agryl P-17 and control (no cover). Row covers promoted leaf area development and resulted in a harvest that was 3–4 days earlier than in controls. Both covers increased total and marketable yields of main heads, with the two types producing similar results. Mean head weight and diameter were smallest in the control. At the earlier planting date, the yields from plastic film-covered plots were slightly higher than from Agryl plots. Growing under plastic decreased the vitamin C (ascorbic acid) concentration of green broccoli compared with Agryl and the control. Covered plants showed good control of cabbage root fly (*D. radicum*), which destroyed a large number of plants in the control plots. There can, however, be considerable unforeseen ancillary difficulties. For example, in the USA nutsedge (*Cyperus* spp.) are problematic in plastic-mulched vegetable production cabbages because of the weed's rapid reproduction and ability to penetrate the mulch (Randell *et al.*, 2020). Vegetable growers rely heavily on halosulfuron to manage nutsedge species; however, the herbicide cannot be applied over mulch before vegetable transplanting due to potential crop injury. Experimental results confirm that halosulfuron binds to the plastic mulch, remaining active, and is slowly released from the mulch over a substantial period, during rainfall or overhead irrigation events, damaging the crop. The use of crop covers for insect control is treated in Chapter 7 section, Important Pests and Pathogens of *Brassica* Species.

BIOSTIMULANTS

The effects of plants and microbes on the growth and development of other organisms (allelopathy) has been the subject of scientific study for many years, see, for example, Molisch (1922) and Rice (1984). As interest in and the importance of sustainable cropping has increased over the last 20 years it has resulted in the appreciation of the properties of biostimulant compounds and mixtures as part of ICM (Awan and Storer, 2017; di Stasio *et al.*, 2017). Primary sources of biostimulants are seaweed (see Fig. 6.14) extracts as discussed by Dixon and Walsh (1998, 2003).

Research has demonstrated properties of seaweed extracts which include increasing the tolerance of abiotic and biotic stresses, additional root and tuber growth, enhanced seed germination, greater production of chlorophyll and better fruit quality. Plant biostimulants, such as seaweed extracts, provide very effective elements for ICM strategies (di Stasio *et al.*, 2017). The aim of their study was to determine growth, yield, photosynthetic rate, single-photon avalanche diode index, mineral composition and leaf quality of a greenhouse-grown *B. rapa* ssp. *sylvestris* var. *esculenta*, also known as friariello campano, either untreated or treated (via root application) with *Ecklonia maxima* seaweed extract. Root application of seaweed extract increased fresh and dry yield by 13.5% and 16.6%, respectively, as well as hydrophilic antioxidant activity in comparison to untreated plants. The better crop performance of friariello campano ecotype plants treated with seaweed extract was associated with an improvement of their nutritional status (higher phosphate and potassium and lower sodium concentrations), higher photosynthetic rate and higher chlorophyll content in plants, independent of nutrient concentration treatment. The positive effect of seaweed extract application could be attributed to the composition of brown seaweed extracts, especially the presence of polysaccharides and osmoprotectants, such as betaines and betaine-like compounds.

Fig. 6.14. Seaweed kelp (*Ascophyllum nodosum*). (Geoff Dixon).

NUTRIENT MANAGEMENT AND WEED CONTROL

Cropping systems require careful nitrogen management to increase the sustainability of agricultural production (van Bueren and Struik, 2017). One important route towards enhanced sustainability is to increase nitrogen use efficiency. Improving nitrogen use efficiency encompasses increasing nutrient uptake, utilization efficiency and harvest index, each involving many physiological mechanisms and agronomic traits of crops (Blackshaw *et al.*, 2003). Head-forming crops, such as cabbage, depend on the prolonged photosynthesis of outer leaves to provide the carbon sources for the continued nitrogen supply and growth of the photosynthetically less active, younger inner leaves. Short-cycle vegetable crops benefit from early below-ground vigour. Discriminative traits related to nitrogen use efficiency are expressed more effectively under low input rather than under high input. Testing, however, under both low and high input can yield cultivars that are not only adapted to low-input conditions but also respond to high-input conditions, huge genotype-by-environment interaction and the complex behaviour of nitrogen in the cropping system.

The effects of nitrogen fertilizer on herbicide dose-response of weeds were investigated by measuring weed biomass after growth at a range of nitrogen levels and treatment with a range of herbicide doses (Kim *et al.*, 2006). Increasing weed biomass with no herbicide treatment and the response rate of the dose-response curve with increasing nitrogen were successfully described by a linear model and an exponential model, respectively. Conversely, decreasing the ED_{50} (median effective dose) values correlated with increasing nitrogen application rates and followed trends predicted by a logistic model. These models could be combined, providing a standardized dose-response model to describe the interactive effects of herbicide dose and nitrogen levels on weed biomass for the groundkeepers of *B. napus*, *Matricaria perforata*, *Papaver rhoeas* and *Galium aparine*. This provided a decision support system for optimizing the uses of nitrogen fertilizer and herbicide.

Rhizosphere microbial communities are known to be highly diverse and strongly dependent on various attributes of the host plant, such as species, nutritional status and growth stage (O'Brien *et al.*, 2018). Fertilizer type and plant growth were found to result in significantly different rhizosphere bacterial communities, while there was no effect of aphid herbivory. Several operational taxonomic units were identified as varying significantly in abundance between the treatment groups and age cohorts. These included members of the sulfur-oxidizing genus *Thiobacillus*, which was significantly more abundant in organically fertilized 12-week-old cabbages, and the nitrogen-fixing cyanobacteria *Phormidium*, which apparently declined in synthetically fertilized soils relative to controls. These responses may be an effect of accumulating root-derived glucosinolates in the *B. oleracea* rhizosphere and increased nitrogen availability, respectively.

BIOFUMIGATION

Sustainable crop production under changing climatic conditions is crucial for feeding the increasing population of the world (Rehman *et al.*, 2019). Allelopathy is a direct or indirect and positive or negative effect of plant species on other plant species and microorganisms, through the release of secondary metabolites known as allelochemicals. Brassica species are well recognized for their allelopathic potential as most of them endogenously produce potent allelochemicals such as glucosinolates, allyl isothiocyanates and brassinosteroids. Glucosinolates are one of the major bioactive compounds in Brassicaceae plants (Nugroho *et al.*, 2020). They play a role in defence against microbes as well as chemo-preventative activity against human cancer. The temporal relationship between the production of aliphatic glucosinolate compounds and the expression profile of their related genes during growth and development was studied in radish (*R. sativus*), Chinese cabbage (*B. rapa* ssp. *pekinensis*) and their intergeneric hybrid, baemoochae. Over the complete life cycle, glucoraphasatin and glucoraphanin predominated in radish, whereas gluconapin, glucobrassicanapin and glucoraphanin abounded in Chinese cabbage. Baemoochae contained intermediate levels of all glucosinolates studied, indicating inheritance from both radish and Chinese cabbage. Expression patterns of *BCAT4*, *CYP79F1*, *CYP83A1*, *UGT74B1*, *GRS1*, *FMOgs-ox1* and

AOP2 genes showed a correlation to their corresponding encoded proteins in radish, Chinese cabbage and baemoochae. *GRS1* was strongly expressed during leaf development, while both of *FMOgs-ox1* and *AOP2* were manifested high in floral tissues. Furthermore, expression of the *GRS1* gene, which is responsible for glucoraphasatin production, was predominantly expressed in leaf tissues of radish and baemoochae, whereas it was only slightly detected in Chinese cabbage root tissue, explaining why radish has an abundance of glucoraphasatin compared to other brassica plants. Aliphatic glucosinolates biosynthesis is dynamically and precisely regulated in a tissue- and development-dependent manner in Brassicaceae family members.

These allelochemicals are highly phytotoxic to target species when released at high concentrations and, therefore, affect their growth and development (Rehman *et al.*, 2019). They can be used for weed management as cover crops, companion crops, and intercrops, for mulching and residue incorporation, or simply by including them in crop rotations. Residue incorporation of brown mustard (*B. juncea*) frequently uses Caliente mustard cv. Rojo© (see Fig. 6.15). Incorporation is a three-stage operation of cutting down the mustard crop and chopping it into small fragments which helps release the glucosinolates and bed-forming for the ware crop (see Fig. 6.16).

Most of the allelochemicals can also act as crop growth promoters when released or applied at low concentrations. The use of ecological options like allelopathy may help in achieving global food security sustainably.

Fig. 6.15. Caliente mustard (*Brassica juncea*) Rojo© crop. (Courtesy of Tozer Seeds).

Fig. 6.16. Soil incorporation of Caliente mustard (*Brassica juncea*) Rojo© crop. (Courtesy of Tozer Seeds).

Soil-less cultivation systems are increasingly used to produce high-quality salad rocket (*Eruca sativa*), appreciated by consumers for its pungent taste, due to the content of glucosinolates. Studies compared the profiles of *Diplotaxis erucoides, Diplotaxis tenuifolia* and *E. sativa* grown side-by-side under a protected environment in a conventional soil and in a soil-less system. The latter cultivation may be adopted to increase the glucosinolate content and enhance the nutritional quality of leafy brassica vegetables (di Gioia *et al.*, 2018).

Technological advancements in light-emitting diode (LED) technology have led to the production, under controlled indoor conditions, of value-added crops that are high in nutritional or nutraceutical contents (Tan *et al.*, 2020). In this study, the growth and glucosinolate profiles of *B. rapa* var. *parachinensis* (choy sum or caisin) were determined under LED lighting. A shift from a high proportion of aliphatic glucosinolates in one-leafed seedlings to indolic and aromatic glucosinolates in three-leafed seedlings and adult plants was attributed to an increase in the proportion of glucobrassicin and/ or 4-hydroxy-glucobrassicin and gluconasturtiin, which are known to have anti-cancer properties.

Little is known about the effects of different light sources on the growth and quality of non-heading Chinese cabbage (*B. rapa*) (Li *et al.*, 2012). The objective of their study was to evaluate the effects of LED light sources (blue, blue plus red, red), fluorescent lamps and sunlight on growth and vitamin C, soluble protein, sucrose, soluble sugar, starch and pigment concentrations in non-heading Chinese cabbage seedlings. Results demonstrated that LED light sources are more effective than fluorescent lamps for the vegetative and reproductive growth of non-heading Chinese cabbage. Moreover, blue LEDs benefit vegetative growth, while red LEDs and blue plus red LEDs support reproductive growth in non-heading Chinese cabbage. In the artificial cultivation and subsequent transplanting of the life cycle of plants, the light source can be selected to meet the requirements of different growth stages of plants and be used to promote the subsequent process in the industrial production of non-heading Chinese cabbage.

CHEMICAL MANAGEMENT OF WEEDS

For a generation (1960–1990), the 'Chemical Revolution' discovered more and more sophisticated synthetic, organic active ingredients that were then developed and marketed. Significantly, increases in their costliness resulting from more complex manufacturing pathways and greater product scrutiny for human and environmental toxicity shifted agrochemical companies' business focus primarily towards the needs of large-scale (broadacre) arable crops. These included first, barley, then later, wheat, rice, soya, maize, sugarbeet, potatoes and grapes, and now includes oilseed rape. Initially, these crops required substantial volumes of spray applied from increasingly large, heavy mechanical sprayers, justifying the use of expensive chemicals. Consequently, field vegetable crops, and brassicas in particular by force of circumstances, have where appropriate used and modified equipment and agrochemicals initially developed for broadacre crops (see Chapter 7 section, Introduction). Few, if any, new agrochemicals are now either researched or marketed specifically for the needs of vegetable crops since these are considered as 'small acreage' crops with minimal profitability for their manufacturers. None the less, market demand continues for fresh produce, including vegetable brassicas that are of uniform high quality and with blemish-free freshness. Consequently, there are continuing requirements for the use of agrochemicals. the development of mechanical and automated weeders, and more sophistica ted pest and pathogen controls, especially in brassica vegetables as discussed a generation ago by Hoyt *et al.* (1996). In some instances, forms of biological control are either used or being developed (see also Chapter 7 section, Introduction).

Herbicides may be selective or non-selective in their mode of action. Selective herbicides destroy specific plant species and may thus be used to eliminate them from populations of other species. Probably the most widely used selective herbicide for use in brassica crops has been trifluralin (α,α,α-trifluoro-2,6-dinitro-*N*,*N*-dipropyl-*p*-toluidine). Non-selective herbicides, such as glyphosate (*N*-(phosphonomethyl)glycine) and paraquat (1,1'-dimethyl-4, 4'-bipyridinium), will destroy all green tissues that they come into contact with. They have been used, where permitted, for destroying weeds that germinate in stale seed beds prior to drilling crops. These two chemicals are inactivated once they come into contact with soil. Neither kills all the weed species commonly present in brassica crops so their spectra of activity are largely complementary and they can be applied in combination (Roberts and Bond, 1975). Herbicides are also classified according to the timing of their application in relation to crop growth stage. These are broadly the pre-emergence and post-emergence categories. Those with pre-emergence characteristics are applied prior to or after sowing the seeds of brassica crops and prior to their emergence. Post-emergence herbicides are applied once the crop has emerged to destroy competing weed species.

PRECISION SPRAY APPLICATION

Reducing spray volumes and the areas treated are now a major goals for agrochemical and spraying machinery manufacturers, regulators and producers. The objectives are the distribution of minimal quantities of active ingredient commensurate with maximum effect, preventing the establishment or progress of weeds, pests or pathogens (Dixon, 1984). Brassica crops which are grown either in rows or beds are efficient candidates for precision control of weeds, pests and pathogens by targeting unwanted organisms and avoiding areas of healthy crop. Interest in inter-row band spraying and intra-row spot treatment was driven by the loss of post-emergent herbicides and calls for reducing overall pesticide load. Increasing application of IPM strategies involves this approach (Clayton, 2014). Precision farming techniques, involving GPS and other guidance techniques applying crop protection products accurately, encourage band spraying. The benefits of band spraying include weeds being targeted more precisely and wider ranges of active ingredients being used, particularly in the early season, reducing dependence on selective herbicides. Band spraying can also deliver fertilizers, fungicides or insecticides directly into the row. Current commercial use in brassica crops estimates that up to 60 ha could be treated per day. Hooded sprayers will accurately deliver minimal quantities of crop protection chemicals directly between crop rows (see Fig. 6.17).

Reducing spray volumes cuts the high costs of transporting water, minimizes crop and soil structural damage and is environmentally beneficial. Formulations which maximize the retention of active ingredient on the target reduce damaging spray drift. Monitoring and control equipment which electronically measures sprayer performance against previously set targets started emerging in the late 1970s (Givelet, 1981). Precision spraying has four objectives: operational efficiency, optimum chemical use, operator safety and environmental care. Recent commercial achievements were described by Alun James (Sagentia Innovation, Cambridge, UK) at the 2020 British Crop Protection Council's Virtual Congress (O'Driscoll, 2021). Developments he described included variable rate control which allows changes in volume depending on tractor speed. Flow rates can be varied in different sections of the boom reducing, for example, the chances of double dosing. This report

Fig. 6.17. Robocrop-guided hooded sprayer complete with band sprayer. (Courtesy of Garford Farm Machinery Ltd).

further highlighted the John Deere Company's Exact Apply system which allows control of dosages at individual nozzles on a boom and their monitoring and documentation. Also reported were computerized vision and machine learning, in conjunction with individualized nozzles, which permit targeted spray applications down to 5 cm using Blue River's See & Spray developments. This has the potential for reducing agrochemical use by 90%.

LIMITED AVAILABILITY OF HERBICIDES FOR *BRASSICA* CROPS

In the USA a collaborative project was established between state agricultural research stations, the US Department of Agriculture–Cooperative States Research Service (USDA–CSRS), chemical manufacturers and growers' organizations to identify and register herbicides for use with speciality crops such as the brassicas (Hopen, 1995). This initiative was driven by the lack of suitable chemicals being placed on the market by the chemical industry for these small area crops. The project (IR-4), with its headquarters at Rutgers University, New Jersey, aimed to assist in gaining label recommendations for use on highly intensive vegetable crops (Baron *et al.*, 2004). Labels were of national, regional and state need, and state emergency categories. The programme succeeded in expanding the availability of herbicides for brassica crops. In particular, the participants attempted to find molecules that were safe and effective for use as selective herbicides applied to established crops.

Weeds such as hairy galinsoga (*Galinsoga ciliata*), wild proso millet (*Panicum miliaceum*), velvet leaf (*Abutilon theophrasti*) and yellow nutsedge

(*Cyperus esculentus*) are difficult to control in standing crops and cause considerable loss of yield and quality. Similar schemes are under development in Canada and Australia.

A similar scheme established in the UK was operated by the Agricultural and Horticultural Development Board (AHDB) and known as the Extension of Authorisation for Minor Use (EAMU) scheme. Recently, this scheme has been privatized and is now managed by producer organizations. Parallel schemes are operated in the European Union (EU). In the UK, AHDB assembled a dossier of information for a particular active ingredient chemical that might be valuable for growers in formulations available from agrochemical companies following field testing over several seasons and locations. The dossier was then submitted to the Chemicals Safety Directorate of the Health and Safety Executive (CRD-HSE) for approval and authorization.

Opportunities for the use of chemicals in vegetable crop production are now severely limited by national governments and international organizations such as the EU. Over the last 30 years more than 60% of products have been removed from the market as their authorizations have either expired or were revoked. As an early example, withdrawal of nitrofen (2,4-dichlorophenyl *p*-nitrophenyl ester) made the control of broad-leaved weeds in all cole crops difficult (Bhowmik and McGlew, 1986) and this chemical reduction process has increased and accelerated. Additionally, most brassicas are traded through the supermarkets who impose their own restrictions on the use of agrochemicals in crop production based on their aspirations for appearing to provide very low or residue-free fresh produce. Currently, the larger manufacturers are seeking active ingredients originating from or exploiting forms of biological control. These are environmentally sustainable and leave no residues but require considerable and detailed understanding of the aerial and edaphic conditions essential for successful commercial use.

PHYSICAL CONTROL METHODS

Weed control can be achieved by physical as well as chemical methods. Automatically detecting the position and density of weeds using computer vision and applying a herbicide treatment, using chemical and mechanical methods, is described by Blasco *et al.* (1998). Image analyses provided a real-time machine vision system that passed information to a moving robot which then applied an electric discharge, destroying the weed. This system is based on a Bayesian algorithm for segmenting images. It requires prior training by an expert who selects areas of different images, trying to represent the colour variability of the plants, the soil and the weeds. After segmentation, pixels belonging to soil classes are correctly classified and morphological operations are applied, discriminating between plants and weeds. This system was able to locate more than 90% of weeds with very little confusion with the crop (1%) in lettuce in under 500 ms.

Studies of GPS technology in Germany concluded that weed control based on GPS guidance of machinery and implements is faster, more cost-effective and more precise (Schwarz and Hege, 2014). Sensors and mapping techniques support intelligent cultivators which target the intra-row spaces within crop rows (Rasmussen *et al.*, 2012). The Danish study by Rasmussen *et al.* (2012) investigated first, an intelligent rotor tine cultivator (the cycloid hoe) for crop and weed selection and second, the synergistic effect between punch planting and post-emergence weed harrowing, in terms of improved crop-weed selectivity. This combination did not give sufficient selectivity between crop and weeds. Use of cutting implements, instead of tine implements is advocated.

WEED CONTROL IN PRACTICE

In practice, there are now very few agrochemicals suitable for post-emergence weed control in brassicas (Andrew Richardson, 2021, Allium and Brassica Centre, Kirton, UK, personal communication). Consequently, after a pre- or post-transplanting residual herbicide applications, inter-row cultivations are used for the control of germinating weeds. There are problems with this approach as crops may become too large and their quality is damaged by the passage of machinery. Additionally, wet weather makes weed control difficult or impossible and attendant soil disturbance produces the germination of another flush of weeds. Garford Farm Machinery Ltd (https://garford.com) manufactures a precision guided hoe which controls weeds between the rows and uses vision guidance for within-row working. Practical problems with mechanical weed control remain however, such as difficulties of working wet soil, crops becoming too large and consequent damaging of quality.

An electrical weeder that trails electrodes just above the soil and discharges 15,000 V that travel from the electrode down through the weed foliage, stems and roots before returning to an earthing wheel completing the circuit has been tested in practice. Discharges of 15,000 V destroy plant tissue. These machines operate most effectively on weeds with a low dry matter content. Eradicating woody perennial weeds, such as established nettles (*Urtica* spp.), proved more difficult. This control method does not disturb the soil. Consequently, further weed germination is not encouraged and the machine can operate in wet conditions. Small lightweight robots are in their developmental stages with capabilities for undertaking a range of tasks, including autonomous non-chemical weed control (O'Driscoll, 2021).

WEED CONTROL IN *BRASSICA* SEED CROPS

Weed competition may reduce seed yield dramatically in cabbage, by more than 50% (Al-Khatib and Libbey, 1992). The presence of weeds at seed harvest increases mechanical damage to cabbage seed and reduces harvest efficiency. Weeds also reduce seed quality by interfering with the processing operations.

Weeds such as wild mustard (*Brassica kaber*) and bedstraw (or cleavers) (*G. aparine*) are of similar size to the cabbage seed, making separation during cleaning operations difficult. Crop seed contaminated by weeds will be rejected during seed testing and certification as regulated by the International Seed Testing Association (ISTA) and controlled by national and international statutory legislation. Weed control in seed crops is made more difficult because of the extended growing seasons required. The seed production cycle is at least 12 months in extent and in some crops even longer. Some herbicides are available for the control of grass weeds but the control of broadleaf weeds is limited by the lack of effective registered chemicals (Al-Khatib *et al.*, 1995).

REFERENCES

A'Brook, J. (1964) The effect of planting date and spacing on the incidence of groundnut rosette disease and of the vector, *Aphis craccivora* Koch, at Mokwa, northern Nigeria. *Annals of Applied Biology* 54(2), 199–208. DOI: 10.1111/j.1744-7348.1964.tb01184.x.

A'Brook, J. (1968) The effect of plant spacing on the numbers of aphids trapped over the groundnut crop. *Annals of Applied Biology* 61(2), 289–294. DOI: 10.1111/j.1744-7348.1968.tb04533.x.

Al-Khatib, K. and Libbey, C. (1992) Weed control in vegetable seed crops. *Proceedings of the Washington Horticultural Association* 83, 30–35.

Al-Khatib, K., Libbey, C. and Kadir, S. (1995) Broadleaf weed control and cabbage seed yield following herbicide application. *HortScience* 30(6), 1211–1214. DOI: 10.21273/HORTSCI.30.6.1211.

Altieri, M.A., Wilson, R.C. and Schmidt, L.L. (1985) The effects of living mulches and weed cover on the dynamics of foliage- and soil-arthropod communities in three crop systems. *Crop Protection* 4(2), 201–213. DOI: 10.1016/0261-2194(85)90018-3.

Andow, D.A., Nicholson, A.G., Wien, H.C. and Willson, H.R. (1986) Insect populations on cabbage grown with living mulches. *Environmental Entomology* 15(2), 293–299. DOI: 10.1093/ee/15.2.293.

Anon (2019) *Principles of Soil Management*. Agriculture and Horticulture Development Board (AHDB) and British Beet Research Organisation (BBRO), Stoneleigh, UK, 20 pp.

Arnhold, S., Lindner, S., Lee, B., Martin, E., Kettering, J. *et al.* (2014) Conventional and organic farming: soil erosion and conservation potential for row crop cultivation. *Geoderma* 219–220, 89–105. DOI: 10.1016/j.geoderma.2013.12.023.

Awan, S. and Storer, K. (2017) *Crop Biostimulants: Factsheet*. Agriculture and Horticulture Development Board (AHDB), Stoneleigh, UK.

Baker, H.G. and Stebbins, G.L. (eds) (1965) *The Genetics of Colonizing Species; Proceedings of the First International Union of Biological Sciences Symposium on General Biology*. Academic Press, New York.

Barnes, J.P. and Putnam, A.R. (1983) Rye residues contribute weed suppression in no-tillage cropping systems. *Journal of Chemical Ecology* 9(8), 1045–1057. DOI: 10.1007/BF00982210.

Baron, J., Kunkel, D. and Holm, R. (2004) The role of the IR-4 project in pest management for fruit, vegetable and otherspeciality crops in the United States.

In: *Advances in Applied Biology: Providing New Opportunities for Consumers and Producers in the 21st Century. Association of Applied Biologists Meeting 15–17 December 2004.* Abstracts obtainable from Association of Applied Biologists, Warwick, UK and Horticulture Research International, Wellesbourne, UK. (unpaginated).

Barré, K., Le Viol, I., Julliard, R. and Kerbiriou, C. (2018) Weed control method drives conservation tillage efficiency on farmland breeding birds. *Agriculture, Ecosystems & Environment* 256, 74–81. DOI: 10.1016/j.agee.2018.01.004.

Batte, M.T. and Ehsani, M.R. (2006) The economics of precision guidance with auto-boom control for farmer-owned agricultural sprayers. *Computers and Electronics in Agriculture* 53(1), 28–44. DOI: 10.1016/j.compag.2006.03.004.

Bhowmik, P.C. and McGlew, E.N. (1986) Effects of oxyfluorfen as a pretransplant treatment on weed control and cabbage yield. *Journal of the American Society for Horticultural Science* 111(5), 686–689. DOI: 10.21273/JASHS.111.5.686.

Blackshaw, R.E., Brandt, R.N., Janzen, H.H., Entz, T., Grant, C.A. *et al.* (2003) Differential response of weed species to added nitrogen. *Weed Science* 51(4), 532–539. DOI: 10.1614/0043-1745(2003)051[0532:DROWST]2.0.CO;2.

Blasco, J., Benlloch, J.V., Agusti, M. and Molto, E. (1998) Machine vision for precise control of weeds. In: Meyer, G.E. and DeShazer, J.A. (eds) *Precision Agriculture and Biological Quality. Proceedings of SPIE 3543.* SPIE, Boston, MA, pp. 336–343.

Brandsæter, L.O. (1996) Alternative strategies for weed control: plant residues and living mulch for weed management in vegetables. Doctor Scientiarum Thesis, 25. Norges Landbrukshøgskole (Agricultural University of Norway), Ås, Norway.

Brandsæter, L.O., Netland, J. and Meadow, R. (1998) Yields, weeds, pests and soil nitrogen in a white cabbage-living mulch system. *Biological Agriculture & Horticulture* 16(3), 291–309. DOI: 10.1080/01448765.1998.10823201.

Bronick, C.J. and Lal, R. (2005) Soil structure and management: a review. *Geoderma* 124(1–2), 3–23. DOI: 10.1016/j.geoderma.2004.03.005.

Calegari, F., Tassi, D. and Vincini, M. (2013) Economic and environmental benefits of using a spray control system for the distribution of pesticides. *Journal of Agricultural Engineering* 44(2s), 163–165. DOI: 10.4081/jae.2013.274.

Camarotto, C., Piccoli, I., Dal Ferro, N., Polese, R., Chiarini, F. *et al.* (2020) Have we reached the turning point? Looking for evidence of SOC increase under conservation agriculture and cover crop practices. (Special issue: Opportunities and challenges in no-till farming.) *European Journal of Soil Science* 71(6), 1050–1063. DOI: 10.1111/ejss.12953.

Chancellor, R.J. (1982) Weed seed investigations. *Advances in Research and Technology of Seeds* 7, 9–29.

Clayton, J.S. (2014) The Varidome precision band sprayer for row crops. *Aspects of Applied Biology* 122, 55–62.

Coaker, T.H. (1980) Insect pest management in Brassica crops by inter-cropping. *SROP/WPRS Bulletin* 111(1), 117–125.

Coaker, T.H. (1988) Insect pest management by intracrop diversity: potential and limitations. In: Cavalloro, R. and Pelerents, C. (eds) *Progress on Pest Management in Field Vegetables.* A.A. Balkema, Rotterdam, pp. 281–288. DOI: 10.1201/9781003079347.

Costello, M.J. (1994) Broccoli growth, yield and level of aphid infestation in leguminous living mulches. *Biological Agriculture & Horticulture* 10(3), 207–222. DOI: 10.1080/01448765.1994.9754669.

Cousens, R. (1985) A simple model relating yield loss to weed density. *Annals of Applied Biology* 107(2), 239–252. DOI: 10.1111/j.1744-7348.1985.tb01567.x.

Cutulle, M., Campbell, H., Couillard, D.M., Ward, B. and Farnham, M.W. (2019) Pre transplant herbicide application and cultivation to manage weeds in south-eastern broccoli production. *Crop Protection* 124, 104862. DOI: 10.1016/j.cropro.2019.104862.

Dabney, S.M., Delgado, J.A. and Reeves, D.W. (2001) Using winter cover crops to improve soil and water quality. *Communications in Soil Science and Plant Analysis* 32(7–8), 1221–1250. DOI: 10.1081/CSS-100104110.

Dempster, J.P. and Coaker, T.H. (1974) Diversification of crop ecosystems as a means of controlling pests. In: Jones, D.P. and Soloman, M.E. (eds) *Biology in Pest & Disease Control; 13th Symposium of the British Ecological Society.* Blackwell Scientific Publications, Oxford, pp. 106–114.

di Gioia, F., Avato, P., Serio, F. and Argentieri, M.P. (2018) Glucosinolate profile of *Eruca sativa, Diplotaxis tenuifolia* and *Diplotaxis erucoides* grown in soil and soilless systems. *Journal of Food Composition and Analysis* 69, 197–204. DOI: 10.1016/j.jfca.2018.01.022.

di Stasio, E., Rouphael, Y., Colla, G., Raimondi, G., Giordano, M. *et al.* (2017) The influence of *Ecklonia maxima* seaweed extract on growth, photosynthetic activity and mineral composition of *Brassica rapa* L. subsp. *sylvestris* under nutrient stress conditions. *European Journal of Horticultural Science* 82(6), 286–293. DOI: 10.17660/eJHS.2017/82.6.3.

Dixon, G.R. (1984) *Plant Pathogens and Their Control in Horticulture.* Macmillan, London. DOI: 10.1007/978-1-349-06923-1.

Dixon, G.R. and Walsh, U.F. (1998) Suppression of plant pathogens by organic extracts a review. *Acta Horticulturae* 469, 383–390. DOI: 10.17660/ActaHortic.1998.469.41.

Dixon, G.R. and Walsh, U.F. (2003) Suppressing *Pythium ultimum* induced damping-off in cabbage seedlings by biostimulation with proprietary liquid seaweed extracts. In: *Symposium S8, 26th International Horticulture Congress, 11–16 August 2002.* Toronto, Canada.

Donald, C.M. (1958) The interaction of competition for light and for nutrients. *Australian Journal of Agricultural Research* 9(4), 421–425. DOI: 10.1071/AR9580421.

Egley, G.H. and Williams, R.D. (1990) Decline of weed seeds and seedling emergence over five years as affected by soil disturbances. *Weed Science* 38(6), 504–510. DOI: 10.1017/S0043174500051389.

Elmstrom, K.M., Andow, D.A. and Barclay, W.W. (1988) Flea beetle movement in a broccoli monoculture and diculture. *Environmental Entomology* 17(2), 299–305. DOI: 10.1093/ee/17.2.299.

Fenner, M. (1985) *Seed Ecology.* Chapman & Hall, London. DOI: 10.1007/978-94-009-4844-0.

Fennimore, S.A. and Cutulle, M. (2019) Robotic weeders can improve weed control options for specialty crops. *Pest Management Science* 75(7), 1767–1774. DOI: 10.1002/ps.5337.

Fogg, J.M. (1975) The silent travellers. *Brooklyn Botanic Garden Records* 31, 12–15.

Fracchiolla, M., Renna, M., D'Imperio, M., Lasorella, C., Santamaria, P. *et al.* (2020) Living mulch and organic fertilization to improve weed management, yield and quality of broccoli raab in organic farming. *Plants (Basel, Switzerland)* 9(2), 177. DOI: 10.3390/plants9020177.

Freyman, S., Brookes, V.R. and Hall, J.W. (1992) Effect of planting pattern on intrarow competition between cabbage and shepherd's purse (*Capsella bursa-pastoris*). *Canadian Journal of Plant Science* 72(4), 1393–1396. DOI: 10.4141/cjps92-172.

Furness, N.H., Jolliffe, P.A. and Upadhyaya, M.K. (2005) Competitive interactions in mixtures of broccoli and *Chenopodium album* grown at two UV-B radiation levels under glasshouse conditions. *Weed Research* 45(6), 449–459. DOI: 10.1111/j.1365-3180.2005.00476.x.

Garrison, A.J., Miller, A.D., Ryan, M.R., Roxburgh, S.H. and Shea, K. (2014) Stacked crop rotations exploit weed-weed competition for sustainable weed management. *Weed Science* 62(1), 166–176. DOI: 10.1614/WS-D-13-00037.1.

Gée, Ch., Bossu, J., Jones, G. and Truchetet, F. (2008) Crop/weed discrimination in perspective agronomic images. *Computers and Electronics in Agriculture* 60(1), 49–59. DOI: 10.1016/j.compag.2007.06.003.

Givelet, M.P. (1981) Electronic control systems in pesticide application machinery. *Outlook on Agriculture* 10(7), 357–360. DOI: 10.1177/003072708101000710.

Gols, R., Wagenaar, R., Poelman, E.H., Kruidhof, H.M., van Loon, J.J.A. *et al.* (2015) Fitness consequences of indirect plant defence in the annual weed, *Sinapis arvensis*. *Functional Ecology* 29(8), 1019–1025. DOI: 10.1111/1365-2435.12415.

Gonzales, D. and Rawlins, W.A. (1968) Aphid sampling efficiency of Möericke traps affected by height and background12. *Journal of Economic Entomology* 61(1), 109–114. DOI: 10.1093/jee/61.1.109.

Graham, M.B. and Crabtree, G. (1987) Management of competition for water between cabbage (*Brassica olerace*) and perennial ryegrass (*Lolium perenne*) living mulch. *Proceedings of the Western Society for Weed Science* 40, 113–117.

Grundy, A.C., Mead, A. and Burston, S. (1999) Modelling the effect of cultivation on seed movement with application to the prediction of weed seedling emergence. *Journal of Applied Ecology* 36(5), 663–678. DOI: 10.1046/j.1365-2664.1999.00438.x.

Heydecker, W. (ed.) (1972) *Seed Ecology: Proceedings of the 19th Easter School in Agricultural Science*. Butterworth, London.

Hill, T.A. (1977) *The Biology of Weeds. The Institute of Biology's Studies in Biology No. 79*. Edward Arnold, London.

Hofsvang, T. (1991) The influence of Intercropping and weeds on the oviposition of the Brassica root flies (*Delia radicum* and D. *floralis*). *Norwegian Journal of Agricultural Sciences* 5, 349–356.

Hopen, H.J. (1995) Herbicides available for commercial cabbage producers during 1965-94. *HortTechnology* 5(1), 25–26. DOI: 10.21273/HORTTECH.5.1.25.

Hoyt, G.D., Bonanno, A.R. and Parker, G.C. (1996) Influence of herbicides and tillage on weed control, yield, and quality of cabbage (*Brassica oleracea* L. var. *capitata*). *Weed Technology* 10(1), 50–54. DOI: 10.1017/S0890037X00045693.

Jaffuel, G., Blanco-Pérez, R., Büchi, L., Mäder, P., Fließbach, A. *et al.* (2017) Effects of cover crops on the overwintering success of entomopathogenic nematodes and their antagonists. *Applied Soil Ecology* 114, 62–73. DOI: 10.1016/j.apsoil.2017.02.006.

Jokela, D. and Nair, A. (2016) No tillage and strip tillage effects on plant performance, weed suppression, and profitability in transitional organic broccoli production. *HortScience* 51(9), 1103–1110. DOI: 10.21273/HORTSCI10706-16.

Kim, D.S., Marshall, E.J.P., Caseley, J.C. and Brain, P. (2006) Modelling interactions between herbicide and nitrogen fertiliser in terms of weed response. *Weed Research* 46(6), 480–491. DOI: 10.1111/j.1365-3180.2006.00531.x.

Knavel, D.E. and Herron, J.W. (1981) Influence of tillage system, plant spacing, and nitrogen on head weight, yield, and nutrient concentration of spring cabbage. *Journal of the American Society for Horticultural Science* 106(5), 540–545. DOI: 10.21273/JASHS.106.5.540.

Kunicki, E., Cebula, S., Libik, A. and Siwek, P. (1996) The influence of row cover on the development and yield of broccoli in spring production. *Acta Horticulturae* 407, 377–384. DOI: 10.17660/ActaHortic.1996.407.48.

Kurstjens, D.A.G. and Perdok, U.D. (2000) The selective soil covering mechanism of weed harrows on sandy soil. *Soil and Tillage Research* 55(3–4), 193–206. DOI: 10.1016/S0167-1987(00)00128-8.

Laber, H. and Stützel, H. (2000) Nebenwirkungen mechanischer unkrautregulationsmašnahmen im gemusebau. *Zeitschrift für Pflanzenkrankheiten und Pflanzenschutz* 17, 653–660.

Lawson, H.M. (1972) Weed competition in transplanted spring cabbage. *Weed Research* 12(3), 254–267. DOI: 10.1111/j.1365-3180.1972.tb01215.x.

Lekasi, J.K., Woomer, P.L., Tenywa, J.S., Bekunda, M.A. and Bekunda, M.A. (2001) Effect of mulching cabbage with banana residues on cabbage yield, soil nutrient and moisture supply, soil biota and weed biomass. *African Crop Science Journal* 9(3), 499–506. DOI: 10.4314/acsj.v9i3.27596.

Li, H., Tang, C., Xu, Z., Liu, X. and Han, X. (2012) Effects of different light sources on the growth of non-heading chinese cabbage (*Brassica campestris* L.). *Journal of Agricultural Science* 4(4), 262–273. DOI: 10.5539/jas.v4n4p262.

Mangan, F., DeGregorio, R., Schonbeck, M., Herbert, S., Guillard, K. *et al.* (1995) Cover cropping systems for brassicas in the northeastern United States. *Journal of Sustainable Agriculture* 5(3), 15–36. DOI: 10.1300/J064v05n03_04.

Miller, A.B. and Hopen, H.J. (1991) Critical weed-control period in seeded cabbage (*Brassica oleracea* var. *capitata*). *Weed Technology* 5(4), 852–857. DOI: 10.1017/S0890037X00033972.

Mohler, C.L. (1993) A model of the effects of tillage on emergence of weed seedlings. *Ecological Applications* 3(1), 53–73. DOI: 10.2307/1941792.

Molisch, H. (1922) *Der Einfluss einer Pflanze auf die Andere, Allelopathie.* Fischer, Jena, Germany.

Morse, R. (1999) Cultural weed management methods for high-residue/no-till production of transplanted broccoli (*Brassica oleracea* var. *italica*). *Acta Horticulturae* 504, 121–128. DOI: 10.17660/ActaHortic.1999.504.13.

Morse, R.D. and Seward, D.L. (1986) No-tillage production of broccoli and cabbage. *Applied Agricultural Research* 1, 96–99.

Moss, S.R. (1985) The survival of *Alopecurus myosuroides* Huds. seeds in soil. *Weed Research* 25(3), 201–211. DOI: 10.1111/j.1365-3180.1985.tb00636.x.

Nicholson, A.G. and Wien, H.C. (1983) Screening turfgrass and clovers for use as living mulch cropping system. MS Thesis, Cornell University, Ithaca, NY. DOI: 10.21273/JASHS.108.6.1071.

Nieto, H.J., Brondo, M.A. and Gonzalez, J.T. (1968) Critical periods of the crop growth period for competition from weeds. *PANS (Pest Articles and News Summaries) Section C* 14(2), 159–166. DOI: 10.1080/05331856809432576.

Nugroho, A.B.D., Han, N., Pervitasari, A.N., Kim, D.-H. and Kim, J. (2020) Correction to: differential expression of major genes involved in the biosynthesis of aliphatic glucosinolates in intergeneric baemoochae (Brassicaceae) and its parents during development. *Plant Molecular Biology* 102(4–5), 569. DOI: 10.1007/s11103-020-00970-8.

O'Brien, F.J.M., Dumont, M.G., Webb, J.S. and Poppy, G.M. (2018) Rhizosphere bacterial communities differ according to fertilizer regimes and cabbage (*Brassica oleracea* var. *capitata* L.) harvest time, but not aphid herbivory. *Frontiers in Microbiology* 9, 1620. DOI: 10.3389/fmicb.2018.01620.

O'Donnell, M.S. and Coaker, T.H. (1975) Potential of intra-crop diversity for the control of Brassica pests. *Proceedings 8th British Insecticide and Fungicide Conference* 1, 101–107.

Odoux, J.F., Aupinel, P., Gateff, S., Requier, F., Henry, M. *et al.* (2014) ECOBEE: a tool for long-term honey bee colony monitoring at the landscape scale in West European intensive agroecosystems. *Journal of Apicultural Research* 53(1), 57–66. DOI: 10.3896/IBRA.1.53.1.05.

O'Driscoll, C. (2021) Pesticide dilemma. *Chemistry & Industry* 85(1), 22–25. DOI: 10.1002/cind.851_7.x.

Palhano, M.G., Norsworthy, J.K. and Barber, T. (2018) Evaluation of chemical termination options for cover crops. *Weed Technology* 32(3), 227–235. DOI: 10.1017/wet.2017.113.

Panetsos, C.A. and Baker, H.G. (1986) The origin of variation in "wild" *Raphanus sativus* (*Cruciferae*) in California. *Genetica* 38(1), 243–274. DOI: 10.1007/BF01507462.

Partel, V., Charan Kakarla, S. and Ampatzidis, Y. (2019) Development and evaluation of a low-cost and smart technology for precision weed management utilizing artificial intelligence. *Computers and Electronics in Agriculture* 157, 339–350. DOI: 10.1016/j.compag.2018.12.048.

Peteinatos, G.G., Weis, M., Andújar, D., Rueda Ayala, V. and Gerhards, R. (2014) Potential use of ground-based sensor technologies for weed detection. *Pest Management Science* 70(2), 190–199. DOI: 10.1002/ps.3677.

Pfeiffer, A., Silva, E. and Colquhoun, J. (2016) Living mulch cover crops for weed control in small-scale applications. *Renewable Agriculture and Food Systems* 31(4), 309–317. DOI: 10.1017/S1742170515000253.

Qasem, J.R. (2009) Weed competition in cauliflower (*Brassica oleracea* L. var. *botrytis*) in the Jordan Valley. *Scientia Horticulturae* 121(3), 255–259. DOI: 10.1016/j.scienta.2009.02.010.

Qasem, J.R. and Hill, T.A. (1993) A comparision of the competitive effects and nutrient accumulation of fat-hen and groundsel. *Journal of Plant Nutrition* 16(4), 679–698. DOI: 10.1080/01904169309364566.

Randell, T.M., Vance, J.C. and Culpepper, A.S. (2020) Broccoli, cabbage, squash and watermelon response to halosulfuron preplant over plastic mulch. *Weed Technology* 34(2), 202–207. DOI: 10.1017/wet.2019.75.

Rasmussen, J., Griepentrog, H.W., Nielsen, J. and Henriksen, C.B. (2012) Automated intelligent rotor tine cultivation and punch planting to improve the selectivity of mechanical intra-row weed control. *Weed Research* 52(4), 327–337. DOI: 10.1111/j.1365-3180.2012.00922.x.

Rehman, S., Shahzad, B., Bajwa, A.A., Hussain, S., Rehman, A. *et al.* (2019) Utilizing the allelopathic potential of brassica species for sustainable crop production: a review. *Journal of Plant Growth Regulation* 38(1), 343–356. DOI: 10.1007/s00344-018-9798-7.

Rice, E.L. (1984) *Allelopathy*. Academic Press, New York.

Risch, S.J. (1981) Insect herbivore abundance in tropical monocultures and polycultures: an experimental test of two hypotheses. *Ecology* 62(5), 1325–1340. DOI: 10.2307/1937296.

Roberts, B.W. and Cartwright, B. (1991) Alternative soil and pest management prac-
tices for sustainable production of fresh-market cabbage. *Journal of Sustainable Agriculture* 1(3), 21–35. DOI: 10.1300/J064v01n03_03.

Roberts, H.A. and Bond, W. (1975) Combined treatments of propachlor and trifluralin
for weed control in cabbage. *Weed Research* 15(3), 195–198. DOI: 10.1111/
j.1365-3180.1975.tb01122.x.

Roberts, H.A., Bond, W. and Hewson, R.T. (1976) Weed competition in drilled
summer cabbage. *Annals of Applied Biology* 84(1), 91–95. DOI: 10.1111/j.1744-
7348.1976.tb01732.x.

Röhrig, M. and Stützel, H. (2001) A model for light competition between vegetable
crops and weeds. *European Journal of Agronomy* 14(1), 13–29. DOI: 10.1016/
S1161-0301(00)00079-4.

Root, R.B. (1973) Organization of a plant-arthropod association in simple and diverse
habitats: the fauna of collards (*Brassica oleracea*). *Ecological Monographs* 43(1),
95–124. DOI: 10.2307/1942161.

Russell, P.E. (2005) A century of fungicide evolution. *The Journal of Agricultural Science*
143(1), 11–25. DOI: 10.1017/S0021859605004971.

Ryan, J., Ryan, M.F. and McNaeidhe, F. (1980) The effect of interrow plant cover on
populations of the cabbage root fly, *Delia brassicae* (Wiedemann). *The Journal of
Applied Ecology* 17(1), 31. DOI: 10.2307/2402961.

Salisbury, Y.E. (1961) *Weeds and Aliens. New Naturalist Series*. Collins, London.

Schellhorn, N.A. and Sork, V.L. (1997) The impact of weed diversity on insect population dynamics and crop yield in collards, *Brassica oleraceae* (Brassicaceae). *Oecologia*
111(2), 233–240. DOI: 10.1007/s004420050230.

Schonbeck, M., Herbert, S., DeGregorio, R., Mangan, F., Guillard, K. *et al.* (1993) Cover
cropping systems for Brassicas in the Northeastern United States. 1. Cover crop
and vegetable yields, nutrients and soil conditions. *Journal of Sustainable Agriculture*
3(3–4), 105–132.

Schwarz, H.-P. and Hege, D. (2014) GPS-based weed control in field vegetable growing.
[German]. *Landtechnik* 69(2), 68–71.

Shaner, D.L. and Beckie, H.J. (2014) The future for weed control and technology. *Pest
Management Science* 70(9), 1329–1339. DOI: 10.1002/ps.3706.

Smith, J.G. (1976a) Influence of crop background on aphids and other phytophagous
insects on Brussels sprouts. *Annals of Applied Biology* 83(1), 1–13. DOI: 10.1111/
j.1744-7348.1976.tb01689.x.

Smith, J.G. (1976b) Influence of crop background on natural enemies of aphids on
Brussels sprouts. *Annals of Applied Biology* 83(1), 15–29. DOI: 10.1111/j.1744-
7348.1976.tb01690.x.

Spitters, C.J.T. (1990) Weeds: population dynamics, germination and competition. In:
Rabbinge, R., Ward, S.A. and van Laar, H.H. (eds) *Theoretical Production Ecology:
Reflections and Prospects*. Pudoc, Wageningen, Netherlands, pp. 182–216.

Stivers-Young, L. (1998) Growth, nitrogen accumulation and weed suppression by fall
cover crops following early harvest of vegetables. *HortScience* 33(1), 60–63.

Stoate, C., Leake, A., Jarvis, P. and Szczur, J. (2017) *Field for the Future: 25 Years of the
Allerton Project – A Winning Blueprint for Farming, Wildlife and the Environment*.
Game and Wildlife Conservation Trust, Loddington, UK.

Tahvanainen, J.O. and Root, R.B. (1972) The influence of vegetational diversity on the
population ecology of a specialized herbivore, *Phyllotreta cruciferae* (Coleoptera:
Chrysomelidae). *Oecologia* 10(4), 321–346. DOI: 10.1007/BF00345736.

Tan, W.K., Goenadie, V., Lee, H.W., Liang, X., Loh, C.S. *et al.* (2020) Growth and glucosinolate profiles of a common Asian green leafy vegetable, *Brassica rapa* subsp. *chinensis* var. *parachinensis* (choy sum), under LED lighting. *Scientia Horticulturae* 261, 108922. DOI: 10.1016/j.scienta.2019.108922.

Theunissen, J. and Ouden, H. (1980) Effects of intercropping with *Spergula arvensis* on pests of brussels sprouts. *Entomologia Experimentalis et Applicata* 27(3), 260–268. DOI: 10.1111/j.1570-7458.1980.tb02973.x.

Theunissen, J., Booij, C.J.H. and Lotz, L.A.P. (1995) Effects of intercropping white cabbage with clovers on pest infestation and yield. *Entomologia Experimentalis et Applicata* 74(1), 7–16. DOI: 10.1111/j.1570-7458.1995.tb01869.x.

Tukahirwa, E.M. and Coaker, T.H. (1982) Effect of mixed cropping on some insect pests of brassicas; reduced *Brevicoryne brassicae* infestations and influences on epigeal predators and the disturbance of oviposition behaviour in *Delia brassicae*. *Entomologia Experimentalis et Applicata* 32(2), 129–140. DOI: 10.1111/j.1570-7458.1982.tb03193.x.

Unger, P.W. and McCalla, T.M. (1980) Conservation tillage systems. *Advances in Agronomy* 33, 1–58. DOI: 10.1016/S0065-2113(08)60163-7.

van Bueren, E.T.L and Struik, P.C. (2017) Diverse concepts of breeding for nitrogen use efficiency:a review. *Agronomy for Sustainable Development* 37(5), 50. DOI: 10.1007/s13593-017-0457-3.

Waisen, P., Sipes, B.S. and Wang, K.H. (2019) Potential of biofumigant cover crops as open-end trap crops against root-knot and reniform nematodes. *Nematropica* 49(2), 254–264.

Walker, R.L., Warson, C.A., Armstrong, G. and Ball, B. (2011) The effects of crop rotation and soil management on worm activity and soil physical quality. *Aspects of Applied Biology* 113, 149–154.

Warneke, B.W., Zhu, H., Pscheidt, J.W. and Nackley, L.L. (2021) Canopy spray application technology in specialty crops: a slowly evolving landscape. *Pest Management Science* 77(5), 2157–2164. DOI: 10.1002/ps.6167.

Weaver, S.E. (1984) Critical period of weed competition in three vegetable crops in relation to management practices. *Weed Research* 24(5), 317–325. DOI: 10.1111/j.1365-3180.1984.tb00593.x.

White, C.M., DuPont, S.T., Hautau, M., Hartman, D., Finney, D.M. *et al.* (2017) Managing the trade off between nitrogen supply and retention with cover crop mixtures. *Agriculture, Ecosystems & Environment* 237, 121–133. DOI: 10.1016/j.agee.2016.12.016.

Zijlstra, C., Lund, I., Justesen, A.F., Nicolaisen, M., Jensen, P.K. *et al.* (2011) Combining novel monitoring tools and precision application technologies for integrated high-tech crop protection in the future (a discussion document). *Pest Management Science* 67(6), 616–625. DOI: 10.1002/ps.2134.

PESTS AND PATHOGENS

7

GEOFFREY R. DIXON*

School of Agriculture, Policy and Development, Earley Gate, Whiteknights Road, PO Box 237, University of Reading, Reading, Berkshire RG6 6EU and GreenGene International, Hill Rising, Horsecastles Lane, Sherborne, Dorset DT9 6BH, UK

Abstract

Pests and pathogens cause diverse damage for brassica crops ranging from reduced land values to direct losses in yield and quality as analysed in this chapter. Monitoring and disease forecasting are increasingly precise with sensing and recording using physical, molecular and artificially intelligent diagnostic systems. These guard against shifting spectra of pests and pathogens which, at least partially, respond to climatic changes. Integrated pest management strategies are critically examined including host resistance, biological control, rotations, tillage methods, altering plant densities and geometries, physical barriers, nutrition, natural products and statutory controls. Smarter and more precise application of agrochemicals is discussed which enables effective use of fewer synthetic molecules.

In the second part of this chapter are discussions and descriptions of specific pests, such as insects, nematodes, molluscs, birds and mammals. In parallel are analyses of the impact of aerial and soil-borne fungal, bacteria, virus and mycoplasma pathogens.

INTRODUCTION

Pests and pathogens are the biggest constraint for efficient, healthy and cost-effective vegetable brassica production. From field to fork they cause 25–30% of loss and wastage. The range and intensity of pest and pathogen invasions is

*geoffrdixon@gmail.com

© Geoffrey R. Dixon and Rachel Wells 2024. *Vegetable Brassicas and Related Crucifers,* 2nd edition. (G.R. Dixon and R. Wells)
DOI: 10.1079/9781789249170.0007

increasing, as discussed by Brown and Hovmøller (2002), partially as a result of climatic changes which make environments more conducive for them (Dixon *et al.*, 2014).

This chapter, first, examines factors determining the success of integrated pest and pathogen management (Theunissen, 1984). The interaction of resistance, host and herbivore chemistry, environmental factors (such as temperature, precipitation and soil structure), the presence of beneficial and benign predators and microbes, and cover and companion vegetation all affect the successful outcome of pest and pathogen control. Biological and cultural control techniques minimizing pest and pathogen damage are becoming more commonly applied in brassica husbandry. Biostimulant products (Rehman *et al.*, 2019) are now regularly used as part of these programmes.

Second, specific damaging organisms of greatest significance worldwide for both ware and seed vegetable brassicas are outlined in the latter part of this chapter. Further detailed information in, for example: Dixon (1981, 1984), Jones and Jones (1984), Finch and Thompson (1992), Koike *et al.* (2007), and Dixon and Tilston (2010), amplifies these descriptions.

In common with Chapter 6, Competitive Ecology and Sustainable Production, consideration of specific synthetic agrochemical controls is limited. Manufacturing companies now direct their very expensive research and development at active ingredients suitable for broadacre crops such as oilseed rape (*Brassica napus*). Where feasible, these may eventually receive 'on-label' clearance for vegetable brassicas. Additionally, the substantial value of 'off-label' forms of synthetic and natural agrochemical (Payne *et al.*, 2004) continues, but economic, environmental and social considerations increasingly weigh against pesticide use because they:

- exert intense selection pressures on pest and pathogen populations resulting in the appearance of tolerant strains in typically 'boom and bust' cycles (see Fig. 7.1);
- adversely affect surrounding populations of non-target organisms, widely diminishing forms of natural biological control and biodiversity; and
- result in undesirable residues and by-products reaching human and other food chains when used in unauthorized and unregulated applications.

FORMS OF CROP LOSS CAUSED BY PESTS AND PATHOGENS

Land capital value loss

Land areas used for vegetable brassica crops tend to be relatively small although in specific locations they may appear as substantial concentrations by virtue of favourable soils, climates, cultural traditions and consequently the availability of skilled manual labour. Intensive cropping without adequate rotations causes rampant infestation by soil-borne pests and pathogens.

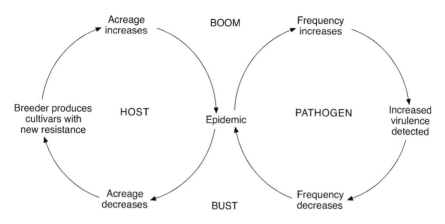

Fig. 7.1. Boom and bust cycle. (Geoff Dixon).

These diminish crop yields rendering the land unfit for economic production and, in the longer term, reducing asset values of the land. Classic examples of such soil-borne pathogens and pests include: clubroot disease caused by *Plasmodiophora brassicae*, white rot caused by *Sclerotinia sclerotiorum* and the brassica cyst nematode, *Heterodera cruciferae*.

Direct yield loss

Each vegetable brassica plant has a high cash value. Pests and pathogens damaging even small proportions of crops can devastate quality, yield and economic returns. Even slight blemishes wreck quality resulting in rejection by customers at fresh vegetable markets and lost value an outcome which differs from broadacre agricultural crops. Forms of crop impairment include:

- failure of growth and reproduction by individual plants in the field;
- damage to the saleable portions of the plant in the field;
- distortion of the harvesting schedule causing failed market delivery at optimal times and maximum values;
- rejection during harvesting, grading, chain handling or display;
- development of blemishes during transit, storage or retail sale;
- downgrading in the marketing chain; and
- total rejection at any point from field to fork.

Consumers react adversely where there are signs of pest activity, eggs, larvae, frass or areas of disease symptoms. Integrated crop husbandry aiming for blemish- and pesticide-free brassicas is driving the successful use of biologically based controls.

INVASION AND INFECTION FORECASTING AND MANAGEMENT

The philosophy of 'forewarned is forearmed', whereby pest and pathogen invasion or infection are anticipated and crops monitored, is not novel. New technologies, however, are making protection more efficient, effective and less wasteful and part of integrated, environmentally benign strategies.

Pest and pathogen forecasting

All forms of control require knowledge of which pests and pathogens are potential causes of damage and capabilities for the identification of incipient invasions. This is achieved using forecasting and monitoring programmes (Collier *et al.*, 2020a). Comprehensive programmes for forecasting risks of aphid and caterpillar invasions, in the UK for example, use relatively simple prediction methods while more complex models are needed for cabbage root fly (*Delia radicum*) and blossom beetle (*Meligethes aeneus*) (Phelps *et al.*, 1993). Accurate weather data is an essential component of these systems and is drawn from a nationwide series of meteorological stations.

Aphids and other flying insects are monitored by the Rothamsted Research Insect Survey comprising a national network of suction traps (https://rotham-sted.ac.uk/insect-survey (accessed 16 August 2023)) primarily serving arable crops but with relevance for brassica vegetables. Samples are analysed for the identification of pest and beneficial species at Rothamsted Research. Data relevant for fresh produce crops are extracted, collated and have been published as both a blog (Collier *et al.*, 2020b) and in the Agricultural and Horticultural Development Board (AHDB) Pest Bulletin (https://horticulture.ahdb.org.uk/knowledge-library/ahdb-pest-bulletin (accessed 16 August 2023)). The mechanics of this systematic approach are summarized by Collier *et al.* (2020b). They utilize software containing forecasting models, colloquially known as MORPH (Methods of Research Practice in Horticulture), which contains more than 20 predictive models. Most models use weather data and MORPH can access data from a number of commercially available 'in-field' weather stations (Murphy *et al.*, 2007). The mathematical model links seasonal weather and the rate of cabbage aphid (*Brevicoryne brassicae*) arrival, predicting when damaging insect populations will expand and are most susceptible to control. These British systems are now, however, likely to change in view of the removal of horticulturally related levy-funded research and development work from the levy board's (AHDB) remit.

Quantifying a developing pest infestation can take advantage of the basic biological processes of pests and utilize them for their control. The first such process is that of their ability to detect odours (Cobb, 1999). Insects may be drawn to odours emitted by the crops. The cabbage seed weevil (*Ceutorhynchus assimilis*), for example, responds positively to extracts of oilseed rape (*B. napus*) and to the colour of the crop plants. The active attractants are α-farnesene (a component of flower scent) and 3-butenyl and 4-pentenyl isothiocyanates (analogous to components of foliage) which are present in brassicas (Ridgway *et al.*, 1990; Evans and Allen-Williams, 1998).

Studies building on and improving prediction and monitoring considered invasions by the diamondback moth (*Plutella xylostella*), a major but sporadic pest of brassica crops worldwide. Two approaches were compared:

- first, a network of pheromone traps, fitted with small cameras which transmitted images on to a website providing remote and timely information but which failed to capture many moths; and
- second, sightings by citizen scientists who recorded their observations on a website and on social media. This approach provided a more effective series of timely warnings (Wainwright *et al.*, 2020).

Earlier, in western USA, more than 1000 weather stations formed a privatized disease warning system (Thomas and Gubler, 2000) in collaboration with universities that developed, validated and inserted algorithms. Weather data are collected and linked with pest and disease warnings by radio reporting. Data are analysed centrally, assessing risks correlated with irrigation, planting, harvesting and nutrient-timing information and distributed daily to farms by internet, fax, proprietary servers or personal visits. Combining with remote sensing, the Global Positioning System (GPS) and geographic information systems (GIS), weather forecasting and scouting technologies constitute an efficient decision making tool financed on a fee-paying basis. These services reduced fungicide use by one application and improved crop quality and yields by 10–50%. Electronic systems for weather monitoring in crops have become increasingly available (see Fig. 7.2).

Brassica Alert system

The Brassica Alert system (www.syngenta.co.uk/brassica-alert/ (accessed 16 August 2023)) provides forecasting for key brassica diseases in collaboration with the AHDB (Kennedy *et al.*, 1999; Kennedy and Gilles, 2003). This system uses suction traps from which samples of captured spores are identified with lateral flow test kits. From July to December, five weather recording centres in Lincolnshire, UK, north of Boston, produce data which identify the potential risks of disease occurrence. Aligning disease conducive weather conditions with the presence of pathogen spores results in disease risk warnings being issued. Knowledge of the presence or absence of specific pathogenic fungal spores ensures that controls are not wasted. Text alerts go out to growers giving high-, medium- and low-risk assessments. Ringspot (*Mycosphaerella brassicicola*) is the main disease targeted and, fortunately, dark leaf spot (*Alternaria* spp.) is also eradicated with currently used ringspot sprays. Powdery mildew (*Erysiphe cruciferarum*) and light leaf spot (*Pyrenopeziza brassicae*) are

Fig. 7.2. Meteorological monitoring station. (Courtesy of Skye Instruments Ltd).

increasing problems in Lincolnshire, for which monitoring and forecasting are needed. Lateral flow test kits for these pathogens are available and those for white rust or white blister (*Albugo candida*) are in development (Mr Simon Jackson, previously at the Allium and Brassica Centre, Kirton, Lincolnshire, UK, personal communication). As with pest monitoring, these British systems are now likely to change in view of the removal of horticulturally related work from AHDB's remit.

Soil-borne pathogens

Identifying the presence of soil-borne pathogens, such as *P. brassicae*, the cause clubroot disease, and understanding the distribution of pathotypes previously required time-consuming assays using *in planta* glasshouse tests with differential host series (Williams, 1966; Toxopeus *et al.*, 1986). Outbreaks of this disease in Canadian oilseed rape (*B. napus*) crops, the source of a very valuable and essential world-traded commodity (canola oil), stimulated intense and very effective research producing routine molecular polymerase chain reaction (PCR) and Quick PCR (qPCR) tests, which identify the pathogen and its virulence spectra, and quantify the proportions of viable and dead spores (Strelkov *et al.*, 2018). Changing from a glasshouse-based *in planta* test to rapid, automated, laboratory-based biochemical assays significantly increased the speed and applicability at which advisory information is available and its resultant effectiveness.

Warning and identification

Initially, the presence of specific pathogens was detected using traps which collected spores and these were identified by detailed laboratory assays (Hirst, 1952). Visual estimation of disease symptoms initially provided a very effective means for assessing disease development and severity, especially in the field using graphical keys (Dixon, 1977). This approach required well-trained, patient and skilful staff whose evaluations were accurate and reproducible. The technique was, however, very time-consuming, laborious and limited in scale and the volume of assessments which were possible was very low (see Figs 7.3 and 7.4 demonstrating field assessment keys for powdery mildew (*E. cruciferarum*) on the buds and leaves of Brussels sprouts (*Brassica oleracea* var. *gemmifera*)) (Dixon and Doodson, 1970, 1971).

Integrated management still requires regular walking and inspection, and this is of considerable relevance for vegetable brassica crops. Drones can now provide monitoring data, which can be accessed by smartphone or laptop. None the less, experienced human visual assessment of quality and evenness of growth is still vital in successful brassica husbandry.

Automatic plant disease detection is proving useful in monitoring broadacre crops and will eventually become used for vegetable brassicas (see Fig. 7.5). Detection of incipient foliar symptoms is also becoming possible (West *et al.*, 2008; Bauer *et al.*, 2011).

Plant disease recognition is based on plant leaf images. First, the lesion is segmented and the disease feature vector is extracted. Then, the extracted features are provided for the *K*-nearest-neighbour classifier recognizing the plant diseases. Experimental results show the effectiveness of this approach (Zhang *et al.*, 2015). Automated in-field methods which assess crops for pest and pathogen risk are being developed initially for broadacre arable crops but with emerging opportunities for fresh produce crops such as brassicas. Optical sensing evaluates the reflection of light coming from crops and detects changes in spectral composition (West *et al.*, 2017). Diseased canopies reflect more light in the visible range and less in infrared and far-red, permitting the measurement of disease symptoms across fields. Rothamsted Research

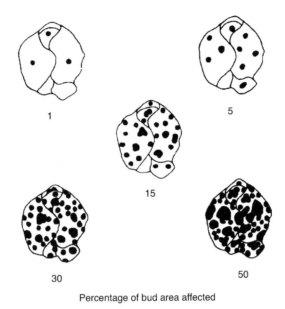

Percentage of bud area affected

Fig. 7.3. Disease assessment key field assessment key for powdery mildew (*Erysiphe cruciferarum*) on the buds of Brussels sprouts (*Brassica oleracea* var. *gemmifera*). (Geoff Dixon).

developed a tractor-mounted system which monitors disease, giving detailed maps of incidence. Currently, this is being evaluated by comparing manual and machine estimations of disease incidence. Further research is speeding up recognition, permitting rapid and effective remedial action. An alternative system detects changes in fluorescence emitted by crops when illuminated with ultraviolet (UV) light. Crop 'pinkness' is an indication of its health status. This system identified powdery mildew (*Blumeria graminis*) in cereals only 4 days after invasion; well before manually visible symptoms appeared.

Diagnostic tests

A prerequisite for automated identification and quantification is the availability of effective molecular diagnostic tests, such as described by King *et al.* (2018) for light leaf spot of brassicas (*P. brassicae*) and Kaczmarek *et al.* (2019) for sugarbeet rust (*Uromyces betae*) using loop-mediated isothermal amplification (LAMP) technologies. Rapid on-site identification of specific pests and pathogens is achieved by analysing their DNA. This assay is becoming much simpler and quicker following adaption of advances in human medical research. Combining in-crop spore sampling with automated, robotic field analysis of DNA and reporting the results via wifi systems is a feasible prospect. 'The laboratory in a box' is a robotic system whereby spore traps collect samples, their

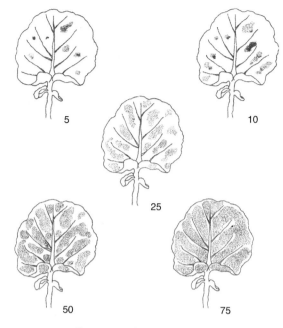

Percentage of leaf area affected

Fig. 7.4. Disease assessment key for powdery mildew (*Erysiphe cruciferarum*) on the leaves of Brussels sprouts (*Brassica oleracea* var. *gemmifera*). (Geoff Dixon).

DNA is extracted and analysed, and results reported to a remote platform. These systems may also detect chemicals which are uniquely produced by plants as part of their responses following invasion by pests or pathogens, such as jasmonates. Analysis and identification of these chemicals can be automated. The automated Blight Alert system for potatoes, as an example, offers a step change in effective crop monitoring and forecasting. Risk assessments sent by automatic text will more quickly place decisions of whether to 'spray or not' in growers' or crop consultants' hands. This is because automated DNA analysis obviates the need for manual spore identification and quantification. Similar automated monitoring of pest invasions is also feasible. Digital cameras will capture images of aphids and then transmit these for identification by artificial intelligence using standardized images and the results transmitted on to growers' or consultants' mobile platforms. In the long term, these systems will reduce the need for both manual field inspections and for laboratory analyses. It is recognized that automation must be practical, robust, readily reliable, available and cost-effective. Aerial crop observations using drones or satellites have a considerable role to play in determining crop health, not least for soil-borne pathogens such as clubroot (*P. brassicae*) and white mould and stem rot (*S. sclerotiorum*). Both pathogens are responsible for major losses in a wide range of brassica vegetables and oilseeds, where, in maturing crops,

Fig. 7.5. Automated spore collection, analysis and reporting equipment. (Courtesy of Dr Jon West).

disease foci in the middle of large fields cannot be monitored because of the density of plant growth and the difficulties of gaining physical access on foot. Aerial observation using drones overcomes these difficulties. Recently, environmental DNA metabarcoding and metagenomics analyses have been applied in aerobiology targeting bacteria, fungi and plants in airborne samples (Banchi *et al.*, 2020). These developments promise complementarity with more traditional biomonitoring frameworks.

Where prediction and forecasting systems indicate an invasion or infection is developing, the decisions rests on whether control is economically sensible. The deciding factor is whether the saleable part of the plant is damaged. For example, from the beginning of September onwards the yield of Brussels sprout (*B. oleracea* var. *gemmifera*) plants is unaffected by defoliation of up to 60% of leaf area. It is unlikely that powdery mildew will cause quantitative damage since infection by *E. cruciferarum* does not start until late August–early September. Only if disease spreads on to the buds and affects their quality will control measures be required.

THE CHANGING SPECTRA OF CROPS, PATHOGENS AND PESTS

Increased production of agricultural oilseed brassicas (mainly *B. napus*) worldwide has raised the incidence of some brassica pests and pathogens on horticultural crops (Lamb, 1989). The economic threshold for damage caused by pests and pathogens of oilseeds is higher than for vegetable brassicas and, therefore, fewer controls are used. Expanding acreages of overwintering oilseed rape (*B. napus*) provided a substantial 'green bridge' for light leaf spot (*P. brassicae*, originally *Gloeosporium concentricum*) (Dixon, 1975). As a result, *P. brassicae* has become the major foliar disease of oilseed rape and a major threat for vegetable brassicas. Subsequently, black leg (*Leptosphaeria maculans*) and dark leaf spot (*Alternaria* spp.) spread from oilseed rape on to vegetable brassicas threatening quality and value.

Worldwide, the spectrum of vegetable brassicas grown has expanded dramatically, embracing crops originally used in Asian countries (Larkcom, 2007) and forms grown as baby leaves for salads. As a result, pests and pathogens affecting these crops are also being more widely spread around the world. Also, the worldwide homogeneity of genetic backgrounds of brassica vegetable F_1 hybrid cultivars used in Asia, Europe and North America offers pests and pathogens very uniform and extensive host platforms for invasion, infection and damage.

Climatic global warming is also favouring the expansion of pest and pathogen populations worldwide (Root *et al.*, 2003; Dixon, 2012). Consequently, the relative importance of individual organisms is changing. The diamondback moth (*P. xylostella*), for example, appears to be increasing in importance as a pest in parts of Europe where previously it was viewed as solely an entomological novelty. Similarly, black rot, caused by the bacterium *Xanthomonas campestris* pv. *campestris*, has become a major disease of northern European crops.

The feeding, growth and reproduction of insects and other ectothermic animals is strongly dependent on their body temperature. Prevailing weather conditions determine the temperature microclimate of the crops on which the insects live. There is strong selection pressure for rapid growth and development which decreases the time during which caterpillars of, for example, the small cabbage white butterfly (imported or European cabbage worm) (*Pieris rapae*) are exposed to predators. Field experiments with this insect showed that slower growth was correlated with higher mortality due to the parasitoid wasp *Cotesia glomerata* (*Apanteles glomeratus*) (Benrey and Denno, 1997). Consumption of collard leaves (*B. oleracea* var. collards) by small cabbage white butterfly, and their caterpillars' growth rates, increased by six- to eightfold as temperatures rose from 10°C to 35°C (see Fig. 7.6) (Kingsolver, 2000) and declined rapidly at higher temperatures.

Herbivores, such as the small cabbage white butterfly, are able to sustain growth over the wide and fluctuating range of body temperatures that they

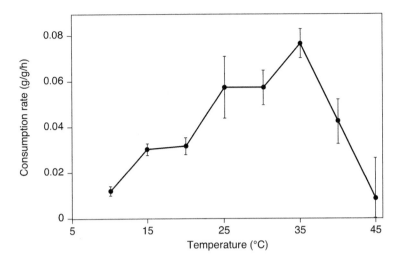

Fig. 7.6. Food consumption by small cabbage white butterfly (*Pieris rapae*) larvae of the fourth instar (g/g/h) as related to temperature (°C) over 24 h. (After Kingsolver, 2000).

routinely experience in the field. This insect thrives in a wide range of weather and climatic conditions throughout its larger geographical range (Gilbert and Raworth, 1996).

STRATEGIES FOR PEST AND PATHOGEN CONTROL

Host resistance

Genetic resistance against pests and pathogens is the principal means of control that is both economically and environmentally advantageous. Resistance against insects, for example, results from:

- antixenosis or a dislike of the plant or non-preference (Kogan and Ortman, 1978);
- antibiosis or antagonistic reaction resulting from the presence of harmful compounds; or
- tolerance and combinations of these attributes.

The mechanisms of resistance have a direct influence on the durability and ultimate success of a pest- or pathogen-resistant cultivar. For example, the integration of low-to-moderate levels of antixenosis, antibiosis and tolerance can be effective in controlling a resident pest population that invades early in the development of a crop and increases gradually during the growing season (Kennedy *et al.*, 1987). Also, combinations of both antixenosis and antibiosis can decrease the likelihood that pest populations will overcome that resistance, compared with either antibiosis or antixenosis alone, as long as the preferred host plant is available.

Resistance to pathogen-induced diseases was developed rapidly in the second half of the 20th century following the use of a single gene, or a few genes, of major dominant effect. Some very notable successes were achieved with particular pathogens of brassica crops, such as the development of resistance to the soil-borne organism *Fusarium oxysporum* f. sp. *conglutinans*, the causal agent of cabbage yellows, in Wisconsin, USA. This retained its effectiveness for many years until a new pathogenic strain of the fungus emerged in California (Williams *et al.*, 1968).

By contrast developing cultivars resistant to *P. brassicae* (the cause of clubroot disease) is more difficult. There are numerous variants of *P. brassicae*, e.g. as analysed by Toxopeus *et al.* (1986), and few sources of resistance, particularly in *B. oleracea* (Dixon and Robinson, 1986). Extensive research and development resulting from a *P. brassicae* epidemic in Canadian oilseed rape (*B. napus*) has produced resistant cultivars but the pathogen has reacted by evolving tolerant variants rapidly (Hollman *et al.*, 2021). Studies of both brassica vegetables and oilseed rape in Canada have identified the very substantial capacity for variant evolution in this pathogen (Mcdonald *et al.*, 2021). Plant breeding companies have, however, devoted significant resources over many years seeking the development of resistant vegetable brassica. Success has been achieved recently in a range of *B. oleracea* crops, including green broccoli (calabrese), Brussels sprouts, Savoy cabbage, white head cabbage and cauliflower (Anon, 2020). Valuable resistance such as this requires careful management ensuring wide rotational distances between these cultivars and the avoidance of land which has been cropped with resistant oilseed rape. This is because common genes for resistance have been deployed in both vegetable and broadacre crops.

Saliently, the understanding of interactions between the pathogen and its host are being elucidated. Messengers, such as calcium, that stimulate resistance responses are being identified and the nature of the response itself unravelled. Much progress has been made with the model brassica *Arabidopsis thaliana* especially with foliar pathogens such as downy mildew (*Hyaloperonospora parasitica*), white blister/white rust (*Albugo candida*) and light leaf spot (*P. brassicae*). This holds out the prospect of more robust resistance being developed in crop plants. Reviews of the basic science are given by Keen *et al.* (2001) and Leong *et al.* (2002).

Reductions in the availability of crop protection chemicals, especially for the control of cabbage root fly (*D. radicum*) have stimulated intense interest in searching for sources of resistance. Some cultivated brassicas have moderate levels of resistance to cabbage root fly, notably red-headed cabbage, glossy brassicas, kale and swedes. None of the levels of resistance are considered sufficiently effective on their own to form the basis of a breeding programme aiming to produce cabbage root fly resistant cultivars. Consequently, there has been a search for resistance sources in allied cruciferous species. To be most easily exploited such sources of resistance should belong to the same genome as the cultivated crop. Ellis *et al.* (1998) found very high levels of cabbage root fly resistance in *Brassica incana*, *Brassica insularis* and *Brassica villosa* and moderate resistance in *Brassica macrocarpa*. Each of these species is part of the CC genome grouping (see Chapter 1 section, Origins and Diversity of *Brassica* Crops) in which the *B. oleracea* horticultural brassicas are included (2*n*=18). Very high levels of antibiosis resistance have also been identified in *Brassica fruticulosa* and *Brassica spinescens*. They belong to the FF genome (2*n*=16), which is closely related to the B genome of *Brassica nigra* (Mizushima, 1980).

Indications are that *Sinapis alba* (white mustard) may be resistant to cabbage root fly or cabbage root maggot as a result of reduced oviposition, diminished weights and survival of larvae, pupae and adults, and lessened damage to host plants. *Sinapis alba* also possesses tolerance to crucifer flea beetle (*Phyllotreta cruciferae*), cabbage seed weevil (*C. assimilis*), cabbage aphid (*B. brassicae*) and diamondback moth (*P. xylostella*) (Jyoti *et al.*, 2001). These effects may be due to the structural properties of the root and stem of *S. alba*, thus the breeding accession Cornell Alt 543 tends to have a more profuse root structure and harder main stem than others tested. Reduced body weight of cabbage root fly could be the result of antibiosis coming from the rapid growth of *S. alba* Cornell Alt 543. In vegetable brassicas it is most important to prevent damage by cabbage root fly during the seedling period or soon after the plants are transplanted.

The *Brassica* species possessing resistance to cabbage root fly also carried resistance to cabbage aphid (*Brevicoryne brassica*) and cabbage whitefly (*Aleyrodes proletella*). Multiple resistance of this nature is especially valuable. Cultivars of swede (*Brassica napus* ssp. *rapifera*) with moderate levels of resistance to cabbage root fly have been developed in Scotland and Scandinavia. These combine this resistance with that to turnip root fly (*Delia floralis*). But to achieve commercially acceptable control these cultivars need to be used in an integrated programme which may include some synthetic organic insecticides.

Antixenosis can provide a promising form of resistance for use against flea beetles, such as crucifer flea beetle and striped flea beetle (*Phyllotreta striolata*) which are oligophagous herbivores feeding principally on brassicas. Related species, such as white mustard (*S. alba*), field pennycress (*Thaspis arvense*) and honesty (*Lunaria annua*), appear unacceptable to striped flea beetles because of the compounds they possess, while *Camelina sativa* fails to offer the pest the necessary cues that initiate feeding and the leaves of *B. villosa* ssp. *drepanensis* have high densities of hairs (>2172 cm^{-2}) and these act as a physical barrier to flea beetle feeding by preventing the insects from settling firmly on the leaf surface (Gavloski *et al.*, 2000). Insect feeding preferences may change with leaf type or growth stage as well as host species. Thus, the true leaves of *B. oleracea* have lower preference for striped flea beetles than those of *S. arvensis* whereas the opposite attraction was observed for cotyledons (Palaniswamy and Lamb, 1992). One avenue for producing resistance to crop pests is the development of plants which express endotoxins from the bacterium *Bacillus thuringiensis* (Bt). Opportunities created by gene-editing techniques might make this possible.

Much of the resistance that has been developed in vegetable brassicas utilizes one or possibly more genes of large effect. This effect is frequently expressed throughout the life of the resistant plant and can be identified by testing seedlings exposed to the pathogen when they are infected artificially. The drawback to this form of plant breeding is the 'boom and bust' cycle referred to in the Introduction section (see Fig. 7.1) where a resistant cultivar becomes used in large proportions of the crops and then, as the size of the tolerant pathogen population increases concomitantly, the host resistance is eroded.

One route suggested to overcome this phenomenon is the development of field resistance (Kocks and Ruissen, 1996). For example, plant breeders seeking resistance to black rot (*X. campestris* pv. *campestris*) concluded that major gene (hypersensitive) resistance could be overcome and that incorporation of additional levels of field resistance might be developed as supplements. Field resistance is defined as 'any resistance that effects epidemics in the field but which is not immediately apparent in laboratory or glasshouse tests' (Robinson, 1969).

MANIPULATION OF HUSBANDRY SYSTEMS

Integrated control strategies have the initial benefit of being less open to erosion by the development of tolerant strains of the pest. Consequently, the life expectancy of such control strategies could be extended. These include the use of husbandry systems as central to achieving control. The crop is manipulated by sowing and maturity date so as to avoid those periods when the pest or pathogen is most active, this can be supported by changing the nutritional regime used for the crop.

Rotation and tillage

Rotation with non-host crops for a sufficient period of time allows the decomposition of infested crop residues and/or a reduction in the viability of pathogen survival structures, eliminating one of the sources of primary reinfection. By increasing the diversity of crops grown within a rotation, the pathogens residing in the soil or on residues from the previous crop may fail to infect the subsequent brassica crop. Understanding the impact of crop residues on survival and carry-over is important in determining the efficacy of rotations. In particular, it is necessary to know if the pathogen survives as a saprophyte as well as a pathogen. Additionally, the roles of weeds and volunteer crop plants as hosts and reservoirs of infection should be established. Knowledge of the extent of the host range is also important. Pathogens with a wide host range and robust survival mechanisms, such as *S. sclerotiorum* (white rot), are unaffected or may even be encouraged by the inclusion of some rotational elements. Similar benefits are achieved for aerial pests and pathogens. For example, rotational diversity increased cabbage aphid (*B. brassicae*) parasitism by parasitoid animals (Scheiner and Martin, 2020) conserving fertile land for crop production. The Allerton Project demonstrates very effectively the environmental and commercial benefits of diversity in the cropped landscape (www.allertontrust.org.uk (accessed 16 August 2023); Stoate *et al.*, 2017) (see Fig. 7.7).

Field margins are valuable habitats for natural enemies of pests and pathogens in agricultural landscapes (Bakker *et al.*, 2021). The presence of flower strips significantly reduced overall injury in the adjacent cabbage (*B. oleracea* var. *capitata*) crop. Diversification can simultaneously support the production ecosystem services by maintaining fresh marketable weight per cabbage plant, and productivity per unit area, and providing pest control (Juventia *et al.*, 2021).

Increasing natural pest control in agricultural fields is an important aim of ecological intensification. The combined effects of landscape context and local placement of agroenvironmental schemes on natural pest control and within-field distance functions of natural pest control agents is untested. It is currently unknown whether ecosystem services provided by adjacent field edges are consistent for different crop types during crop rotation. Densities of carabid beetles and staphylinid beetles, as well as crop yields, increased towards the field centres. Increasing distance to the field edge also increased the effects of crop rotation on carabid beetle assemblages, indicating a source habitat function for field edges (Boetzl *et al.*, 2020). A network of semi-natural

Fig. 7.7. Borage (*Borago officinalis*) field margin supporting predators of crop pests. (Courtesy of the Allerton Project).

habitats and spatially optimized habitats can benefit pest control in agricultural landscapes, but constraints as a result of crop type need to be addressed by annually targeted, spatially shifting agro-environment schemes for different crops.

Substantial efforts are being applied to understand how cropping practices may be designed to encourage benign organisms and discourage pests and pathogens. The most basic husbandry practice is primary cultivation by ploughing, an activity that has taken place since the dawn of agriculture. There may be some disadvantages such as:

- increasing water loss due to run off;
- leaving the soil exposed to heat and wind increases drying by evaporation; and
- increasing the loss of organic matter and carbon, which is required for an active benign microbial population. Carbon loss is associated with rising temperatures and global warming (Dixon *et al.*, 2014).

Studies have shown that soil tillage can be deleterious to ground dwelling predators. The main deleterious effect of soil tillage was on the emergence of those carabid species that overwinter partly as larvae. Consequently, spring tillage disrupts their emergence and could reduce pest control effects (Mesmin *et al.*, 2020). Reducing tillage (see Chapter 6 section, Non-tillage and Reduced Tillage) under American conditions enhanced microbial species diversity with larger populations in the upper soil horizons. Tillage, however, redistributes microbes throughout the upper and lower soil horizons. Several higher fungi produce ergosterol that can be utilized as a biological marker to measure the activity of soil microbial populations. Ergosterol content was greater under non-tillage conditions (Monreal *et al.*, 2000; Krupinsky *et al.*, 2002) indicating increased microbial activity. Conversely, the advantages of tillage lay in the burial of crop residues, levelling, consolidation and warming of seed beds in spring, reducing surface compaction, breaking up soil pans, incorporation of pesticides and fertilizers, and reducing weed competition.

Biological control

Biological control of pests and pathogens has been sought for many years as an alternative to using synthetic pesticides. Biological control of pests using single predatory species is particularly successful with protected glasshouse crops for pests such as aphids (see Fig. 7.8).

Similar success is now being achieved with field crops. For pathogens, the initial approach is finding individual microbes that inhibit their growth and reproduction. Most fungi used in biocontrol are Hyphomycetes, especially the genera *Trichoderma*, *Penicillium* and *Gliocladium*, and the bacteria *Pseudomonas* and *Bacillus* (Bello *et al.*, 2002).

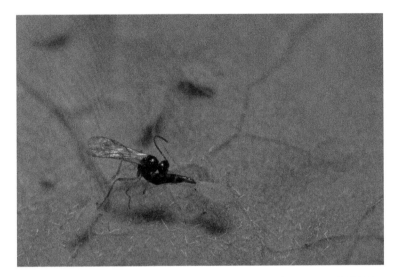

Fig. 7.8. The wasp *Aphidius ervi* parasitizing a green peach-potato aphid (*Myzus persicae*). (Courtesy of Mr Tony Girard, Koppert Biological Systems).

In geographical regions with high ambient temperatures solarization techniques will diminish soil-borne pest and pathogen populations. Solarization exploits the heating effects gained by placing plastic sheeting over soil. Population counts of *F. oxysporum* f. sp. *conglutinans* were greatly reduced and cabbage yellows disease undetected in plots treated with solar heating and cabbage amendments at 1% weight for weight. The sheeting traps fungitoxic gases derived from the cabbage residues that are concentrated; these together with high temperatures are responsible for the diminution in the viability of pathogen propagules. Ramirez-Villapudua and Munnecke (1987) postulated that these result from changes to the biodiversity of soil microflora, diminished soil fungistasis and the release of nutrients that may induce chlamydospore germination leading to biological control of *F. oxysporum* f. sp. *conglutinans*. Biosolarization combines biotoxic products from organic matter decomposition and heat from solarization, offering an effective option for soil pest control (Ney *et al.*, 2019). The fungistatic effect of brassica residues was reported earlier by Kocks *et al.* (1998) for black rot (*X. campestris* pv. *campestris*). When black rot infested plants were ploughed back in after harvest and the plants broken up before incorporation, presumably releasing sulfurous compounds, the infection risks diminished in the following season. Heat can also be used for the control of seed-borne pathogens (thermotherapy) but this can impair germination. Tests with green broccoli (*B. oleracea* var. *italica*) seeds held at 55°C should not be exposed for longer than 20 min (Soriano *et al.*, 2019).

There may also be opportunities to load seed with biological control organisms, a process termed bacterization (Boruah and Kumar, 2002). Seed bacterization improved: germination, shoot height, root length, fresh and dry mass, enhanced yield and chlorophyll content of leaves in the field. Plant growth promotion resulted from the presence of siderophore and disease suppression from the effects of an antibiotic. Siderophores are high-affinity iron (Fe^{3+})-chelating molecules that enhance the acquisition of iron in a deficient environment, thereby making iron available to crop plants but unavailable to pathogens. Identified strains of *Pseudomonas fluorescens* are capable of enhancing plant growth and diminishing the effects of fungal and bacterial plant pathogens. Attempts are being made to utilize this approach with *Bacillus polymyxa* (Pichard and Thouvenot, 1999), isolated from cauliflower (*B. oleracea* var. *botrytis*) seeds, to control black rot and damping-off of cauliflower.

Maintenance of a beneficial microbial community, especially in the rhizosphere, is indispensable for plant growth and agricultural sustainability. Plant growth-promoting rhizobacteria, e.g. *Proteus vulgaris* JBLS202 strain, promoted the growth of kimchi Chinese cabbage and altered the relative abundance of total bacteria and *Pseudomonas* spp. in the treated rhizosphere. Pyrosequencing data revealed an increase in total bacteria abundance, including specific groups such as Proteobacteria, Acidobacteria and Actinobacteria, in the treated rhizosphere. Time-course quantitative PCR analysis confirmed the increase in the abundance of Acidobacteria, Actinobacteria, Alphaproteobacteria and Betaproteobacteria. Genes involved in nitrogen cycling were up-regulated by JBLS202 treatment, indicating changes in the ecological function of the rhizosphere soil. Introduction of JBLS202 alters both the composition and function of the rhizosphere bacterial community, which can have direct and indirect effects on plant growth (Bhattacharyya *et al.*, 2018). Soft rot caused by *Pectobacterium carotovorum* ssp. *carotovorum* is a major constraint in the production of Chinese cabbage. *Bacillus amyloliquefaciens* KC-1 survives and suppresses the growth of *P. carotovorum* ssp. *carotovorum* in Chinese cabbage and its rhizosphere, protecting the host from the pathogen (Cui *et al.*, 2019).

The 'push–pull' strategy aims at manipulating insect pest behaviour using a combination of attractive and repulsive stimuli using either plant-derived volatile organic compounds or insect host plant preferences (Lamy *et al.*, 2018). Green broccoli field crops were exposed to 'push–pull' factors which included the oviposition deterrent effect of dimethyl disulfide and the attractive effect of a Chinese cabbage strip enhanced with Z-3-hexenyl-acetate. The push component dimethyl disulfide reduced cabbage root fly (*D. radicum*) oviposition on green broccoli by nearly 30%. Conversely, applying Z-3-hexenyl-acetate in the pull component of Chinese cabbage increased it by 40%.

Cabbage whitefly (*A. proletella*) causes severe economic damage on cabbage crops, especially in organic vegetable production. Parasitoids and predators were sampled from organic Brussels sprouts (*B. oleracea* var. *gemmifera*) fields that were highly infested with cabbage whitefly. The wasp *Encarsia tricolor* was the dominant parasitoid species and comprised 99.5% of all parasitoids encountered. Others wasps included *Encarsia inaron* and *Euderomphale chelidonii.* Most sampled predators were hoverfly larvae (49.6%) followed by spiders (33.8%), ladybeetles (14.2%) and lacewing larvae (0.8%) (Laurenz *et al.*, 2019).

Induced systemic resistance has been demonstrated in different plant species against a broad spectrum of pathogens. The mechanisms by which rhizobacteria species induce resistance vary. Some trigger the salicylic acid-dependent resistance pathway by producing this compound at the root surface. When appropriately stimulated, plants systemically enhance their defensive capacity against pathogen attack. This induced resistance is generally characterized by a restriction of pathogen growth and a reduction in disease severity. The development of a broad-spectrum, systemic acquired resistance (SAR) after primary pathogen infection is characterized by an early increase in endogenously synthesized salicylic acid. This is an essential compound in the SAR signalling pathway, because transgenic plants that are unable to accumulate salicylic acid are incapable of developing SAR. Further applications of salicylic acid are associated with the systemic activation of SAR genes. These include genes that encode for pathogenesis-related proteins. Some of these proteins have *in vivo* antifungal activity and may contribute to the resistance state of SAR (Ton *et al.*, 2001).

Most recently, it has become apparent that selected strains of non-pathogenic, root-colonizing rhizobacteria are capable of inducing disease resistance in crop plants. This is referred to as rhizobacteria-mediated induced systemic resistance. Rhizobacteria are present in large numbers on the root surface where plant exudates and lysates provide nutrients. Besides inducing resistance rhizobacterial strains are reported to directly antagonize soil-borne pathogens and to stimulate plant growth.

Plant defences often mediate whether competing chewing and sucking herbivores indirectly benefit or harm one another (Blubaugh *et al.*, 2018). Dual-guild herbivory muddles plant signals used by specialist natural enemies to locate prey, further complicating the net impact of herbivore–herbivore interactions in naturally diverse settings. Results suggest that enemy-mediated apparent commensalism may override constraints to growth induced by competing herbivores in field environments, and emphasize the value of placing chemically mediated interactions within their broader environmental and community contexts. In nature, herbivorous insects and plant pathogens are generally abundant when plants are flowering (Chretien *et al.*, 2018). Plants are exposed to a diversity of attackers during their reproductive phase. Plant responses to one attacker can interfere with responses to a second attacker, and phytohormones that orchestrate plant reproduction are also involved in resistance to insect and pathogen attack. Caterpillars were the main inducers of jasmonates in inflorescences, and the phytohormonal profile of leaves was not affected by either insect or pathogen attack. Results demonstrate that plants prioritize resistance of reproductive over vegetative tissues, and that a chewing herbivore species is the main driver of responses in flowering *B. nigra*.

Secreted lipopeptide fractions from *Bacillus* UCMB5113, together with synthetic peptide mimics, were evaluated for their effects on fungal phytopathogens and *A. thaliana*. The structures of secreted lipopeptides were analysed using mass spectrometry. In plate tests, *Bacillus* UCMB5113 and lipopeptide extracts suppressed growth of several fungal pathogens infecting brassica plants. Application of the lipopeptide extracts on *A. thaliana* roots provided systemic protection against *Alternaria brassicicola* on leaves. *Arabidopsis thaliana* signalling mutants and PDF1.2 and VSP2 promoter-driven β-glucuronidase marker lines indicated that the lipopeptide fraction involved jasmonic acid-dependent host responses for suppression of fungal growth, indicative of induced systemic resistance (Asari *et al.*, 2017).

Endophytes are fungi or bacteria which live within plants without causing disease stresses. Recently, they have been demonstrated as having a range of benefits such as:

- improving and promoting plant growth;
- increasing yield;
- reducing disease symptoms caused by plant pathogens;
- reducing herbivory from insect pests;
- removing contaminants from soil;
- improving plant performance under extreme conditions of temperature and water availability; and
- solubilizing phosphate and contributing assimilable nitrogen to their hosts (Card *et al.*, 2015).

Crop density

Stand density and crop plant architecture will affect the impact of diseases and their risks. Stand density affects air movement, shading and moisture retention within the crop. Root (1973) proposed the 'resource concentration hypothesis' whereby 'many herbivores, especially those with a narrow host range are more likely to find hosts that are concentrated'. This hypothesis predicts that the density of herbivores per host plant is higher in denser stands of their host plants. The density of herbivores per host plant in the field is determined by: (i) the number of eggs laid by the individuals that enter from the surrounding area and (ii) the survival rate of individuals growing inside the field. Yamamura and Yano (1999) argue that in small areas the density of herbivores per plant is dominated by the frequency of entry but in large areas of crop the density will be dominated by the survival rate of individuals inside the plot or crop (see Chapter 6 section, Intercropping and Cover Cropping).

Host nutrition

Plant nutrition affects the rate at which the signs and symptoms of pest and pathogen invasion develop. Balanced fertility is more easily maintained in a diverse cropping rotation. Each crop species has a varying requirement for optimal growth, development and reproduction and so utilizes different nutrients at different rates. The suitability of food plants influences the ovipositional responses. Small cabbage white butterfly (*P. rapae*) females can detect the physiological status of plants (Myers, 1985). The behavioural events prior to oviposition by small cabbage white butterflies include searching flight, landing and contact evaluation on a potential host plant. These factors may be influenced by the presence of non-host plants. Chlorophyll levels and the percentage composition of nitrogen, sulfur and other macronutrients could affect host selection (Hooks and Johnson, 2001). Plant water content may also be an important indicator of the nutritional status of plants to ovipositing butterflies.

Female small cabbage white butterflies, for instance, assess the nutritional value of plants and tend to lay more eggs on plants that are richer in nutrients. This insect will also respond to other host characteristics such as size, leaf water content, host colour (green and blue-greens which relate to chlorophyll content) and plant growth stage.

Changing the nutritional regime used for the crop offers avenues for pest and pathogen management. It is well recognized that use of some fertilizers is associated with increased pest incidence while others tend to have the effect of diminishing pest impact. Integral to this approach is the use of crop walking and assessment. Excessive use of nitrogen fertilizers, especially where this results in an imbalance between ammonium and nitrate ions in the soil, causes stress which then allows pests and pathogens to cause damage more easily. The use of environmentally benign nitrogen sources, such as calcium cyanamide or calcium nitrate, is associated with diminished crop stress and in consequence less disease. This topic is discussed by Engelhard (1996), Hall (1996) and Dixon (2002, 2009, 2017). Appreciation that some fertilizers and other compounds are associated with the diminution of pest and pathogen impact has led to the development of the concept of 'biostimulation'. This identifies that in addition to their direct nutritional properties some compounds are capable of stimulating benign microbial populations associated with plant roots and the diminution of populations of pathogenic organisms.

The crop is inspected at regular intervals to determine whether pest problems are emerging. If there is a developing infestation then growers are advised to apply chemical remedies that may be used at a reduced application rate. Kurppa and Ollula (1993) suggest that limiting the quantities of nitrogen applied to brassica crops could result in fewer herbivores affecting them, but this work was done with summer turnip rape and the blossom beetle (*M. aeneus*).

Physical barriers

Crop covering involves laying transparent plastic sheets over drilled or transplanted crops. The growing plants support the cover as they develop underneath. Alternatively, opaque plastic sheets are placed on the soil surface and the crop. The crop either grows through holes in the sheet or is transplanted through them. Both covers and mulches offer barriers between the developing crop and pests (Ester *et al.*, 1994). For brassica crops, covers have more universal value in the control of pests. In early studies, the effectiveness of two polyethylene nets and two non-woven polypropylene fleeces in controlling cabbage root fly (*D. radicum*), cabbage moth (*Mamestra brassicae*) and small cabbage white butterfly (*P. rapae*) on cabbages was studied in Belgium during 1990–1991. Good control of cabbage root fly was achieved by covers removed 4–5 weeks before harvest (van de Steene *et al.*, 1992).

Polythene sheets were often perforated to allow ventilation and ease of handling. Polypropylene sheets tend to heat up less quickly on hot, sunny days and hence can be left on the crops longer than was the case with polythene sheeting. These sheets are either 'woven' or 'non-woven'. Fibres in the non-woven forms are bonded together by heat (thermal spun bonding). These fabrics are lighter and cheaper than the woven forms and often described as 'fleeces'. They are permeable to air and water. Woven polypropylene is sold as netting with extremely small gaps between the warp and weft fibres (see Fig. 7.9).

Crop covers exclude many major pests from brassica crops and have become important ingredients in pest control. They have become the most

Fig. 7.9. Woven crop cover barrier against pests, CosyTex. (Courtesy of Mr Owen Lane, XL Horticulture).

reliable way of producing swede (*B. napus* ssp. *rapifera*) crops free from damage caused by cabbage root fly. Mesh size in woven netting governs the species of insect that are excluded by netting. Mesh of 0.6 mm² excludes leaf miner and cabbage whitefly (*A. proletella*), 0.36 mm² excludes aphids and 0.06 mm² excludes thrips (*Thrips tabaci*). Brassica pests that have been controlled satisfactorily with crop covers include: cabbage root fly, cabbage aphid (*B. brassicae*) and caterpillars of cabbage white butterflies, small (*P. rapae*) and large (*Pieris brassicae*). A 40-mesh-size repellent net treated with α-cypermethrin irritant repelled the green peach-potato aphid (*Myzus persicae*) and its parasitoid, *Aphidius colemani*. Under field conditions, there were no pests on protected cabbage (Martin *et al.*, 2013). Trials over years and locations showed that fences made of insect netting 180 cm in height, with a downward fold of 50 cm toward the outside, gave significant reductions in attack by root flies (*Delia* spp.) (Meadow and Johansen, 2005).

Photoselective nets have proved effective for aphid control as they limit their dispersal ability. UV-absorbing nets have an effect on the beneficial hover fly (*Sphaerophoria rueppellii*), while immature hover fly density was higher, their dispersion was reduced by the UV-absorbing nets. The use of photoselective nets and the release of predators, such as hover flies, are compatible aphid control strategies (Amoros-Jimenez *et al.*, 2020).

Natural products

Numerous plant extracts or plant-derived compounds can potentially be incorporated into an alternative and novel strategy to control various insects, such as diamondback moth (*P. xylostella*). Extracts can come from tropical plants, such as the rhizomes of galangal (*Alpinia galanga*), the tubers of purple nutsedge (*Cyperus rotundus*) and the seeds of globe amaranth (*Gomphrena globosa*) (Ohsawa and Ohsawa, 2001). The active compounds in purple nutsedge and galangal are α-cyperone and 1′-acetoxychavicol, respectively, which reduced the larval density of the diamondback moth. Glucosinolates are secondary metabolites present in Brassicaceae species and are implicated in their defence against plant pathogens as reviewed by Madloo *et al.* (2019). When a pathogen causes tissue damage, the enzyme myrosinase hydrolyses glucosinolates into diverse products with antimicrobial activity. The effects of glucosinolates were apparently dependent on the pathogen and the type. Thus, the aliphatic sinigrin was inhibitory to infection by *S. sclerotiorum* and the indolic glucobrassicin was inhibitory to infection by black rot (*X. campestris* pv. *campestris*).

STATUTORY CONTROL

The international network of legislation is designed to ensure growers receive reliable seed and mainly provides for the determination of germination, purity and trueness to type (see Chapter 3 section, Seed Quality). There are also requirements for freedom from pests and pathogens where these are seed borne. Relatively few brassica pests and pathogens fall into this category with the prime exception of the cause of black rot (*X. campestris* pv. *campestris*). Normally seed batches are tested for the presence of this bacterium and the presence declared. The International Seed Testing Association (ISTA) has evaluated methods for testing for seed-borne organisms. These tests are sufficiently robust for direct drilled crops where the seed is going straight into its cropping stations. Much of the European and Asian cauliflower (*B. oleracea* var. *botrytis*) and green broccoli (*B. oleracea* var. *italica*) crops, however, are now transplanted and in consequence even minute traces of microbe infection can be devastating at the time of propagation. The seedlings are raised in warm moist protected glasshouse environments that are ideal for the reproduction and spread of this bacterial pathogen. The rate of spread can reach 1% increase per day. This position is complicated by the capacity of black rot for symptomless spread (Shigaki *et al.*, 2000). In consequence, this pathogen has become a major source of loss for the vast majority of world vegetable brassica propagators who supply transplants for field production. Legislation normally prohibits the sale of diseased transplants. This applies to bacterial diseases, such as black rot, and, in some countries, also to plants infected with the cause of clubroot disease (*P. brassicae*). While this pathogen is not transmitted by seed it can be carried in propagation media, especially those that are peat based. There have been several cases where peat infested with *P. brassicae* has been used for the propagation of brassica transplants with devastating consequences when planted out into previously pathogen-free land, which then becomes infested with a pathogen which is very difficult to eradicate.

IMPORTANT PESTS AND PATHOGENS OF BRASSICA CROPS

Aphids

Cabbage aphid – *Brevicoryne brassicae* and green peach-potato aphid – *Myzus persicae*

Aphids (see Fig. 7.10) are destructive pests worldwide, damaging agricultural and horticultural crops (Li *et al.*, 2019b).

The intensity of aphid epidemics is regulated by prevailing weather conditions. In warm dry conditions there is rapid and extensive colony growth, while cool damp weather inhibits population expansion (Blackman and Eastop, 1984; Minks *et al.*, 1987). Initial spring invasions rely on the proximity of overwintered crops to new plantations. Consequently, field hygiene by ploughing in or rotavating residues is of paramount significance, depriving these pests of sources of hibernation cover. Spring-planted vegetable brassicas should not be sited close to overwintered oilseed rape (*B. napus*) crops. The attraction of new crops may be diminished by the use of inter-row mulches of

Fig. 7.10. Green peach-potato aphid (*Myzus persicae*) invasion of Brussels sprout (*Brassica oleracea* var. *gemmifera*) heads. (Geoff Dixon).

transparent or blue plastic or straw. Recent studies suggest that the UV components of sunlight influence aphid flight initiation, dispersal, host finding and establishment and subsequent population build-up (Gulidov and Poehling, 2013).

Breeding for resistance to cabbage aphid (*B. brassicae*) is rendered difficult because of the large number of biotypes. There has been extensive selection and some breeding work in several parts of the world, notably in England and in France. It is suggested that useful dominant gene resistance with some maternal effects is present in some cabbages. Some F_1 hybrids between susceptible and resistant lines failed to provide protection (Metz *et al.*, 1995), whereas Ellis *et al.* (1996) reported partial levels of antixenosis in red-coloured accessions of cvs 'Ruby Ball', 'Yates Giant Red', Brussels sprout cv. 'Rubine' and a strain of 'Italian Red Kale'. Glossy accessions of cabbage and cauliflower (*B. oleracea* var. *botrytis*) possessed antixenosis and antibiosis resistance that lasted throughout the growing season of the crop. In kale, Ellis *et al.* (1998) found they were more likely than other forms of *B. oleracea* to be resistant, while green broccoli (*B. oleracea* var. *italica*) was a poor source of resistance. Several Portuguese kales have resistance to the green peach-potato aphid (*M. persicae*) especially selections of 'Crista-de-Galinha', followed by 'Joenes' and 'Roxa' (Leite *et al.*, 1996). Seven *Brassica* species were tested by Ellis *et al.* (2000) who found *B. fruticulosa*, *B. spinescens*, *B. incana* and *B. villosa* had high levels of antibiosis; these were the same species reported to possess resistance to cabbage root fly (*D. radicum*). There is a suggestion that light-coloured foliage is associated with resistance, as in some of the Australian cauliflower cultivars. Possibly, this relates to a failure of these aphids to locate such coloured crops.

Early use of IPM involves monitoring and forecasting, sampling and decision making, biological control, host plant resistance and cultural control (Collier and Finch, 2007). Minimizing nitrogen fertilizer applications and the use of mixed cropping have been advocated as part of IPM. Similar principles and problems apply to the control of the green peach-potato aphid by the use of pesticides as for the cabbage aphid (*B. brassicae*). The development of strains of these pests with insecticide resistance is a major problem, especially with the green peach-potato aphid because of its wide host range and consequently increased exposure of aphid populations to sprays applied to many host crops. Integrated control systems using combinations of husbandry, chemical and genetic host resistance currently offer the most environmentally attractive means for control but this demands regular crop monitoring in order to evaluate the build-up of pest populations (Finch, 1987). The concept of 'control thresholds' has been used whereby the build-up of aphid colonies is monitored and applications varied according to crop growth stage. Thus, with Brussels sprouts (*B. oleracea* var. *gemmifera*) fewer aphids are required to trigger the application of remedial sprays in the early stages after transplanting than in the main growing season. Tolerance levels are reduced again as the axillary buds form and zero tolerance is permitted close to harvest (Theunissen, 1984). Aphids are generally controlled in nature by parasites, pathogens and predators. This provides opportunities for the development of biological control systems.

Details of host–pest relationships are becoming clearer. Specialist herbivores exploit the chemical defences of their food plants to their own advantage (Goodey *et al.*, 2015). Brassica plants produce glucosinolates (see Chapter 6 section, Biofumigation) that are broken down into defensive toxins when tissues are damaged. Evidently, some aphids use these chemicals against their natural enemies by becoming a 'walking mustard-oil bomb'. Aphids sequester glucosinolates selectively, accumulating sinigrin to high concentrations while preferentially excreting a structurally similar glucosinolate, progoitrin. Surveys of aphid infestation in wild populations of *B. oleracea* show that this pattern of sequestration and excretion is associated with host plant selection. Above-ground and below-ground herbivore species modify plant defence responses differently. Simultaneous attack leads to non-additive effects on primary and secondary metabolite composition in roots and shoots (Hol *et al.*, 2013). The population growth of cabbage aphid on *B. oleracea* was reduced where plants were infested with sugarbeet cyst nematodes (*Heterodera schachtii*) 4 weeks prior to aphid infestation. Since nematodes reduce amino acid and sugar concentrations in the phloem, this limited subsequent aphid population growth and disturbed feeding relations between host plants and aphids. Volatile-mediated plant-to-plant communication has been assessed in multitrophic systems in different plant and pest species by Kang *et al.* (2018). β-Ocimene is recognized as a herbivore-induced plant volatile that plays an important role in the chemical communication between plants and pests. Treatment of Chinese cabbage with β-ocimene inhibited the growth of the green peach-potato aphid, negatively influencing its feeding behaviour by shortening the total feeding period and phloem ingestion and increasing the frequency of stylet puncture. This suggests that β-ocimene can activate the defence response of Chinese cabbage against the green peach-potato aphid.

Aphid damage can extend to stored crops, where they feed on and excrete faeces (honeydew) which is an excellent growth medium for microbes (Zhan *et al.*, 2020). The dominant bacterium in stored Chinese cabbage infested by aphids was *Staphylococcus aureus* which increased significantly after prolonged storage.

Elevated concentrations of atmospheric carbon dioxide (CO_2) may interfere with tritrophic interactions (Klaiber *et al.*, 2013). These authors used a combination of brassica plants, the cabbage aphid and the endoparasitoid, *Diaeretiella rapae*, which specializes on aphids feeding on brassicas, and compared the effects of elevated CO_2 (800 ppm) vs ambient CO_2 (400 ppm). Parasitoid progeny emerged earlier but offspring adults were shorter lived. Plant glucosinolate concentrations were higher under elevated CO_2 compared to ambient CO_2, and aphid glucosinolate concentrations were significantly lower. Aphid body mass was 20% lower under elevated CO_2 compared to ambient CO_2. Elevated CO_2 concentrations enhance direct defence of plants as well as encouraging biological controls.

Cabbage whitefly – *Aleyrodes proletella*

Historically the cabbage whitefly (*A. proletella*) was a minor pest of brassica crops, but recently it has become an increasingly important pest in Europe, particularly of Brussels sprouts and kale (Collins, 2014). Cabbage whitefly colonize vegetable brassicas primarily from oilseed rape. The flight behaviour of cabbage whitefly morphs shows phototactic attraction but this is only effective during the later phases of flight. Therefore, cabbage whitefly ignore host plants close to their origin and disperse as much as several kilometres away (Ludwig *et al.*, 2019). The colonization of Brussels sprouts (*B. oleracea* var. *gemmifera*) occurs before the emergence of the first generation of adults on late season plants, suggesting that overwintering females were the first colonizers. Hondelmann *et al.* (2020) assessed resistance in Brussels sprout cultivars, finding that it was not strong or consistent enough for use alone but might be useful as a component of integrated management strategies.

This pest inhabits the undersurface of leaves and consequently chemical control is effective only when insecticides are directed at high volume on to this target with suitable drop-leg spray booms (Mound and Halsey, 1978; Byrne and Bellows, 1991). A German crop inspection and application routine required sprays when 20 adults or 50 larvae per plant were found, this routine substantially reduced the number of applications required for Savoy and red-headed cabbage. There may be opportunities for biological control (Hulden, 1986). *Encarsia tricolor* is the dominant parasitoid wasp species of the cabbage whitefly. The latter finds sufficient overwintering habitats to appear in masses on cabbage crops during cultivation periods, whereas habitats with suitable overwintering hosts for *E. tricolor* are less common (Laurenz *et al.*, 2017). Studies show that *E. tricolor* successfully survived winter as immature stages, but no adults were found during late winter months. The first adult *E. tricolor* were found on yellow sticky traps in the field in early May.

Beetles and weevils

The order Coleoptera is the largest grouping of all insects, containing crop pests and many species that are valuable predators attacking other organisms. Beetles are essential members of soil communities and many are very beneficial controllers of soil pests. Rove beetles, such as the devil's coach-horse beetle (*Ocypus olens*), have an omnivorous diet, embracing other soil inhabitants and any organic remains but not living plants. This would make them prime candidates for biological control strategies but so far this has not been successfully achieved. Brassica husbandry should aim at preserving and encouraging populations of these beneficial insects. Studies by Depalo *et al.* (2020) showed that rove beetles were more abundant where the cover crops were terminated by the roller–crimper machinery. Using crop fencing can limit or delay surface-active predatory beetles from accessing green broccoli (*B. oleracea* var. *italica*) fields (Renkema *et al.*, 2016). The performance of herbivorous pest insects depends on a balance of their nutrient uptake and toxin avoidance (Tremmel and Müller, 2014). High concentrations of defensive plant metabolites impair both generalist and specialist pests; but generalists are less well adapted to particular hosts and thus more badly affected by plant defence traits.

Blossom beetles – *Meligethes* spp.

Trap crops may be employed which divert these pests from summer cauliflower (*B. oleracea* var. *botrytis*). Success has been achieved with flowering crops of Chinese cabbage and green broccoli (*B. oleracea* var. *italica*). Crop monitoring minimizes the use of pesticides. Population densities of adult beetles (see Fig. 7.11) in autumn crops should be 15–20 beetles and, in spring, three beetles per plant, before applications are needed.

Once crops have reached the flowering stage, however, most of the damage has been done and spraying is ineffective. Computer-based forecasting systems are available for some crops.

Flea beetles – *Phyllotreta* spp. and cabbage stem flea beetle – *Psylliodes chrysocephala*

Dry, warm spring weather favours these insects and seedlings under water stress are more vulnerable. Under sowing or mulching, brassica crops diminish their attractiveness for flea beetles (*Phyllotreta* spp.) (see Fig. 7.12).

Fig. 7.11. A blossom beetle (*Meligethes* spp.). (Courtesy of Professor Rosemary Collier).

Fig. 7.12. Flea beetles (*Phyllotreta* spp.). (Courtesy of Professor Rosemary Collier).

There is some suggestion that the use of organic, as opposed to soluble, fertilizers discourages flea beetles. Unsuccessful attempts at biological control have involved using nematodes pathogenic to flea beetles. The invasive predator *Pterostichus melanarius* could offer an avenue for biological control of crucifer flea beetles (*P. cruciferae*) (Blubaugh *et al.*, 2021). But manipulating plant communities that increase parasitoid evenness and promote pest control would require significant husbandry expertise. Opportunities for using chemical controls against flea beetles are extremely limited since neonicotinoid compounds have been removed from use.

Mustard leaf beetle – *Phaedon cochleariae*

The insect family of leaf beetles (Chrysomelidae) is one of the largest and most successful in the animal kingdom with an estimated 35,000–50,000 species. The natural host of the mustard leaf beetle (*P. cochleariae*) is the aquatic crucifer watercress (*Nasturtium officinale*). They can also cause considerable damage to cabbage crops. The food plant type and quality influences the feeding preferences and various life history traits of these herbivorous insects (Muller and Muller, 2017). Beetles experimentally fed on cabbage became more active with age, bolder, reproduced more quickly and developed a different behavioural phenotype compared to beetles fed on watercress, expressing a faster pace of life.

Cabbage stem weevil – *Ceutorhynchus quadridens* and cabbage seed weevil – *Ceutorhynchus assimilis*

Control measures include avoiding growing susceptible horticultural brassicas adjacent to oilseed rape (*B. napus*) crops. Chemical treatments, provided they are available, applied to control blossom beetles (*M. aeneus*) will also help to control cabbage seed weevil (*C. assimilis*), while chemicals used for cabbage root fly (*D. radicum*) control may give some control of cabbage stem weevil (*C. quadridens*). Seven *Brassica* species were evaluated for susceptibility to infestation by the cabbage seedpod weevil (*Ceutorhynchus obstrictus*). *Brassica rapa* was the most susceptible and intermediate susceptibility was observed for *B. napus*, *Brassica napus* × *S. alba*, *Brassica tournefortii* and *Brassica juncea*, although the last species displayed some antixenotic resistance. *Sinapis alba*, *B. nigra* and *Crambe abyssinica* were least susceptible (Kalischuk and Dosdall, 2004)

Flies

Cabbage root fly (also known as cabbage root maggot in North America) – *Delia radicum* and turnip root fly – *Delia floralis*

Cabbage root fly (*D. radicum*) (see Fig. 7.13) is predated by a range of other flies, mites, ants, beetles and fungi at several stages in the life cycle. So far,

Fig. 7.13. Cabbage root fly (*Delia radicum*) larva. (Courtesy of Professor Rosemary Collier).

no commercially acceptable system of biological control has been developed. Several beetles will consume the eggs of turnip root fly (*D. floralis*) while the fungus *Strongwellsea castrans* sterilizes the females but only after egg laying has commenced. Neither the beetles nor the fungus have yet been formulated into biocontrol systems.

Husbandry may be manipulated to avoid infestation by cabbage root fly through crop rotation, the destruction of infected hosts and the avoidance of autumn sowing which provides overwintering sites. Spatially separating ware production of vegetable brassicas from seed crops is of major importance. Forming physical barriers by mulching with finely netted, non-woven or fleece covers, or using high-level barriers have become standard means of commercial control. Placing collars around the stem bases of transplants prevents the adult flies from laying eggs adjacent to host plants and may be practiced with small-scale market garden or organically grown crops.

No commercially resistant cultivars have been produced so far. Screening for resistance in root brassicas and cauliflower (*B. oleracea* var. *botrytis*) has been a main breeding aim and some tolerant strains are known. Resistance is reported in *B. fructiculosa*, *B. incana*, *B. villosa* and *B. spinescens* to cabbage root. All *Brassica* species were attractive to egg-laying by cabbage root fly and therefore lacked antixenosis resistance. Freuler *et al.* (2000) found resistance to cabbage root fly in the cauliflower cvs 'Imperator Nouveau', 'Panda', 'Asterix', 'Belot', 'Talbot' and 'Lara' by counting eggs laid around the plant stem and estimating root damage. Shelton and Dickson (unpublished information) identified resistance to cabbage root fly in cauliflower by response to egg placement at the stem base in the glasshouse and, in the field, by damage assessment following baiting with offal. Dosdall *et al.* (2000) found *S. alba* had the greatest resistance to infestation by cabbage root fly maggots and in crosses of *S. alba* × *B. napus* some hybrids had high levels of resistance, similar to *S. alba*, which persisted from year to year. Scottish swede cvs 'Angus' and 'Melfort' apparently have some resistance to turnip root fly (*D. florialis*). Resistance in a range of brassica types has been described by Alborn *et al.* (1985).

Butterflies and moths

Large cabbage white butterfly – *Pieris brassicae* and small cabbage white butterfly (also known as the imported or European cabbageworm in North America) – *Pieris rapae*

Herbivorous insects and plant pathogens are usually more abundant when plants are flowering (Chretien *et al.*, 2018). Generally, plant responses against one attacker can interfere with responses to a second one. Phytohormones that orchestrate plant reproduction are also involved in resistance against insect and pathogen attack. Jasmonates, a class of oxylipins, are signalling molecules involved with the development of host resistance reactions (Kurakin, 2022).

Natural control of cabbage white caterpillars (see Fig. 7.14) is provided by birds, carabid beetles, the parasitoid wasp *Cotesia glomerata* and a baculovirus.

Several predators from wasp parasitoids (*Apanteles* spp., *Pteromalus puparum* and *Meteorus versicolor*) and nematodes (*Heterorhabditis* spp., *Poinax* spp. and *Neoaplectana* spp.) have been tested as potential commercial

Fig. 7.14. Caterpillars of the cabbage white butterfly (*Pieris* spp.). (Geoff Dixon).

biological control agents (Feltwell, 1982; Carter, 1984). Spiders are important suppressers of lepidopteran pests in brassica crops, particularly early in the season before parasitoids become established (Senior *et al.*, 2016). The most numerous of these spiders were from the family Theridiidae, which were more strongly associated with cauliflower (*B. oleracea* var. *botrytis*) and Chinese cabbage (*B. rapa* ssp. *pekinensis*), and less so European cabbage (*B. oleracea* var. *capitata*). Experimentally, they have proved capable of feeding on larvae of the diamondback moth (*P. xylostella*), and cabbage cluster caterpillar (*Crocidolomia pavonana*). Commercially, preparations of the bacterium *B. thuringiensis* (Bt) may provide biological control applied at 14-day intervals. Green *et al.* (2015) suggest that although leaf colour affects both levels of host defences and herbivore abundance in the field, the responsiveness of herbivores is limited. Wind may influence plant–herbivore interactions (Chen *et al.*, 2018). Windiness delayed flowering, decreased host plant height and increased leaf concentrations of amino acids and glucosinolates in black mustard (*B. nigra*). Wind-exposed adult large cabbage white butterflies (*P. brassicae*) were significantly larger, revealing a trade-off between development time and adult size. But the avian predator, the great tit (*Parus major*), captured significantly more caterpillars on still than on wind-exposed plants.

Cabbage moth – *Mamestra brassicae*

Several wasps colonize the eggs (*Trichogramma evanescens*), larvae (*Cotesia glomerata* and *Hyposoter ebenius*) and pupae (*Pteromalus puparum* and *Pimpla instigator*). Additionally, the flies *Phryxe vulgaris* and *Compsilura concinnata* are parasites of the cabbage moth (*M. brassicae*). Studies indicate that several fungal and viral parasites offer potential biological control systems; the larvae are not susceptible to current commercial strains of the bacterium, *B. thuringiensis* (Bt), unlike other lepidopteran pests, e.g. cabbage white butterflies (*Pieris* spp.). Reputedly, there is resistance in the cauliflower (*B. oleracea* var. *botrytis*) line PI 234599. This was the first line associated with glossy leaves and resistance to lepidopteran pests. Invasions by cabbage moths, resulting in larvae developing deeply inside cabbage heads, can cause substantial loss of quality, being difficult to detect until the product is used.

Diamondback moth – *Plutella xylostella*

Several wasp predators reportedly attack the larvae of the diamondback moth (*P. xylostella*) (see Fig. 7.15).

The larvae are generally susceptible to the bacterial biological control agent *B. thuringiensis* (Bt) (Talekar *et al.*, 1985; Talekar, 1986; Entwistle *et al.*, 1993; Talekar and Shelton, 1993). Genes from two strains of *B. thuringiensis* (Bt) have been transferred into green broccoli (*B. oleracea* var. *italica*) by *Agrobacterium tumefaciens*-mediated transformation. This bacterial control is effective against the diamondback moth (Cao *et al.*, 1999) invading Chinese

Fig. 7.15. Diamondback moth (*Plutella xylostella*) adult. (Courtesy of Professor Rosemary Collier).

cabbage (Cho *et al.*, 2001). Plants with high levels of Cry1C protein cause rapid and complete mortality of three types of diamondback moth larvae without defoliation (Earle and Knauf, 1999). Because of the rapid ability of the diamondback moth to mutate and overcome *B. thuringiensis* (Bt), Shelton *et al.* (2000) have shown that it is necessary to provide refuges which conserve susceptible larvae and thus reduce the numbers of homozygous offspring. At the same time, the refuges with high levels of *B. thuringiensis* (Bt) toxic plants resulted in a reduction of predators (Riggin-Bucci and Gould, 1997). The diamondback moth was monitored using pheromone traps (Reddy and Guerrero, 2001), by counting immature stages on plants, and by accessing citizen science data (records of sightings of moths) from websites and X (formerly Twitter). Studies confirmed that the diamondback moth is a sporadic but important pest. Very large numbers of moths can arrive suddenly, most often in the UK's early summer (Wainwright *et al.*, 2020). Sightings by citizen scientists appears to be an effective way of providing timely warnings.

   Considerable effort has been invested into developing resistance to the diamondback moth and small cabbage white butterfly (*P. rapae*) in *B. oleracea*. Some glossy-leafed lines have shown good consistent resistance (Lin *et al.*, 1983) in many parts of the world. But the glossy leaf is unacceptable horticulturally as it reduces yield and delays maturity and, so far, no adequate resistance has been found in non-glossy lines. Resistance was found in cauliflower (*B. oleracea* var. *botrytis*) PI 234599, cabbage G8329 and lines of Chinese cabbage. There are, however, reports of good resistance from a commercial breeding programme in South Korea. Pheromone traps are important elements in integrated control, limiting the use of pesticides, where permitted. Overuse of pesticides in Asia has resulted in the emergence of resistant insect populations. White mustard (*S. alba*) may be used as a trap crop in cauliflowers (George *et al.*, 2009). Benefit is maximized by using trap crop plants that are larger than the main commercial ones.

Garden pebble moth – *Evergestis forficalis*

Intercropping. especially in Brussels sprouts (*B. oleracea* var. *gemmifera*), reduces the amount of insect damage but yields are adversely affected because the population density of the main crop is diminished.

Turnip moth (cutworm) – *Agrotis* spp., especially *Agrotis segetum*

A spatial barrier of 500 m between new and old season seed crops is frequently sufficient to prevent infestation because adult turnip moths (*Agrotis segetum*) are weak fliers (see Figs. 7.16 and 7.17).

Brassica crops vary in their attraction to this pest with fewer eggs being laid in mustard crops, such as black mustard (*B. nigra*), brown mustard (*B. juncea*) and Ethiopian mustard (*Brassica carinata*), compared with forms of *B. rapa* and *B. napus*. Forecasting using computerized prediction systems allows control by targeted applications of insecticide, where permissible, based on increasing populations relative to prevailing and predicted weather conditions. This pest only causes measurable damage when cabbage seed weevil (*C. assimilis*) damages the host pods in advance and provides entry to them.

Fig. 7.16. Adult turnip moth (*Agrotis segetum*). (Courtesy of Professor Rosemary Collier).

Fig. 7.17. Cutworm larva stage of turnip moth (*Agrotis segetum*). (Courtesy of Professor Rosemary Collier).

Cabbage leaf miner – *Phytomyza rufipes*

The options for control are very limited, hence this pest can be the cause of significant loss of quality and yield in leafy brassicas.

Thrips

Onion thrips – *Thrips tabaci*

Onion thrips (*Thrips tabaci*) are an important pest of cabbage. Two reproductive modes, arrhenotoky (parthenogenesis forming males) and thelytoky (parthenogenesis forming females), are known and co-occur in the field (Li *et al.*, 2014). Within populations attacking onions (*Allium* spp.), arrhenotokous onion thrips performed better than thelytokous onion thrips, while on cabbage the opposite was reported. Pest management strategies should reflect these differences.

Breeding programmes for resistance against onion thrips showed quantitative and recessive resistance are available (Stoner *et al.*, 1989; Shelton, 1995). Storage cabbage crops in the Netherlands are usually harvested around mid-October and may be severely damaged by onion thrips (Voorrips *et al.*, 2010). Populations and the more severe symptoms develop mostly during September and October and subsequently in cold storage symptoms continue developing. Large differences exist between cabbage cultivars in their susceptibility to onion thrips damage. Regression studies showed that more advanced plant development at the end of August increased onion thrips damage at the final harvest. Other plant traits affecting onion thrips damage were Brix (sugar content) and the amount of leaf surface wax. No single plant trait, however, explained more than 45% of the variation in onion thrips damage at the final harvest.

Cultural control is possible when the crop maturity is scheduled to avoid periods during which onion thrips populations are at their highest. Pest monitoring using yellow-coloured sticky traps is essential, identifying periods of highest risk. The lacewing predator (*Chrysopa* spp.) has been used for biological control in India and integrated with pesticide applications.

Nematodes

Brassica cyst nematode – *Heterodera cruciferae* and sugarbeet cyst nematode – *Heterodera schachtii*

Control is largely obtained by crop rotation, avoiding brassica overcropping. Where a rotation of brassicas (including oilseed rape (*B. napus*)) of no greater frequency than once every 5 years is maintained, the nematode populations are eroded. Sugarbeet cyst nematode (*H. schachtii*) has recently been detected as a severe pest of Chinese cabbage fields in South Korea (Kabir *et al.*, 2017). Maximum nematode multiplication rates were recorded on older plants, hence transplanting of Chinese cabbage in modules is recommended for the management of sugarbeet cyst nematodes, as compared with direct drilling.

Molluscs

Slugs – *Deroceras* spp., *Arion hortensis* and *Milax* spp.

Studies at Long Ashton Research Station, UK (Hass *et al.*, 1999) identified the nematode *Phasmarhabditis hermaphrodita* as a potential biological control for slugs. This research demonstrated that Chinese cabbage (*B. rapa* ssp. *pekinensis*) seedlings could be protected from serious damage by the grey field slug (*Deroceras reticulatum*). Elsewhere, slugs are serious pests of a range of crops worldwide (Rae *et al.*, 2009). *Phasmarhabditis hermaphrodita* is a slug-parasitic nematode that can be an effective means of control. Trials showed that *P. hermaphrodita* persisted in soil resulting in significant slug control 38 days after initial application. Three low-dose applications of *P. hermaphrodita* provided slug control comparable to one broadcast spraying. Iron phosphate molluscicides significantly reduce damage caused by grey field slugs and black slugs (*Arion ater*) and are replacing previous environmentally unacceptable products.

Warmer, more moist seasons are encouraging increasingly large populations of slugs which damage the quality and attractiveness of brassica crops (see Fig. 7.18).

Fig. 7.18. Slugs on Brussels sprout (*Brassica oleracea* var. *gemmifera*) buds. (Geoff Dixon).

Growing crops on soils that are heavily infested with slugs should be avoided. Baits are now being formulated which are compatible with wildlife populations. For example, iron phosphate molluscicides significantly reduce damage caused by grey field slugs and black slugs. The use of some organic fertilizers, such as calcium cyanamide, appears to be associated with reductions in slug populations, possibly by encouraging beneficial microbes in the soil which are antagonistic towards slugs. Studies of molluscs as crop pests are reviewed by Barker (2002).

Vertebrates

Wood pigeon (ring dove) – *Columba palumbus palumbus*, rabbit – *Oryctolagus cuniculus*, hare – *Lepus europaeus occidentalis*, deer (e.g. roe deer – *Capreolus capreolus*, fallow deer – *Dama dama*, Japanese deer – *Sika sika*, Indian muntjac – *Muntiacus muntjac*) and rat (brown rat – *Rattus norvegicus*, black rat – *Rattus rattus*)

Management of vertebrates requires combinations of methods and now tends to be controlled by environmental and conservation legislation, and social pressures. Shooting has short-term benefit but will only slightly reduce the populations of larger mammals such as deer, hares and rabbits. Scaring by various methods can be effective but needs constant changes as pigeons, for example, will quickly learn to ignore anything that presents no real danger and return to browse crops (see Fig. 7.19).

The use of toxic baits has now been largely abandoned, except for rats and mice where the baits are placed in containers specifically designed to catch these animals and exclude others. The loss of some active ingredient molecules for use as baits is resulting in increasing levels of damage caused by rats, especially in maturing crops of autumn brassicas (see Fig. 7.20).

This raises very significant human health risks, both to those harvesting the crops and ultimately the consumer, from disease-causing microbes carried

Fig. 7.19. Pigeon damage to rocket baby leaf crops. (Geoff Dixon).

Fig. 7.20. Rat damage on Brussels sprouts (*Brassica oleracea* var. *gemmifera*). (Geoff Dixon).

in rat excreta and urine. This can lead to Weil's disease, caused by forms of *Leptospira* bacteria, resulting in serious cases of meningitis and kidney failure, and possibly leading to fatalities. The faeces of dogs and foxes can contaminate crops with parasitic roundworms (*Toxicana* spp.) which pose a significant danger for humans when present on crops. Domestic dogs should be excluded from all brassica crops.

Transplants raised in seed beds may be protected most effectively by netting or the use of vibrating tapes. Browsing mammals can be diverted from young brassica crops by the presence of adjacent sowings of clover (*Trifolium* spp.) and grasses. Depending on the geographical location, mammal pests can vary from the small to elephants and other large mammals in tropical and subtropical countries.

Airborne microbes

White rust or white blister – *Albugo candida*

The incidence of white rust or white blister disease (*A. candida*) is increasing, particularly on late-summer and autumn brassica crops as a result of warmer and moister weather (see Fig. 7.21).

Race-specific resistance governed by single dominant genes has been identified in several *Brassica* and *Raphanus* spp. Quantitative inheritance of disease reaction type was demonstrated in *B. rapa*. Resistance to race 2 of white rust is controlled by a single dominant gene, and resistance was associated with leaf pubescence, which is also governed by a single dominant allele (Kole *et al.*, 1996). The resistance locus *ACA1* and pubescence locus *PUB1* were mapped using restriction fragment length polymorphisms (RFLPs) to linkage group 4. Portuguese landraces of *B. oleracea* were screened for white rust resistance and 'Couve Agarvia', 'Couve Gloria de Portugal' and 'Couve Portuguese' were the most resistant (Santos *et al.*, 1996). Some cultivars of *B. carinata* have resistance to both white rust and downy mildew (*H. parasitica*), while dominant resistance is present in *B. juncea* and in *B. napus*. Resistance

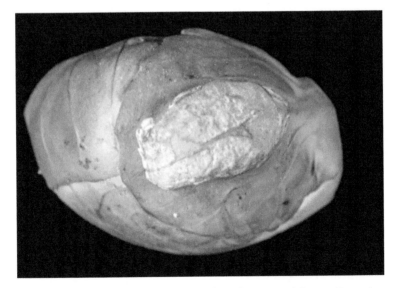

Fig. 7.21. White rust or white blister caused by *Albugo candida* on a Brussels sprout (*Brassica oleracea* var. *gemmifera*) bud. (Courtesy of Phillip Effingham).

to white rust is present in interspecific crosses between *B. carinata* and *B. juncea*. Mass selection and selection between and within half-sib families were effective methods of accumulating minor genes conferring reduced pathogen sporulation in rapid-cycling populations of *B. rapa* (Edwards and Williams, 1987). Similar genes have been identified and mapped by Kole *et al.* (2002) in *A. thaliana*.

Rapid diagnosis is an essential part of integrated disease management. An interactive, iterative smartphone application which uses colour images that can distinguish diseased from healthy plant tissues and calculate the percentage of disease severity has been developed by Pethybridge and Nelson (2015). Estimates of disease severity were highly accurate with the advantages of rapid image processing, low cost, ease of use and the ability for reporting digitally. Botanically, white rust provides a model for exploring how plant biodiversity has channelled speciation of biotrophic parasites (Borhan *et al.*, 2008). The causal agents of white rust infest across a wide breadth of cruciferous hosts and are named currently as variants of a single oomycete species, *A. candida*.

Dark leaf spot – *Alternaria* spp.

Dark leaf spot of cruciferous vegetables, incited by different species of *Alternaria*, is an increasing threat to crops worldwide. The pathogens have a wide host range, such as head cabbage, Chinese cabbage, cauliflower, green broccoli and other crucifers, including cultivated and wild-grown plants. *Alternaria* spp. usually cause damping-off of seedlings, leaf spotting of cabbages, blackleg of cabbage heads, and spotting of cauliflower curds and green broccoli florets. There is considerable risk for horticultural brassicas posed by the development of *Alternaria* spp. on agricultural crops. Humpherson-Jones (1989) found that seed crops of marrow-stem kale (*B. oleracea* var. *acephala*) liberated 50 spores/m^3 of air at cutting and 3200 spores/m^3 of air when windrowed crops were harvested.

Resistance is dominant and quantitative to this pathogen. Selections from species such as *B. tournefortii*, *Camelina sativa*, *S. alba*, *Capsella bursa-pastoris* and *B. carinata* appear to be highly resistant. Successful molecular fusion of *C. sativa* with *B. oleracea* transferred a high level of resistance to *B. oleracea* (Sigareva and Earle, 1995). Subsequently, Hansen and Earle (1997), fused *S. alba* with *B. oleracea* and obtained resistance to *Alternaria brassicae* equal to that in *S. alba*, although developing true breeding lines with high levels of resistance has been difficult. Resistance to *Alternaria brassicicola* does not necessarily, however, result in resistance to *A. brassicae* (Ryschka *et al.*, 1996). Variations in resistance in oilseed rape (*B. napus*) and field mustard (*B. rapa*) have been attributed to differences in the thickness of epicuticular waxes.

Subsequent studies using conventional breeding techniques have shown that strong cross-incompatibility and polygenic background of the resistance (additive and dominant gene interactions), as well as the differences in ploidy between the *Brassica* species, render the transfer of resistance from the wild species into the cultivated forms difficult (Nowicki *et al.*, 2012). Cultivar trials of cauliflowers (*B. oleracea* var. *botrytis*) evaluated susceptibility to *A. brassicicola* at Geneva, NY, USA (Kreis *et al.*, 2016). The purple-curded cvs 'Graffiti' and 'Violet Queen' had significantly less *Alternaria* leaf spot than others. The rate of initial pathogen development on leaves varies with different *Brassica* cultivars but by 5 days after infection all showed similar leaf necrosis (Macioszek *et al.*, 2018). An ability for tracking disease development from invasion is, therefore, valuable for this pathogen (Pethybridge and Nelson, 2015). Brassinin is an antifungal compound induced in brassica plants after infection (Srivastava *et al.*, 2013). The ability of *A. brassicicola* to detoxify brassinin appears necessary for successful infection. Biological control may be developed using phylloplane fungi, such as *Aureobasidium pulliulans* and *Epicoccum nigrum*, which are pathogenic to *A. brassicicola*.

Powdery mildew – *Erysiphe cruciferarum*

In the UK, powdery mildew (*E. cruciferarum*) on Brussels sprouts (*B. oleracea* var. *gemmifera*) increased in importance following the intensification of production in the 1970s onwards (Dixon, 1974, Dixon, 1981, Dixon, 1984) (see Fig. 7.22).

The use of very susceptible F_1 hybrid cultivars that matured uniformly and rapidly and are aimed at supplying the processing industry encouraged

Fig. 7.22. Powdery mildew (*Erysiphe cruciferarum*) on Brussels sprout (*Brassica oleracea* var. *gemmifera*) foliage. (Geoff Dixon).

disease incidence. Originally, *E. cruciferarum* was found sporadically in Bedfordshire and the Vale of Evesham in the UK – now it is an international problem. Currently, it has become a major cause of lost quality at harvest worldwide. Considerable efforts are being made to control *E. cruciferarum* using resistant cultivars. Rutabaga (*B. napus* ssp. *rapifera*) plants in Ontario, Canada were reported as severely infected with *E. cruciferarum* (Shattuck, 1993). As an example of increasing damage caused by this disease, Chinese cabbage (*B. rapa* ssp. *pekinensis*) seedlings in a glasshouse in Suwon, South Korea, showed typical powdery mildew symptoms. Initially as circular or irregular white colonies, which developed into abundant growth on both leaf surfaces. Severe infections caused leaf distortion, withering and premature senescence (Jee *et al.*, 2008). Powdery mildew (*E. cruciferarum*) is recorded on broccoli raab (*B. rapa* ssp. *ruvo*; a leafy vegetable known in Europe as rapini) in California, USA.

Resistance in cabbage is attributed to a single dominant gene that is influenced by numbers of minor genes, since in parental generations resistance is incompletely dominant. Thus, under conditions of heavy inoculum, a heterozygotic host may support limited fungal growth. Multiple disease-resistant cabbages were produced in the northern USA for sauerkraut production, and the fresh market trade in Florida during winter. Inheritance studies involving interspecific crosses between *B. juncea* and *B. carinata* found resistance from *B. carinata* was dominant. The resistance gene was on the C genome, a diploid progenitor of *B. carinata* (Singh *et al.*, 1997). When assessing infection of Brussels sprouts leaves, all stem positions must be sampled as younger top leaves tend to be much less infected (de Jong and Hasper, 1996).

Husbandry control by late sowing of swedes and turnips avoids the worst epidemics of *E. cruciferarum* but there is a significant penalty as yield is reduced and this technique cannot be used with most brassica crops.

Black leg, stem and leaf canker – *Leptosphaeria maculans* (previously named *Phoma lingam*)

Blackleg disease (*L. maculans* (anamorph *P. lingam*)) of Brussels sprouts, cabbage and cauliflower was described in an authoritative early primary source by Clayton (1927). It is now a most ubiquitous pathogen of brassica crops, mainly oilseed brassicas, e.g. oilseed rape, canola (*B. napus*), causing the devastating 'stem canker' or 'black leg'. Primary disease symptoms are greyish-green collapse of cotyledon or leaf tissue, without a visible margin, and bearing tiny black spots (pycnidia). The fungus then develops an endophytic symptomless growth for many months. Secondary symptoms, at the end of the growing season, are dry necroses of the crown tissues with occasional blackening (stem canker or blackleg) causing lodging of the plants (see Fig. 7.23).

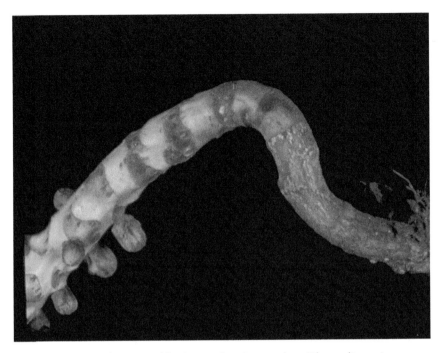

Fig. 7.23. Stem canker caused by *Leptosphaeria maculans* (*Phoma lingam*) on a Brussels sprout (*Brassica oleracea* var. *gemmifera*) stem. (Geoff Dixon).

Pseudothecia differentiate on leftover residues. Seedling damping-off and premature ripening are also reported under certain environmental conditions (Rouxel and Balesdent, 2005).

Breeding for resistance has been made difficult because of the widespread variability in the pathogen. Crop types can be categorized as: *very susceptible* – red and green cabbage, Savoy cabbage, Chinese cabbage, Brussels sprouts, some radish cultivars, some swede cultivars, white mustard and kohlrabi; *medium susceptible* – cauliflower, green broccoli, rape, kale, collards, some turnip cultivars, wild radish and black mustard; *mildly susceptible* – some turnip and swede cultivars, Chinese mustard, garden cress and many strains of mustard; and *resistant* – horseradish, penny-cress, ball mustard, yellow rocket, shepherd's purse and pepper cress. Swede (*B. napus* ssp. *rapifera*) cultivars often appear to be a mixture of susceptible and resistant populations, but the Wilhelmsburger types contain a high proportion of resistant material. *Camelina sativa, S. alba* and *Sinapis arvensis* and *B. nigra* are good sources of resistance and attempts are being made to transfer this resistance to *B. napus* and *B. oleracea* through both embryo rescue and fusion, and resistance to toxins by protoplasts. Some Savoy cabbage, kale, turnip rape and white mustard expressed quantitative resistance and are being used to transfer resistance to white cabbage.

Husbandry controls include the use of rotations that ensure breaks of 4–5 years between brassica crops. Legumes, such as lucerne (alfalfa) (*Medicago sativa*) or clover (*Trifolium spp.*), are particularly effective break crops. Provision of adequate soil drainage and manipulation of plant stand density ensuring rapid air movement are essential, thereby discouraging the build-up of a moist microclimate within the crop. Ploughing should be done deeply in the autumn, which hastens the decay of infested debris.

Hot water treatment can be used to free seed of *L. maculans* infestation, by soaking for 25 min at 50°C for cabbage (*B. oleracea* var. *capitata*) and Brussels sprouts (*B. oleracea* var. *gemmifera*) and 20°C minimum for cauliflower (*B. oleracea* var. *botrytis*) and green broccoli (*B. oleracea* var. *italica*). But the technique is unreliable, often impairing subsequent germination. This pathogen has become a major cause of losses in oilseed rape, resulting in significant research effort.

Ringspot – *Mycosphaerella brassicicola*

Ringspot (*M. brassicicola*) infecting Brussels sprouts (*B. oleracea* var. *gemmifera*) causes substantial yield loss as the sprout buttons are downgraded (see Fig. 7.24).

Resistance to *M. brassicicola* is claimed to exist in Roscoff-type cauliflower (*B. oleracea* var. *botrytis*), which have been selected for centuries against this

Fig. 7.24. Ringspot (*Mycosphaerella brassicicola*) on Brussels sprout (*Brassica oleracea* var. *gemmifera*) buds. (Source unknown).

pathogen in Brittany, France. In India, differences in disease severity have also been noted with the cauliflower cv. 'Improved Japanese', exhibiting greatest resistance. Subsequent breeding of F_1 hybrids has sought improved resistance, particularly in cauliflower cultivars.

Studies of the influence of temperature and wetness duration on infection processes correlated ($r = 0.92$) temperature with leaf wetness periods. An optimal wind run (the product of the average wind speed and the period over which that average speed was measured) of 250–500 km/day was required for the dissemination of ringspot inoculum (Wakeham and Kennedy, 2010). These conditions also encourage the development of leaf spot (*A. brassicicola* and *A. brassicae*) and black rot (*X. campestris* pv. *campestris*). Infections of the outer leaves of Brussels sprouts causes severe losses in quality. Crop residues can be a major primary inoculum source of the pathogens. The observed variation in population sizes of the pathogens between individual pieces of crop residues indicates a stochastic spread of pathogens (Kohl *et al.*, 2011).

Cultural measures are of paramount significance for the control of *M. brassicicola*. Infection leads to premature leaf abscission, particularly in overwintered Angers-type cauliflower; growers encourage rapid production of new foliage in the following spring by applications of nitrogenous top dressings. Autumnal applications of potassium fertilizers (400 kg/ha) are considered as a further means of combating infection for overwintered cauliflower. Hot water treatment may be used to eradicate seed-borne infection.

Light leaf spot – *Pyrenopeziza brassicae*

Originally this pathogen damaged vegetable brassicas grown in mild, moist environments, such as the south-western counties of England. The development of extensive populations of oilseed rape (*B. napus*) provided 'green bridges' resulting in light leaf spot becoming a major disease of that crop (Dixon, 1975) and posing significant threats to adjacent brassica vegetables (see Fig. 7.25).

Pyrenopeziza brassicae (anamorph *Cylindrosporium concentricum*) is an ascomycete fungus and is spread by both asexual splash-dispersed conidia and sexual wind-dispersed ascospores. Inoculum can be detected with existing qualitative and quantitative PCR diagnostics, but these require laboratory-based processing. Loop-mediated isothermal amplification (LAMP) assays, targeting internal transcribed spacer or β-tubulin DNA sequences, offers fast and specific detection of *P. brassicae* isolates from a broad geographical range (throughout Europe, and Australia and New Zealand) and multiple brassica host species (*B. napus*, *B. oleracea* and *B. rapa*). Assays consistently detected DNA in amounts equivalent to 100 *P. brassicae* conidia per sample within 30 min, although the β-tubulin assay was more rapid. *In planta* application of the β-tubulin sequence-based LAMP assay to individual oilseed rape leaves collected from the field found no statistically significant difference in the amount of pathogen DNA present in parts of leaves, either with or without visible

Fig. 7.25. Light leaf spot (*Pyrenopeziza brassicae*) on a cabbage leaf. (Geoff Dixon).

light leaf spot symptoms. This offers an automated real-time monitoring of pathogen inoculum (King *et al.*, 2018).

Downy mildew – *Hyaloperonospora parasitica*

Downy mildew caused by the oomycete *H. parasitica* (formerly *Peronospora parasitica*) is a worldwide foliar disease of brassica vegetables. It causes seedling loss during propagation (see Fig. 7.26) and damages adult plants in the field.

Disease symptoms start from the lower leaves and progress upwards. Immature leaves may be more resistant than adjacent, fully expanded mature leaves, demonstrating that susceptibility increases with leaf age (Coelho *et al.*, 2009). Where crops are grown intensively for the leaf production, as with turnips (*B. rapa*) in California, USA, *H. parasitica* is a major disease problem.

Attempts to control *H. parasitica* by resistance have been restricted to the use of highly specific major genes. Several studies suggest, however, that more generalized resistance is available which could be exploited. There are many physiological races of the pathogen and they tend to be specific to single species of brassica. Resistance was considered to be due to a single dominant gene for seedling resistance. More recently, the seedling resistance has been shown to be quantitative and there is a continuous variation in the level of resistance and susceptibility, as well as variation in the seedling and mature plant resistance (after seven leaves). As with other downy mildews, the pathogen evolves new physiological races quite readily. Portuguese studies reported no accessions with higher resistance at the cotyledon stage than at the adult stage, but that the reverse did not apply (Coelho *et al.*, 1997).

Fig. 7.26. Downy mildew (*Hyaloperonospora parasitica*) on the seedling leaf of a cauliflower (*Brassica oleracea* var. *botrytis*). (Geoff Dixon).

Evaluation of physiological race collections and sources of resistance have been reported along with efforts to find a universal and wide-ranging source of resistance. In most cases, resistance appears to be a dominant trait as reported by (Soursa *et al.*, 1997). Mitchell-Olds *et al.* (1995), using *B. rapa*, showed a positive correlation for resistance to *H. parasitica* and *L. maculans*. This suggests some resistance genes provide defence against two very different pathogens which might not be easily circumvented by rapidly evolving physiological races. Resistance genes *br8* and *br9* reduced conidia production on cotyledons by 50–70% compared with some susceptible lines. Gene *br9* was the strongest resistance source (Jensen *et al.*, 1999).

Studies of induced resistance pre-inoculated susceptible brassica plants with avirulent isolates of *H. parasitica*, attempting to prevent disease from virulent forms. Induced resistance was systemic and lasted at least 15 days (Monot *et al.*, 2002). Plant density should be regulated to prevent overcrowding. Deliberate 'checks-to-growth' often given to seedlings in order to prevent them becoming too large before transplanting are often associated with attacks by *H. parasitica*. For direct-drilled crops, wider spacing should be employed where downy mildew is known to be a hazard. Growth stimulation by fertilizer and biostimulant applications enable seedlings to outgrow infections. Potassium phosphite may be usefully applied to collards to prevent yielding loss caused by downy mildew (Keinath, 2010). Crop debris must be removed from seed beds since this pathogen perennates as oospores in old foliage.

Soil-borne microbes

Damping-off and other diseases – *Thanetophorus cucumeris* (also known as *Rhizoctonia solani, Corticium solani, Hypochnus cucumeris, Hypochnus solani* and *Pellicularia filamentosa*), *Pythium* spp.

Suppressive soils containing a high percentage of bacteria may inhibit the fungus *Thanetophorus cucumeris*. *Lysobacter, Streptomyces* and *Pseudomonas* spp. were suppressive of the pathogen in soil from cauliflower (*B. oleracea* var. *botrytis*) crops. Successive cauliflower plantings can cause a decline of the damage caused by *T. cucumeris* (Postma *et al.*, 2010). Damping-off is a good example of a disease where the modification of cultural practices involving the avoidance of excessively moist environments and splash transmission of the pathogen during propagation is of major importance. Typical symptoms include the 'pinching' of seedling stems close to the soil or compost surface and the subsequent collapse of the plant (see Fig. 7.27).

Alteration of the sowing date will reduce the risk of infection from soil-borne inoculum for crops where the seed germination requires high temperatures. Shallow seeding, encouraging rapid emergence, is essential. Movement of soil around the hypocotyls of emerging seedlings should be avoided, this

Fig. 7.27. Damping-off disease syndrome, showing collapsed seedlings and pinched stems caused by several water moulds. (Geoff Dixon).

usually occurs during irrigation. Crop sequence can influence the extent of *T. cucumeris* infection by increasing or reducing inoculum levels in soils. Growing lucerne (alfalfa) (*M. sativa*) or clover (*Trifolium* spp.) increased the level of infection in subsequent crops, whereas cropping with cereals or incorporation of straw into the soil reduced the *T. cucumeris* population. This effect possibly results from increased CO_2 concentrations, reduced nitrogen levels in the soil solution (*T. cucumeris* is favoured by soil nitrogen) and stimulation of antagonistic soil microorganisms. The antagonist *Trichoderma lignorum*, for instance, produces a toxin that destroys *T. cucumeris*. The search for biological forms of control is especially active in Japan and other parts of Asia where radish and Chinese cabbage are essential crops and seriously damaged by damping-off. An ingenious alternative control method is the stimulation of hypovirulent forms of *T. cucumeris* that compete for infection sites with pathogenic forms, rendering their recognition and occupation unavailable to virulent strains.

The oomycete *Pythium ultimum* caused significantly less damping-off symptoms in cabbage (*B. oleracea* var. *capitata*) seedlings when the peat-based potting compost was treated with liquid seaweed extract prior to the infection (Dixon and Walsh, 2004). Biostimulants of this type provide a structural basis for sustainable control. Formulating integrated disease control strategies with a large range of components enhances their effectiveness making them less prone to erosion following the development of tolerance in pathogen populations. Host plants and soil-borne pathogens that attack them exist within an ecological matrix populated by numerous microbial species that may influence the access of pathogenesis. These events are moderated by the physical and chemical components of the soil. Organic nutrients that have associated biostimulant properties will reduce the development of *Brassica–P. ultimum* combinations and subsequent damping-off disease (Dixon, 2002).

Cabbage yellows – *Fusarium oxysporum* f. sp. *conglutinans*

Cabbage yellows, caused by *F. oxysporum* f. sp. *conglutinans*, is a major disease of brassicas, particularly in warm environments. In studies using inoculated and control plants of resistant and susceptible lines of Chinese cabbage (*B. rapa* ssp. *pekinensis*), Miyaji *et al.* (2017) identified differentially expressed genes, most of which were up-regulated by pathogen inoculation. This activated SAR in resistant lines and tryptophan biosynthetic processes and responses to chitin and ethylene in susceptible lines.

Control of *F. oxysporum* f. sp. *conglutinans* is largely achieved by the use of resistant cultivars of cabbages and is a classic example of the success of this approach, which has remained effective for decades in the USA. Although the pathogen is extremely variable in its growth characteristics *in vitro*, it is stable in pathogenicity towards its hosts. Initially, field selection for resistance within diseased crops resulted in the release of cv. 'Wisconsin Hollander'. Subsequently, this cultivar was found to have incomplete resistance at high temperatures, when pathogen growth is most vigorous, and some yellowing developed. Further work, using inbred lines produced by selfing plants using bud pollination to overcome host self-incompatibility mechanisms, led to the development of cultivars possessing resistance unaffected by temperatures below 26°C. The cabbage cultivars bearing major gene resistance to cabbage yellows, internal tipburn, *Rhizoctonia* head rot, powdery mildew and cabbage mosaic resulted (Williams *et al.*, 1968). Cultivars of Chinese cabbage are increasingly being developed with heat tolerance as a major attribute (traditionally, it is grown in the cool seasons when *Fusarium* is not a problem) and hence wilt is now developing as a constraint for production as a result of global warming (Shattuck, 1992).

Husbandry controls include use of high levels of potassium fertilizers. Soil solarization has been used successfully to reduce populations of this pathogen. In Germany, soil treatment with steam, alone and in combination with calcium cyanamide, encourages suppressive soil-borne microbes which reduce *F. oxysporum* f. sp. *conglutinans*-induced symptoms. Rotation is an effective control measure since *F. oxysporum* f. sp. *conglutinans* survives for only limited periods in the absence of host tissues.

Biological control may be developed since cross-protection against *F. oxysporum* f. sp. *conglutinans* followed prior inoculation with other formae speciales of *F. oxysporum*. Some soils are suppressive to *F. oxysporum* f. sp. *conglutinans*. This characteristic may be related to the presence of siderophore-producing antagonists, such as the bacterium *Pseudomonas putida*. Other antagonistic organisms, e.g. the fungi *Penicillium oxalicum* and *Trichoderma viride*, reduce the rate of hyphal development. Composted hardwood bark, amended with the fungus *Trichoderma hamatum* and the bacterium *Flavobacterium balustinum*, restored soil suppressive characteristics.

Clubroot – *Plasmodiophora brassicae*

Clubroot is a scourge of brassica crops worldwide, identified by the presence of severely swollen 'cancerous' clubs or galls, which destroy the functions of host root systems as conduits for water and nutrient uptake, and anchorage (see Fig. 7.28).

The invasion of clubroot (*P. brassicae*) into oilseed rape (*B. napus*), particularly in Canada and China, has stimulated substantial research efforts which are providing valuable information relevant for disease control in vegetable brassicas (Chai *et al.*, 2014; Strelkov and Hwang, 2014). This research has transformed the situation; for evidence compare the reports of Dixon and Doodson (1970) with Dixon (2006) and Dixon (2014).

Most resting spores of *P. brassicae* in field soils are concentrated in the upper profiles, decreasing gradually with depth, soil type, tillage practice and cropping history. Spore dissemination is via drainage and irrigation water,

Fig. 7.28. Clubroot symptoms caused by the protist *Plasmodiophora brassicae* on green broccoli (*Brassica oleracea* var. *italica*), destroying root functioning. (Geoff Dixon).

wind, organic material derived from infested roots, wind-borne soil particles and diseased transplanting material, not least soil carried on seed potato tubers and bulbs grown in contaminated soils (Rennie *et al.*, 2011). Contaminated transplants are a major means of long-range spread. Spores have been identified in peat used in composts for container-grown brassica transplants (see, e.g. Donald and Porter, 2014).

Early studies of wild *Brassica* spp. revealed that there are very few sources of major gene resistance to *P. brassicae*, posing a substantial problem for plant breeders. Resistance in *B. rapa* ssp. *rapifera* (turnip) and *B. napus* ssp. *rapifera* (swede) is of a monogenic dominant type while that of *B. oleracea* (cole crops) is polygenic and recessive. Several lines of *B. rapa* were selected in the Netherlands that carried strong resistance to *P. brassicae*. These were developed from turnip landraces identified in Belgium and southern Netherlands. Resistance results from the presence of three genes: A, B, C. In *B. napus* there are four resistance factors which may be inherited as single independent genes with resistance being dominant to susceptibility. Resistant swedes are mostly derived from either the green-skinned 'Wilhelmsburger' or bronze-skinned 'Bangholm' types. Clubroot resistance is rarely encountered in *B. oleracea*. The cabbage cv. 'Badger Shipper' was developed from selfed progeny of a chance cross of kale × cabbage hybrid and originally thought to possess polygenic resistance. Subsequent research indicated that susceptibility is dominant to resistance and probably only results from a limited number of genes. Resistance to *P. brasssicae* in the landraces of the cabbage 'Bindsachsener' and kale 'Verheul' results from the interactive effects of several common genes of quantitative effect. Further resistance has been identified in the landrace 'Bohmerwaldkohl', which shares common resistance genes with 'Bindsachsener'. In general, resistance in *B. oleracea* is quantitatively expressed, recessively inherited and dependent on the environment. For example, the cv. 'Resista', developed from 'Bohmerwaldkohl' and landraces from Shetland, UK, exhibits a high level of variation in reaction to *P. brassicae*. The combined characteristics of resistance and tolerance to *P. brassicae* are inherited recessively with probably only a few genes involved. Similar characteristics are present in landrace material from northern, eastern and central Europe. Incomplete dominance for tolerance to *P. brassicae* is linked to storage quality and in favour of extended growing periods for white cabbage (*B. oleracea* var. *capitata*) types. Attempts have been made in the USA and Japan to transfer resistance to Brussels sprouts, cauliflower and green broccoli from *B. oleracea* sources via interspecific hybrids with *B. napus*. Little resistance was recorded in an extensive range of *B. oleracea* cultivars (Dixon, 1976; Dixon and Robinson, 1986). Commercial breeding programmes (Anon, 2020) have produced resistant cultivars of *B. oleracea* utilizing the dominate genes in *B. rapa*, which have also been used in developing resistant oilseed rape cultivars (Diederichsen *et al.*, 2014). Interest centres particularly on transferring resistance from *B. rapa* to *B. napus* ssp. *oleifera* (one of the oilseed crops). The disease outbreak in Canadian oilseed rape (canola) has resulted in significant efforts there aimed at developing resistant lines and cultivars by both conventional breeding and molecular transfers.

Development of resistance to *P. brassicae* in Chinese cabbage and pak choi (*B rapa* ssp. *pekinensis* and *B. rapa* var. *chinensis*) is of major economic significance in Asia. The cv. 'Michehili' possesses one dominant gene while at low inoculum densities several other hosts express resistance. Elsewhere, cytoplasmic male sterility has been utilized to develop F_1 hybrids possessing clubroot resistance. Use may also be made of resistance in *Raphanus* (radish) by forming the intergeneric hybrids *Raphanobrassica* and *Brassicoraphanus*, both of which are resistant to *P. brassicae*.

The kales are generally the most resistant of vegetable brassicas, together with some rutabagas (swedes) and turnips. Each has been used in resistance breeding programmes, an early example being the cabbage cv. 'Badger Shipper' which was selected in Wisconsin (the Badger State), USA from kale in the 1930s. Tolerant green broccoli, cauliflower and Chinese cabbage have been developed using resistance derived from *B. napus* and *B. rapa*.

The Italian *B. oleracea* lines have displayed forms of tolerance, possibly reflecting selection in crops grown for several human generations on pathogen infested land and possibly reflecting close proximity to the centres of origin of these crops. Most resistance, especially in *B. oleracea*, is partially recessive and partially additive traits governed by a few major genes. There is a debate as to the presence or absence of gene-for-gene relationships between *Brassica* species and this pathogen. Work is in progress to develop genetic maps for resistance to clubroot in various European and Asian host species and includes *A. thaliana* (Dixon, 2007). The situation is complicated by the occurrence of variants in *P. brassicae* populations as described by Toxopeus *et al.* (1986) and analysed recently in detail by Hollman *et al.* (2021) and Mcdonald *et al.* (2021).

Controlling clubroot in horticultural brassicas requires carefully formulated integrated disease management strategies (Dixon, 2020). These involve management of land use, such as rotations (particularly with rented facilities), sustainable soil care ensuring adequate drainage and effective structure, and the application of lime and fertilizers, especially those which encourage the activities of benign microbes.

Husbandry control using lime, which increases soil pH above 7.0, is a traditional method of clubroot control over several centuries. Liming will control the pathogen if the spore load is low, but even heavy applications are ineffective if soil is heavily contaminated. Applying sufficient lime, which raises the pH to 7.3–7.5, over several seasons will reduce inoculum potential. A more rapid effect is achieved with calcium oxide (hot lime) or sugarbeet lime. The fertilizer calcium cyanamide provides sources of nitrogen and calcium and is associated with a diminution in the severity of clubroot disease, especially when used over several seasons. It probably achieves this by stimulating populations of microbes antagonistic to *P. brassicae* (Dixon, 2017). Calcium nitrate fertilizer has a similar effect and offers a readily soluble source of calcium. Calcium, boron, nitrogen and pH have been demonstrated to affect the processes of resting spore germination, host invasion and colonization in the *Brassica*–*P. brassicae* combination that results in clubroot disease (Dixon, 2010).

Rotations with at least 5-year breaks between cruciferous crops are effective as control methods, but frequently they are not economically feasible for horticultural brassicas. Breaks in cropping are equally important where resistant brassicas are grown. This is because the root hairs of resistant types are invaded, during the primary stages of pathogenesis, to a similar extent to those of susceptible forms. Weeds, such as grasses, can also harbour the primary stages of *P. brassicae* and should be used with caution in rotations. Even longer rotations are required for spring- and summer-growing brassica crops that occupy the soil during optimal environmental conditions for the growth of *P. brassicae*. Some break crops, such as maize (maize) and lucerne (alfalfa) (*M. sativa*), diminish the inoculum potential of *P. brassicae*.

Green manuring, following ploughing, reduces infestation, as will trap cropping with summer rape for 5 weeks prior to planting cauliflower. Organic soil amendments have been demonstrated to reduce clubroot in Taiwan, where intensive brassica production is common. Soil solarization using plastic sheeting is also an effective control as reported from Australia, California and Florida in the USA, and Japan. Direct soil heating reduces the viability of *P. brassicae* spores, rather than any induced form of biological control being responsible for the effects of solarization. This technique was used effectively for clubroot control in Florida, controlling clubroot affecting watercress (*N. officinale*) (Dixon and Strelkov, unpublished). Suppressive soils have been identified which inhibit the development of *P. brassicae*. The suppressive potential remained unchanged after steam sterilization but was increased with alkaline pH and calcium content. Development of biological control has been suggested by Arie *et al.* (1998).

White rot (white mould in the USA) – *Sclerotinia sclerotiorum*

Control of this pathogen is difficult because of its wide host range and formation of robust and persistent sclerotial resting bodies. All brassica ware and seed crops are attacked (see Fig. 7.29); for example, of the latter, brown mustard (*B. juncea*) sustains considerable damage (Sharma and Sharma, 2001).

Fig. 7.29. White rot (white mould) caused by *Sclerotinia sclerotiorum* on the foliage of Brussels sprouts (*Brassica oleracea* var. *gemmifera*). (Geoff Dixon).

Resistance is recessive and quantitative. The introduction line PI 206942, a non-heading cabbage from Turkey, has shown superior levels of resistance (Dickson and Petzoldt, 1996). Resistance was transferred to cabbage, green broccoli and cauliflower. Sharma *et al.* (1995) reported cauliflower cv. 'Early Winter Adam's White Head' (EWAWH) and EC 162587 were highly resistant, and RSK 1301 and MRS1 were moderately resistant. Singh and Kalda (1995) indicated EC 103576, EWAWH and EC 177283 were resistant. Ethylene response factors are members of the APETALA2/ERF transcription factor family, having important roles in plant growth, development and multiple environmental stress responses. An ethylene response transcription factor gene was designated as *BoERF1* and regulates in resistance against salt stress and *Sclerotinia* stem rot, which could be used in breeding green broccoli (*B. oleracea* var. *italica*) (Jiang *et al.*, 2019).

Cultural controls, such as soil cultivation, aeration and flooding, are helpful in combating this pathogen. Forms of nitrogen fertilizer, such as calcium cyanamide, are also associated with diminished pathogen activity. Control of storage rots incited by *S. sclerotiorum* is aided by retaining fresh produce in a turgid state and reducing the temperature to 5°C. Some encouraging results have been achieved using biological control, particularly with hyperparasites, such as the fungi *Coniothyrium* spp. that attack the sclerotia of *S. sclerotiorum*.

Sclerotial survival was reduced significantly by plastic covering and amendment with brassica residues, and in one instance by mycoparasitism with the fungus *Coniothyrium minitans* (Thaning and Gerhardson, 2001). The controlling effects of brassica residues result from sulfur compounds derived from glucosinolates. These are secondary metabolites present in Brassicaceae species implicated in their defence against plant pathogens. When a pathogen causes tissue damage, the enzyme myrosinase hydrolyzes glucosinolates into compounds with antimicrobial activity. The effects of glucosinolates are dependent on the pathogen and the form of glucosinolate. Aliphatic sinogrin is inhibitory to infection by *S. sclerotiorum* and the indolic glucobrassicin is inhibitory to infection by black rot (*X. campestris* pv. *campestris*) (Madloo *et al.*, 2019). Soil solarization is widely practised in Mediterranean countries. However, in cooler, more northerly areas a combination of plastic covering and crop residues containing glucosinolates is a useful control (Kirkegaard *et al.*, 1998).

Bacterial pathogens

Black rot – *Xanthomonas campestris* pv. *campestris*

Black rot, caused by *X. campestris* pv. *campestris*, is a serious disease of vegetable crucifers worldwide (Griffiths *et al.*, 2009). Periodic epidemics of black rot follow the introduction of susceptible cultivars, careless use of contaminated seeds and seedlings, and weather conditions favourable to the disease outbreaks (see Fig. 7.30).

Fig. 7.30. Black rot caused by *Xanthomonas campestris* pv. *campestris* on cabbage. (Geoff Dixon).

It has been suggested that there are new, highly aggressive variants of the pathogen and breeding has been done in the absence of recognition of the existence of these races.

This pathogen induces two types of symptoms, namely, black rot and blight. Black rot symptoms are V-shaped lesions and black veins on the leaf, and blight symptoms are a sudden collapse of interveinal tissues following the lack of veinal necrosis at early stages of infection. These two symptoms can occur simultaneously (Shigaki *et al.*, 2000). However, the tendency to induce either symptom type is strain-dependent. Symptomless responses to races 1 and 4 were observed in the US Department of Agriculture Plant Introductions of mustard species (*B. juncea, B. carinata* and *B. nigra*). Resistance was identified in several accessions of *B. carinata*, and four accessions of *B. nigra* and *S. arvensis* (PI 296079).

Since the early 1990s, diseases caused by *X. campestris* have been spreading on to new hosts and into new regions. *Brassica oleracea* appears the most susceptible host. Resistance genes have been identified and the gene-for-gene studies indicate there are different physiological races with unrelated resistance being identified in South-east Asian cabbage and in Portuguese Penca kale (Ignatov *et al.*, 1998; Ignatov *et al.*, 1999). The origin of the Asian cabbage was traced to the Flat Dutch group of varieties and heading Mediterranean kale. Novel resistance may be available in Chinese kale, green broccoli and cabbage. Pathogen-resistant cabbage lines are available – the mechanism of resistance appears to be present in the hydathode. Co-inoculation of cabbages (*B. oleracea* var. *capitata*) with *Xanthomonas campestris* pv. *carotae* and *X. campestris* pv. *campestris* increased resistance (Cook and Robeson, 1986). Determinants that trigger host resistance responses may reside on the bacterial cell surfaces (Taylor *et al.*, 2002). Sources of resistance have been found in two accessions of *B. carinata* (PIs 199947 and 199949). These are thought to result from a single dominant gene. In *B. rapa*, resistance is quantitative and of moderate effect. Resistance to black rot was found initially in the Japanese cabbage cvs 'Early Fuji' and 'Hugenot' in the early 1950s. Later, PI 436606 from China, another cabbage, provided further sources of resistance. Williams *et al.* (1972) demonstrated that resistance in 'Early Fuji' was due to one dominant major gene *f*, modified by one dominant and one recessive gene where *f* is in the heterozygous condition. There is, however, an almost continuous variation in the level of resistance, from high resistance (but far from immunity) to extreme susceptibility. There is a seedling resistance due to an additional recessive gene (Vicente *et al.*, 2000). The relationship between *X. campestris* pv. *campestris* and *Brassica* spp. in the 'Triangle of U' (see Chapter 1) is shown in Fig. 7.31.

Screening can best be done at temperatures of 25–30°C. The source of resistance in *B. carinata* PI 199947 and also in PI 199949 gives quasi-immunity. This has been transferred into green broccoli (*B. oleracea* var. *italica*) by fusion (Hansen and Earle, 1995). Following this, several generations of rigorous selection were necessary to obtain a true breeding line with the immunity level of resistance found in the *B. carinata* parent. The *B. carinata* resistance is dominant, but back crosses are still segregated. Guo *et al.* (1991) transferred resistance from *B. carinata* to *B. rapa* by classical breeding. Ignatov *et al.* (1998) collected isolates from the UK, Japan and Russia and identified five races whose inheritance for resistance was controlled by dominant genes in races 1–4 and race 5 by a recessive gene *r5*. A gene-for-gene relationship best explained the data. Kamoun *et al.* (1992) identified five races (0–4). Vicente *et al.* (2000) modified the numbering of the races. They withdrew race 3, separated Kamoun's race 1 into three new races (1, 3 and 5) and also

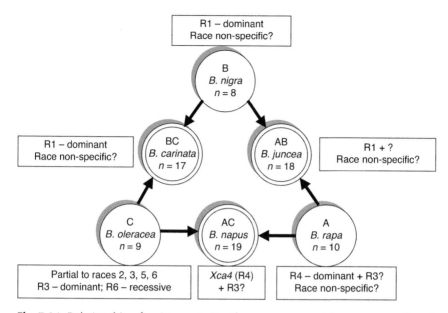

Fig. 7.31. Relationship of resistance to *Xanthomonas campestris* pv. *campestris* and between *Brassica* spp. (After Vicente *et al.*, 2000).

proposed a gene-for-gene model to explain interactions between *B rapa* and *Brassica* differentials. In the UK, race one predominates, and worldwide races 1 and 4 predominate.

No line, so far, has been found to be resistant to all six races. The cv. 'Wrosa' (*B. oleracea*) was susceptible to all six races, the cvs 'Cobra' (*B. napus*) and 'Just Right' (*B. nigra*) were resistant to race 4, the cv. 'Seven Top Turnip' (*B. rapa*) to races 2 and 4, PI 199947 (*B. carinata*) to races 1, 3 and 4, the cv. 'Florida Broad Leaved Mustard' (*B. juncea*) to races 1, 3 and 4, and the cv. 'Miracle' (*B. oleracea*) to races 2, 3 and 5. Jamwal and Sharma (1986) reported that resistance in cauliflower was governed by dominant polygenes. The USA PI 436606, originating from China, is resistant in both juvenile and adult stages, these effects apparently being controlled by a single recessive allele (Dickson and Hunter, 1987; Hunter *et al.*, 1987). Two ecotypes of *A. thaliana* have demonstrated differential response to *X. campestris* pv. *campestris* suggesting that resistance genes may be present which could be of value in breeding programmes for crop brassicas.

Quantitative trait loci for juvenile and adult resistance in *B. oleracea* have been mapped (Camaago *et al.*, 1995). Two regions on linkage groups 1 and 9 were associated with both juvenile and adult plant resistance. Malvas *et al.* (1999) mapped 900 F_2 individuals from line 'Badger 16' × 'LC201' cross and six random amplified polymorphic DNA markers were associated with resistance linked to petal colour. Alleles from the susceptible 'LC201' contributed towards resistance. Chitinase genes cloned from plants and microorganisms have been inserted in the genome of a number of plant species including brassicas in order to achieve resistance against a range of fungal pathogens. Mora and Earle (2001) utilized chitinase genes in attempts to develop resistance.

Xanthomonas campestris pv. *campestris* is principally a seed-borne bacterium. A real-time PCR assay based on a dual-labelled hydrolysis TaqMan® probe has been developed for the rapid and sensitive detection of the bacterium and related pathovars that affect mainly Brassicaceae crops and ornamentals. Primers were designed which specifically amplified a 152 bp fragment of the zinc-specific regulator (*Zur*) gene sequenced from *X. campestris* (Eichmeier *et al.*, 2019), although the real-time PCR assay based on detection of *Zur* also detected *X. campestris* pvs *raphani*, *armoraciae* and *incanae* and a strain of *X. campestris* pv. *carotae*, Tests of brassica tissues and seeds artificially inoculated with *X. campestris* pv. *campestris* showed that the real-time PCR based on detection of *Zur* is an efficient and robust assay. The pathogen can survive in soil, and even on plant debris, for only one to two growing seasons. Survival on contaminated seeds and on weed crucifers is considered essential for the cycle of disease.

Several pathovars of *X. campestris* are prevalent in South Korea, of which pv. *campestris*, causing black rot, threatens the production of Chinese cabbage in one of the world's major producing countries. Rapid and sensitive detection of the pathovar is an essential prerequisite for any plant protection programme (Rubel *et al.*, 2019). Sequence characterized amplified region primer sets were designed, of which three sets, namely Xcc_48 F/R, Xcc_53 F/R and Xcc_79 F/R, specifically amplified all *X. campestris* pv. *campestris* strains and did not amplify other pathovars and bacterial strains. This approach produced sensitive, specific and reliable markers offering efficient and inexpensive detection for quarantine and disease forecasting by early detection from asymptomatic field samples.

Plant morphology and life cycle play important roles in the degree of black rot development in the field. Hydathodes are water pores found on leaves of a wide range of vascular plants and are the sites of guttation (Cerutti *et al.*, 2017). These are natural infection routes for several pathogens such as *X. campestris* pv. *campestris*. Six days post-inoculation there was high bacterial multiplication forming an initial niche for subsequent vascular systemic dissemination to distant plant tissues. The rate of guttation is important in determining the susceptibility of cultivars. The ability of a pathogen to multiply in the vascular system is also an important factor in determining the eventual level of black rot symptoms. Leaf resistance and stem resistance are governed by different genes. Race-specific resistance exists at the site of inoculation as a hypersensitive response. Resistance in *B. oleracea* that is severely damaged by black rot is very low, with no true resistance in many botanical varieties and landraces. Resistance developed in Japan is related to the use of heading Mediterranean kale of the 'Penca de Mirandella' type.

Hot water treatment of fresh plump seed at 50°C for 25–30 min for cabbage helps eradicate black rot bacteria. Other primary control measures include phytosanitary inspections, accompanied by rogueing and seed certification, crop rotation, avoidance of excessive irrigation and location of seed crops well clear of ware crops.

Soft rots – *Erwinia carotovora* ssp. *carotovora, Pseudomonas marginalis* and *Pseudomonas fluorescens*

Soft rots are especially damaging for green broccoli (*B. oleracea* var. *italica*) crops when symptoms are expressed in the heads close to harvest, destroying quality and preventing marketing (see Fig. 7.32).

Of 101 bacterial isolates from green broccoli crops affected with spear rot in Scotland, 72 were fluorescent *Pseudomonas* spp., two were *Pectobacterium carotovorum* spp. *carotovorum* (*E. carotovora* spp. *carotovora*) and 27 were unidentified non-fluorescent Gram-negative organisms (Harling *et al.*, 1994). Research indicates that additive genetic effects predominated in controlling inheritance of resistance to bacterial soft rot. Estimated narrow-sense heritability averaged 42%. It is suggested that resistance might be controlled by two major partially dominant genes plus minor genes (Ren and Dickson, 1992). Green broccoli cvs 'Shogun', 'Green Defender' and related hybrids are quite tolerant, and some cauliflowers (*B. oleracea* var. *botrytis*) had higher levels of

Fig. 7.32. Soft rot symptoms expressed in green broccoli (*Brassica oleracea* var. *italica*) at harvest. (Originally from Dr Nicola Holden, The James Hutton Institute).

tolerance, both in seedling and mature plants. When cv. 'Shogun' was crossed with the best Chinese cabbage lines using molecular fusion, higher levels of resistance were obtained in the back cross than in either fusion parent (Ren *et al.*, 2000). Improved resistance to bacterial soft rot following fusion between *B. rapa* and *B. oleracea* has been recorded. Cultivars vary in susceptibility to soft rots, the green broccoli cvs 'Marathon', 'Headline' and 'Trixie' appear to have some tolerance to *P. flourescens*, while there are major efforts to breed for resistance to soft rot in Chinese cabbage in Japan, Taiwan and other parts of Asia. In studies of over 800 accessions only low levels of tolerance were found, but three cycles of recurrent selection, starting with the best 23 accessions, improved this tolerance significantly (Ren *et al.*, 2001). These authors suggested that resistance might be controlled by two major partially dominant genes plus associate minor genes. It is also suggested that *Erwinia* spp. was more virulent than *Pseudomonas* spp.

An increase in the amount of nitrogen fertilizer supplied (from 0 to 196 kg N/ha) to green broccoli (*B. oleracea* var. *italica* cv. Emperor) led to an increased incidence of head rot, probably caused by a number of *Pseudomonas* and *Erwinia* species, and this resulted in a decrease of the amount of marketable produce (Everaarts, 1994). Consequently, cultural controls should involve the avoidance of excessive applications of nitrogenous fertilizers and encourage aeration by wider spacing which minimizes the impact of these pathogens. With green broccoli in particular, however, crop spacing is used to determine head size according to market demands and this leaves little option for change. Straw or plastic mulch placed around the base of plants minimizes the splash transference of bacteria from the soil on to foliage and helps reduce disease incidence.

Virus pathogens

Brassicas suffer from a range of abiotic (see Chapter 8, Table 8.6) and biotic stresses that do not occur in isolation but often simultaneously. Productivity of natural and agricultural systems is frequently constrained by water limitation, and the frequency and duration of drought periods will be likely to increase due to global climate change. Phytoviruses represent a highly prevalent biotic threat in wild and cultivated plants (Berges *et al.*, 2018). Under water deficit, transmission rate and virulence were negatively correlated. Changes in the rate of transmission under water deficit were not related to changes in viral load. Transmission–virulence trade-off is highly dependent on the environment and growth traits of the host.

Elimination of the vector, where known, is the main route whereby virus induced diseases in brassicas are controlled. Mostly, the vectors are insects and their control is outlined earlier in this chapter. The principle viruses affecting brassica crops include: cauliflower mosaic virus (CaMV), beet western yellows virus (BWYV) (also known as turnip mild yellows virus) (see Fig. 7.33),

Fig. 7.33. Symptoms of mottling and mosaicing resembling cauliflower mosaic virus and beet western yellows virus infesting a Brussels sprout (*Brassica oleracea* var. *gemmifera*) leaf. (Geoff Dixon).

broccoli necrotic yellows virus (BNYV), radish mosaic virus (RaMV), radish yellow edge virus (RaYEV), turnip crinkle virus (TuCV), turnip mosaic virus (TuMV), turnip yellow mosaic virus (TuYMV) and turnip rosette mosaic virus (TuRMV). Turnip mosaic virus (Fjellstrom and Williams, 1997) is a serious disease of *B. oleracea* and *R. sativus* causing substantial losses to *B. rapa* and *B. juncea* as well (Sako, 1981). Resistance to TuMV is scarce (Shattuck, 1992).

Turnip mosaic virus is one of the most damaging pathogens of field-grown vegetables, particularly affecting brassicas in Asia, North America and Europe (Walkey and Pink, 1988; Walsh, 1997). The biology, epidemiology and control of TuMV was reviewed by Shattuck (1992) and the first complete sequence published by Nicolas and Laliberté (1992) (GenBank accession D10927). A further review by Walsh and Jenner (2002) concentrated on host resistance.

Turnip mosaic virus is the only potyvirus known to infect brassicas. It threatens brassica crops, including cabbage, Chinese cabbage, oilseed rape and mustard. Turnip mosaic virus disease was first discovered in the USA. Infection by TuMV results in a yield loss of up to 70%. This virus is also characterized by high pathotype diversity because of its highly variable genome structure and has been divided into 12 pathotypes. These characteristics, as well as its non-persistent transmission mode by as many as 89 aphid species, means the disease is difficult to prevent (Li *et al.*, 2019a). During the last decade, extensive studies have investigated inheritance, mapping and cloning of the TuMV resistance genes, and several nucleotide binding domains – leucine-rich repeats or eukaryotic initiation factor-encoding loci with divergent molecular mechanisms have been uncovered. Resistance has been transferred to desirable types and is available. Turnip mosaic virus can be a problem in *B. oleracea* and the cabbage cv. 'Globelle' exhibits quantitative resistance to races 1 and 2 of TuMV. In China, 19 isolates were screened for resistance on 3000 *B. rapa* accessions and fell into seven strain groupings. Eight accessions were immune to all seven TuMV isolates (Liu *et al.*, 1996). Rusholme *et al.* (2007) suggested a mode of action for the resistance that is based on denying the viral RNA access to the translation initiation complex of the plant host. The gene *retr01* is the first reported example of a recessive resistance gene mapped in a brassica species.

Use of resistant or tolerant cultivars is recommended such as white cabbage (*B. oleracea* var. *capitata*). Watercress (*N. officinale*) crops should be replaced annually from seed-raised stock. Sources of resistance controlled by up to four genes in Brussels sprouts have been identified. Rutabaga cvs. 'Calder', 'Sensation', 'Vogesa', line 165 and a line developed by recurrent selection from the cross 'Laurentian' × 'Macomber' have shown a form of resistance to Canadian strains of TuMV which may be determined by a single dominant gene (Shattuck and Stobbs, 1987). White cabbage cvs 'Decema Extra', 'Vitala' and 'Winter White III' have high levels of resistance to the UK strain of TuMV. In comparison with other European isolates, those from the UK caused the highest level of external necrosis while German isolates were associated with high levels of internal necrosis. Both caused greater yield reductions than Danish or Greek strains. Differences in necrotic symptom expression may be utilized as markers in breeding for combined resistance to both German and UK strains. Virus infection had no effect on the incidence of pepper spot necrosis. Differences in the TuMV symptoms on many cvs of swede (*B. napus* ssp. *rapifera*) were assessed on a 0–9 scale. The least affected cvs were 'Bangholm Dima', 'Bangholm Sahna' and 'Ruta OEtofte' (Dixon *et al.*, 1975). Provvidenti (1980, 1981) found four isolates and appropriate resistance in Chinese cabbage (*B. rapa* ssp. *pekinensis*) from China. Later, a fifth isolate was found in Taiwan. The introduction PI 418957 is resistant to races 1–4, PI 391560 to races 1–3, and other PIs and cultivars were sources of resistance to one or two races. Resistance to race five was found at the Asian Vegetable Research and Development Centre (AVRDC), Taiwan, and has been incorporated into some advanced lines.

Turnip mosaic virus incidence is usually greatest where seed beds are sited near to old infected crops. Consequently, these seed beds should be placed well away from overwintered brassica crops. Extracts from trees and shrubs, such as Malabar plum (*Syzygium cumini*), wattle (*Acacia arabica*) and bottlebrush (*Callistemon lanceolatus*), decrease local lesion production by TuMV in goosefoot (*Chenopodium amaranticolor*) (Pandey and Mohan, 1986).

Physiological changes overlapping, at least partially, the defence pathways elicited both by viruses and their herbivore vectors. van Munster *et al.* (2017) assessed the effect of a severe water deficit on the efficiency of aphid transmission of CaMV or TuMV. For both viruses, results demonstrated that the rate of vector transmission is significantly increased from water-deprived source plants. CaMV transmission reproducibility increased by 34% and that of TuMV by 100%. The evidence shows that infected plants subjected to drought are much better virus sources for insect vectors.

Losses caused by CaMV result from insufficient protection of curds due to reduced leaf area and diminished size of inflorescence that can make crops unmarketable. Although losses are generally limited, between 20% and 50% reductions in yield have been noted for some cultivars. Apparently, CaMV is not seed-transmitted so that protection of new brassica crops is achieved by the destruction of the old season's infected plants prior to the emergence of new seedlings. There should be a break of at least one month between destruction of old crops and braiding of new ones. Seed beds require geographical isolation from older brassica crops or to be surrounded by trap crops of kale (*B. oleracea* var. *acephala*) or oilseed rape (*B. napus*), either to produce virus-free seed or transplants. The most satisfactory form of control is raising transplants outside the areas of crop production. Virus-free (or more accurately virus-tested) plants may be raised by *in vitro* culture from inflorescence tissue.

Resistance to CaMV in Brussels sprouts (*B. oleracea* var. *gemmifera*) has been discussed by Walsh and Jenner (2002) as being controlled by at least one dominant gene and possibly also a recessive gene. Field studies of summer and autumn maturing cauliflowers (*B. oleracea* var. *botrytis*) showed that cultivars varied in their reaction to virus infection with significant differences in the effects of virus on yield, marketability and time of maturity. Genotype and environmental interactions were established for these characteristics. The relative susceptibilities and symptom responses of several *Brassica* species infected with CaMV have been compared and related to the molecular events of the virus multiplication cycle. A major component of the susceptibility of brassica plants, and probably other hosts of CaMV, is at the level of the expression of the viral mini-chromosome and this is influenced by the host genotype.

Control of RaMV by the development of resistant cultivars may be possible since 'Tender Green Mustard' (*B. juncea*) is heterozygous for resistance that has also been identified in *B. rapa* cv. 'Purple Top White Globe', whereas cvs 'Early White Flat Dutch' and 'Shogoin' are susceptible. Resistance is apparently present in radish lines 5-white, 5-24-2a and 'L. R. Bombay'.

Mycoplasma-like organisms

Aster yellows complex

Phytoplasmas were detected by the combined use of fluorescence (4′,6-diamidino-2-phenylindole) staining and transmission and scanning electron microscopy in cabbage (*B. oleracea* var. *capitata*), sprouting broccoli (*B. oleracea* var. *italica*), turnip (*B. rapa* var. *rapifera*) and palm kale (*B. oleracea* var. *palmifolia*) collected from several vegetable growing areas in Lazio (Latium), Campania, Basilicata and Calabria (southern Italy). Diseased plants were characterized by malformations of floral parts involving virescence and phyllody, and by yellowing, stunting, proliferation of axillary shoots and deformation of the leaves. Fluorescence and electron microscopy were consistent and revealed a great number of phytoplasmas only in mature sieve elements of diseased plants. The RFLP analysis indicated they belonged to the aster yellows phytoplasma group (Marcone and Ragozzino, 1995). Antibiotics have provided effective control of this pathogen in North America. But their use is prohibited in Europe because of fears that this will erode the usefulness of antibiotics in the treatment of human diseases. Hot water treatment is used with perennial planting material (Lee *et al.*, 1998).

REFERENCES

Alborn, H., Karlsson, H., Lundgren, L., Ruuth, P. and Stenhagen, G. (1985) Resistance in crop species of the genus *Brassica* to oviposition by the turnip root fly, *Hylemya floralis*. *Oikos* 44(1), 61. DOI: 10.2307/3544044.

Amoros-Jimenez, R., Plaza, M., Montserrat, M., Marcos-García, M.Á. and Fereres, A. (2020) Effect of UV-absorbing nets on the performance of the aphid predator *Sphaerophoria rueppellii* (Diptera: Syrphidae). *Insects* 11(3), 166. DOI: 10.3390/insects11030166.

Anon (2020) Controlling Clubroot through resistance plant breeding. *The Vegetable Farmer* August, 19.

Arie, T., Kobayashi, Y., Okada, G., Kono, Y. and Yamaguchi, I. (1998) Control of soil-borne clubroot disease of cruciferous plants by epoxydon from *Phoma glomerata*. *Plant Pathology* 47(6), 743–748. DOI: 10.1046/j.1365-3059.1998.00298.x.

Asari, S., Ongena, M., Debois, D., De Pauw, E., Chen, K. *et al.* (2017) Insights into the molecular basis of biocontrol of Brassica pathogens by *Bacillus amyloliquefaciens*

UCMB5113 lipopeptides. *Annals of Botany* 120(4), 551–562. DOI: 10.1093/aob/mcx089.

Bakker, L., van der Werf, W. and Bianchi, F.J.J.A. (2021) No significant effects of insecticide use indicators and landscape variables on biocontrol in field margins. *Agriculture, Ecosystems & Environment* 308, 107253. DOI: 10.1016/j.agee.2020.107253.

Banchi, E., Ametrano, C.G., Tordoni, E., Stanković, D., Ongaro, S. *et al.* (2020) Environmental DNA assessment of airborne plant and fungal seasonal diversity. *Science of the Total Environment* 738, 140249. DOI: 10.1016/j.scitotenv.2020.140249.

Barker, G.M. (ed.) (2002) *Molluscs as Crop Pests*. CAB International, Wallingford, UK. DOI: 10.1079/9780851993201.0000.

Bauer, S.D., Korč, F. and Förstner, W. (2011) The potential of automatic methods of classification to identify leaf diseases from multispectral images. *Precision Agriculture* 12(3), 361–377. DOI: 10.1007/s11119-011-9217-6.

Bello, D., Mónaco, C.I. and Simón, M.R. (2002) Biological control of seedling blight of wheat caused by *Fusarium graminearum* with beneficial rhizosphere microorganisms. *World Journal of Microbiology and Biotechnology* 18(7), 627–636. DOI: 10.1023/A:1016898020810.

Benrey, B. and Denno, R.F. (1997) The slow-growth–high-mortality hypothesis: a test using the cabbage butterfly. *Ecology* 78(4), 987–999. DOI: 10.1890/0012-9658(1997)078[0987:TSGHMH]2.0.CO;2.

Berges, S.E., Vile, D., Vazquez-Rovere, C., Blanc, S., Yvon, M. *et al.* (2018) Interactions between drought and plant genotype change epidemiological traits of cauliflower mosaic virus. *Frontiers in Plant Science* 9, 703. DOI: 10.3389/fpls.2018.00703.

Bhattacharyya, D., Duta, S., Yu, S.-M., Jeong, S.-C. and Lee, Y.-H. (2018) Taxonomic and functional changes of bacterial communities in the rhizosphere of Kimchi cabbage after seed bacterization with *Proteus vulgaris* JBLS202. *The Plant Pathology Journal* 34(4), 286–296. DOI: 10.5423/PPJ.OA.03.2018.0047.

Blackman, R.L. and Eastop, V.F. (1984) *Aphids on the World's Crops. An Identification Guide*. John Wiley, Chichester, UK.

Blubaugh, C.K., Asplund, J.S., Eigenbrode, S.D., Morra, M.J., Philips, C.R. *et al.* (2018) Dual-guild herbivory disrupts predator-prey interactions in the field. *Ecology* 99(5), 1089–1098. DOI: 10.1002/ecy.2192.

Blubaugh, C.K., Asplund, J.S., Judson, S.M. and Snyder, W.E. (2021) Invasive predator disrupts link between predator evenness and herbivore suppression. *Biological Control* 153(1), 104470. DOI: 10.1016/j.biocontrol.2020.104470.

Boetzl, F.A., Schuele, M., Krauss, J., Steffan-Dewenter, I. and Marini, L. (2020) Pest control potential of adjacent agri-environment schemes varies with crop type and is shaped by landscape context and within-field position. *Journal of Applied Ecology* 57(8), 1482–1493. DOI: 10.1111/1365-2664.13653.

Borhan, M.H., Gunn, N., Cooper, A., Gulden, S., Tör, M. *et al.* (2008) WRR4 encodes a TIR-NB-LRR protein that confers broad-spectrum white rust resistance in *Arabidopsis thaliana* to four physiological races of *Albugo candida*. *Molecular Plant-Microbe Interactions* 21(6), 757–768. DOI: 10.1094/MPMI-21-6-0757.

Boruah, H.P.D. and Kumar, B.S.D. (2002) Biological activity of secondary metabolites produced by a strain of *Pseudomonas fluorescens*. *Folia Microbiologica* 47(4), 359–363. DOI: 10.1007/BF02818690.

Brown, J.K.M. and Hovmøller, M.S. (2002) Aerial dispersal of pathogens on the global and continental scales and its impact on plant disease. *Science* 297(5581), 537–541. DOI: 10.1126/science.1072678.

Byrne, D.N. and Bellows, T.S. (1991) Whitefly biology. *Annual Review of Entomology* 36(1), 431–457. DOI: 10.1146/annurev.en.36.010191.002243.

Camaago, L.E.A., Williams, P.H. and Osborn, T.C. (1995) Mapping of quantitative trait Loci controlling resistance of *Brassica oleracea* to *Xanthomonas campestris* pv. *Campestris* in field and greenhouse. *Phytopathology* 85(10), 1296–1300. DOI: 10.1094/Phyto-85-1296.

Cao, J., Tang, J.D., Strizhov, N., Shelton, A.M. and Earle, E.D. (1999) Transgenic broccoli with high levels of *Bacillus thuringiensis* Cry1C protein control diamondback moth larvae resistant to Cry1A or Cry1C. *Molecular Breeding* 5(2), 131–141. DOI: 10.1023/A:1009619924620.

Card, S.D., Hume, D.E., Roodi, D., McGill, C.R., Millner, J.P. *et al.* (2015) Beneficial endophytic microorganisms of *Brassica* – a review. *Biological Control* 90, 102–112. DOI: 10.1016/j.biocontrol.2015.06.001.

Carter, D.J. (1984) *Pest Lepidoptera of Europe: With Special Reference to the British Isles. Series Entomologica, Vol. 31*. Junk, Dordrecht, Netherlands.

Cerutti, A., Jauneau, A., Auriac, M.C., Lauber, E., Martinez, Y, *et al.* (2017) Immunity at cauliflower Hydathodes controls systemic infection by *Xanthomonas campestris* pv campestris. *Plant Physiology* 174(2), 700–716. DOI: 10.1104/pp.16.01852.

Chai, A.L., Xie, X.W., Shi, Y.X. and Li, B.J. (2014) Research status of clubroot (*Plasmodiophora brassicae*) on cruciferous crops in China. *Canadian Journal of Plant Pathology* 36(suppl. 1), 142–153. DOI: 10.1080/07060661.2013.868829.

Chen, C., Biere, A., Gols, R., Halfwerk, W., van Oers, K. *et al.* (2018) Responses of insect herbivores and their food plants to wind exposure and the importance of predation risk. *Journal of Animal Ecology* 87(4), 1046–1057. DOI: 10.1111/1365-2656.12835.

Cho, H.S., Cao, J., Ren, J.P. and Earle, E.D. (2001) Control of Lepidopteran insect pests in transgenic Chinese cabbage (*Brassica rapa* ssp. *pekinensis*) transformed with a synthetic *Bacillus thuringiensis cry1C* gene. *Plant Cell Reports* 20(1), 1–7. DOI: 10.1007/s002990000278.

Chretien, L.T.S., David, A., Daikou, E., Boland, W., Gershenzon, J. *et al.* (2018) Caterpillars induce jasmonates in flowers and alter plant responses to a second attacker. *The New Phytologist* 217(3), 1279–1291. DOI: 10.1111/nph.14904.

Clayton, E.E. (1927) Black-leg disease of Brussels sprouts, cabbage, and cauliflower. *New York (Geneva) Agricultural Experiment Station Bulletin* 550, 1–27.

Cobb, M. (1999) What and how do maggots smell? *Biological Reviews of the Cambridge Philosophical Society* 74(4), 425–459. DOI: 10.1017/S0006323199005393.

Coelho, P., Leckie, D., Bahcevandziev, K., Valério, L., Astley, D. *et al.* (1997) The relationship between cotyledon and adult plant resistance to downy mildew (*Peronospora Parasitica*) in *Brassica Oleracea*. *Acta Horticulturae* 459, 335–342. DOI: 10.17660/ActaHortic.1998.459.39.

Coelho, P.S., Valerio, L. and Monteiro, A.A. (2009) Leaf position, leaf age and plant age affect the expression of downy mildew resistance in *Brassica oleracea*. *European Journal of Plant Pathology* 125(2), 179–188. DOI: 10.1007/s10658-009-9469-4.

Collier, R.H. and Finch, S. (2007) IPM case studies: brassicas. In: Van Emden, H.F. and Harrington, R. (eds) *Aphids as Crop Pests*. CAB International, Wallingford, UK, pp. 549–559. DOI: 10.1079/9780851998190.0000.

Collier, R., Mazzi, D., Schjøll, A.F., Schorpp, Q., Thöming, G. *et al.* (2020a) The potential for decision support tools to improve the management of root-feeding fly pests of vegetables in Western Europe. *Insects* 11(6), 369–384. DOI: 10.3390/insects11060369.

Collier, R., Elliott, M., Wilson, D., Teverson, D. and Cowgill, S. (2020b) Phenology and abundance of pest insects of vegetable and salad crops in Britain: decision support for growers. In: Meadow, R. (ed.), *Proceedings of the Working Group 'Integrated Protection in Field Vegetables' Meeting, 13–16 October 2019. IOBC-WPRS Bulletin Vol. 153*, pp. 5–10.

Collins, S. (2014) Biology of the cabbage whitefly, *Aleyrodes proletella*. *IOBC/WPRS Bulletin* 107, 131–141.

Cook, D.R. and Robeson, D.J. (1986) Active resistance of cabbage (*Brassica oleracea*) to *Xanthomonas campestris* pv. *carotae* against the causal agent of black rot by co-inoculation. *Physiological and Molecular Plant Pathology* 28(1), 41–52. DOI: 10.1016/S0048-4059(86)80006-6.

Cui, W., He, P., Munir, S., He, P., He, Y, *et al.* (2019) Biocontrol of soft rot of Chinese cabbage using an endophytic bacterial strain. *Frontiers in Microbiology* 10, 1471. DOI: 10.3389/fmicb.2019.01471.

de Jong, P.D. and Hasper, G.A. (1996) Threshold values for chemical control of powdery mildew (*Erysiphe cruciferarum*) on Brussels sprouts. *European Journal of Plant Pathology* 102(2), 205–207. DOI: 10.1007/BF01877108.

Depalo, L., Burgio, G., Magagnoli, S., Sommaggio, D., Montemurro, F. *et al.* (2020) Influence of cover crop termination on ground dwelling arthropods in organic vegetable systems. *Insects* 11(7), 445. DOI: 10.3390/insects11070445.

Dickson, M.H. and Hunter, J.E. (1987) Inheritance of resistance in cabbage seedlings to black rot. *HortScience* 22(1), 108–109. DOI: 10.21273/HORTSCI.22.1.108.

Dickson, M.H. and Petzoldt, R. (1996) Breeding for resistance to *Sclerotinia sclerotiorum* in *Brassica oleracea*. *Acta Horticulturae* 407, 103–108. DOI: 10.17660/ActaHortic.1996.407.11.

Diederichsen, E., Frauen, M. and Ludwig-Müller, J. (2014) Clubroot disease management challenges from a German perspective. *Canadian Journal of Plant Pathology* 36(suppl. 1), 85–98. DOI: 10.1080/07060661.2013.861871.

Dixon, G.R. (1974) Field studies of powdery mildew (*Erysiphe crucifearum*) on Brussels sprouts. *Plant Pathology* 23(3), 105–109. DOI: 10.1111/j.1365-3059.1974.tb02917.x.

Dixon, G.R. (1975) The reaction of some oil rape cultivars to some fungal pathogens. *Proceedings 8th British Insecticide and Fungicide Conference* 2, 503–506.

Dixon, G.R. (1976) Disease assessment keys for clubroot (*Plasmodiophora brassicae*) on brassicas, powdery mildew (*Erysiphe cruciferarum*) on swedes and brussels sprouts, pea wilt (*Fusarium oxysporum f. sp. pisi*). In: *Manual of Plant Growth Stages and Disease Assessment Keys*. Ministry of Agriculture, Fisheries and Food, London.

Dixon, G.R. (1977) Standardised visual assessment keys for clubroot (*Plasmodiophora brassicae*) and brussels sprout powdery mildew (*Erysiphe cruciferarum*). In: Ministry of Agriculture, Fisheries and Food, London.

Dixon, G.R. (1981) *Vegetable Crop Diseases*. Macmillan, London. DOI: 10.1007/978-1-349-03704-9.

Dixon, G.R. (1984) Plant Pathogens and their Control in Horticulture. In: *Plant Pathogens and Their Control in Horticulture*. Macmillan, London, in collaboration with the Horticultural Education Association, Sutton Bonington, UK and Royal Horticultural Society, London. DOI: 10.1007/978-1-349-06923-1.

Dixon, G.R. (2002) Interactions of soil nutrient environment, pathogenesis and host resistance. *Plant Protection Science* 38, S87–S94. DOI: 10.17221/10326-PPS.

Dixon, G.R. (2006) The biology of *Plasmodiophora brassicae* Wor. – a review of recent advances. *Acta Horticulturae* 706, 271–282. DOI: 10.17660/ActaHortic.2006.706.32.

Dixon, G.R. (2007) *Vegetable Brassicas and Related Crucifers*, 1st edn. CAB International, Wallingford, UK. DOI: 10.1079/9780851993959.0000.

Dixon, G.R. (2009) Husbandry – the sustainable means of controlling soil borne pathogens: a synoptic review. *Acta Horticulturae* 817, 233–242. DOI: 10.17660/ActaHortic.2009.817.24.

Dixon, G.R. (2010) Calcium and pH as parts of a coherent control strategy for clubroot disease (*Plasmodiophora brassicae*). *Acta Horticulturae* 867, 151–156. DOI: 10.17660/ActaHortic.2010.867.19.

Dixon, G.R. (2012) Climate change, plant pathogens and food production. *Canadian Journal of Plant Pathology* 34(3), 362–379. DOI: 10.1080/07060661.2012.701233.

Dixon, G.R. (2014) Clubroot (*Plasmodiophora brassicae* Woronin) – an agricultural and biological challenge worldwide. *Canadian Journal of Plant Pathology* 36(suppl. 1), 5–18. DOI: 10.1080/07060661.2013.875487.

Dixon, G.R. (2017) Managing clubroot disease (caused by *Plasmodiophora brassicae* Wor.) by exploiting the interactions between calcium cyanamide fertilizer and soil microorganisms. *The Journal of Agricultural Science* 155(4), 527–543. DOI: 10.1017/S0021859616000800.

Dixon, G.R. (2020) Controlling clubroot. *Vegetable Farmer* July, 26–27.

Dixon, G.R. and Doodson, J.K. (1970) Club root of Brassica crops. *Agriculture* 77, 500–503.

Dixon, G.R. and Doodson, J.K. (1971) Assessment keys for some diseases of vegetable, fodder and herbage crops. *Journal of the National Institute of Agricultural Botany* 12(2), 299–307.

Dixon, G.R. and Robinson, D.L. (1986) The susceptibility of *Brassica oleracea* cultivars to *Plasmodiophora brassicae* (clubroot). *Plant Pathology* 35(1), 101–107. DOI: 10.1111/j.1365-3059.1986.tb01987.x.

Dixon, G.R. and Tilston, E.L. (2010) *Soil Microbiology and Sustainable Crop Production*. Springer Science + Business Media BV,. Dordrecht. DOI: 10.1007/978-90-481-9479-7.

Dixon, G.R. and Walsh, U.F. (2004) Suppressing *Pythium ultimum* induced damping-off in cabbage seedlings by biostimulation with proprietary liquid seaweed extracts. *Acta Horticulturae* 635, 103–106. DOI: 10.17660/ActaHortic.2004.635.13.

Dixon, G.R., Hague, J.M. and Braybrooks, J. (1975) Field reaction of swede cultivars to turnip crinkle virus. *Plant Pathology* 24(1), 31–32. DOI: 10.1111/j.1365-3059.1975.tb01854.x.

Dixon, G.R., Collier, R.H. and Bhattacharya, I. (2014) An assessment of the effects of climate change on horticulture. In: Dixon, G.R. and Aldous, D.E. (eds) *Horticulture: Plants for People and Places. Volume 2: Environmental Horticulture*. Springer, Dordrecht, Netherlands, pp. 817–858. DOI: 10.1007/978-94-017-8581-5.

Donald, E.C. and Porter, I.J. (2014) Clubroot in Australia: the history and impact of *Plasmodiophora brassicae* in *Brassica* crops and research efforts directed towards its control. *Canadian Journal of Plant Pathology* 36(suppl. 1), 66–84. DOI: 10.1080/07060661.2013.873482.

Dosdall, L.M., Good, A., Keddie, B.A., Ekuere, U. and Stringam, G. (2000) Identification and evaluation of root maggot (*Delia* spp.) (Diptera: Anthomyiidae) resistance within Brassicaceae. *Crop Protection* 19(4), 247–253. DOI: 10.1016/S0261-2194(00)00015-6.

Earle, E.D. and Knauf, V.C. (1999) Genetic engineering. In: Gómez–Campo, C. (ed.) *Biology of Brassica Coenospecies.* Elsevier, Amsterdam, pp. 287–313.

Edwards, M.D. and Williams, P.H. (1987) Selection for minor gene resistance to *Albugo candida* in a rapid-cycling population of *Brassica campestris. Phytopathology* 77(4), 527–532. DOI: 10.1094/Phyto-77-527.

Eichmeier, A., Peňázová, E., Pokluda, R. and Vicente, J.G. (2019) Detection of *Xanthomonas campestris* pv. *campestris* through a real-time PCR assay targeting the *Zur* gene and comparison with detection targeting the *hrpF* gene. *European Journal of Plant Pathology* 155(3), 891–902. DOI: 10.1007/s10658-019-01820-0.

Ellis, P.R., Singh, R., Pink, D.A.C., Lynn, J.R. and Saw, P.L. (1996) Resistance to *Brevicoryne brassicae* in horticultural brassicas. *Euphytica* 88(2), 85–96. DOI: 10.1007/BF00032439.

Ellis, P.R., Pink, D.A.C., Phelps, K., Jukes, P.L., Breeds, S.E, *et al.* (1998) Evaluation of a core collection of *Brassica oleracea* accessions for resistance to *Brevicoryne brassicae*, the cabbage aphid. *Euphytica* 88, 85–96. DOI: 10.1023/A:1018342101069.

Ellis, P.R., Kift, N.B., Pink, D.A.C., Jukes, P.L., Lynn, J. *et al.* (2000) Variation in resistance to the cabbage aphid (*Brevicoryne brassicae*) between and within wild and cultivated *Brassica* species. *Genetic Resources and Crop Evolution* 47(4), 395–401. DOI: 10.1023/A:1008755411053.

Engelhard, A.W. (ed.) (1996) *Soilborne Plant Pathogens: Management of Diseases with Macro – and Microelements.* American Phytopathological Society, St Paul, MN.

Entwistle, P.F., Cory, J.S., Bailey, M.J. and Higgs, S. (1993) *Bacillus Thuringiensis: An Environmental Biopesticide.* John Wiley, New York.

Ester, A., van der Zande, J.C. and Frost, A.J.P. (1994) Crop covering to prevent pest damage to field vegetables, and the feasibility of pesticide application through polyethylene nets. In: *Proceedings of the Brighton Crop Protection Conference (Pests and Diseases)*, British Crop Protection Council, Farnham, UK, pp. 761–766.

Evans, K.A. and Allen-Williams, L.J. (1998) Response of cabbage seed weevil (*Ceutorhynchus assimilis*) to baits of extracted and synthetic host-plant odor. *Journal of Chemical Ecology* 24(12), 2101–2114. DOI: 10.1023/A:1020741827544.

Everaarts, A.P. (1994) Nitrogen fertilization and head rot in broccoli. *Netherlands Journal of Agricultural Science* 42(3), 195–201. DOI: 10.18174/njas.v42i3.597.

Feltwell, J. (1982) *Large White Butterfly: The Biology, Biochemistry and Physiology of Pieris brassicae (Linnaeus). Series Entomologica, Vol. 18.* Kluwer, Alphen aan den Rijn, Netherlands.

Finch, S. (1987) Horticultural crops. In: Burn, A.J., Coaker, T.H. and Jepson, P.C. (eds) *Integrated Pest Management.* Academic Press, London, pp. 257–293.

Finch, S. and Thompson, A.R. (1992) Pests of cruciferous crops. In: McKinley, R.G. (ed.) *Vegetable Crop Pests.* Macmillan, London, pp. 87–138. DOI: 10.1007/978-1-349-09924-5.

Fjellstrom, R.G. and Williams, P.H. (1997) Fusarium yellows and turnip mosaic virus resistance in *Brassica rapa* and *B. juncea. HortScience* 32(5), 927–930. DOI: 10.21273/HORTSCI.32.5.927.

Freuler, P., Fischer, A.J., Gagnebin, S., Parko, F., Granges, J. *et al.* (2000) Resistance of cauliflower against cabbage root fly. *Revue Suisse de Viticulture, d'Arboriculture et d'Horticulture* 32, 109–113.

Gavloski, J.E., Ekuere, U., Keddie, A., Dosdall, L., Kott, L. *et al.* (2000) Identification and evaluation of flea beetle (*Phyllotreta cruciferae*) resistance within Brassicaceae. *Canadian Journal of Plant Science* 80(4), 881–887. DOI: 10.4141/P99-164.

George, D.R., Collier, R. and Port, G. (2009) Testing and improving the effectiveness of trap crops for management of the diamondback moth *Plutella xylostella* (L.): a laboratory-based study. *Pest Management Science* 65(11), 1219–1227. DOI: 10.1002/ps.1813.

Gilbert, N. and Raworth, D.A. (1996) Insects and temperature – a general theory. *The Canadian Entomologist* 128(1), 1–13. DOI: 10.4039/Ent1281-1.

Goodey, N.A., Florance, H.V., Smirnoff, N. and Hodgson, D.J. (2015) Aphids pick their poison: selective sequestration of plant chemicals affects host plant use in a specialist herbivore. *Journal of Chemical Ecology* 41(10), 956–964. DOI: 10.1007/s10886-015-0634-2.

Green, J.P., Foster, R., Wilkins, L., Osorio, D. and Hartley, S.E. (2015) Leaf colour as a signal of chemical defence to insect herbivores in wild cabbage (*Brassica oleracea*). *PLoS ONE* 10(9), e0136884. DOI: 10.1371/journal.pone.0136884.

Griffiths, P.D., Marek, L.F. and Robertson, L.D. (2009) Identification of crucifer accessions from the NC-7 and NE-9 Plant Introduction Collections that are resistant to black rot (*Xanthomonas campestris* pv. *campestris*) races 1 and 4. *HortScience* 44(2), 284–288. DOI: 10.21273/HORTSCI.44.2.284.

Gulidov, S. and Poehling, H.M. (2013) Control of aphids and whiteflies on Brussels sprouts by means of UV-absorbing plastic films. *Journal of Plant Diseases and Protection* 120(3), 122–130. DOI: 10.1007/BF03356463.

Guo, H., Dickson, M.H. and Hunter, J.E. (1991) *Brassica napus* sources of resistance to black rot in crucifers and inheritance of resistance. *HortScience* 26(12), 1545–1547. DOI: 10.21273/HORTSCI.26.12.1545.

Hall, R. (ed.) (1996) *Principles and Practice of Managing Soilborne Plant Pathogens.* The American Phytopathological Society, St Paul, MN.

Hansen, L.N. and Earle, E.D. (1995) Transfer of resistance to *Xanthomonas campestris* pv *campestris* into *Brassica oleracea* L. by protoplast fusion. *Theoretical and Applied Genetics* 91(8), 1293–1300. DOI: 10.1007/BF00220944.

Hansen, L.N. and Earle, E.D. (1997) Somatic hybrids between *Brassica oleracea* L. and *Sinapis alba* L. with resistance to *Alternaria brassicae* (Berk.) Sacc. *Theoretical and Applied Genetics* 94, 1078–1085. DOI: 10.1007/s001220050518.

Harling, R., Kellock, L.J. and Chard, J. (1994) Pathogenicity of bacteria causing spear (head) rot of Calabrese (broccoli) in Scotland. *Mededelingen-Faculteit Landbouwkundige en Toegepaste Biologische Wetenschappen, Universiteit Gent* 59(3b), 1189–1191.

Hass, B., Glen, D.M., Brain, P. and Hughes, L.A. (1999) Targeting biocontrol with the slug-parasitic nematode *Phasmarhabditis hermaphrodita* in slug feeding areas: a model study. *Biocontrol Science and Technology* 9(4), 587–598. DOI: 10.1080/09583159929541.

Hirst, J.M. (1952) An automatic volumetric spore trap. *Annals of Applied Biology* 39(2), 257–265. DOI: 10.1111/j.1744-7348.1952.tb00904.x.

Hol, W.H.G., De Boer, W., Termorshuizen, A.J., Meyer, K.M., Schneider, J.H.M. *et al.* (2013) *Heterodera schachtii* nematodes interfere with aphid-plant relations on

Brassica oleracea. Journal of Chemical Ecology 39(9), 1193–1203. DOI: 10.1007/s10886-013-0338-4.

Hollman, K.B., Hwang, S.F., Manolii, V.P. and Strelkov, S.E. (2021) Pathotypes of *Plasmodiophora brassicae* collected from clubroot resistant canola (*Brassica napus* L.) cultivars in western Canada in 2017–2018. *Canadian Journal of Plant Pathology* 43(4), 622–630. DOI: 10.1080/07060661.2020.1851893.

Hondelmann, P., Paul, C., Schreiner, M. and Meyhöfer, R. (2020) Importance of antixenosis and antibiosis resistance to the cabbage whitefly (*Aleyrodes proletella*) in Brussels sprout cultivars. *Insects* 11(1), 56. DOI: 10.3390/insects11010056.

Hooks, C.R.R. and Johnson, M.W. (2001) Broccoli growth parameters and level of head infestations in simple and mixed plantings: impact of increased flora diversification. *Annals of Applied Biology* 138(3), 269–280. DOI: 10.1111/j.1744-7348.2001.tb00112.x.

Hulden, L. (1986) The Whiteflies (Homoptera, Aleyrodidae) and their parasites in Finland. *Notulae Entomologicae* 66, 1–40.

Humpherson-Jones, F.M. (1989) Survival of *Alternaria brassicae* and *Alternaria brassicicola* on crop debris of oilseed rape and cabbage. *Annals of Applied Biology* 115(1), 45–50. DOI: 10.1111/j.1744-7348.1989.tb06810.x.

Hunter, J.E., Dickson, M.H. and Ludwig, J. (1987) Source of resistance to black rot of cabbage expressed in seedlings and adult plants. *Plant Disease* 71(3), 263. DOI: 10.1094/PD-71-0263.

Ignatov, A., Hida, K. and Kuginuki, Y. (1998) Black rot of crucifers and sources of resistance in Brassica crops. *Japan Agricultural Research Quarterly* 32(3), 167–172.

Ignatov, A., Kuginuki, Y. and Hida, K. (1999) Vascular stem resistance to black rot in *Brassica oleracea. Canadian Journal of Botany* 77(3), 442–446. DOI: 10.1139/b98-224.

Jamwal, R.S. and Sharma, P.P. (1986) Inheritance of resistance to black rot (*Xanthomonas campestris* pv. *campestris*) in cauliflower (*Brassica oleracea* var. *botrytis*). *Euphytica* 35(3), 941–943. DOI: 10.1007/BF00028603.

Jee, H.J., Shim, C.K., Choi, Y.J. and Shin, H.D. (2008) Powdery mildew caused by *Erysiphe cruciferarum* is found for the first time on Chinese cabbage in Korea. *Plant Pathology* 57(4), 777–777. DOI: 10.1111/j.1365-3059.2007.01807.x.

Jensen, B.D., Vaerbak, S.V., Munk, L. and Andersen, S.B. (1999) Characterization and inheritance of partial resistance to downy mildew, *Peronospora parasitica*, in breeding material of broccoli, *Brassica oleracea* convar. *botrytis* var. *italica. Plant Breeding* 118(6), 549–554. DOI: 10.1046/j.1439-0523.1999.00409.x.

Jiang, M., Ye, Z.-H., Zhang, H.-J. and Miao, L.-X. (2019) Broccoli plants over-expressing an ERF transcription factor gene *BoERF1* facilitates both salt stress and sclerotinia stem rot resistance. *Journal of Plant Growth Regulation* 38(1), 1–13. DOI: 10.1007/s00344-018-9799-6.

Jones, F.G.W. and Jones, M.G. (1984) *Pests of Field Crops*. Edward Arnold, London.

Juventia, S.D., Rossing, W.A.H., Ditzler, L. and van Apeldoorn, D.F. (2021) Spatial and genetic crop diversity support ecosystem service delivery: a case of yield and biocontrol in Dutch organic cabbage production. *Field Crops Research* 261, 108015. DOI: 10.1016/j.fcr.2020.108015.

Jyoti, J.L., Shelton, A.M. and Earle, E.D. (2001) Identifying sources and mechanisms of resistance in crucifers for control of cabbage maggot (Diptera: Anthomyiidae). *Journal of Economic Entomology* 94(4), 942–949. DOI: 10.1603/0022-0493-94.4.942.

Kabir, M.F., Lee, J.-K. and Lee, D.-W. (2017) Effect of plant age and Nematode inoculation density on final population of *Heterodera schachtii* on Chinese cabbage. *Korean Journal of Applied Entomology* 56(4), 413–420. DOI: 10.5656/KSAE.2017.11.0.025.

Kaczmarek, A.M., King, K.M., West, J.S., Stevens, M., Sparkes, D. *et al.* (2019) A loop-mediated isothermal amplification (LAMP) assay for rapid and specific detection of airborne inoculum of *Uromyces betae* (sugar beet rust). *Plant Diseases* 103(3), 417–421. DOI: 10.1094/PDIS-02-18-0337-RE.

Kalischuk, A.R. and Dosdall, L.M. (2004) Susceptibilities of seven Brassicaceae species to infestation by the cabbage seedpod weevil (Coleoptera: Curculionidae). *The Canadian Entomologist* 136(2), 265–276. DOI: 10.4039/n03-058.

Kamoun, S., Kamdar, H.V., Tola, E. and Cado, C.I. (1992) Incompatible interaction between crucifers and *Xanthomonas Campestris* involves a vascular hypersensitive response, role of the *htpX* locus. *Molecular Plant-Microbe Interactions* 5(1), 22–33. DOI: 10.1094/MPMI-5-022.

Kang, Z.-W., Liu, F.-H., Zhang, Z.-F., Tian, H.-G. and Liu, T.-X. (2018) Volatile beta-ocimene can regulate developmental performance of peach aphid *Myzus persicae* through activation of defense responses in chinese cabbage Brassica pekinensis. *Frontiers in Plant Science* 9, 708. DOI: 10.3389/fpls.2018.00708.

Keen, N.T., Mayama, S., Leach, J.E. and Tsuyuma, S. (eds) (2001) *Delivery and Perception of Pathogen Signals in Plants*. The American Phytopathological Society, St Paul, MN.

Keinath, A.P. (2010) Potassium phosphite mixed with other fungicides reduces yield loss to downy mildew on collard. *Plant Health Progress* 11(1). DOI: 10.1094/PHP-2010-0823-01-RS.

Kennedy, G.G., Gould, F., Deponti, O.M.B. and Stinner, R.E. (1987) Ecological, agricultural, genetic, and commercial considerations in the deployment of insect-resistant germplasm. *Environmental Entomology* 16(2), 327–338. DOI: 10.1093/ee/16.2.327.

Kennedy, R. and Gilles, T. (2003) Brassica[spot]: a forecasting system for diseases of vegetable brassicas. In: *Proceedings of the 8th International Conference in Plant Pathology (ICPP), 2–7 February 2003*. Christchurch, New Zealand, Vol. 2, p. 131.

Kennedy, R., Wakeham, A.J. and Cullington, J.E. (1999) Production and immunodetection of ascospores of *Mycosphaerella brassicicola*: ringspot of vegetable crucifers. *Plant Pathology* 48(3), 297–307. DOI: 10.1046/j.1365-3059.1999.00341.x.

King, K.M., Krivova, V., Canning, G.G.M., Hawkins, N.J., Kaczmarek, A.M. *et al.* (2018) Loop-mediated isothermal amplification (LAMP) assays for rapid detection of *Pyrenopeziza brassicae* (light leaf spot of Brassicas). *Plant Pathology* 67(1), 167–174. DOI: 10.1111/ppa.12717.

Kingsolver, J.G. (2000) Feeding, growth, and the thermal environment of cabbage white caterpillars, *Pieris rapae* L. *Physiological and Biochemical Zoology* 73(5), 621–628. DOI: 10.1086/317758.

Kirkegaard, J.A., Sarwar, M. and Matthiessen, J.N. (1998) Assessing the biofumigation potential of crucifers. *Acta Horticulturae* 459, 105–112. DOI: 10.17660/ActaHortic.1998.459.10.

Klaiber, J., Najar-Rodriguez, A.J., Dialer, E. and Dorn, S. (2013) Elevated carbon dioxide impairs the performance of a specialized parasitoid of an aphid host feeding on *Brassica* plants. *Biological Control* 66(1), 49–55. DOI: 10.1016/j.biocontrol.2013.03.006.

Kocks, C.G. and Ruissen, M.A. (1996) Measuring field resistance of cabbage cultivars to black rot. *Euphytica* 91(1), 45–53. DOI: 10.1007/BF00035275.

Kocks, C.G., Ruissen, M.A., Zadoks, J.C. and Duijkers, M.G. (1998) Survival and extinction of *Xanthomonas campestris* pv. *campestris* in soil. *European Journal of Plant Pathology* 104(9), 911–923. DOI: 10.1023/A:1008685832604.

Kogan, M. and Ortman, E.F. (1978) Antixenosis – a new term proposed to define Painter's "nonpreference" modality of resistance. *Bulletin of the Entomological Society of America* 24(2), 175–176. DOI: 10.1093/besa/24.2.175.

Kohl, J., Vlaswinkel, M., Groenenboom-de Haas, B.H., Kastelein, P., van Hoof, R.A. *et al.* (2011) Survival of pathogens of Brussels sprouts (*Brassica oleracea* Gemmifera Group) in crop residues. *Plant Pathology* 60(4), 661–670. DOI: 10.1111/j.1365-3059.2010.02422.x.

Koike, S.T., Gladders, P. and Paulus, A.O. (2007) *Vegetable Diseases: A Colour Handbook.* Manson, London.

Kole, C., Teutonico, R., Mengistu, A., Williams, P.H. and Osborn, T. (1996) Molecular mapping of a locus controlling resistance to *Albugo candida* in *Brassica rapa*. *Phytopathology* 86(4), 367–369. DOI: 10.1094/Phyto-86-367.

Kole, C., Williams, P.H., Rimmer, S.R. and Osborn, T.C. (2002) Linkage mapping of genes controlling resistance to white rust (*Albugo candida*) in *Brassica rapa* (syn. *campestris*) and comparative mapping to *Brassica napus* and *Arabidopsis thaliana*. *Genome* 45(1), 22–27. DOI: 10.1139/g01-123.

Kreis, R.A., Lange, H.W., Reiners, S. and Smart, C.D. (2016) Cauliflower yield and susceptibility to Alternaria leaf spot under New York field conditions. *HortTechnology* 26(4), 542–546. DOI: 10.21273/HORTTECH.26.4.542.

Krupinsky, J.M., Bailey, K.L., McMullen, M.P., Gossen, B.D. and Turkington, T.K. (2002) managing plant disease risk in diversified cropping systems. *Agronomy Journal* 94(2), 198–209. DOI: 10.2134/agronj2002.1980.

Kurakin, G. (2022) Chemical attraction. *Biologist* 68(4), 26–29.

Kurppa, S. and Ollula, A. (1993) Optimizing insect pest control and nitrogen fertilizing of the summer turnip rape (*Brassica campestris sutiva*). *Agricultural Science Finland* 2(2), 149–160. DOI: 10.23986/afsci.72644.

Lamb, R.J. (1989) Entomology of oilseed *Brassica* crops. *Annual Review of Entomology* 34(1), 211–229. DOI: 10.1146/annurev.en.34.010189.001235.

Lamy, F., Dugravot, S., Cortesero, A.M., Chaminade, V., Faloya, V. *et al.* (2018) One more step toward a push-pull strategy combining both a trap crop and plant volatile organic compounds against the cabbage root fly *Delia radicum*. *Environmental Science and Pollution Research International* 25(30), 29868–29879. DOI: 10.1007/s11356-017-9483-6.

Larkcom, J. (2007) *Oriental Vegetables: The Complete Guide for the Gardening Cook.* Francis Lincoln, London.

Laurenz, S., Brun, A. and Meyhofer, R. (2017) Overwintering of *Encarsia tricolor* on the cabbage whitefly. *IOBC/WPRS Bulletin* 122, 156–159.

Laurenz, S., Schmidt, S., Balkenhol, B. and Meyhofer, R. (2019) Natural enemies associated with the cabbage whitefly *Aleyrodes proletella* in Germany. *Journal of Plant Diseases and Protection* 126(1), 47–54. DOI: 10.1007/s41348-018-0194-0.

Lee, I.-M., Gundersen-Rindal, D.E. and Bertaccini, A. (1998) Phytoplasma: ecology and genomic diversity. *Phytopathology* 88(12), 1359–1366. DOI: 10.1094/PHYTO.1998.88.12.1359.

Leite, G.L.D., Picanco, M., Bastos, C.S., de Araujo, J.M. and Azevedo, A.A. (1996) Resistance of kale clones to the green peach aphid. *Horticultura Brasileira* 14, 178–181.

Leong, S.A., Allen, C. and Triplett, E.W. (eds) (2002) *Biology of Plant-Microbe Interactions, Volume 3. Proceedings of the 10th International Plant-Microbe Interactions Congress, Madison, Wisconsin, 10–14 July 2001.* The International Society for Plant-Microbe Interactions, St Paul, MN.

Li, G., Lv, H., Zhang, S., Zhang, S., Li, F. *et al.* (2019a) TuMV management for brassica crops through host resistance: retrospect and prospects. *Plant Pathology* 68(6), 1035–1044. DOI: 10.1111/ppa.13016.

Li, R., Wang, R.-J., Xie, C.-J., Liu, L., Zhang, J. *et al.* (2019b) A coarse-to-fine network for aphid recognition and detection in the field. *Biosystems Engineering* 187, 39–52. DOI: 10.1016/j.biosystemseng.2019.08.013.

Li, X.-W., Fail, J., Wang, P., Feng, J.-N. and Shelton, A.M. (2014) Performance of arrhenotokous and thelytokous *Thrips tabaci* (Thysanoptera: Thripidae) on onion and cabbage and its implications on evolution and pest management. *Journal of Economic Entomology* 107(4), 1526–1534. DOI: 10.1603/ec14070.

Lin, J., Eckenrode, C.J. and Dickson, M.H. (1983) Variation in *Brassica oleracea* resistance to diamondback moth (Lepidoptera: Plutellidae). *Journal of Economic Entomology* 76(6), 1423–1427. DOI: 10.1093/jee/76.6.1423.

Liu, X.P., Lu, W.C., Liu, Y.K., Wei, S.Q., Xu, J.B. *et al.* (1996) Occurrence and strain differentiation of turnip mosaic potyvirus and sources of resistance in chinese cabbage in China. *Acta Horticulturae* 407, 431–440. DOI: 10.17660/ActaHortic.1996.407.55.

Ludwig, M., Ludwig, H., Conrad, C., Dahms, T. and Meyhöfer, R. (2019) Cabbage whiteflies colonise *Brassica* vegetables primarily from distant, upwind source habitats. *Entomologia Experimentalis et Applicata* 167(8), 713–721. DOI: 10.1111/eea.12827.

Macioszek, V.K., Lawrence, C.B. and Kononowicz, A.K. (2018) Infection cycle of *Alternaria brassicicola* on *Brassica oleracea* leaves under growth room conditions. *Plant Pathology* 67(5), 1088–1096. DOI: 10.1111/ppa.12828.

Madloo, P., Lema, M., Francisco, M. and Soengas, P. (2019) Role of major glucosinolates in the defense of kale against *Sclerotinia sclerotiorum* and *Xanthomonas campestris* pv. *campestris*. *Phytopathology* 109(7), 1246–1256. DOI: 10.1094/PHYTO-09-18-0340-R.

Malvas, C.C., Coelho, R.M.S. and Camargo, L.E.A. (1999) Identification of resistance genes in *Brassica oleracea* to black rot by selective genotyping. *Fitopatologia Brasileira* 24, 143–148.

Marcone, C. and Ragozzino, A. (1995) Detection of phytoplasmas in *Brassica* spp. in Southern Italy and their characterization by RFLP analysis. *Zeitschrift Fur Pflanzenkrankheiten und Pflanzenschutz* 102(5), 449–460.

Martin, T., Palix, R., Kamal, A., Deletre, E., Bonafos, R. *et al.* (2013) A repellent net as a new technology to protect cabbage crops. *Journal of Economic Entomology* 106(4), 1699–1706. DOI: 10.1603/ec13004.

McDonald, M.R., Al-Daoud, F., Sedaghatkish, A., Moran, M., Cranmer, T.J. *et al.* (2021) Changes in the range and virulence of *Plasmodiophora brassicae* across Canada. *Canadian Journal of Plant Pathology* 43(2), 304–310. DOI: 10.1080/07060661.2020.1797882.

Meadow, R. and Johansen, T.J. (2005) Exclusion fences against brassica root flies (*Delia radicum* and *D. floralis*). *Bulletin OILB/SROP* 28(4), 39–43.

Mesmin, X., Cortesero, A.-M., Daniel, L., Plantegenest, M., Faloya, V. *et al.* (2020) Influence of soil tillage on natural regulation of the cabbage root fly *Delia radicum* in brassicaceous crops. *Agriculture, Ecosystems & Environment* 293, 106834. DOI: 10.1016/j.agee.2020.106834.

Metz, T.D., Roush, R.T., Tang, J.D., Shelton, A.M. and Earle, E.D. (1995) Transgenic broccoli expressing *Bacillus thuringiensis* insecticidal crystal protein: implications for pest resistance management strategies. *Molecular Breeding* 1(4), 309–317. DOI: 10.1007/BF01248408.

Minks, A.K., Harrewijn, P. and Helle, W. (eds) (1987) *Aphids: Their Biology Natural Enemies and Control. World Crop Pests, Vol. 2A*. Elsevier, Amsterdam.

Mitchell-Olds, T., James, R.V., Palmer, M.J. and Williams, P.H. (1995) Genetics of *Brassica rapa* (syn. *campestris*). 2. Multiple disease resistance to three fungal pathogens: *Peronospora parasitica, Albugo candida* and *Leptosphaeria maculans*. *Heredity* 75(4), 362–369. DOI: 10.1038/hdy.1995.147.

Miyaji, N., Shimizu, M., Miyazaki, J., Osabe, K., Sato, M. *et al.* (2017) Comparison of transcriptome profiles by *Fusarium oxysporum* inoculation between Fusarium yellows resistant and susceptible lines in *Brassica rapa* L. *Plant Cell Reports* 36(12), 1841–1854. DOI: 10.1007/s00299-017-2198-9.

Mizushima, U. (1980) Genome analysis in *Brassica* and allied genera. In: Tsunoda, S., Hinata, K. and Gómez-Campo, C. (eds) *Brassica Crops and Wild Allies*. Japan Scientific Society Press, Tokyo, pp. 89–106.

Monot, C., Pajot, E., Le Corre, D. and Silué, D. (2002) Induction of systemic resistance in broccoli (*Brassica oleracea* var. *botrytis*) against downy mildew (*Peronospora parasitica*) by avirulent isolates. *Biological Control* 24(1), 75–81. DOI: 10.1016/S1049-9644(02)00006-3.

Monreal, M.A., Derksen, D.A., Watson, P.R. and Monreal, C.M. (2000) Effect of crop management practices on soil microbial communities. In: *Proceedings of the 43rd Annual Manitoba Society of Soil Science Meeting, 25–26 January 2000*. Winnipeg, Manitoba, Canada, pp. 216–228.

Morra, A.A. and Earle, E.D. (2001) Resistance to *Alternaria brassicicola* in transgenic broccoli expressing a *Trichoderma harzianum* endochitinase gene. *Molecular Breeding* 8(1), 1–9. DOI: 10.1023/A:1011913100783.

Mound, L.A. and Halsey, S.H. (1978) *Whitefly of the World: A Systematic Catalogue of the Aleyrodidae (Homoptera) with Host Plant and Natural Enemy Data*. British Museum (Natural History) and John Wiley, Chichester, UK. DOI: 10.5962/bhl.title.118687.

Muller, T. and Muller, C. (2017) Host plant effects on the behavioural phenotype of a Chrysomelid. *Ecological Entomology* 42(3), 336–344. DOI: 10.1111/een.12389.

Murphy, J., Reader, R., Phelps, K., Fellows, J. and Collier, R.H. (2007) Delivering science to the horticultural industry: learning from past mistakes. In: *Proceedings of the 6th Conference of the European Federation of IT in Agriculture and the 5th World Congress on Computers in Agriculture EFITS/WCCA, 2–5 July 2007*. Glasgow, UK.

Myers, J.H. (1985) Effect of physiological condition of the host plant on the ovipositional choice of the cabbage white butterfly, *Pieris rapae. Journal of Animal Ecology* 54(1), 193. DOI: 10.2307/4630.

Ney, L., Franklin, D., Mahmud, K., Cabrera, M., Hancock, D. *et al.* (2019) Rebuilding soil ecosystems for improved productivity in biosolarized soils. *International Journal of Agronomy* 2019, 5827585. DOI: 10.1155/2019/5827585.

Nicolas, O. and Laliberté, J.F. (1992) The complete nucleotide sequence of turnip mosaic virus RNA. *Journal of General Virology* 73(11), 2785–2793. DOI: 10.1099/0022-1317-73-11-2785.

Nowicki, M., Nowakowska, M., Niezgoda, A. and Kozik, E.U. (2012) Alternaria black spot of crucifers: symptoms, importance of disease, and perspectives of resistance breeding. *Vegetable Crops Research Bulletin* 76(1), 5–19. DOI: 10.2478/v10032-012-0001-6.

Ohsawa, K. and Ohsawa, D. (2001) Efficacy of plant extracts for reducing larval populations of the diamondback moth, *Plutella xylostella* L.(Lepidoptera: Yponomeutidae) and cabbage webworm, *Crocidolomia binotalis* Zeller(Lepidoptera: Pyralidae), and evaluation of cabbage damage. *Applied Entomology and Zoology* 36(1), 143–149. DOI: 10.1303/aez.2001.143.

Palaniswamy, P. and Lamb, R.J. (1992) Host preferences of the flea beetles *Phyllotreta cruciferae* and *P. striolata* (Coleoptera: Chrysomelidae) for crucifer seedlings. *Journal of Economic Entomology* 85(3), 743–752. DOI: 10.1093/jee/85.3.743.

Pandey, B.P. and Mohan, J. (1986) Inhibition of turnip mosaic virus by plant extracts. *Indian Phytopathology* 39, 489–491.

Payne, C., Battey, N., Cooke, R., Dixon, G.R., Ijpelaar, J. *et al.* (2004) *Economic Evaluation of the Horticultural Development Council (HDC). A report commissioned by the Department for Environment, Food and Rural Affairs (Defra), London.* Centre for Agricultural Strategy, The University of Reading, Reading, UK, 153 pp.

Pethybridge, S.J. and Nelson, S.C. (2015) Leaf doctor: a new portable application for quantifying plant disease severity. *Plant Disease* 99(10), 1310–1316. DOI: 10.1094/PDIS-03-15-0319-RE.

Phelps, K., Collier, R.H., Reader, R.J. and Finch, S. (1993) Monte Carlo simulation method for forecasting the timing of pest insect attacks. *Crop Protection* 12(5), 335–342. DOI: 10.1016/0261-2194(93)90075-T.

Pichard, B. and Thouvenot, D. (1999) Effect of *Bacillus polymyxa* seed treatments on control of black-rot and damping-off of cauliflower. *Seed Science and Technology* 27, 455–465.

Postma, J., Scheper, R.W.A. and Schilder, M.T. (2010) Effect of successive cauliflower plantings and *Rhizoctonia solani* AG 2-1 inoculations on disease suppressiveness of a suppressive and a conducive soil. *Soil Biology and Biochemistry* 42(5), 804–812. DOI: 10.1016/j.soilbio.2010.01.017.

Provvidenti, R. (1980) Evaluation of Chinese cabbage cultivars from Japan and the people's republic of China for resistance to turnip mosaic virus and cauliflower mosaic virus. *Journal of the American Society for Horticultural Science* 105(4), 571–573. DOI: 10.21273/JASHS.105.4.571.

Provvidenti, R. (1981) Sources of resistance to turnip mosaic virus in Chinese cabbage. In: Talekar, N.S. and Griggs, T.D. (eds) *Chinese Cabbage: Proceedings of the First International Symposium.* Asian Vegetable Research and Development Centre, Shanhua, Taiwan.

Rae, R.G., Robertson, J.F. and Wilson, M.J. (2009) Optimization of biological (*Phasmarhabditis hermaphrodita*) and chemical (iron phosphate and metaldehyde) slug control. *Crop Protection* 28(9), 765–773. DOI: 10.1016/j.cropro.2009.04.005.

Ramirez-Villapudua, J. and Munnecke, D.E. (1987) Control of cabbage yellows (*Fusarium oxysporum* f. sp. *conglutinans*) by solar heating of field soils amended with dry cabbage residues. *Plant Disease* 71(3), 217–221. DOI: 10.1094/PD-71-0217.

Reddy, G.V.P. and Guerrero, A. (2001) Optimum timing of insecticide applications against diamondback moth *Plutella xylostella* in cole crops using threshold catches in sex pheromone traps. *Pest Management Science* 57(1), 90–94. DOI: 10.1002/1526-4998(200101)57:1<90::AID-PS258>3.0.CO;2-N.

Rehman, S., Shahzad, B., Bajwa, A.A., Hussain, S., Rehman, A. *et al.* (2019) Utilizing the allelopathic potential of *Brassica* species for sustainable crop production: a review. *Journal of Plant Growth Regulation* 38(1), 343–356. DOI: 10.1007/s00344-018-9798-7.

Ren, J.P and Dickson, M.H. (1992) Inheritance of resistance to bacterial soft rot in *Brassica rapa. Cruciferae Newsletter* 1999(21), 131–132.

Ren, J.P., Dickson, M.H. and Earle, E.D. (2000) Improved resistance to bacterial soft rot by protoplast fusion between *B. rapa* and *B. oleracea. Theoretical and Applied Genetics* 100(5), 810–819. DOI: 10.1007/s001220051356.

Ren, J.P., Petzoldt, R. and Dickson, M.H. (2001) Genetics and population improvement of resistance to bacterial soft rot in Chinese cabbage. *Euphytica* 117(3), 197–207. DOI: 10.1023/A:1026541724001.

Renkema, J.M., Evans, B.G., House, C. and Hallett, R.H. (2016) Exclusion fencing inhibits early-season beetle (Coleoptera) activity-density in broccoli. *Journal of the Entomological Society of Ontario* 147, 15–28.

Rennie, D.C., Manoli, V.P., Cao, T., Hwang, S.F., Howard, R.J. *et al.* (2011) Direct evidence of surface infestation of seeds and tubers by *Plasmodiophora brassicae* and quantification of spore loads. *Plant Pathology* 60(5), 811–819. DOI: 10.1111/j.1365-3059.2011.02449.x.

Ridgway, R.L., Silverstein, R.M. and Inscoe, M.N. (1990) *Behaviour-Modifying Chemicals for Insect Management.* Marcel Dekker, New York.

Riggin-Bucci, T.M. and Gould, F. (1997) Impact of intraplot mixtures of toxic and nontoxic plants on population dynamics of diamondback moth (Lepidoptera: Plutellidae) and its natural enemies. *Journal of Economic Entomology* 90(2), 241–251. DOI: 10.1093/jee/90.2.241.

Robinson, R.A. (1969) Disease resistance terminology. *Review of Applied Mycology* 48, 593–605.

Root, R.B. (1973) Organization of a plant-arthropod association in simple and diverse habitats: the fauna of collards (*Brassica oleracea*). *Ecological Monographs* 43(1), 95–124. DOI: 10.2307/1942161.

Root, T.L., Price, J.T., Hall, K.R., Schneider, S.H., Rosenzweig, C. *et al.* (2003) Fingerprints of global warming on wild animals and plants. *Nature* 421(6918), 57–60. DOI: 10.1038/nature01333.

Rouxel, T. and Balesdent, M.H. (2005) The stem canker (blackleg) fungus, *Leptosphaeria maculans,* enters the genomic era. *Molecular Plant Pathology* 6(3), 225–241. DOI: 10.1111/j.1364-3703.2005.00282.x.

Rubel, M.H., Natarajan, S., Hossain, M.R., Nath, U.K., Afrin, K.S. *et al.* (2019) Pathovar specific molecular detection of *Xanthomonas campestris* pv. *campestris,* the causal agent of black rot disease in cabbage. *Canadian Journal of Plant Pathology* 41(3), 318–328. DOI: 10.1080/07060661.2019.1570973.

Rusholme, R.L., Higgins, E.E., Walsh, J.A. and Lydiate, D.J. (2007) Genetic control of broad-spectrum resistance to turnip mosaic virus in *Brassica rapa* (Chinese cabbage). *Journal of General Virology* 88(11), 3177–3186. DOI: 10.1099/vir.0.83194-0.

Ryschka, U., Schumann, G., Klocke, E., Scholze, P. and Neumann, M. (1996) Somatic hybridization in Brassicaceae. *Acta Horticulturae* 407, 201–208. DOI: 10.17660/ActaHortic.1996.407.24.

Sako, N. (1981) Virus diseases of Chinese cabbage in Japan. In: Talekar, N.S. and Griggs, T.D. (eds) *Chinese Cabbage: Proceedings of the First International Symposium*. Asian Vegetable Development Centre, Shanhua, Taiwan, pp. 129–142.

Santos, M.P., Dias, J.S. and Monteiro, A.A. (1996) Screening portuguese cole landraces for resistance to white rust [*Albugo candida* (Pers.) Kuntze]. *Acta Horticulturae* 407, 453–460. DOI: 10.17660/ActaHortic.1996.407.58.

Scheiner, C. and Martin, E.A. (2020) Spatiotemporal changes in landscape crop composition differently affect density and seasonal variability of pests, parasitoids and biological pest control in cabbage. *Agriculture, Ecosystems & Environment* 301, 107051. DOI: 10.1016/j.agee.2020.107051.

Senior, L.J., Healey, M.A. and Wright, C.L. (2016) The role of spiders as predators of two lepidopteran *Brassica* pests. *Austral Entomology* 55(4), 383–391. DOI: 10.1111/aen.12201.

Sharma, S. and Sharma, G.R. (2001) Influence of white rot (*Sclerotinia sclerotiorum*) on growth and yield parameters of Indian mustard (*Brassica juncea*) varieties. *Indian Journal of Agricultural Sciences* 71(4), 273–274.

Sharma, S.R., Kapoor, K.S. and Gill, H.S. (1995) Screening against sclerotinia rot (*Sclerotinia sclerotiorum*), downy mildew (*Peronospora parasitica*) and black rot (*Xanthomonas campestris*) in cauliflower (*Brassica oleracea* var. *botrytis subvar. cauliflora*). *Indian Journal of Agricultural Sciences* 65(12), 916–918.

Shattuck, V.I. (1992) The biology, epidemiology and control of turnip mosaic virus. *Horticultural Reviews* 14, 199–238. DOI: 10.1002/9780470650523.

Shattuck, V.I. (1993) Powdery mildew-resistant UG3 and UG4 rutabaga germplasm. *Canadian Journal of Plant Science* 73(1), 301–302. DOI: 10.4141/cjps93-046.

Shattuck, V.I. and Stobbs, L.W. (1987) Evaluation of rutabaga cultivars for turnip mosaic virus resistance and the inheritance of resistance. *HortScience* 22(5), 935–937. DOI: 10.21273/HORTSCI.22.5.935.

Shelton, A.M. (1995) Temporal and spatial dynamics of thrips populations in a diverse ecosystem: theory and management. In: Parker, B.L., Skinner, M. and Lewis, T. (eds) *Thrips Biology and Management. NATO ASI Series 276*. Springer, Boston, MA, pp. 425–432. DOI: 10.1007/978-1-4899-1409-5.

Shelton, A.M., Tang, J.D., Roush, R.T., Metz, T.D. and Earle, E.D. (2000) Field tests on managing resistance to Bt-engineered plants. *Nature Biotechnology* 18(3), 339–342. DOI: 10.1038/73804.

Shigaki, T., Nelson, S.C. and Alvarez, A.M. (2000) Symptomless spread of blight-inducing strains of *Xanthomonas campestris* pv. *campestris* on cabbage seedlings in misted seedbeds. *European Journal of Plant Pathology* 106(4), 339–346. DOI: 10.1023/A:1008771217477.

Sigareva, M. and Earle, E.D. (eds) (1995) Intertribal somatic hybrids between *Camelina Sativa* and rapid cycling *Brassica oleracea*. *Cruciferae Newsletter* 19, 49–50.

Singh, D., Naveen, C. and Gupta, P.P. (1997) Inheritance of powdery mildew resistance in interspecific crosses of Indian and Ethiopean mustard. *Annals of Biology* 13, 73–77.

Singh, H. and Kalda, T.S. (1995) Screening of cauliflower germplasm against *Sclerotinia* rot. *Indian Journal of Genetics and Plant Breeding* 55(1), 98–102.

Soriano, F., Claudio, M.T.R. and Cardoso, A.I.I. (2019) Germination of broccoli organic seeds treated with thermotherapy. *Acta Horticulturae* 1249, 73–78. DOI: 10.17660/ActaHortic.2019.1249.14.

Soursa, M.S., Dias, J.S. and Monteiro, A.A. (1997) Screening Portuguese cole landraces for resistance to seven indigenous downy mildew isolates. *Scientia Horticulturae* 68(1–4), 49–58. DOI: 10.1016/S0304-4238(96)00979-X.

Srivastava, A., Cho, I.-K. and Cho, Y.-R. (2013) The *Bdtf1* gene in *Alternaria brassicicola* is important in detoxifying brassinin and maintaining virulence on *Brassica* species. *Molecular Plant-Microbe Interactions* 26(12), 1429–1440. DOI: 10.1094/MPMI-07-13-0186-R.

Stoate, C., Leake, A., Jarvis, P. and Szczur, J. (2017) *Fields for the Future: 25 Years of the Allerton Project – A Winning Blueprint for Farming, Wildlife and the Environment.* Game and Wildlife Conservation Trust, Loddington, UK.

Stoner, K.A., Dickson, M.H. and Shelton, A.M. (1989) Inheritance of resistance to damage by *Thrips tabaci* Lindeman (Thysanoptera: Thripidae) in cabbage. *Euphytica* 40(3), 233–239. DOI: 10.1007/BF00024517.

Strelkov, S.E. and Hwang, S.-F. (2014) Clubroot in the Canadian canola crop: 10 years into the outbreak. *Canadian Journal of Plant Pathology* 36(suppl. 1), 27–36. DOI: 10.1080/07060661.2013.863807.

Strelkov, S.E., Hwang, S.-F., Manolii, V.P., Cao, T., Fredua-Agyeman, R. *et al.* (2018) Virulence and pathotype classification of *Plasmodiophora brassicae* populations collected from clubroot resistant canola (*Brassica napus*) in Canada. *Canadian Journal of Plant Pathology* 40(2), 284–298. DOI: 10.1080/07060661.2018.1459851.

Talekar, N.S. (1986) Diamondback moth management. In: *Proceedings of the First International Workshop, Tainan, Taiwan, 11–15 March 1985.* Asian Vegetable Research and Development Centre, Shanhua, Taiwan.

Talekar, N.S. and Shelton, A.M. (1993) Biology, ecology, and management of the diamondback moth. *Annual Review of Entomology* 38(1), 275–301. DOI: 10.1146/annurev.en.38.010193.001423.

Talekar, N.S., Yong, H.C., Lee, S.T., Chen, B.S. and Sun, L.Y. (1985) *Annotated Bibliography of Diamondback Moth.* Asian Vegetable Research and Development Center, Shanhua, Taiwan, pp. 85–229.

Taylor, J.D., Conway, J., Roberts, S.J., Astley, D. and Vicente, J.G. (2002) Sources and origin of resistance to *Xanthomonas campestris* pv. *campestris* in *Brassica* genomes. *Phytopathology* 92(1), 105–111. DOI: 10.1094/PHYTO.2002.92.1.105.

Thaning, C. and Gerhardson, B. (2001) Reduced sclerotial soil-longevity by whole-crop amendment and plastic covering. *Journal of Plant Diseases and Protection* 108(2), 143–151.

Theunissen, J. (1984) Supervised pest control in cabbage crops: theory and practice. *Mitteilungen Biologische Bundesanstalt für Land und Forstwirtschaft* 218, 76–84.

Thomas, C.S. and Gubler, W.D. (2000) A privatized crop warning system in the USA. *Bulletin EPPO* 30(1), 45–48. DOI: 10.1111/j.1365-2338.2000.tb00849.x.

Ton, J., Davison, S., Van Loon, L.C. and Pieterse, C.M.J. (2001) Heritability of rhizobacteria-mediated induced systemic resistance and basal resistance in *Arabidopsis*. *European Journal of Plant Pathology* (107), 63–68. DOI: 10.1023/A:1008743809840.

Toxopeus, H., Dixon, G.R. and Mattusch, P. (1986) Physiological specialization in *Plasmodiophora brassicae*: an analysis by international experimentation. *Mycological Research* 87(2), 279–287. DOI: 10.1016/S0007-1536(86)80031-6.

Tremmel, M. and Müller, C. (2014) Diet dependent experience and physiological state shape the behavior of a generalist herbivore. *Physiology & Behavior* 129, 95–103. DOI: 10.1016/j.physbeh.2014.02.030.

van de Steene, F., Benoit, F. and Ceustermans, N. (1992) The use of covers to reduce cabbage root fly and caterpillar damage in white cabbage crops. *Bulletin OILB/SROP* 15(4), 155–167.

van Munster, M., Yvon, M., Vile, D., Dader, B., Fereres, A. *et al.* (2017) Water deficit enhances the transmission of plant viruses by insect vectors. *PLoS ONE* 12(5), e0174398. DOI: 10.1371/journal.pone.0174398.

Vicente, J.G., Conway, J., King, G.J. and Taylor, J.D. (2000) Resistance to *Xanthomonas campestris* pv. *campestris* in *Brassica* spp. *Acta Horticulturae* 539, 61–67. DOI: 10.17660/ActaHortic.2000.539.6.

Voorrips, R.E., Steenhuis-Broers, G., Tiemens-Hulscher, M. and van Bueren, E.T.L. (2010) Earliness, leaf surface wax and sugar content predict varietal differences for thrips damage in cabbage. *Acta Horticulturae* 867, 127–132. DOI: 10.17660/ActaHortic.2010.867.16.

Wainwright, C., Jenkins, S., Wilson, D., Elliott, M., Jukes, A. *et al.* (2020) Phenology of the diamondback moth (*Plutella xylostella*) in the UK and provision of decision support for *Brassica* growers. *Insects* 11(2), 118–133. DOI: 10.3390/insects11020118.

Wakeham, A.J. and Kennedy, R. (2010) Risk assessment methods for the ringspot pathogen *Mycosphaerella brassicicola* in vegetable brassica crops. *Plant Disease* 94(7), 851–859. DOI: 10.1094/PDIS-94-7-0851.

Walkey, D.G.A. and Pink, D.A.C. (1988) Reactions of white cabbage (*Brassica oleracea* var. *capitata*) to four different strains of turnip mosaic virus. *Annals of Applied Biology* 112(2), 273–284. DOI: 10.1111/j.1744-7348.1988.tb02063.x.

Walsh, J.A. (1997) *Turnip Mosaic Virus. Data sheet for Commonwealth Agricultural Bureau International Global Crop Protection Compendium*. CAB International, Wallingford, UK.

Walsh, J.A. and Jenner, C.E. (2002) Turnip mosaic virus and the quest for durable resistance. *Molecular Plant Pathology* 3(5), 289–300. DOI: 10.1046/j.1364-3703.2002.00132.x.

West, J.S., Atkins, S.D., Emberlin, J. and Fitt, B.D.L. (2008) PCR to predict risk of airborne disease. *Trends in Microbiology* 16(8), 380–387. DOI: 10.1016/j.tim.2008.05.004.

West, J.S., Canning, G.G.M., Perryman, S.A. and King, K. (2017) Novel technologies for the detection of Fusarium head blight disease and airborne inoculum. *Tropical Plant Pathology* 42(3), 203–209. DOI: 10.1007/s40858-017-0138-4.

Williams, P.H. (1966) A system for the determination of races of *Plasmodiophora brassicae* that infect cabbage and rutabaga. *Phytopathology* 56(6), 624–626.

Williams, P.H., Walker, J.C. and Pound, G.S. (1968) Hybelle and Sanibel, multiple disease-resistant F$_1$ hybrid cabbages. *Phytopathology* 58, 791–796.

Williams, P.H., Staub, T. and Sutton, J.C. (1972) Inheritance of resistance in cabbage to black rot. *Phytopathology* 62(2), 247. DOI: 10.1094/Phyto-62-247.

Yamamura, K. and Yano, E. (1999) Effects of plant density on the survival rate of cabbage pests. *Research on Population Ecology* 41(2), 183–188. DOI: 10.1007/s101440050021.

Zhan, Y., Tian, H., Ji, X. and Liu, Y. (2020) *Myzus persicae* (Hemiptera: Aphididae) infestation increases the risk of bacterial contamination and alters nutritional content in storage Chinese cabbage. *Journal of the Science of Food and Agriculture* 100(7), 3007–3012. DOI: 10.1002/jsfa.10331.

Zhang, S.W., Shang, YJ. and Wang, L. (2015) Plant disease recognition based on plant leaf image. *Journal of Animal and Plant Sciences* 25(3), 42–45.

POSTHARVEST QUALITY, VALUE AND MARKETING

8

GEOFFREY R. DIXON*

School of Agriculture, Policy and Development, Earley Gate, Whiteknights Road, PO Box 237, University of Reading, Reading, Berkshire RG6 6EU and GreenGene International, Hill Rising, Horsecastles Lane, Sherborne, Dorset, DT9 6BH, UK

Abstract

The postharvest period sees the culmination of research-based husbandries requiring that product quality is retained and marketed. Quality factors are defined in this chapter, especially appearance and health benefits for which brassicas are significantly important. Moving quality onto consumers' plates relies on a range of husbandry attributes and the impacts of these attributes are discussed. These attributes particularly influence the freshness of produce, which can be damaged by several physiological disorders. These abiotic stresses result from imbalances between genotype and the husbandry environment, aspects of which are defined. Imbalanced calcium nutrition is a salient factor resulting in the onset of several disorders. Storage blemishes will also result from specific microbial infections.

Harvesting operations for brassicas are being automated, and specialized equipment developed for this purpose is identified in relation to particular crops. There follows an evaluation of storage, grading, food safety and degradation. Aspects of marketing are also considered.

*geoffrdixon@gmail.com

© Geoffrey R. Dixon and Rachel Wells 2024. *Vegetable Brassicas and Related Crucifers,* 2nd edition. (G.R. Dixon and R. Wells) DOI: 10.1079/9781789249170.0008

INTRODUCTION

Research, development and husbandry objectives for vegetable brassicas have changed substantially in the past 25 years. Focus now concentrates on the science and technology that will underpin production strategies which are environmentally sustainable as climatic change becomes more intense, linked with an appreciation of brassicas' dietary and medical values, consumers' worldwide demands for food novelty in Asian and European forms, and plant breeders' increasingly efficient capabilities for offering new hybrids. This chapter concerns factors of transparency and trustworthiness in the transfer of high-quality produce from the field via postharvest handling through supply chains to the ultimate consumers.

Previously, attention was almost exclusively aimed at improving the effectiveness and efficiency of husbandry operations and achieving the highest returns from the units of resource employed, including land, labour and capital. While the priority of these business objectives is still the ultimate basis of company survival, their interpretations are now moderated by social and environmental responsibilities. Underpinning all the changes described in this book is recognition of the imperatives driven by global climatic change and the essential need for reaching carbon net zero at all stages from field to fork by 2050.

Alterations in marketing systems have helped to drive change. The arrival of supermarkets from the 1970s onwards increasingly offered consumers wider ranges of high-quality greengrocery. Regrettably, the dominance of massive retailing companies, offering consumers convenience shopping, and now including remote online trading, resulted in unbalanced and distorted markets at the production end. Fresh vegetables, including the brassicas, are grown by very modest companies compared with the vast international retail empires. As a result, many smaller producers have ceased trading. Larger growers require increasing precision and business efficiency, delivered by science and innovation, as a means for coping with static financial returns from the market. These market inequalities have gone about as far as the supply side can accommodate, particularly when the costs of husbandry resources are inflating very fast.

The alternative is the wholesale markets, which currently supply greengrocery for industrial processing, restaurants and other food outlets. During the COVID-19 pandemic, wholesale markets gained a much wider client base. Additionally, farm shops, street market traders and small shops were favoured by consumers; processes which may be set to continue because of the high level of personal communication between vendor and purchaser (see Fig. 8.1).

Each of these changes demands differing research, development and innovation strategies funded either by the producer, producer organizations, levy boards or public taxpayer funds.

There has been a massive increase in research and development for brassica vegetables on an international scale over the past 30 years. This is driven by the economic and social importance of brassicas in both developed

Fig. 8.1. Market trader – an example of direct communication between vendor and purchaser. (Geoff Dixon).

and developing nations. In the latter, brassica vegetables traditionally form substantial parts of the diet and hence are key supports for burgeoning populations. In all nations, the contribution of brassicas towards 'healthy living' is a key factor in the growth of consumer demand. Research and development for vegetable brassicas also benefits from much larger funding provisions made for industrial oilseed brassicas. Vegetable oils are major agricultural commodities traded on international markets. Hence, they are supported by substantial research and development financing. Scientific and technical information relevant for oilseed crops is deliberately excluded from this book. It is to be found in its own voluminous publications. None the less, the research outcomes benefit vegetable brassicas, particularly in studies of their genetics, plant breeding, pests and pathogens, and physiology.

The quality of fresh vegetable brassicas as they leave the field is determined by the interactions of growth stage, cultivar, environment and husbandry practices. Postharvest quality reflects the effectiveness of crop husbandry and postharvest handling, storage and transport. Attractive packaging can enhance good quality produce but does little for the visual appeal of poor-quality, blemished goods. The recent realization of the environmental damage wrought by plastic packaging is leading to changes in the materials used. In some instances, vegetable brassicas are being sold without any packaging. Ensuring quality and shelf life are safeguarded for each marketing sector requires relevant and novel research and development. Retaining the benefits of modified atmosphere packaging (MAP) is an important consideration with all these changes.

WHAT IS QUALITY?

Quality is a subjective and somewhat nebulous term composed of attributes that vary with different brassica crops and the attitudes of individual consumers. Basically, these attributes fall into four groups:

- sight: colour, gloss, viscosity, size, shape and visual defects;
- touch or texture: in hand, finger and mouth;
- odour: smell, taste or flavour; and
- hidden attributes: such as nutritional value and the presence of either harmless adulterants or toxic substances.

Improving the horticultural quality in regionally adapted green broccoli (calabrese) (*Brassica oleracea* var. *italica*) and other *B. oleracea* crops is challenging because of the complex genetic control of traits affecting morphology, development and yield. Understanding the mechanisms underlying horticultural quality enables multi-trait, marker-assisted selection for improved, resilient and regionally adapted *B. oleracea* germplasm (Stansell *et al.*, 2019).

Quality is linked with environmental impact, yet along the whole supply chain this topic has scarcely begun being studied (Frankowska *et al.*, 2019). This group found that cabbage and Brussels sprouts (*B. oleracea* var. *gemmifera*) are generally environmentally most sustainable. Overall, in the UK, for instance, vegetables generate 20.3 t carbon dioxide (CO_2) equivalent, consume 260.7 PJ of primary energy and use 253 t equivalent of water. The environmental impacts of airfreighted fresh vegetables are around five times higher than of those from domestic production. Even processed products have lower impacts than fresh airfreighted produce. Packaging contributes significantly to environmental impacts, especially the glass jars and metal cans used for processed vegetables. Other significant hotspots of environmental impact are the vast ranks of open chill cabinets used by the large retailers and vegetable cooking.

Colour and gloss

Colour and gloss are products of light reflected from the brassica in question. They result from the amount of light reflected in proportion to that absorbed by plant pigments. Thus, whiteness in cauliflower (*B. oleracea* var. *botrytis*) curds results from an almost total absence of pigmentation, leading to a very high proportion of light being reflected into the eyes of the observer. Pigment such as the green chlorophyll are present in high quantities in most brassica crops, contributing to increased visual appeal by conveying the impression of freshness. This attribute can be measured using chlorophyll fluorescence as the parameter (see Fig. 8.2).

Fig. 8.2. Measuring chlorophyll fluorescence as an indicator of freshness. (Courtesy of Professor Debbie Rees).

With cabbage, Brussels sprouts and green broccoli spears the consumer requires an appearance of freshness, which is associated with a green and glossy appearance. Some crops, such as red-headed cabbage (*B. oleracea* var. *capitata*) and coloured cauliflower curds, are regarded as of good quality due to the presence of attractive red, orange or yellow anthocyanin pigments. In coloured cauliflower, there may be up to 100-fold increase in vitamin A compared with white curds. Coloured cauliflower has become very popular worldwide since research started in England and North America 50 years ago. By contrast, where there is yellowing of the foliage and flowering it is associated with senescence and is not acceptable to the consumer.

A clean, glossy appearance much enhances the apparent quality of brassica vegetables. This appeal is increased by a film of or droplets of moisture on the surface of leaves or buds. Moisture films or droplets increase the directional reflection of light as compared with light being reflected evenly at all angles, which conveys a dull, lifeless finish. Glossiness can be associated genetically with resistance to insect damage, such as that caused by the diamondback moth (*Plutella xylostella*), but has the disadvantage of causing reductions in growth rate.

Viscosity

Viscosity describes high levels of internal friction in semi-fluid substances. It is not an attribute normally associated with fresh brassica vegetables. It only applies to crops, such as swedes (*Brassica napus* ssp. *rapifera*) and turnips (*Brassica rapa* ssp. *rapifera*), when they are processed into soups and purees or 'smoothies', such as those prepared from green broccoli (*B. oleracea* var. *italica*) which are now advocated as part of some 'healthy-eating' diets.

Size and shape

Botanically, improved predictions of fitness and yield may be obtained by characterizing the genetic controls and environmental dependencies of organismal ontogeny. Baker *et al.* (2015) modelled leaf growth and allometry as valued traits and examined genetic correlations between these traits and aspects of phenology, physiology, circadian rhythms and fitness. Leaf shape is associated with venation features that affect desiccation resistance. The genetic independence of leaf shape from other leaf traits may, therefore, enable crop optimization for this trait without negative effects on characters such as size, growth rate, duration or gas exchange.

Size and shape are qualities of major significance for brassica crops. Retail consumers' size requirements for fresh vegetables have decreased sharply in parallel with the reduction in the size of family units and the rising dominance of single persons catering for themselves. Consumers demand produce that is easily and quickly prepared and can be consumed at a single sitting with no wasted residues. Few retail purchasers require products larger than 0.5 kg and frequently even smaller portions are demanded (see Fig. 8.3).

Fig. 8.3. A green broccoli (*Brassica oleracea* var. *italica*) spear suitable for a single meal for a small family unit. (Geoff Dixon).

Consequently, growers are required to divide large cauliflower (*B. oleracea* var. *botrytis*) heads into florets before packaging or provide single heads and spears of cabbages or green broccoli (*B. oleracea* var. *italica*). An ability to present brassica products at peak freshness and turgidity is increasing as their use in uncooked, salad-style foods rises.

Traits related to head shape, including Hvd (head vertical diameter), Htd (head transverse diameter) and Hsi (head shape index, the ratio of Hvd/Htd), are very important agronomic traits associated with both yield and quality in cabbage (*B. oleracea* var. *capitata*). There is little information describing inheritance analysis and quantitative trait locus mapping of these traits remains rare. Work by Zhang *et al.* (2016) indicated that Htd was controlled by two major independent genes plus polygenes with recessive-epistatic effects. Head shape was controlled by two major linkage genes and polygenes with cumulative effects.

Shape is of especial importance in crops such as cauliflower and green broccoli. The market demands domed heads that differ markedly from the flat-headed types produced by the now largely obsolete open-pollinated cultivars. Well-formed spears of green broccoli or the densely packed heads of Chinese cabbage, white (head) cabbage or leafy greens are essential quality characters. Chinese cabbage (*Brassica rapa* ssp. *pekinensis*) plants, for example, pass through seedling and rosette stages before forming their leafy head. Chinese cabbage plants resemble pak choi plants in their seedling stage, but by their rosette stage the leaves of Chinese cabbage begin differentiating as they increase in size with shorter petioles (Sun *et al.*, 2019). Comparison of the transcriptome between leaves of two very different Chinese cabbages with pak choi during plant development allowed the identification of specific gene categories associated with leafy head formation. There is a correlation between rosette leaf traits and both head traits and heading capacity in Chinese cabbage. However, the leaf number of the mature head is not correlated to heading degree or head shape (Sun *et al.*, 2018). The correlation between rosette leaf and heading traits provides an insight into the leaf requirements for head forming.

Radish is an important source of root, leafy and fruit vegetables, oil crops, and cover plants. The economic importance and characteristics of radishes differ between the Asia and Europe. In Asia, radish cultivars may form large roots with various shapes called 'Asian big radish' and those types grown for production of immature pods or oilseeds. In Europe, radish is a small, rapidly maturing vegetable (Nishio, 2017). In California, USA, wild radish (*Raphanus raphanistrum*) has developed introgressed populations after hybridization with its cultivated counterpart (*Raphanus sativus*). Hybridization between various *Brassica* and *Sinapis* species is also possible. However, the origin of morphological diversity in wild radish is unclear – is it native or due to gene flow from the cultivated radish or other Brassicaceae? Significant morphological divergence was found that could have impacts on plant ecology and adaptation (Liu and Darmency, 2019).

Defects

Crop defects may have genetic, physiological, pathological or mechanical origins, or alternatively result from the presence of extraneous organic or inorganic items. Defects inherent in the crop resulting from interactions between genotype, environment and microbial pathogens are dealt with in Chapter 7.

Surface blemishes, insect deposits, fungal growth, necrotic zones and virus-induced yellowing can all constitute defects of brassica crops. Mechanical damage includes cuts, bruises and discoloration resulting from defective harvesting processes. Cauliflowers (*B. oleracea* var. *botrytis*) and green broccoli (*B. oleracea* var. *italica*) are probably the most easily damaged crops since rough handling will result in bruising of the flowering organ and subsequently the downgrading or rejection of heads and spears. Where Brussels sprout (*B. oleracea* var. *gemmifera*) buds are harvested mechanically the failure to sharpen and adjust the stripper's blades correctly can ruin previously high-quality crops.

The effects of mechanical stress (dropping, compression and trimming) on Chinese cabbage (*B. rapa* ssp. *pekinensis*) cv. 'Yuki' were tested by Porter *et al.* (2004). Dropping and compression did not affect marketable quality where the produce was sold immediately. But storage of damaged produce for 9 weeks (at 2°C) impaired quality. Heads that were repeatedly trimmed produced less ethylene at the end of the storage period compared with the start. This reflected the removal of outer senescing and rotting leaves. Marketable yield was not improved by trimming, emphasizing the point that treatments after harvest cannot improve quality, they can only maintain what has already been established in the field.

The presence of extraneous items will also ruin crop quality. Prepackaging by the grower for the supermarkets presupposes that there will be strict adherence to quality standards set as part of the contract. These are easily forfeited where, for example, leaves are packed with Brussels sprout buds or soil is allowed into the package along with cauliflower florets.

Texture

Texture is an interaction of those physical characteristics that are sensed by the feeling or touch. Purchasers still consider they can judge quality by the outer feel of vegetables, especially cabbage heads. Sensing firmness or softness by hand is used as a guide to maturity and quality by experienced crop technologists for determining the start of harvesting. In brassica crops, firmness is more important since softness indicates incipient enzyme-induced breakdown resulting from over maturity or pathogen-related rotting. This attribute can be quantified by using standardized penetrometer tests that determine the rate at which needles pierce the produce when driven by a standardized force. Compression can be determined numerically by exposing the produce to standardized mass.

Texture in the mouth is generally sensed as chewiness, fibrousness, grittiness, mealiness, stickiness, oiliness or dryness. For brassica crops which are cooked prior to consumption, the retention of flavour appeal and crispness is important. These are aspects of quality that cannot be controlled by the grower since it is very easily lost by overcooking. Increasingly, both leaf and root brassica crops are being consumed in the raw state where the retention of crispness and flavour are of major importance. Here the application of MAP for the produce retains these qualities and is of prime importance.

Flavour and taste

Flavour is a composite of taste and odour. Taste is a four-dimensional character which distinguishes sweetness, sourness, saltiness and bitterness. Odour comes from the combination and interaction of a range of chemical constituents. Consumers would not expect odours from brassica vegetables and their presence will reduce their value. Taste is very much a combination of physical and chemical components. The chemical constituents can be identified by detailed analytical techniques, such as mass spectroscopy and liquid chromatography. Such analytical techniques now allow plant breeders to select for their presence or absence.

The main sensory sensations related to brassicas are their characteristic sharp and bitter taste, and unique aroma (Wieczorek *et al.*, 2018). But there is considerable variability between individuals in bitter taste perception and sensibility. The intensities of sweetness, sourness, bitterness, umami, saltiness and fattiness are key factors in the acceptability of brassicas after preparation by the consumer (van Stokkom *et al.*, 2016). Conditions prior to purchase affect subsequent acceptability; typically, green broccoli (*B. oleracea* var. *italica*) arrives at the store between 7 and 14 days of harvest and is kept refrigerated until purchased or considered as waste. Information is limited as to how far shelf life or storage temperature affects the sensory attributes that contribute to green broccoli purchases or repurchases (Pellegrino *et al.*, 2019). Tests showed that future purchases were affected by tastes, colours and flavours such as grassy, musty or 'dirt-like' from previous purchases.

Kale (*B. oleracea* ssp. *acephala*) is a recently popularized commodity in Western markets, but there is limited information regarding the underlying sensory characteristics motivating this trend (Swegarden *et al.*, 2019). Initially, consumers preferred familiar kale types (i.e. curly leaved types), while accepting newer hybrids for flavour once tested. Plant breeders have developed new Bimi® broccoli which is a natural hybrid between Chinese kale (*B. oleracea* var. *alboglabra*) and conventional green broccoli (Martínez-Hernández *et al.*, 2013). It has a tender stem (similar to asparagus) and a small floret with mild flavour, and is ideal for fresh-cut purposes with nutritional benefits. However, it is easily perishable as indicated by yellowing, stem bending, off-odours and off-flavours. It requires controlled atmosphere storage with 10–15 kPa CO_2 plus 2 kPa oxygen (O_2) and balanced with nitrogen.

Root brassicas, such as fresh-cut and diced swede (*B. napus* ssp. *rapifera*) and turnip (*B. rapa* ssp. *rapifera*), can be frozen at −2.67°C and −1.97°C, respectively (Helland *et al.*, 2016). Increasing storage temperature from 0°C to 5°C or 10°C did not change appearance in swede, but in turnip the hue and colour intensity increased, and colour evenness and whiteness decreased. Differences in odour, taste and flavour developed with storage at 5°C or 10°C. Deterioration may be slowed by using cyclodextrins which are macromolecules used by the food industry as health-promoting compounds, flavour stabilizers or to eliminate undesirable tastes and browning (Martinez-Hernandez *et al.*, 2019). Potentially, for example, they may preserve glucoraphanin in green broccoli juice during industrial processing, which is later transformed by endogenous myrosinase after ingestion into sulforaphane.

The extreme forms of processing of brassica, as in the production of regional delicacies, such as sauerkraut from head cabbage (*B. oleracea* var. *capitata*) in Germany and North America and kimchi from Chinese cabbage in Korea, accentuate saltiness and bitterness as essential indicators of quality. A sensory assessment of sauerkraut after a 7-day fermentation with various salt concentrations used a non-numerical approach based on several criteria and person aggregations (Fadhil *et al.*, 2021). Overall, the salt concentration variations in sauerkraut resulted in different acceptance patterns based on sensory assessment. Cabbage kimchi is a popular side dish in Korean cuisine which produces several fermentation by-products. Kimchi is praised by its consumers for flavour, taste and texture when fermented at 0.7–0.9% total acidity, or a pH of approximately 4.1 (Moon *et al.*, 2020).

Microgreens are an emerging brassica production system, along with other food crops grown in enclosed factory systems. They have some promise for sustainably diversifying global food systems, facilitating adaptations to urbanization, mitigating global climate change and promoting human health (Michell *et al.*, 2020). This study evaluated consumer sensory perception and acceptability of microgreens species green broccoli and red-headed cabbage (*B. oleracea* var. *capitata*) and potential factors affecting consumer acceptance. Sensory perception and food neophobia affected their acceptability. Participants in the study indicated that factors, such as knowledge, access and availability, cost, freshness and shelf life, regulate purchasing of microgreens.

HEALTH AND WELFARE BENEFITS

Food containing brassica vegetables has high nutritional and potentially health-promoting properties (Perez-Balibrea *et al.*, 2011). The consumption of brassica vegetables may be associated with reducing the incidence of human cancer and coronary diseases when consumed over periods of several years as part of a balanced diet (Mazza, 2004) (see Table 8.1).

In this respect the brassica crops, such as green broccoli (*B. oleracea* var. *italica*), have especial interest because of their sulforaphane content which

Table 8.1. Examples of the constituents of brassicas that may enhance human health. (After Mazza, 2004).

Component	Potential benefit
Lutein	Contributes to healthy vision
Sulforaphane	Neutralizes free radicals, may reduce cancer risk
Lignans	May protect against heart disease and some cancers, lowers low-density lipoprotein cholesterol, total cholesterol and triglycerides
Allyl methyl trisulfide, dithiothiones	Lowers low-density lipoprotein cholesterol, maintains healthy immune system

is associated with the reduction of active oxygen in tissues and may provide protection from cancer and coronary diseases. Green broccoli has now become a very popular convenience vegetable worldwide with potentially significant benefits for long-term human health. It has been recommended as a dietary additive for cancer prevention since the early 1980s (Nestle, 1998). There is now growing evidence that the glucosinolate derivative, sulforaphane, in these green leafy vegetables may be effective in preventing and treating various cancers, such as prostate, breast, colon, skin, bladder and oral cancer (Nandini *et al.*, 2020). The potential use of green broccoli-derived sulforaphane as a functional food ingredient may attenuate the inflammation-induced hepatic hepcidin and reduce liver cancer risk (Al-Bakheit and Abu-Qatouseh, 2020).

Glucosinolates are a large group of sulfur-containing compounds found in most brassicas. Wild and domesticated brassicas contain over 100 different forms of glucosinolate molecule. They all have a common basic structure composed of three parts: a β-D-thioglucose group, a sulfonated oxime moiety and a variable side chain. The latter can be a straight-chained alkyl or alkenyl structure, a ring-shaped aromatic group or an indolyl formation. Many common vegetable brassicas, such as Brussels sprouts or green broccoli, contain all three moieties. They remain inactive in intact cells but damage releases myrosinase enzymes which break down the glucosinolate into several products, notably nitriles and isothiocyanates. These are the source of hot and bitter flavours in brassicas, especially condiments. Their original function may be to provide natural deterrents active against some insect and vertebrate pests.

Some glucosinolate breakdown products have been thought to produce toxic or anti-nutritional effects in grazing animals. Consequently, plant breeders aimed to reduce their content in crops such as forage rape (*B. napus*) and fodder kale (*B. oleracea* ssp. *acephala*). They have also been ascribed goitrogenic effects in humans but more recent evidence offers a contrary view of their value. There is a growing body of epidemiological and experimental evidence showing that the consumption of brassica vegetables specifically reduces the risk of cancer in the human lung and alimentary tract. Evidence suggests, for instance, that sulforaphane aids in the detoxification of carcinogens. Others, such as those present in watercress (*Nasturtium officinale*), help in the excretion of toxic agents from tobacco smoke, while yet another group inhibits the proliferation of cancerous cells. The relative importance of specific health knowledge and taste on acceptance of brassica vegetables (green broccoli, red and green cabbages, broccolini, cauliflower, and Brussels sprouts) is reported by Cox *et al.* (2012).

Since glucosinolates are highly labile, their health benefits depend on many variables related to intake and metabolism and influenced by a range of factors which affect their concentrations following their harvesting, storing and processing. In some crops, such as red and white cabbage (*B. oleracea* var. *capitata*), concentrations of glucosinolates remain stable for several months after harvest. Where postharvest processing causes physical damage this is likely to lead to the activation of myrosinase and the release of glucosinolate breakdown products. Thus, the use of brassicas in chopped, ready-to-use salads containing cabbage or green broccoli increases the concentration of anticarcinogenic agents. Where the crops are subjected to heat treatment, as in the blanching of chopped green broccoli, there will be reductions in the concentrations of glucosinolates. There is evidence that fermentation, as in the production of sauerkraut or kimchi, causes a complete loss of glucosinolates. These putative health promoting effects of brassica vegetables have substantial implications for producers, retailers and consumers. Breeding strategies are being developed that aim to produce new 'healthier' cultivars utilizing the abundant genetic variation present in *B. oleracea* (Johnson, 2000). The glucoraphanin content in green broccoli seedlings decreases before first foliage leaf emergence (Kim *et al.*, 2020). Treatment with methyl jasmonate significantly increased the glucosinolate content in green broccoli, as did magnesium sulfate applications.

Kale (*B. oleracea* ssp. *acephala*) has become popular recently for its reputed anti-inflammatory, antigenotoxic, gastro-protective activities, inhibition of carcinogenic compounds and positive effects on gut microbe populations (Satheesh *et al.*, 2020). Curly leaved kale is a robust, cold-tolerant plant with a high content of health-promoting compounds that grows at a range of latitudes (Steindal *et al.*, 2015) (see Fig. 8.4).

Cold acclimatization increases the soluble sugar content improving taste, although unsaturated fats and glucosinolates may decrease. Curly kale stored at 1°C for 3 and 6 weeks was compared with plants remaining in the field (Hagen *et al.*, 2009). In plants left in the field for 6 weeks, flavonols, total phenols and antioxidant capacity were reduced by 25–35% and the vitamin C content by more than 50%, whereas soluble sugars and dry matter increased by roughly 20% and 30%. Antioxidant capacity was positively correlated with the level of total phenols ($r = 0.73$, $P < 0.001$) and total flavonols ($r = 0.70$, $P < 0.001$). Black broccoli (*B. oleracea* var. *italica*) is a particular landrace, grown on the slopes of Mount Etna in Sicily (Terzo *et al.*, 2020), which appears particularly capable of modifying gut microbiota.

Fig. 8.4. Curly leaved kale crop. (Geoff Dixon).

Kale has become known as a 'superfood' in the USA (Samec *et al.*, 2019) due to its association with health benefits, mainly because of polyphenol, glucosinolate and carotenoid content. As in many countries, childhood obesity continues increasing as a critical health concern in the USA (DeJesus and Venkatesh, 2020). Interventions that focus on delivering verbal lessons about food and health to children in preschool classrooms have had only modest effects. There are associations between children's food intake and pickiness that provide growing evidence of alignment between parent assessments of their children's typical eating behaviour and children's food choices. Campaigns such as 'Love Your Greens' aim at increasing public awareness of the health and nutritional value of brassica vegetables (see Fig. 8.5).

TRANSFERRING HIGH QUALITY FROM FIELD TO PLATE

Shelf life

Development, pre-maturation, maturation, ripening and senescence are the five phases that vegetables and fruit pass through during production, harvesting, storage and into the distribution chain to the ultimate retail consumer. Saleable brassicas do not ripen or senesce. Indeed, if they do, then they become unmarketable and worthless. Development of brassicas, in terms of product quality, begins with the initiation and subsequent growth of the edible portion. This may be the flowering curds and spears of cauliflower and green broccoli, the buds of Brussels sprouts, the leaves of cabbage, including Chinese cabbage,

Fig. 8.5. The 'Love Your Greens' campaign, raising awareness of the health and nutritional value of brassica vegetables. (Courtesy of British Growers Ltd).

or the roots of swede (*B. napus* ssp. *rapifera*), turnip (*B. rapa* ssp. *rapifera*) and radish (*R. sativus*). Development ends when there is a change in growth pattern or when natural enlargement ceases (see Chapter 4). Maturation may be interpreted as the final development of size and quality in the field and is the point where commercial crops are harvested. The maturation period for brassica crops is short, providing only a brief period when the highest quality is available for harvest.

Green broccoli (*B. oleracea* var. *italica*) is a perishable brassica vegetable with a limited shelf life (Hagen and Larsen, 2020). Spears may be washed immediately after harvesting in order to reduce field heat and ensure cleanliness (see Fig. 8.6).

Studies demonstrate the importance of packaging that prevents weight loss and retains firmness in green broccoli, but this does not prevent yellowing or decay development during the period at room temperature. Hence, shelf life, is prolonged by cold-chain handling in the distribution chain and in retail display. Typically, green broccoli will arrive at a retail store within 7–14 days of harvest and is kept chilled until purchased (Pellegrino *et al.*, 2019). Quality degrades during storage, reducing the attractiveness of green broccoli, regardless of the holding temperature and cultivar. The actual quality and storability of a single green broccoli curd after harvest remains uncertain before changes are noticed by the human eye (Kabakeris *et al.*, 2015a).

Controlled atmosphere storage, using low O_2 and high CO_2 concentrations, effectively extends the shelf life of brassicas. This technique retained the freshness of green broccoli heads by depressing the induction of alternative oxidase enzymes (Makino *et al.*, 2020). Green broccoli heads usually aged by yellowing or producing an off-flavour after harvesting, and during storage

Fig. 8.6. Washing freshly harvested green broccoli (*Brassica oleracea* var. *italica*) spears. (Source unknown).

under improper conditions (Wang *et al.*, 2019c). Controlled O_2 and CO_2 atmospheres increased the energy production efficiency, improved stress resistance and markedly extended the shelf life of green broccoli heads (Guo *et al.*, 2013). Excessively high CO_2 proportions, however, promoted glycolysis and fermentation, reduced energy generation efficiency, altered sulfur-component metabolism and promoted the accumulation of an offensive odour in green broccoli heads. Developing a blend of sulfonated polyether ether ketone with polyvinylidene fluoride produced films suitable as a MAP for green broccoli (He and Xiao, 2018). The use of MAP also reduced loss of green colour and maintained texture and chemical composition of green broccoli. Quality is further maintained by using 1-methylcyclopropene, which inhibits ethylene formation and extends storage quality for green broccoli. Predicting storability requires early indicators of postharvest changes in chlorophyll and its derivatives in the florets of green broccoli curds (Kabakeris *et al.*, 2015b). Green broccoli harvested throughout the year at four planting sites in Germany showed different hue angle values ranging from 96° to 114°, resulting in different periods of satisfactory storage. Green broccoli shelf life is reduced by physicochemical losses and microbiological degradation (Ben-Fadhel *et al.*, 2018). Combining pre-treatment with calcium and alginate coating reduced the weight loss and respiration rate of green broccoli, maintained its colour and texture, and extended its shelf life by 6 days, limiting postharvest losses.

Several packaging strategies have been tested for fresh-cut broccoli raab (*B. rapa* var. *ruvo*). These have included combinations of polypropylene/polyethy lene terephthalate and microperforated polypropylene/polyamide in modified atmospheres and using CO_2 absorbing sachets, which reached an equilibrium condition very close to optimal for broccoli raab (5% O_2 and 5% CO_2). Broccoli raab was stored for 8 days showing negligible reduction of appearance and odour scores, and product was comparable to the fresh samples (Cefola *et al.*, 2016). The influence of passive MAP using perforated polyethylene packages and three different types of O_2 transmission rate active packaging was tested on the quality maintenance of fresh Chinese kale (*B. oleracea* var. *alboglabra*) stored at 5°C (Boonyakiat and Boonprasom, 2012). Active packaging resulted in longer storage life compared to passive polypropylene packaging. The visual quality of Chinese kale and caisin (choy sum) (*B. rapa* var. *parachinensis*) was best maintained at 0°C, at which shelf life greater than 21 days was achieved (Zong *et al.*, 1998). When stored at 5°C or 10°C, marketable quality was only maintained for 7–14 and 3–4 days, respectively. Exposure to ethylene (1 ppm) at 5°C reduced the shelf life of both products by approximately 30%. Several antimicrobial agents, organic salicylic acid and inorganic nano-sized titanium dioxide (TiO_2) were added into polylactic acid producing biodegradable packaging films (Yin and Huang, 2020). These antimicrobial films reduced the respiration rate of Chinese rape (*B. napus*) and extended the shelf life. These antimicrobial films could be suitable for the preservation of other brassica vegetables. Fresh-cut swede and turnip could be stored at temperatures below 0°C (Helland *et al.*, 2016). Storage in antimicrobial acid film resulted in higher weight loss than the biaxially oriented polypropylene film for both vegetables.

Several essential oils active against pathogens were tested using a disc vola-tilization method (Hyun *et al.*, 2015). Combining essential oils in vapour phase and MAP (100% CO_2) using lemongrass-2 oil inactivated bacterial pathogens. Combining MAP with a high-voltage electrostatic field prolonged the shelf life of fresh-cut cabbage by up to 60 days (Huang *et al.*, 2021). Supercooling during preservation has the potential for increasing the shelf life of brassicas at sub-zero temperatures without freezing. The survival ratio of cut cabbage (*B. oleracea* var. *capitata*) increased when subjected to sub-zero temperatures (−5°C or −10°C) by use of ethanol brine for 12 h (Koide *et al.*, 2019). Survival percentages of supercooled samples were 100% and over 30% for −5°C and −10°C, respectively. The need for cool storage environments is exemplified by a report from the Philippines where cabbage is the leading leafy vegetable (Valida *et al.*, 2018). Yellowing and wilting of outer leaves limited the retail shelf life, and at 21–25°C, bacterial soft rot caused further quality deterioration. Similar problems are reported with kale (*B. oleracea* ssp. *acephala*) where consumption in fresh-cut salads is increasing (Albornoz and Cantwell, 2016). Leaves may be harvested at different stages of maturity and the resultant heterogeneity reduces fresh-cut salad quality and shelf life. Loss of composition and visual quality increased with rising temperatures, days of storage and leaf maturity. Increasing the concentration of abscisic acid (ABA) (or its partial agonist pyrabactin) from the start of postharvest suppresses senescence of stored leaves, changes the transcriptional regulation of glucosinolate metabolism and down-regulates biotic stress defence mechanisms (Miret *et al.*, 2018). It is suggested that manipulating ABA signalling could offer a route for improving postharvest quality of leafy brassicas stored at ambient temperatures.

Recently introduced small, conically headed cabbage cultivars (*B. oleracea* var. *capitata* f. *alba*) have gained considerable market favour since they are suitable for single persons or limited family units (Gajewski *et al.*, 2015). Controlled atmosphere storage produced a higher percentage of marketable product, with less mass loss of the heads and better quality, including higher vitamin C content compared with storage at ambient atmospheres.

Genetic variation

The great variability inherent in *Brassica* species has led to the enormous spectrum of crop types (see Chapter 1 section, Origins and Diversity of *Brassica* Crops). Such high variability has ultimately led to a spread of maturity periods within crops that has caused much difficulty for growers in achieving efficient harvesting of high-quality produce. The traditional means of overcoming variation in the maturity of brassica crops was by the use of multiple harvests in which there was picking-over by hand with the subjective visual assessment of quality and maturity. This technique (see Fig. 8.7) is very labour intensive, laborious and inefficient and makes scheduling for marketing extremely difficult. Increasing development of automated robotic harvesting demands crops which mature uniformly.

Fig. 8.7. Hand harvesting pak choi (*Brassica rapa* var. *chinensis*) in Canada. (Geoff Dixon).

Minor changes in the weather, giving a few hot days or a more prolonged cold period, will disrupt the most carefully prepared harvesting schedules. The development of F$_1$ hybrid brassica crops offered a considerable advance in their uniformity and reliability of maturation and allowed the development of fully mechanized harvesting for Brussels sprouts (*B. oleracea* var. *gemmifera*) and cabbage (see Chapter 5 section, New Technologies), and the use of gantry systems for crops such as cauliflower (*B. oleracea* var. *botrytis*). Fully automated harvesting of fragile crops, such as cauliflower, is now in the early phases of development.

The move towards raising of transplants in modular containers by specialist propagators has added considerably to the improvements in the uniformity of maturation gained from genetic improvement. Propagation (see Chapter 3 section, Transplanting) has been separated from crop growing and the growing of seedlings prior to transplanting into the field is carefully controlled and regulated by specialists. Standardized transplants are spaced mechanically into their field stations with accuracy and consistency. Provided soil fertility has been carefully adjusted to suit brassica crops, then the subsequent transplants will establish quickly and uniformly. Both the breeding of F$_1$ hybrids and the production of modularized transplants have been major advances, reducing the variation of brassica crops in their maturity stage and increasing their quality postharvest.

Soil fertility, structure and moisture content

Over the last two decades, the importance of soil conservation and improving its quality, health and fertility have become driving forces for sustainable commercial crop production. Maintaining soil fertility and structure are cardinal factors in achieving high brassica quality within the field and preventing its subsequent deterioration after harvesting. Provided soil fertility is adjusted, which removes deficiencies in major- and micronutrients, then the availability of sufficient but not excessive nitrogen during brassica growth and development is of paramount significance in achieving brassicas with consistent, even quality and freedom from pathogens and pests (see Chapter 5 section, Soil Health and Quality). Luxury consumption of nitrogen, or alternatively shortage of nutrients, both contribute greatly towards the loss of brassica quality and pest and disease susceptibility (see Table 8.2).

There are no immediate solutions for the presence of excess nitrogen in the root zone, but incipient nitrogen deficiencies can be corrected before visible stress symptoms develop. Diagnosis of incipient nutrient stress is achieved by regular crop monitoring using foliar analyses. Traditionally, this is achieved by laboratory analysis of leaves and sap where several days may elapse between collecting the samples and the availability of results. Field test kits for such analysis are steadily coming on to the market. Automated and non-invasive systems for the measurement of incipient stress in crops

Table 8.2. The effect of increasing nitrogen fertilizer applications on quality characteristics of Brussels sprouts. (After Babik *et al.*, 1996).

Nitrogen rate (kg/ha)	Dry matter (%)	Total sugars (%)	Vitamin C (mg %)	Chlorophyll (mg/l)
100 + 0[a]	18.4a	6.0b	148d	5.0
200 + 0	18.7a	6.4a	170a	5.6b
100 + 100	17.7b	6.0b	157b	5.8b
400 + 0	17.1b	6.0b	154c	5.9b
300 + 100	17.1b	5.8c	139e	6.6a
600 + 0	17.0b	5.4d	137e	6.9a
500 + 100	16.9b	5.2d	134f	7.1a

Data followed by different letters are significantly different using the Newman–Keuls test at $\alpha = 0.05$.
[a]Applications split between pre- and post-planting.

are also now readily available. Mostly, these utilize changes to chlorophyll fluorescence as the criterion (Wydrzynski *et al.*, 1995). This allows remedial action to be taken before obvious stress takes place. Once brassica crops show visual yellowing, stress damage has become irreversible and losses in quality and maturity follow. Top dressings with highly soluble and readily available fertilizers, such as calcium nitrate or calcium cyanamide, may prevent nutrient stress-induced variability from developing and maintain even growth towards maturity, thereby retaining the harvesting quality and scheduling. Increasing nitrogen-based dressings to Brussels sprouts (*B. oleracea* var. *gemmifera*) beyond 400 kg/ha reduced bud sweetness, vitamin C content and dry matter content (Babik *et al.*, 1996). Initially, nitrogen fertilizer use in Brussels sprouts was designed to boost and increase yields, especially for processing and freezing crops. However, it resulted in increased acidity and reduced tastefulness, which had adverse effects on retail consumer acceptability. For organically grown brassicas, nitrogen-fixing green manure provides a source of nutrient supply (Koller *et al.*, 2008). Tests showed that at harvest, 15% of the nitrogen contained in the borage (*Phacelia*) biomass was present in the cabbage plants. Banded soil steaming is a valuable means of weed control and soil structural conservation for organic brassica crops but consumes significant energy (Melander and Kristensen, 2011). Soil temperatures of 80°C result in satisfactory weed control under moist soil conditions, especially if soils are cultivated first. Frequently, brassica residues are left in the field after harvesting and may cause nitrous oxide and ammonia emissions because of their high nutrient and water content (Nett *et al.*, 2016). Ploughing residues into the soil apparently produces the highest nitrous oxide emissions in the range of soils tested while surface decomposition primarily increases ammonia emissions in coarse-textured soils.

The effects of soil compaction and loosening on soil physical conditions and crop yield are explained by Fisher *et al.* (1975) and Gooderham (1977). Compaction of soil increased the mechanical resistance encountered by penetrating roots. Wet-ploughing Kentish silt loams increased bulk density by 4% and mechanical resistance by 10% to depths of 8 cm. Yields of kale were reduced by 7–17%. Cultivation of soil in wet conditions causes deterioration of soil structure and reduces crop yields. Compaction could be alleviated by using single-tined subsoilers. Significant differences in soil structure and physical soil properties were detected in Australian red earth (Oxic Paleustalf) soils (Chan and Heenan, 1996). Crops can modify soil biological and chemical properties, such as microbial biomass, dissolved carbon and soluble cation concentrations. Rotations improve water stability and soil strength but the effects are transient.

Moisture influences both nutrient uptake and the innate qualities of the brassica products after harvest. Moisture stress increases the thickness of cell walls, relative dry weight, and deposition of lignin, suberin and cellulose, resulting in a fibrous or woody texture. The development of fibrous sclerenchyma deposits is unacceptable to retail consumers of fresh vegetables, especially those purchased for their leaves, flowers or root and hypocotyl organs, as is the case with most brassica crops. Quality decreases following erratic irrigation regimes that fail to sustain continuous development and impose stresses at all crop growth stages, and this effect becomes critical as maturity approaches.

Temperature

Temperature is not normally controlled in open field crops. But the advent of plastic film covers, nets and mulching will allow modest manipulation of temperature. Applying crop covers in spring increases soil and air temperature and retains heat in the autumn and early winter. These techniques will allow growers to produce four or more crops of green broccoli (*B. oleracea* var. *italica*) on a single area in 1 year with substantial reductions in fixed costs. The planting date is advanced and growth accelerated, leading to earlier maturity and possibly higher market prices (see Chapter 5 section, Transplanting). The timing of covering removal becomes critical in late spring or early summer to avoid overheating in crops, such as green broccoli, leafy and hearted cabbage and early turnip production. Low or freezing field temperatures cause damage and reduce quality, especially when crops are approaching maturity. Rapid changes in temperature alter growth rates and induce stress with resultant losses in quality, and applying late season covers helps avoid these hazards. Head formation of green broccoli is greatly reduced under high temperatures (22–27°C). Green broccoli inbred lines, capable of producing heads at high temperatures in summer, are unique to Taiwan (Lin *et al.*, 2019). The expression pattern of high-temperature-associated signalling genes suggests they become involved in stress defence instead of transitioning to the reproductive phase in response to heat stress. Transcriptome profiling of high temperature (HT) and high stress (HS) green broccoli helps in understanding the molecular mechanisms underlying head-forming capacity and in promoting functional marker-assisted breeding. Seasonal variations occur in the antioxidant metabolism of komatsuna (*Brassica rapa* var. *perviridis*) leaves harvested in summer and winter seasons. The level of hydrogen peroxide, that prevents oxidase reactions, was significantly higher in the summer leaves (Imahori *et al.*, 2016). It is suggested that the antioxidant metabolism of komatsuna leaves responds differently to the high summer temperature and low winter temperature stresses. Plants are exposed to biotic disorders caused by pathogen attack and complex abiotic stresses including heat and drought resulting from climate changes (Shin *et al.*, 2019). *Bacillus aryabhattai* H26-2 and *Bacillus siamensis* H30-3 could offer opportunities for plant growth promotion and mitigation of heat and drought stresses in Chinese cabbage. Foliar application of zinc may mitigate the adverse effects of heat stress on pak choi (*B. rapa* var. *chinensis*) plants (Han *et al.*, 2020) by maintaining superoxide dismutase activity and membrane stability and protecting photosynthesis against heat damage.

A wild species, *Brassica fruticulosa*, found in south-east Spain, is adapted to both heat and high light intensity in its natural habitat, while cultivated *B. oleracea* has lost this attribute (Díaz *et al.*, 2007). It is suggested that the chloro-respiration processes are involved in adaptation to heat and high illumination in wild brassica. As global warming becomes more apparent, there is a need for the development of high temperature and high humidity-tolerant cultivars (Song *et al.*, 2020). Suitable genes are being identified using molecular markers which explain the genetic basis of tolerances in at least 80% of the cabbage genotypes tested. Long non-coding RNAs mediate epigenetic regulation of heat tolerance and highlight a set of candidate genes in non-heading Chinese cabbage *B. rapa* (Wang *et al.*, 2019a).

Integrated crop management

There are still valuable fungicides, herbicides and insecticides used in the production of vegetable brassicas worldwide. Increasingly, however, their use as a principal husbandry strategy is being reduced, and discontinued completely in many cases. They are being replaced by integrated crop management (ICM) strategies which often include the use of biological control agents, biopesticides, biostimulant formulations or complete switches into organic cultural methodology (see also the Introduction sections to Chapters 5, 6 and 7). Retail consumers are encouraging these changes along with the major vendors – the supermarkets – using labels such as 'pesticide-free' as marketing and promotional tools. Quite probably, forms of sustainable and environmentally friendly practices will have beneficial effects on soil microbial communities (Liao *et al.*,

2019) with resultant improvements of soil quality and microbial community composition.

The husbandry processes of soil cultivation, addition of fertilizers, transplanting, application of agrochemicals, irrigation and harvesting are now regarded as an integrated continuum. Collectively, they are aimed at gaining the highest possible quality in the product compatible with minimum use of resources, and are summarized as ICM (this is referred to in several chapters, notably the Introductions to Chapters 6 and 7). The formulation of standardized husbandry protocols well in advance of crop production and their review in the light of performance in preceding seasons has become known as 'smart scheduling'. This approach to crop husbandry has reached a high level of sophistication with all vegetable crops and is now often regulated by the wholesale purchasing system through the supermarkets, which collectively have established their own quality control systems as part of the purchasing contract. Red Tractor Assurance in the UK and the more widely applied GLOBALG.A.P. scheme offer standardized protocols for crop production that are acceptable to the supermarket buyers and should provide food safety and security for the consumer. As an example of changes in methodology, thermotherapy has significant potential to control seed-borne pathogens in seeds produced in an organic system (Soriano *et al.*, 2019). The aim of this research was to study the effect of thermotherapy on green broccoli (*B. oleracea* var. *italica*) seed germination. These authors concluded that thermotherapy of green broccoli seeds at 55°C should not be performed for over 20 min, if the initial seed vigour is low. However, 60 min can be used for seed lots which are more vigorous, especially when the storage of seeds is not necessary.

The first edition of *Vegetable Brassicas and Related Crucifers* contained detailed references to the use of agrochemicals. This second edition does not because use of these materials, while still essential in many instances, is being reduced significantly.

Retaining freshness

Mechanization of harvesting becomes an urgent task nowhere more importantly than with Chinese cabbage (*B. rapa* var. *pekinensis*), which is a major part of the diet in Asia. The total output of Chinese cabbage in China, for example, is the biggest worldwide and requires substantial manual labour. Mechanized harvesting is becoming urgent (Zhang *et al.*, 2020) because rural manual labour supplies are diminishing. Measuring the physical and mechanical properties of Chinese cabbage produced a mathematical curve of the cutting force needed for the cabbage stem and the displacement of the cabbage head, and provided a theoretical basis for designing a cabbage harvester.

Freshness is a very desirable attribute for all brassicas, but especially leafy greens. Loss of freshness may occur due to postharvest factors such as time, temperature abuse and mechanical injury (Cantwell *et al.*, 2019). At low temperatures, loss of quality is due both to colour change from green to yellow and the onset of rotting induced by microorganisms. Practices that minimize the accumulation of ethylene in the storage environment or inhibit endogenous ethylene development can be effective in extending storage life and quality. Green broccoli (*B. oleracea* var. *italica*) is a crop where the product must be moved to the consumer as rapidly as possible, whereas head cabbage offers opportunities for storage for periods of up to several months. Cabbage in storage is affected by a range of physiological disorders and fungal pathogen problems, especially infection by grey mould (*Botrytis cinerea*). Prevention of grey mould was most effective at low temperatures as compared with controlled (modified) atmosphere conditions. Genetic resistance to grey mould would be a very valuable attribute and it is suggested that this may be a character linked with resistance to white mould (*Sclerotinia sclerotiorum*).

Check-all-that-apply (CATA) analysis reveals the qualities governing consumer acceptance of green broccoli, such as taste, colour and flavour such as grassy, musty and 'dirt-like'. This information helps determine shelf-life esti-mation and the physiological processes involved with deterioration (Pellegrino *et al.*, 2019). Senescence commences as the physiological changes in flavour, composition and structure are initiated by the cessation of growth at harvest-ing. The harvested organ ceases receiving supplies of energy as nutrients, water from the roots and photo-assimilates from the younger leaves. The term 'deterioration' includes senescence plus the effects of pests, pathogens, disorders and mechanical damage before and after harvest. Deterioration in the brassica crops can take on many forms but is most widely characterized by chlorosis or yellowing and eventually necrosis, resulting from a breakdown of chlorophyll pigments within the tissues.

The edible portions of the crop, while attached to the growing plant, derive their constituents of quality in relation to its rates of nutrient uptake, photosynthesis, respiration, transpiration and other metabolic processes. Once harvested each portion becomes an independent entity where quality is controlled by its own rates of respiration and transpiration. Excessive tran-spiration is the greatest source of postharvest damage to quality. All brassica produce needs to be maintained in a turgid state, but free surface water should be absent since that provides conditions for the multiplication of bacteria and fungi that can rapidly cause degradation. At harvest, the water source of the crop product is severed. Respiration continues through the stomata and cut surfaces of detached organs.

The rate of water loss rises with increasing temperature and is exacer-bated by decreasing relative humidity (RH) and atmospheric pressure. Loss rates are highest where there is a large area available for transpiration relative to unit weight, hence leafy crops lose water to the atmosphere more rapidly than densely packed produce such as cabbage heads and turnip (*B. rapa* ssp. *rapifera*) or swede (*B. napus* ssp. *rapifera*) 'roots'. For most vegetables a water loss of between 5% and 10% causes visual deterioration.

After harvest, changes to carbohydrate, organic acid and secondary metabolite content takes place and each of these will affect product quality. Probably, the release of ethylene from tissues has the most critical effect on quality. Ethylene is involved in the acceleration of ripening and consequent senescence of tissues. For leafy or flower crops (cabbage, Chinese cabbage, cauliflower and green broccoli), the presence of even small amounts of ethylene will accelerate senescence and increase deterioration. Respiration rates of vegetables vary widely and increase markedly in response to rising temperatures. High rates of respiration are characteristic of young and immature tissues and developing flower tissues, such as leaves or cauliflower (*B. oleracea* var. *botrytis*) heads, preventing their storage for more than a few days. The rates of heat loss in brassica vegetables due to respiration are illustrated in Table 8.3.

Some crops, such as cabbage heads, may be stored for 6–8 months provided the temperature is reduced effectively. Use is made of controlled atmosphere storage with increased concentrations of CO_2 and reduced O_2 levels minimizing postharvest deterioration. Uncontrolled accumulation of CO_2 concentrations and rising temperatures accelerate respiration in storage or transit leading to lost quality. Detailed protocols for storage of brassica crops are given by Thompson (1998).

Physiological changes preceding and accompanying visual deterioration of harvested green broccoli floral tissue are discussed by Downs *et al.* (1997). There is a loss of sucrose in the head following harvesting – over the first 6 h sucrose content declines by 50%. Treatment of florets with 6-benzylaminopurine (6-BA) delayed the associated large increases in the amides, asparagine and glutamine that usually accompany sucrose loss in the first 48 h after harvesting. Also, the decline in concentrations of amino

Table 8.3. Examples of heat loss caused by respiration in brassica vegetables postharvest. (After Ryall and Lipton, 1972).

Crop type	Heat loss (BTU/t/day)		
	0°C	10°C	20°C
Low rates			
Cabbage	n/a	4000	9500
Turnip	1300	4300	7000
Moderate rates			
Brussels sprouts	5300	18,600	n/a
Cauliflower	n/a	7400	17,600
High rates			
Green broccoli	5800	20,300	n/a
Watercress	5800	20,300	n/a

n/a = not available.
BTU = British thermal unit.

acids and soluble proteins and increase in ammonia normally associated with postharvest deterioration were delayed by application of 6-BA. Breakdown of chlorophyll in harvested green broccoli florets and the subsequent accumulation of degradation products are discussed by Yamauchi *et al.* (1997).

Floret maturity has the greatest effect on the rate of yellowing in green broccoli and Tian *et al.* (1995) showed that differing patterns of ethylene production, ethylene sensitivity, respiratory activity and chlorophyll loss occurred in florets of varying maturity. Floret maturity could be assessed by an examination of the developmental stage of the pollen. Dipping florets of cv. Shogun in 6-BA stimulated ethylene production, depressed respiration rate and delayed floret yellowing. The effect of this cytokinin analogue on green broccoli was related to the concentration used and the postharvest age of the floret at the time of treatment. Application of a hormetic dose of ultraviolet C (UV-C) can be beneficial in maintaining not only the quality of green broccoli florets, but also in enhancing the glucosinolates, phenolic acids and their precursor amino acids during the low-temperature storage at 4°C and 90–95% RH (Duarte-Sierra *et al.*, 2019).

Fresh green broccoli is harvested when the flowering heads (spears or sprouts) are immature and hence it is a highly perishable product, with a storage life of 2–3 days at 20°C and 3–4 weeks at 0°C (Makhlouf *et al.*, 1989). The major limitation in storage at ambient temperature is rapid yellowing of the flower buds due to chlorophyll breakdown, ethylene production and subsequent flower opening. Yellowing was further delayed where green broccoli was treated at 20°C for 6 h with 1 μl/l of 1-MCP at the same temperature. This doubled storage life where the atmosphere contained 0.1 μl/l ethylene. Storage life was extended even further when the treatment temperature was retained at 20°C and storage temperature reduced to 5°C. Treatment at a high temperature followed by storage at a lesser temperature provided the most satisfactory combination (Ku and Wills, 1999). Measurement of chlorophyll fluorescence offers a technique for the assessment of quality in green broccoli florets under commercial conditions (Toivonen and DeEll, 1998).

Ammonia, a product of protein catabolism associated with senescence of leafy greens, is toxic to plant cells and accumulates during postharvest handling. Ammonia accumulation could be a useful freshness indicator. Ammonia can be determined in the leafy tissues or as a volatile in the packaged product (Cantwell *et al.*, 2019). Increasing the concentration of ABA (or its partial agonist pyrabactin) from the start of postharvest suppressed senescence of stored leaves, changed the transcriptional regulation of glucosinolates metabolism and down-regulated biotic stress defence mechanisms. These results suggest a potential for manipulating ABA signalling for improving postharvest quality of leafy brassicas stored at ambient temperature. For example, in green broccoli, forms of glucosinolate were being retained by suitable (low O_2, high CO_2) gas conditions (Schouten *et al.*, 2008).

Green broccoli is one of the most vulnerable and perishable crops. Predicting postharvest storability can be related to water availability in the last 2 weeks before harvest ($R^2 = 0.71$) (Kabakeris, 2018). Integration of climate and cultivation events with reflectance gauged using hand-held devices during storage, provided predictions and ratings of green broccoli storability. Postharvest ethylene plays an important role in green broccoli floret yellowing and senescence (Wang *et al.*, 2014). Water stress and nutrient deficiency during green broccoli storage affected ethylene production rate by increasing the transcription of gene *BoACO3*. Key enzymes involved included 1-aminocyclopropane-1-carboxylic acid (ACC), malonyl ACC and ACC oxidase. Ethylene encourages the yellowing of florets, which reduces their economic and nutritional value (Aghdam and Luo, 2021). The storage temperature of green broccoli and pak choi (*B. rapa* var. *chinensis*) at 10°C with ethylene reduced to 0.001 μl/l (Li *et al.*, 2017) increased storage life to 14 days and economized on energy requirements. The synergistic effect of ultrasonic treatment and MAP results in effective reduction of peroxidase and polyphenol oxidase activities in pak choi (Zhang *et al.*, 2019) and offers the best method for preserving this brassica for up to 30 days.

Exogenously applying phytosulfokine-α (PSK-α) at 150 nM delayed senescence in florets stored at 4°C for 28 days. This bioactive peptide might provide a means for delaying the senescence of green broccoli florets during cold storage. A 5 mM dip in amino acids such as L-arginine, L-cysteine and L-methionine at 10°C was optimal in delaying senescence. There was increased retention of green colour, vitamin C and antioxidant activity, and less ethylene production, a lower respiration rate, weight loss, phenylalanine ammonia lyase activity and ion leakage, with the benefits being similar for all three amino acids (Sohail *et al.*, 2021). Applications of 1-MCP delayed senescence of green broccoli florets by maintaining higher sugar content (Xu *et al.*, 2016). Irradiation with light-emitting diodes (LEDs) delayed senescence and maintained the quality of pak choi by enhancing the activity and relative gene expression level of antioxidant enzymes and regulating chlorophyll metabolism (Zhou *et al.*, 2020). Packaging green broccoli spears, especially in bags with 1-MCP treatment, delayed chlorophyll degradation and colour changes in green broccoli. In studies by Kasim *et al.* (2007), the hue angle of green broccoli colour and chlorophyll loss were correlated.

Kale (*B. oleracea* ssp. *acephala*) is a very nutritious leafy brassica. Leaves are harvested at different stages of maturity, resulting in a heterogeneity that may be detrimental to fresh-cut salad quality and shelf life (Albornoz and Cantwell, 2016). The visual quality (yellowing, decay, cut-end browning) of fresh-cut kale leaves deteriorated with rising temperatures, length of storage and leaf maturity. The link between leaf structure modifications and nutrient remobilization in oilseed rape (*B. napus*) is discussed by Sorin *et al.* (2016). They proposed that leaf structure monitoring during senescence through nuclear magnetic resonance testing could be valuable in selecting genotypes with high nutrient use efficiencies.

The synergies of MAP and a high-voltage electrostatic field could prolong the shelf life of fresh-cut cabbage by up to 60 days (Huang *et al.*, 2021). LED irradiation is effective for enriching the chlorophyll, vitamin C and polyphenol contents of cabbage stored at a low temperature but LED colours have different effects (Lee *et al.*, 2014). The lightness of minimally processed cabbage decreased linearly from 70.94 ± 6 to $63.8 \pm 8.5–61.3 \pm 8$ units for the chemical treatments during 22 days of storage at 0°C. Hue angle values during storage time were also significantly influenced by ascorbic acid, citric acid and calcium chloride treatments, mainly at 0°C (Manolopoulou and Varzakas, 2011).

Caisin (choy sum) (*B. rapa* var. *parachinensis*) is a dark green leafy vegetable containing high folate (vitamin B9) levels comparable to spinach. Folate is essential for the maintenance of human health and is obtained solely through diet (O'Hare *et al.*, 2012). It is possible to store choy sum for up to 3 weeks at 4°C without significantly affecting the total folate concentration of the edible portion. Red LED illumination for 60 min/day was most effective at suppressing shoot elongation and reducing the loss of photosynthetic ability in cabbage seedlings during low-temperature storage (Sato and Okada, 2014).

PHYSIOLOGICAL DISORDERS

Disorders of brassicas frequently originate from interactions between environmental conditions and forms of nutrient deficiency or excess. Several of these conditions have been associated with a limitation of the availability of calcium within the tissues of rapidly growing organs that form the components of yield in brassicas. Calcium is an immobile element, recognized as mainly being translocated from the roots through the xylem by mass flow. Mass flow results from transpiration, root pressure and diurnal change in water stress. There is some evidence that calcium ions in the vessels do not move primarily by mass flow but by exchange reactions along negatively charged sites on the walls of the xylem vessels. Once, however, calcium has reached its destination in the plant there is little or no subsequent redistribution. Most of the water flux is channelled to leaves exposed to the sun by the transpiration stream that contributes to the cooling of the plant. Increasingly, an understanding of the physiological roles that calcium has for both plants and animals as a secondary signalling substance is emerging. In plants, next to the secondary messengers, lies an array of signal relaying molecules among which calmodulins convey the unequivocal alarms of calcium influxes to calmodulin-binding transcription activators (CAMTAs) (Noman *et al.*, 2021). Upon reception, CAMTA transcription factors decode the calcium signatures by transcribing the genes corresponding to the specific stimulus, thus having direct/indirect engagement in the complex signalling crosstalk. Understanding of how calcium signatures are coded, decoded and translated identifies the multiple-stress-responsive nature of CAMTAs and potentially points towards means by which plant breeders may develop cultivars with increased tolerances towards environmentally induced disorders.

The complexity of causal factors resulting in the expressing of physiological disorders can be glimpsed by considering how developing tissues fed by the phloem are often disadvantaged by competition from transpiring tissues. Competition between sinks, such as buds and developing leaves, and flowering organs, such as green broccoli (*B. oleracea* var. *italica*) spears or cauliflower (*B. oleracea* var. *botrytis*) florets, is high when calcium in the xylem is low and transpiration is high. It is at this point that green broccoli florets, cabbage heads or cauliflower curds are very susceptible to calcium deficiencies (see Fig. 8.8).

Furthermore, the availability of excessive potassium and magnesium is likely to aggravate calcium deficiencies within the plant. Add to this

Fig. 8.8. Calcium deficiency in cauliflower (*Brassica oleracea* var. *botrytis*). (Courtesy of Tom Decamp, YARA).

consideration of the effects of periods of excessively high and low tempera-tures or their fluctuations, and it begins to provide a picture of interactions resulting in physiological/environmental disorders.

These conditions may develop after harvesting and during storage, resulting in substantial loss of quality. Cabbage (*B. oleracea* var. *capitata*), for example, is an important source of dietary antioxidants. Studies of white cabbage during 6 months of commercial storage (Hounsome *et al.*, 2009) revealed changes during long-term storage resulting from processes, posthar-vest senescence, biennial cycling and response pathogens, and physiological disorders. Storage losses of brassica vegetables, such as winter white cabbage and Chinese cabbage, are regularly reported at 10% or higher. Brassicas suffer from a range of physiological disorders that are major causes of these losses after harvesting. Physiological disorders have been attributed to a range of nutrient deficiencies interacting with environmental or climatic conditions and, more recently, virus pathogens. It appears that under specified conditions of nutrient stress the viruses enhance the development of tissue chlorosis and necrosis.

Plant breeding is a major avenue for developing tolerance to physiological disorders. The complexity and long-term nature of this process is illustrated by Li *et al.* (2013). Studies of cauliflower lines showed that six characters were affected by additive and non-additive effects. Maturity, bracts ratio, head height and leaf coverage ratio, significantly affected by an additive gene, were mainly inherited by a general combining ability; head diameter and head weight were mainly inherited by a specific combining ability and were strongly controlled by non-additive genes.

Crop nutrition can be used by growers as a means for reducing the onset and impact of physiological disorders. Excessive applications of synthetic nitrogen may exacerbate these disorders. Cultivation of legume cover crops, such as white clover (*Trifolium repens*), can supply a significant amount of the nitrogen requirements for brassicas. Grown with green broccoli crops (Gaskin *et al.*, 2020) it may, for example, result in acceptable yields, especially when decision support tools are used for predicting nutrient requirements.

Tipburn

Internal tipburn is one of the commonest physiological disorders affecting a wide range of vegetable and fruit crops. A necrotic breakdown affects the internal tissues of leaves in the heads of both Chinese cabbage (*B. rapa* ssp. *pekinensis*) and white cabbage (*B. oleracea* var. *capitata*). It is usually attributed to localized calcium deficiency and related in incidence to genotype, prevailing weather conditions and the availability of nitrogen fertilizer. Large applications of easily assimilated nitrogen at transplanting increases the shoot-to-root ratio, imbalancing the plant and leading to physiological disorders such as tipburn. Magnusson (2002) suggested that rapidly increasing growth and high total nitrogen and nitrate concentrations at harvest increased the incidence of tipburn. He made comparisons of the growth of Chinese cabbage with 'Green Mulch' grass leys used as intercrops (see Chapter 6 section, Intercropping and Cover Cropping) and in combination with mineral fertilizers, in which intercropping decreased the prevalence of tipburn.

An extensive review of tipburn in *Brassica* was published by Everaarts and Blom-Zandstra (2001) who subscribe to the view that a number of interacting factors lead to the expression of tipburn symptoms. This physiological problem is increasing with all forms of European and Asian brassicas. Possibly, this is analogous to the upsurge in pathogenic and pest problems and reflects the increasingly concentrated genetic base from which commercial brassica cultivars are bred. Normally, there are no external symptoms, while internally they vary between genotypes and between storage and fresh market types. In the fresh market types, the symptoms are desiccated, papery-thin, light, broad leaf margins extending in zones from as small as several millimetres to eventually discoloring the entire head. In genotypes intended for storage, the symptoms are dry, papery-thin, dark brown, circular to oval spots with deep brown to black margins ranging from a few millimetres to several centimetres in size.

Localized calcium deficiency in the leaves is found in rapidly growing tissues with low transpiration rates and is exacerbated by constant high RH, which stimulates calcium-related disorders. Since calcium is transported mainly in the xylem, the amount reaching the growing and meristematic tissues is closely related to the rate of the transpiration stream – where this is low then calcium deficiencies will develop.

Calcium accumulates in the outer leaves during the day by mass flow in the transpiration stream and in heads at night when growth takes place and root pressure forces water and calcium into the head. Some genotypes exhibit levels of resistance or tolerance to tipburn. Since calcium is only taken up by the very young unsuberized roots, the number of young roots and their position in the soil of root apices relates to calcium uptake. Consequently, root architecture is an important factor. The large, vigorous and deeply rooting genotypes are less likely to be prone to tipburn. Cultivars producing high yields and having rapid growth are prone to tipburn, but it is discouraged by the use of wider planting distances. Tipburn is most frequent where plants grow rapidly but fail to develop a sufficient root system and, in consequence, there is a high leaf to root ratio. The leaves of field-grown cauliflower (*B. oleracea* var. *botrytis*) plants with tipburn contained less calcium compared to leaves without symptoms, but did not differ in nitrogen, phosphate, potassium or magnesium concentrations (Maynard *et al.*, 1981).

The rate of biomass production in Brussels sprouts (*B. oleracea* var. *gemmifera*) is known to be proportional to the intercepted radiation. High radiation rates will increase growth rate and, in consequence, the amount of tipburn in this crop. The incidence of tipburn is also associated with high levels of nitrogen fertilizer use. This is due to accelerated growth resulting from the availability of nitrogen, not the effect of the element *per se*. Ammonium, as a nitrogen source, reduces the uptake of calcium due to competition for uptake between the two cations. Husbandry factors, such as planting date, also affect the incidence of tipburn. Cabbages intended for long-term storage are normally planted between the end of April and mid-May, with head formation starting 60–75 days later. Later planting is associated with increased tipburn resulting from high growth rates during periods of a limited duration of darkness. Delaying the harvest increases the likelihood of tipburn developing. Probable interacting factors are the age of tissue and storage, which increases tipburn with cool (low temperature) storage.

White cabbage roots can penetrate to 100 cm or more into soil and the root shape is obconical reaching 150 cm deep, but the greatest intensity of the root is in the top 20 cm. Large root size and vigour tends to minimize the incidence of tipburn. Soil characteristics can also affect the incidence of tipburn. Thus, Dutch growers stop liming when soil calcium content exceeds 2%. Normally, they use well-cultivated and drained, moisture-retaining fertile soils for cabbage. Soil, where waterlogging is present, leads to dysfunctional root growth resulting in anaerobiosis, which even when it only lasts for short periods is sufficient to cause root death and, in consequence, reduced calcium uptake and subsequent tip burn. But calcium applications made directly to the crop are unlikely to be of benefit, except possibly sprays of calcium chloride or nitrate which are quickly assimilated. Tipburn is associated with inadequate calcium uptake by young, rapidly growing leaves. Several soil chemical and environmental factors that increase plant growth and decrease calcium mobility and transpiration have been implicated.

Soil cations (calcium, potassium, magnesium and ammonium) play critical roles in the development and prevention of physiological disorders that usually involve competition for plant uptake, resulting in an excess or deficiency of a particular element or elements in the tissues (Cubeta *et al.*, 2000). Rapidity of growth in the presence of excessive nitrogen and minimal calcium appear to be dominating factors in the development of this syndrome and affects all brassica crops that form heads or inflorescences. The association with calcium deficiency is shown in Table 8.4.

Table 8.4. Nutrient composition of field-grown cauliflower leaves showing normal and tipburn growth. (After Maynard *et al.*, 1981).

Nutrient	Nutrient composition (% dry weight)		Probability based on *t*-test
	Normal	Tipburn	
Nitrogen	4.70	4.80 (102%)	0.40
Phosphorus	0.64	0.52 (81%)	0.50
Potassium	1.31	1.58 (121%)	0.20
Calcium	0.50	0.18 (36%)	0.05
Magnesium	0.33	0.26 (79%)	0.30

The association between tipburn and root size was demonstrated by Johnson (1991). The root system of a tipburn-susceptible cultivar was smaller than for a tipburn-resistant one, hence the plant was more susceptible to moisture stress. But, *per se*, this does not necessarily correlate with the development of tipburn. The calcium efficiency ratio (mg of dry matter produced per mg of calcium in the tissue) in young leaves was, however, greater for tipburn-tolerant cultivars than for tipburn-susceptible ones and provides a valuable gauge to the likelihood of symptom expression (see Table 8.5).

Studies of Chinese cabbage (*B. rapa* ssp. *pekinensis*) identified the main husbandry problem as tipburn. Borkowski *et al.* (2016) connected tipburn with low levels of calcium in young leaves and with water deficiency. Applications of calcium nitrate limited the occurrence of tipburn and bacterial rotting of Chinese cabbage. Symptoms developed quickly and became more serious with increased ammonium nitrate in the absence of calcium. Furthermore, tipburn occurred earlier where there were combined calcium and trace element deficiencies (Yu *et al.*, 2004). The *BrCRT2*[R] line appeared to be capable of increasing calcium storage in the *Arabidopsis crt2* mutant and also reduced cell death in leaf tips and margins under calcium-depleted conditions. Su *et al.* (2019) suggest that *BrCRT2* may be a possible candidate gene for controlling tipburn in Chinese cabbage. Currently, it appears that there are no known resistance genes in Chinese cabbage.

Virus infection may exacerbate physiological disorders. Beet western yellows virus (BWYV) infection was associated with collapse of leaf tissue at the margins (tipburn) in heads of stored white cabbage (Hunter *et al.*, 2002). There was a significant association between detection of turnip mosaic virus (TuMV) and the development of internal necrotic spots, while cauliflower mosaic virus (CaMV) reduced cabbage yields.

Table 8.5. Calcium efficiency ratio for collards (*Brassica oleracea* ssp. *acephala*) grown in controlled conditions. (After Johnson, 1991).

| | | Calcium efficiency ratio[a] | | |
| | | Leaves[b] | | |
Cultivar	Tipburn reaction	Young	Old	Total plant
Vates	Susceptible	107b	44a	69a
Blue Max	Tolerant	206a	43a	64a
Heavi Crop	Tolerant	183a	42a	66a

[a]Mean separation among cultivars within a calcium level by least significant difference at $P=0.05$.
[b]Young leaves (blade and midrib) = terminal five leaves that were 2–6 cm long; old leaves (blade and midrib) = all leaves older than fifth leaf from the terminal.

Internal browning

Internal browning of Brussels sprout (*B. oleracea* var. *gemmifera*) buds is seen mainly as a problem in crops destined for 'quick freeze' processing where even a small percentage of blemished buds causes total rejection. Affected buds could not be identified and extracted from the processing line. Consequently, they would end up in the packs sold by supermarkets and found by the retail consumer. Factors that have been implicated in the development of internal browning include: bud size – larger are sprouts more prone to the syndrome; the density of leaf packing in the bud – the greater the density, the greater the likelihood of internal browning; seasonality – incidence is most common in early maturing, rapidly growing types and mid-season (October to December); and genotype – some cultivars are more susceptible than others. Symptoms are characterized by the death and subsequent brown discoloration of leaf tissue in the apical third of the sprout bud. Symptoms are never seen on outer leaves, nor the youngest leaves. In severe cases, the browning can spread downwards along the petiole to the base of the sprout. It has been suggested that water condenses within the sprout bud, which may then restrict calcium transport and lead to marginal leaf necrosis in the bud.

Brown bead

In the 1970s, in California, USA, brown bead (or dead bead) was first seen in green broccoli (*B. oleracea* var. *italica*) as a physiological disorder, causing loss of quality and the abandonment of substantial areas (see Fig. 8.9).

Fig. 8.9. Brown bead (or dead bead) on green broccoli (*Brassica oleracea* var. *italica*). (Courtesy of Phillip Effingham).

This disorder was also found in Canada (Jenni *et al.*, 2001b). There is no correlation with nitrogen fertilization but low calcium content, associated with rapid growth, has been related to brown bead. Also, there is some association with low potassium levels. There is tentative evidence for an association with excessive temperatures (22–38°C) in the 5 days prior to maturity. Some evidence exists suggesting an association with elevated ethylene levels. Experimental studies showed less brown bead in the fastest growing crops but where there is nutritional imbalance, especially between calcium, potassium and magnesium, brown bead developed, particularly when calcium was deficient. The most likely contributory factor is high air temperature, whereas a regular water supply diminishes brown bead. Adequate nitrogen supplies and moderate availability of potassium and magnesium help to prevent this syndrome. Whenever there is a restriction in water supply and calcium or nitrogen shortages associated with high temperatures, the condition is prone to occur. Applications of calcium chloride decreased brown bead.

In general, more solar radiation and less precipitation translated into more green broccoli heads showing brown bead symptoms (Jenni *et al.*, 2001a). For transplanted green broccoli plants, the minimum temperature from the button stage to maturity was a key variable in the prediction of the percentage of heads with brown bead and the corresponding index of severity. Canadian field studies (Jenni *et al.*, 2001b, 2017) demonstrated an interaction between nitrogen fertilization, soil type and the incidence of brown bead. Soil effects became more important where low nitrogen applications were used.

Pepper spotting

Pepper spotting develops inside cabbage (*B. oleracea* var. *capitata*) heads, especially the Dutch or white cabbage that are stored for several months (Cox, 1977). Pepper spotting is seen as clusters of small black spots of less than 1 mm in diameter. The causes of pepper spotting (also known as pepper spot and grey speck) are unknown. The use of controlled atmosphere storage (2.5–3.0% O_2 and 5.0–6.0% CO_2) will extend cabbage storage life for 5–6 months at 0°C, delay yellowing and maintain good quality characteristics. Differences in the expression of pepper spotting by genotypes has been reported (Shipway, 1978). Turnip mosaic virus was found to be a cause of internal necrosis in stored white cabbage. The occurrence of internal necrotic lesions was associated with the presence of severe TuMV symptoms on the outer leaves of plants at harvest. The average head weight of marketable cabbage from TuMV infected plants was significantly less than from healthy plants (Walkey and Webb, 1978; Walkey and Neely, 1980). The aphid-transmitted TuMV is responsible for storage losses in volume and quality. This virus causes either large necrotic leaf spots (0.5–1 cm in diameter) or small necrotic flecks. The common CaMV can similarly cause internal necrosis, particularly of stored white cabbage heads.

Black speck

This non-parasitic disorder was first reported by Strandberg *et al.* (1969) as numerous dark spots of 0.1–2 mm in diameter on the outer leaves of cabbage and throughout the head, progressing rapidly during storage at low temperatures. Development of individual flecks begins with necrosis of stomatal guard cells, followed by darkening and collapse of adjacent epidermal and mesophyll cells. The cause was ascribed to the accumulation of salts to toxic levels in guttation droplets. Closer definition of the syndrome characterized it by small, sharply sunken brown or black specks and similarity to pepper spotting or grey speck disorder (Loughton and Riekels, 1988). Decreased O_2 content in the controlled atmospheres, such as 2–2.5% O_2 at a temperature of 0°C, reduced the disorder (Geeson and Browne, 1980). It is most frequently seen about 1 week after cold storage starts. In many cases, black speck is found on the outer leaves and accompanied by fungal and bacterial infections of leaves 3–10 towards the centre of the head, necessitating substantial post storage trimming to provide a marketable head. On occasion, black speck may extend to the core, making the entire head unmarketable. Black speck in stored green broccoli (*B. oleracea* var. *italica*) is well described. On this crop, black speck is characterized by small sunken black spots on the inflorescence stalk which can coalesce into lesions of 0.5–4 mm in diameter. It is encouraged in green broccoli by the use of high-nitrogen fertilizers, rapid vigorous growth, inadequate mineral uptake and sometimes related to hollow stem or internal browning, as also found in cauliflower (*B. oleracea* var. *botrytis*), which indicates boron deficiency and reduced availability of potassium (see Fig. 8.10).

Black speck is related to genotype, soil type and the use of some postharvest fungicidal dips. The use of controlled atmosphere storage has been suggested to reduce black speck in green broccoli held at 3% O_2 and 5% CO_2 at 1°C.

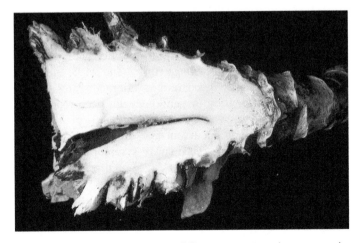

Fig. 8.10. Internal stem breakdown in cauliflower (*Brassica oleracea* var. *botrytis*). (Geoff Dixon).

(Lougheed, 1983). More recently, low temperatures have been associated with black speck. Iizuka *et al.* (2013) concluded that rapid thawing of frozen leaves causes an acute temperature increase inside cabbage head leaves, damages cells and results in internal black spot. Studies of nitrogen fertilization, black speck and storability of kimchi cabbage (Hong *et al.*, 2017) suggested that the occurrence of black speck was related to the amount of nitrogen fertilization.

Broccoli scarring

Broccoli (*B. oleracea* var. *italica*) scarring is a relatively rare disorder develop-ing where the leaves are removed or the stem is cut and the exposed tissue is initially whitish but in store turns grey or blackish. Sodium hypochlorite dips of up to 300 ppm reduced black speck and also scarring discoloration, both in ambient and controlled atmospheres.

Cigar burn

Cigar burn is one of the major internal disorders of cabbage (*B. oleracea* var. *capitata*). It is seen as sunken necrotic spots of 5–10 mm in diameter on inter-nal tissues. Walsh *et al.* (2004) have added evidence that virus pathogens may also be implicated in these syndromes. They propose that cigar burn is caused by TuMV on cvs 'Polinius' and 'Impala'. The condition reached its most severe at 4 months in store and did not progress beyond that date. The presence of CaMV, while not directly causing cigar burn, enhanced the symptoms resulting from TuMV infection. Cauliflower mosaic virus can cause raised pale lesions of 3–8 mm in diameter. Storage conditions can influence the progression of these physiological disorders. High-quality control of temperature and atmos-pheric composition in a sealed store repressed the development of losses due to cigar burn and also the rotting caused by microbial pathogens. In the field, these viruses reduced yield by between 16–76%. The use of an enzyme-linked immunosorbent assay (ELISA) test for the presence of TuMV in cabbage heads at harvest permits an estimate of the risk of cigar burn developing being made and allows those heads at most risk to be marketed more quickly in advance of symptom expression (Walsh and Hunter, 2004).

Several viruses are associated with the development of both tipburn and cigar burn. Beet western yellows virus can be associated with increased sever-ity of tipburn but this is heavily affected by genotype (cv. 'Impala' appears to be very susceptible). All these viruses are aphid spread and BWYV is likely to come from adjacent oilseed rape crops. Walsh and Hunter (2004) suggest that BWYV, spread by the green peach-potato aphid (*Myzus persicae*), encourages the tipburn syndrome. Cigar burn is encouraged by TuMV and CaMV spread by the green peach-potato aphid and the cabbage aphid (*Brevicoryne brassicae*), respectively. Pepper spotting has not, as yet, been associated with a virus disease vector but is linked by these authors with excessive nitrogen fertilizer. They also identify 'veinal streak' which is seen as large black lesions coalesced along the midrib and only visible when the head is split open. No association with virus vectors is proposed for this disorder. As control measures, they advocate cautious site selection that avoids proximity to oilseed rape (*B. napus*) crops. It is considered that waterlogged fields exacerbate the problems of calcium uptake and cultivars vary in susceptibility to the disorder. In all areas where physiological problems are likely to occur, it is important to monitor crops and avoid using those showing field symptoms of virus infection for storage. The use of ELISA testing prior to storage can aid the identification of the presence of viruses and the invasion of crops by aphids should, where feasible, be prevented (Walsh *et al.*, 2004). Table 8.6 summarizes the relationships between physiological disorders and virus infection.

Riceyness

Cauliflower (*B. oleracea* var. *botrytis*) curd quality and the timing of continuous production are closely related to temperature during the early development of the plant (Grevsen and Olesen, 1994). The occurrence of 'bracting' (small leaves growing through the curd surface) was positively correlated with the average temperature during the 10 days after curd initiation but there were large cultivar differences. The incidence of 'riceyness' (premature development of small flower buds on the curd surface) was positively correlated with the temperature from 15 to 25 days after curd initiation. This result for riceyness does not agree with earlier findings and further research is needed before a definitive model is obtained.

Cauliflower curds suffer from a range of defects from discoloration (see Fig. 8.11) and yellowing to riceyness and overmaturity, where the florets begin to turn green.

Broccoli of both the green calabrese and sprouting (white or purple) types can easily become overmature and the florets can start expanding and opening. This may be accelerated during display at the point of sale, where the produce

Table 8.6. Viruses associated with physiological disorders in brassicas. (After Walsh *et al.*, 2004).

Disorder	Cause	Aphid transmission	Storage
Tipburn	BWYV	Slow	Continues to develop in store
Cigar burn	TuMV	Rapid	Does not get worse after 4 months in store
Cigar burn (severe)	TuMV + CaMV	Rapid	Does not get worse after 4 months in store

BWYV = beet western yellows virus; TuMV = turnip mosaic virus;
CaMV = cauliflower mosaic virus.

Fig. 8.11. Cauliflower (*Brassica oleracea* var. *botrytis*) head discoloration. (Courtesy of Phillip Effingham).

is held for longer than its sell-by date. A candidate gene (*Bo4g024850*) and an upstream single nucleotide polymorphism are suggested as associated with the riceyness phenotype as they influence the expression (Zhao *et al.*, 2020).

Cold and heat stress

Heat stress, exacerbated by increases in global temperatures, is posing serious threats across the agricultural sector, not least for brassica crops in many parts of the world. Heat tolerance in brassicas is of major value as production expands, especially for cauliflower (*B. oleracea* var. *botrytis*) and Chinese cabbage. Cauliflower curds tend to initiate in response to temperature values, which vary according to their maturity type. In India, cauliflower has become a popular vegetable, following its introduction by the British in the 1800s, and moderate temperature tolerance has been developed by selection over 200 years. The development of lines capable of being grown, for example, in warmer conditions in Mauritius was reported by Nowbuth and Pearson (1998). In addition, Chinese cabbage breeders have long sought to develop high-temperature-tolerant types for use throughout Asia. In green broccoli (*B. oleracea* var. *italica*), the critical time for heat response is at bud initiation at about 3–4 weeks prior to harvest. There are substantial differences in heat tolerance in green broccoli. Some of the Japanese seed houses have cultivars with extended heat tolerance. Cabbage production is difficult in hot climates since it grows best at 16–18°C. In hot conditions, such as in Israel, production is restricted to the autumn and winter with storage providing supplies in spring and summer months (Chalupowicz *et al.*, 2019).

Wilson *et al.* (2019) studied the mechanism of thermotolerance in 4-day-old seedlings of Chinese cabbage (*B. rapa*, 44 genotypes) on the basis of various enzymatic and nonenzymatic antioxidants. There were significant increases in electrolyte leakage, lipid peroxidation, peroxidase activity, glutathione and proline content compared with control seedlings, and a decline in catalase activity and chlorophyll content. The effects of heat stress on root growth, root antioxidant capacity and cellular ultrastructure (Yuan *et al.*, 2016) reputedly caused decreases in root biomass, relative water content and root vigour in susceptible genotypes.

A seedling disorder of green broccoli characterized by chlorotic cotyledons and delayed seedling growth is described by Mayberry *et al.* (1991). This disorder, termed 'yellow cotyledon', was observed in field sowings under high temperatures with some green broccoli seed batches. Yellow cotyledon disorder is found in laboratory germination tests where they are conducted under illumination. Unaffected seedlings develop green cotyledons and those with yellow cotyledon disorder eventually developed into normal plants, but maturity was delayed and total yields were reduced.

Brassica vegetables are evolutionarily suited for cool to warm temperate conditions. Some of the cabbage, Brussels sprout and kale types can withstand low and freezing temperatures, although not very severe low temperature winter conditions, whereas cauliflower can be badly damaged (see Fig. 8.12).

Fig. 8.12. Freezing damage to cauliflower (*Brassica oleracea* var. *botrytis*) crops. (Geoff Dixon).

Opportunities for improving frost tolerance in cauliflower have not been exploited sufficiently despite some genotypes being regularly used over winter (Deane *et al.*, 1996). Breeders have developed forms of Chinese cabbage capable of withstanding low temperatures such that in Japan and Korea fresh vegetables are now available year-round. Cold tolerance has been utilized from Japanese radish or daikon (*Raphanus sativus* ssp. *longipinnatus*) (Ogura, 1968; Heath *et al.*, 1994). There is a correlation between high dry matter content and cold tolerance. Kale has a dry matter content of over 18% and is the cold-tolerant brassica. Savoy cabbage has dry matter content of around 12% in comparison to summer cabbage, which is about 6–7%. Dry matter content is a useful indicator when selecting for frost tolerance. Winter hardy cabbage will survive −15°C to −20°C. Crosses made between kale, green broccoli and cabbage suggest there are two epistatic genes controlling tolerance to frost. The molecular markers for freezing tolerance that are needed for marker-assisted breeding of freezing-tolerant cabbage are described by Song *et al.* (2018). The expression patterns of genes *BoCCA1-1* and *BoCCA1-2* were similar under normal vs temperature-stressed conditions (low and high temperatures), suggesting a functional difference at the post-transcriptional level.

Since plants are relatively sessile organisms they can suffer serious growth and developmental consequences under cold stress conditions. Glutathione transferases (GSTs) are ubiquitous and multifunctional conjugating proteins, which play a major role in stress responses through preventing oxidative damage by reactive oxygen species (Vijayakumar *et al.*, 2016). Most of these genes were highly expressed at 6 h and 1 h in the cold-tolerant and cold-susceptible lines. Overall, *BoGSTU19*, *BoGSTU24* and *BoGSTF10* are candidate genes which are highly expressed in *B. oleracea*, cabbage and could provide an understanding of the complex mechanisms underlying cold tolerance.

Lipid metabolism plays an important role in the mechanism of frost or cold tolerance in plants (Badea and Basu, 2009). Plant membrane lipids change from a gel to a liquid-crystalline phase in response to low-temperature stress. This process is due to the increased level of lipid desaturation. The responsible components of this process are, among others, the fatty acid desaturases. Controlling the activity of these enzymes affects the amount of polyunsaturated fatty acids on the glycerol backbone and eventually controls the plants sensitivity to low-temperature stress. Curly kale is a robust, cold-tolerant plant with a high content of health-promoting compounds and is grown at a range of latitudes (Steindal *et al.*, 2015). Cold acclimatization increases the content of soluble sugar and improves the taste, although the content of unsaturated fatty and glucosinolate acids may decrease. Heptahelical proteins (HHPs) respond to a variety of abiotic stresses in plants, including cold stress (Wang *et al.*, 2019b). Results indicate that the (*B. rapa* var. *chinensis*) BcHHP5 protein might play a role in the abiotic response of pak choi. Abscisic acid significantly decreased both cold tolerance in acclimatized and non-acclimated cauliflower microshoots, while molybdenum had a highly positive effect on the cold tolerance of cauliflower microshoots (Rihan *et al.*, 2015). Subsequent studies of dehydrin proteins through the maturation stages of cauliflower seeds (Rihan *et al.*, 2017) indicate they had no role in the cold tolerance of artificial seeds.

Salinity

The wild counterparts of *B. oleracea* occur naturally on European sea cliffs (Lema *et al.*, 2019). Domestication and selection processes have led to phenotypic and genetic divergence between domesticated plants and their wild ancestors that are exposed to saline conditions. Salinity is one of the most limiting factors for brassica vegetables but little is known about how salinity affects crop plant productivity and growth. Generally, salt stress significantly affects all major physiological process such as germination, changes in dry matter, water relations and mineral uptake. At early stages of development, wild plants are more succulent than cultivated plants and have a higher capacity to maintain lower sodium concentrations in their shoots in response to increasing levels of salinity. Ahmad *et al.* (2019) suggest that potassium applications may induce tolerance to salt stress in cabbage (*B. oleracea*).

PATHOGENS AND PESTS

Plants have evolved a range of defensive mechanisms against diverse communities of herbivorous insects and pathogens (see also Chapters 6 and 7). The consequences of an investment in defence are determined by its direct metabolic costs and indirect or ecological costs through interactions with other organisms (de Vries *et al.*, 2019). In nature, the value of defence is highly dependent on ecological interactions and is predominantly determined by the outcome of competition for light. Natural populations of brassicas may also harbour internally benign bacteria which themselves can contribute towards defence. These endophytic bacteria are known for their ability in promoting plant growth and defence against biotic and abiotic stress (Tyc *et al.*, 2020). Little is known, however, about the microbial endophytes living in the spermosphere surrounding the seeds. Bacteria were isolated from the seeds of five closely associated populations of wild cabbage (*B. oleracea*) growing on the Dorset coast in England and each population was found to contain a unique microbiome. Sequencing of the 16S rRNA genes revealed that these bacteria belong to three different phyla (Actinobacteria, Firmicutes and Proteobacteria). Volatile organic compounds emitted by the endophytic bacteria had a profound effect on plant development but only a minor effect on resistance against herbivores of brassicas, although they were active against the fungal pathogen *Fusarium culmorum*.

Vegetable brassica cultivars developed for resistance or tolerance against specific pests and pathogens will compromise the value of these traits against other characteristics, not least ultimately marketable product quality or storability. Similar considerations occur where biological control is applied. Control of the larvae of cabbage looper (*Trichoplusia ni*) silver 'Y' moths by entomopathogens was affected by temperature and the nutritional status of the caterpillars (Shikano and Cory, 2015) and hence the health of the cabbage host.

Plant diseases are often thought to be caused by one species or even by a specific microbial strain. Microbes in nature, however, mostly occur as part of complex communities and this has been known since van Leeuwenhoek's discoveries in the 17th century. There are only a few reports of synergistic pathogen–pathogen interactions in plant diseases and the mechanisms of interactions are currently unknown (Lamichhane and Venturi, 2015). Disease complexes with several causal agents are more common than expected and exert considerable costs on the host plant in the resultant damage and crop management difficulties.

Postharvest damage from diseases usually results from either infections that happen in the field or following damage during harvesting and storage followed by the entry of pathogenic microbes. Control of many storage rots was achieved with fungicidal dips applied postharvest but this practice is becoming less acceptable since there is little opportunity for the active ingredients to dissipate from organs which have ceased growth and development. The ubiquitous grey mould fungus (*Botrytis cinerea*) is probably responsible for the most widespread losses worldwide and can limit storage periods. This disease is characterized by causing a soft watery rot of cabbage head tissues with felt-like mats of spreading grey mycelium and spores. In mild infections, only the outer leaves are affected and these may be removed by trimming but the pathogen soon colonizes cut stems and petioles, penetrating deeply into the head and resulting in total loss. Mechanical injury during harvesting has been associated with increased losses. The practice of 'mid-term trimming' during longer term storage has been associated with accelerated disease spread. Possibly, that results from the wounding caused by the trimming or from the greater susceptibility of younger leaves to the pathogen compared with the older wrapper leaves. Freezing damage inflicted by frost in the field or poor control of the storage environment increases disease risk. Good husbandry practice requires the early harvest of crops destined for long-term storage before damage is inflicted by autumnal night frosts. The grey mould pathogen has weak powers of invasion, requiring entry through wounds or as a secondary invader following damage caused by other organisms, such as dark leaf spot (*Alternaria* spp.), ringspot (*Mycosphaerella brassicicola*) or TuMV. Susceptibility to grey mould disease varies with host genotype and previous husbandry practices. There is increasing evidence that manipulation of fertilizer strategy (see Chapter 5 section, Nitrogen Use) will minimize subsequent disease development; excessive applications of nitrogen (especially ammonium forms) have been associated with enhanced disease development. Once infection foci are initiated the rate of spread accelerates due to senescence induced by ethylene formation that may itself encourage symptoms.

Once in store the major factors controlling disease development are temperature and RH. Lowering the temperature below 4°C frequently reduces disease severity. Reductions below −1°C may, however, result in low-temperature damage to the cabbage tissues. Results from studies of the effects of RH are conflicting. Some reports suggest that maintaining the wrapper leaves in a turgid state diminished losses due to grey mould whereas others have identified converse effects. It is likely that some dehydration of the wrapper leaves will slow the rate at which grey mould is able to colonize leaves and improve the longevity of produce in store. Hence, losses are probably increased in ice-bank cooled stores which maintain 98% RH, whereas brine-cooled or direct expansion-cooled stores operating at 90–95% RH are less conducive to fungal spoilage.

Manipulation of storage atmosphere can be used to inhibit pathogen development. Reduced concentrations of O_2 (1.0%) at 0°C delayed yellowing in Chinese cabbage and decreased the incidence of grey mould-induced decay. This approach requires careful testing since higher levels of CO_2 (>6.0%) have been associated with the development of off-odours and off-flavours (Menniti *et al.*, 1997). Fungal inoculum will reside on the walls of stores, on containers and other surfaces ready to cause infection should conditions become favourable. It is essential to maintain a strict sanitary policy involving washing and cleaning stores and containers during the periods between crops.

Three species of *Alternaria* (*A. brassicae*, *A. brassicicola* and *A. alternata*) are responsible for serious losses of cabbage during storage and in the subsequent distribution and marketing chain, causing 'Alternaria spot' and 'dark leaf spot'. All three pathogens have been identified as causing substantial crop losses in north-western Europe and North America. Symptoms are common to all three of these pathogens; invaded tissues are surrounded by chlorotic margins within which are discolored areas that are dark brown or black with a dry or leathery texture, which may show a superficial growth of mycelium and dark spore bodies (conidia). *Alternaria brassicicola* is characterized by causing large lesions bearing uniform, dark, olive-black sporulation. In distinction, *A. alternata* produces mycelial growth that is characteristically dark grey to greyish-black and *A. brassicae* has brown to dark brown sporulation and distinct concentric zonation.

Infection initiated in the field continues to spread mainly on the outer wrapper leaves after harvest. Trimming can remove much infection but areas of damage to leaves, petioles and stem butts offer portals for further invasion and the spread of existing infections. Although lesions caused by *Alternaria* spp. are typically limited in size (<5 cm in diameter), damage may be greater than is initially apparent due to penetration below the wrapper leaves extending deeply into the head tissues. Damage may be increased following secondary invasion by grey mould and soft rotting bacteria.

Primary infection by *Alternaria* spp. can encourage secondary organisms by predisposing the tissues to infection through their release of ethylene. High RH and temperatures above 5°C favour the growth of *Alternaria* spp. Disease risk increases as the length of storage increases and *Alternaria* spp. are frequently damaging pathogens on cabbage stored beyond 6–9 months.

The field-phase pathogen *M. brassicicola*, causing ring spot disease, results in dark grey to black lesions, whereas with *Alternaria* spp. the tissues desiccate becoming dry, leathery and corky in texture. Both pathogens can infect produce simultaneously, causing significant loss of quality. The randomly distributed lesions of *M. brassicicola* are best identified by the presence of small brown pycnidia that exude pale pink or whitish droplets containing pycnospores. Infection is initiated through bruised outer leaves or through cut stems and petioles. Lesions frequently extend deeply into the head tissues and may cause dark discoloration of the stem vascular tissues, but spread between heads is not frequent. In the field, ringspot disease is associated with cool, moist growing periods, hence in store high RH and low temperatures (0–2°C) encourage infection.

Storage losses of cabbage in Scandinavia have been attributed to species of water mould or oomycete (*Phytophthora*) and in one incident in Norway the pathogen *Phytophthora porri* (the cause of white tip disease on leeks (*Allium porrum*) and salad onions (*Allium fistulosum*)) was isolated. Similar infections have now been identified in the UK and the Netherlands. This pathogen can cause substantial losses in store. Infection is characterized by dark brown or grey–brown lesions spreading upwards from the stem base. Diseased tissue remains firm but acquires a distinctly acidic, vinegary odour. No mycelia are evident on the outer surfaces of infected heads and hence the disease may be confused with bacterial rotting or frost injury. Small areas of white hyphal growth are found between the surfaces of heart leaves and within cavities in the stem medulla. It is possible that soil-borne oospores or chlamydospores of *P. porri* enter cuts made to cabbage stems during harvesting, especially in wet, muddy conditions. Disease appears to be more frequent when harvesting takes place in very wet field conditions. Mid-term trimming in the store is associated with increased losses due to the spread of infection on knives. This pathogen can remain viable in soils for 3 years after *Allium* crops. Consequently, rotations should carefully avoid their close cultivation with brassicas.

Fusarium avanaceum is responsible for rapidly spreading soft brown rots characterized by dense pink or whitish-pink woolly mycelial overgrowths. Infection by white rot (*S. sclerotiorum*) leads to watery soft rotting where the head collapses and the pathogen moves swiftly between heads, frequently affecting entire batches of cabbage. The pathogen is identified by black, sclerotial, resting bodies and dense cottony white mycelium found on diseased leaves and heads. Disease spread is retarded by storage close to 0°C.

Dark internal discoloration of cabbage and cauliflower (*B. oleracea* var. *botrytis*) heads can result from early infection at the propagation stage by downy mildew (*Hyaloperonospora parasitica*) or of growing crops. This pathogen becomes latent between seedling infection and harvest, only becoming obvious after cutting, illustrating the impact that early field phase infection can have on subsequent marketable quality.

Black leg disease (*Leptosphaeria maculans* or *Phoma lingam*) invades cabbage heads causing black rotting of the stem and leaf bases following invasion in the field of root collars and stem bases. High temperatures allow the normally saprophytic fungus *Rhizopus stolonifer* (*R. nigricans*) to become parasitic and invade wounded tissues leading to watery soft rotting.

The bacterium *Pseudomonas marginalis* is responsible for wet, slimy, soft rotting and the emission of foul, sour odours in cabbages held under refrigeration. At sub-zero temperatures this organism spreads rapidly and is encouraged by poor ventilation of the store. Barn-stored crops are especially susceptible to this pathogen, other *Pseudomonas* spp. and *Erwinia carotovora* ssp. *carotovora*, particularly when crops have been harvested in a frosted condition.

MARKETING

The rate of the successful introduction of mechanization and automation into brassica harvesting and postharvesting treatments varies depending largely on the robustness of the final product. Delicate crops where eye selection for maturity and quality are paramount have still retained manual harvesting, largely because pools of field staff are available at minimal cost. But worldwide, supplies of rural staff who will accept this type of work are now diminishing (Huffman, 2012). Robotics and automation offer solutions which may ultimately result in better quality produce harvested more cheaply. Key requirements are machines which can perceive produce position, assess its degree of maturity and successfully cut and move brassica heads, buds or flowers (Kviesis and Osadcuks, 2013). Vision-based control can provide means by which this is achieved (Zhao *et al.*, 2016). But the phenotypes of suitable cultivars must be tailored by plant breeders such that they will fit in with the requirements of automated robotic harvesting. Designing, developing and manufacturing robots requires 'smart technologies, which combine sensors and robotics with localized and/or cloud based Artificial Intelligence' (Grieve *et al.*, 2019). Where engineering industries perceive commercial opportunities for developing suitable robotic brassica harvesting equipment then they will rapidly exploit these new markets.

Automated harvesting of robust crops

Automating harvesting of the root brassicas, swedes (*B. napus* ssp. *rapifera*) and turnips (*B. rapa* ssp. *rapifera*), uses machinery which was originally designed for potato crops. Main-season root brassicas have comparable robustness with tuberous crops and once the foliage is removed can be lifted from the soil mechanically and transported in bulk loads for cleaning, grading, storage and packaging (Anon, 2015). Crops are grown using multi-row beds which are straddled by the machinery to avoid damaging the roots. Early, and more sensitive, crops of swedes and turnips may still be harvested by hand.

The robust, mature buds of Brussels sprouts (*B. oleracea* var. *gemmifera*) can be stripped mechanically from stems (Davies and Wheeler, 1968). Originally, strippers were static machines requiring that stems were cut and transported to a central station for bud removal. Mobile machines now move across a crop; operatives harvest stems individually which are then manually fed into rotating, bud-stripping knives. Early, and specialist, crops may still be picked manually, visually selecting mature sprouts on several passes across the field (Anon, 2018). Both manual picking and machine harvesting require field staff capable of working in very inclement weather during the late autumn and winter months.

Automated harvesting of leafy crops

Harvesting of leafy brassicas, such as Dutch white cabbage (*B. oleracea* var. *capitata*), is being mechanized and uptake is predicted to expand substantially in the next 5 years worldwide, reflecting increased global demand and short-ages of rural labour. Self-propelled, pedestrian- or tractor-hauled, single- or double-row machines, suitable for small areas, are available in China (Du *et al.*, 2016) and other countries with small acreage intensive cropping. These machines cut the cabbage heads, remove wrapper leaves and pass the heads into collecting bins. Similar commercial drivers aimed at reducing costs, time and labour in Egypt encouraged the development of machines that harvested 12 t/h, causing less than 4% head damage, and the harvested cabbages could be stored for 4–6 months (El Didamony and El Shal, 2020). This equipment uses principles comparable with those developed for the powered cutting of green broccoli (*B. oleracea* var. *italica*) heads using rotating disc knives (Wilhoit and Vaughan, 1991) and those developed for mechanical harvesting of Chinese cabbage (*B. rapa* ssp. *pekinensis*) (Kanamitsu and Yamamoto, 1996). This latter machinery lifted the heads of Chinese cabbage from the soil, cut off the roots at the stem base, removed the wrapper leaves and passed the product into accompanying bulk bins. Industrial-scale extensive production is stimu-lating the design and development by engineering companies, principally in Europe and North America, of machines capable of harvesting larger areas of cabbage, especially cultivars of Dutch White which are subject to several months of storage and used for processed products such as coleslaw.

Automated harvesting of juvenile 'baby' crops

Baby leaf crops are now immensely popular for consumer salads. The cut leaves are delicate, easily bruised and deteriorate quickly. Automating harvesting for baby leaf brassicas such as mizuna (*B. rapa* var. *nipposinica*) and rocket (*Eruca sativa*) requires machinery which cuts the juvenile crops at controlled heights, preserving leaf integrity and quality and carefully elevating the produce into collecting bins for transport into packaging and storage facilities. Large crop areas are harvested with tractor-powered machines using a reciprocating toothed cutter blade passing through crops. Pedestrian-controlled machines are available for specialist crops and smaller areas. Similar machines are used for harvesting watercress (*N. officinale*) crops.

Automated harvesting of flowering crops

Flower-producing brassicas, such as cauliflower (*B. oleracea* var. *botrytis*), green broccoli (*B. oleracea* var *italica*), the increasingly popular forms of purple sprouting broccoli (*B. oleracea* var *italica*) and their hybrids, require eye selection for maturity and quality. Most crops are still harvested manually with the heads loaded on to gantry conveyors and passed into mobile packing facilities in the field. Introducing mobile gantries and conveyors, on to which harvested heads are loaded, substantially improved productivity (Holt and Sharp, 1989). Automation and robotic harvesting of these crops is currently an area of substantial research and development. Self-learning algorithms, open-source robotic software and generic mechatronic solutions are available and adaptive for automated harvesting of these crops (Pekkeriet *et al.*, 2015). Analysing the tasks involved precedes the development of autonomous robotic systems capable of harvesting and handling delicate produce with care and precision (Duckett *et al.*, 2018). The need is for lightweight machinery which is manufactured at low cost using smart components. These specifications are satisfied by utilizing the electronic components widely available for general retail consumption such as: mobile phones, gaming consoles, mobile computers (laptops, tablets), high-quality cameras and embedded processors. Additionally, robotic platforms can be manufactured using new materials and fabrication techniques, such as three-dimensional (3D) printers, which could offer opportunities for making spare parts on the farm. Key harvesting requirements are abilities for sensing the stages of maturity and then harvesting with sufficient delicacy such that the produce is not damaged (Bac *et al.*, 2014).

A 3D vision system for robotic harvesting of green broccoli using low-cost red, green, blue depth (RGB-D) sensors (Diego-Mas *et al.*, 2017) was tested for field harvesting of green broccoli in the UK and Spain. This equipment detected mature green broccoli heads and their 3D locations relative to a carrying vehicle. Combinations of viewpoint feature histograms, support vector machine classifiers and temporal filters detected green broccoli heads with high precision. Temporal filtering generated 3D maps of green broccoli head positions and size, indicating readiness for harvesting under field conditions. Comparisons of systems trained separately in British and Spanish conditions indicated mutual compatibility for harvesting in either country. This confirmed the value of low-cost 3D imaging for commercial green broccoli harvesting (Kusumam *et al.*, 2017).

Flowering crops offer substantial opportunities for using robotics and artificial intelligence (AI) as substitutes for manual field staff. A prime requirement is that each crop matures uniformly and retains high quality after harvesting, thereby justifying the capital outlays required. Achieving cultivars with these phenotypic traits demands a combination of physiological knowledge and plant breeding expertise. Science underpinning the physiology of growth and uniform maturity in brassicas is described in Chapters 4 and 5, and plant breeding developments in Chapter 2. This is a classical exemplar of the need for combinations of plant science, information technology and industrial-scale investment.

Yield estimation

Effective robotic autonomous harvesting requires the continuous monitoring of crop establishment, development and potential yields as part of husbandry management (Saldana *et al.*, 2006). Spanish studies recorded mass accumulation rates of green broccoli (*B. oleracea* var. *italica*). In these studies, ground-based sensors determined the harvested area and a Global Positioning System defined the field position. Assessments of yield variability within fields and worker performance allowed more precise management of land allocation and labour use. In other studies, crop mapping via satellite-identified productivity zones and data handling, based on Sentinel-2 A/B and Landsat eight multispectral data for oilseed rape (*B. napus*), closely correlated with subsequent data obtained with yield meters fitted to CASE 1H combine harvesters.

Current British intensive research and development (Simon Pearson, 2019, Lincoln University, UK, personal communication) utilizes a wide range of electronic tools, such as computerized data acquisition and analysis, the internet of things (IoT), cloud data management, machine learning, robotics, AI and blockchains, which will potentially transform all aspects of brassica production, harvesting and marketing. Most of the robots being developed are AI-provided and capable of dealing with large data blocks, processed either on the robot or in the cloud. Blockchains offer digital tracing systems that could connect with supply chains and directly with the consumers, who are then able to link with the grower, considerably extending the concepts of produce traceability, safety and transparency. Robotic green broccoli harvesters are in development which incorporate Saga Norwegian hardware (https://sagarobotics.com (accessed 17 August 2023)) with software prepared by Lincoln University (www.lincoln.ac.uk (accessed 17 August 2023)) and partner organizations. Saga's Thorvald robots are small, lightweight and operate autonomously, using advanced navigation systems and AI, and are capable of performing various tasks more effectively and more cheaply than conventional tractors or manual staff. Thorvald robots consist of two picking arms and one work platform. Interdisciplinary collaboration is an essential ingredient for automation. For example, green broccoli cultivars with well exposed spears enabling efficient robotic selection are required for ease of growth stage identification by the robot.

GRADING AND STORAGE

Physiological factors governing postharvest quality and value in brassica vegetables are of major significance in the preparation of produce for marketing. Respiration continues and quality starts declining as soon as vegetables are harvested. It is essential, therefore, that the mechanics of handling, storing and packaging operations are swift and provide environments that minimize losses. Senescence, particularly in harvested flowers, buds and leaves, such as the inflorescences of green broccoli (*B. oleracea* var. *italica*), is a highly regulated process characterized by a massive degradation of chloroplast components. Harvesting, either manually or potentially robotically, may be accompanied by immediate visual grading and packaging using mobile covered facilities, especially for fragile, easily blemished flowering and leafy brassicas. Alternatively, more physically robust root crops, such as swedes (*B. napus* ssp. *rapifera*) and turnips (*B. rapa* ssp. *rapifera*), or the firm heads of Dutch white cabbage (*B. oleracea* var. *capitata*) and Chinese cabbage (*B. rapa* ssp. *pekinensis*), may be harvested mechanically and transported into handling facilities for grading. Both streams of produce will be placed into short- or long-term cool storage with temperature and humidity controls. That reduces field heat, slows loss of quality and permits the alignment of harvesting volumes with contractual requirements and current market demands. A review of vision-based grading systems for a range of horticultural produce is provided by Sivaranjani *et al.* (2022).

Handling and storage facilities

Designing, building and equipping facilities for receiving, handling, grading and short- or long-term storage of brassica crops are commercial construction and engineering decisions and projects backed by business investments. The basic requirements for these facilities resemble those described for potato stores (Pringle *et al.*, 2009; Cunnington, 2019). The principles for retaining quality in stored crops, such as Dutch white cabbage (*B. oleracea* var. *capitata*), swedes (*B. napus* ssp. *rapifera*), turnips (*B. rapa* ssp. *rapifera*) and kohlrabi (*B. oleracea* var. *gongylodes*), for prolonged periods are well-established for potatoes (Cargill, 1976; Cunnington *et al.*, 1992) and broadly applicable for other similarly bulky crops. Storage facilities suitable for brassica vegetables were evaluated in detail by Kumar and Kumar (2007) and Perkins and Tiffin (2008). These authors identified that the first requirement for satisfactory storage is the reduction of field heat, which slows respiration and sustains quality, and is the first link in cold-chain handling systems for both the shorter and longer terms. This is achieved by:

- Vacuum cooling – suitable for leafy cabbage, but not for bulky crops or those with thick waxy surfaces, it is a quick batch system with high power consumption.

- Air cooling and conventional cold stores – produce is cooled by cold air circulated inside an insulated room and heat removed by refrigerated coolers with enhanced positive or forced ventilation. Conventional cold stores fitted with humidifiers will retain produce for 2–3 days. Crops, such as Brussels sprouts (*B. oleracea* var *gemmifera*), may take up to 24 h to lose field heat.
- Wet air cooling with positive ventilation – air is moistened by water cooled by a block of ice (ice bank). This is suitable for Brussels sprouts, cabbage, salads and root brassicas with the advantage that since air is forced through the crop achieving uniform cooling, it does not dehydrate the crop and can therefore use off-peak electricity and minimize costs.
- Hydrocooling – uses similar principles to air cooling but with the water temperature being reduced by a block of ice in a continuous process and it is ideal for crops washed or moved in water, such as Dutch white cabbage, swede, turnip or radish. There will be a requirement for subsequent cool storage for holding treated produce.
- Controlled atmosphere storage – in these systems temperature control combines with restriction of CO_2, derived from produce respiration, from accumulating in the atmosphere while O_2 concentrations decline. Essentially, this requires a gas-tight store fitted with ventilation and monitoring equipment by which the proportions of CO_2 and O_2 are controlled. Flushing with nitrogen can be used to establish the controlled atmosphere when stores are first closed or, if partially opened, for withdrawing some of the contents.

Storage environment

Field conditions influence the keeping quality of brassicas before going into storage. For example, the influence of five nitrogen application rates (60, 90, 120, 150 and 180 kg N/ha) were evaluated using limestone ammonium nitrate on the curd, fresh and dry mass, size, chromaticity coordinates, L^*, a^*, b^* and ascorbic acid content, in cauliflower (*B. oleracea* var. *botrytis*) cultivars (Mashabela *et al.*, 2018). With increasing nitrogen application rates, curds became darker and less intense when yellow in colour in Italian cultivars. Increasing rates of nitrogen reduced the ascorbic acid content. High-intensity white LED illumination of green broccoli (*B. oleracea* var. *italica*) maintained higher levels of ascorbic acid, carotenoids and phenolics at the end of the storage period (Pintos *et al.*, 2020). Exposure to mid-intensity light for 3 h/day reduced dehydration, chlorophyll, sucrose, glucose and fructose losses. White LED illumination during retail may be used to extend the shelf life of refrigerated green broccoli.

Preharvest climatic conditions and genotype affect head quality and postharvest performance of fresh-cut green broccoli (Conversa *et al.*, 2020). Green broccoli grown in southern Italian summer–autumn periods showed a distinctive decay during storage. Winter-grown heads had higher ascorbic acid contents. Yellowing of brassica vegetables is due to chlorophyll decomposition and visually indicates serious deterioration of freshness. It can be assessed by measuring colour space values (Makino and Amino, 2020). This is accompanied by loss of mass resulting from dehydration. A freshness evaluation index for brassicas is suggested which combines the degree of greenness and mass loss. It is considered more accurate than the conventional evaluation index that uses only greenness. Ethylene causes yellowing of green broccoli spears, which can be delayed by use of absorber treatments (Cai *et al.*, 2019). Yellowing first occurred in the base of the bud when its shape had not changed significantly and there was no sign of floret opening.

Preharvest methyl jasmonate and postharvest 1-MCP treatments on green broccoli florets influenced glucosinolate concentrations and quinone reductase (an *in vitro* anti-cancer biomarker) inducing activity (Ku *et al.*, 2013). Sulforaphane, the isothiocyanate hydrolysis product of the glucoraphanin, was found to be the most potent quinone reductase induction agent. Increased sulforaphane formation from the hydrolysis of glucoraphanin was associated with up-regulated gene expression of myrosinase (*BoMyo*) and the myrosinase enzyme co-factor gene, epithiospecifier modifier1 (*BoESM1*). This study demonstrates the combined treatment of methyl jasmonate and 1-MCP increased quinone reductase activity without loss of postharvest quality. The carotenoid isomerase gene (*BoaCRTISO*) of Chinese kale (*B. oleracea* ssp. *alboglabra*) can be edited using the CRISPR (clustered regularly interspersed short palindromic repeats)/Cas9 system (Sun *et al.*, 2020). This affects the expression levels of most carotenoid and chlorophyll biosynthesis-related genes, including *CRTISO*. Consequently, CRISPR/Cas9 might be used for quality improvement of Chinese kale and other brassica vegetables.

Harvesting cauliflowers, and the subsequent processing, can cause severe stress, determining the appearance of accelerated senescence symptoms (Miceli *et al.*, 2018). Florets that were treated with antioxidants before storage showed not only a good shelf life and overall quality maintenance but also a susceptibility to browning of cut zones. Subacute or low doses of stresses might enhance or induce protective mechanisms, a biological phenomenon known as hormesis (Duarte-Sierra *et al.*, 2020). The titres of indole-type glucosinolates and hydroxycinnamates in green broccoli were significantly higher ($P < 0.05$) when exposed to ultraviolet B (UV-B) compared to the non-exposed florets. There is good correlation between gene expression of *CYP79B3*, and the titres of indole glucosinolates in the treated green broccoli florets, suggesting that the target of UV-B could be the branch pathway of indole glucosinolates. Illumination with UV-C reduced weight loss and hue angle of green broccoli florets compared with the control treatment (Dogan *et al.*, 2018). The pH and total phenolics were constantly higher in UV-C illuminated florets than those measured on control florets. Furthermore, in the packages containing UV-C illuminated florets, O_2 levels were always lower while CO_2 levels were always higher compared with control florets. Additionally, UV-C light is a non-thermal method for improving the safety and shelf life of cold-pressed kale juices, with minimal impact on quality and nutrition (Pierscianowski *et al.*, 2021).

Early research (Geeson, 1983) showed that the optimum controlled or modified atmospheres for brassicas are achieved by substantially increasing CO_2 concentration and decreasing O_2 levels. Detailed requirements depend on the field conditions under which the crop is grown, length of storage required and storage facilities used. Broadly, reduced temperatures combined with atmospheres of CO_2 of 2.5–7% and O_2 of 2.5–6% were found to be adequate. For white cabbage (*B. oleracea* var. *capitata*), Geeson (1983) found that at temperatures of 0–1°C for 39 weeks, 92% of heads were marketable after trimming when held in 5% CO_2 and 3% O_2. By contrast, only 70% of heads held in ambient air were marketable. Cabbage and other brassicas (such as Brussels sprouts, green broccoli and Chinese cabbage) held in controlled atmospheres retain green coloration, fresh appearance and texture far longer than those held at ambient conditions. Optimum temperature conditions during green broccoli transport can prevent rapid deterioration of freshly harvested green broccoli for 3 weeks without further benefit from the use of controlled and modified atmospheres (Pliakoni *et al.*, 2015)

Green broccoli sales in the Australian markets, for example, fell due to poor quality at retail and disappointing shelf life (Ekman *et al.*, 2019). Quality could be improved by using the SmartFresh™ InBox, a new delivery system for the ethylene inhibitor 1-MCP, which enhances green broccoli retail appearance. Under Australian conditions, it could replace top icing, especially for long-distance transport and storage, and under high temperatures. This would be particularly useful during retail and consumer handling.

Typically, green broccoli should arrive at retail stores within 7–14 days of harvest, if not sooner, and be kept refrigerated until purchased or considered waste (Pellegrino *et al.*, 2019). Characteristics which govern consumer acceptance of green broccoli include tastes, colours and flavours (e.g. grassy, musty and 'dirt-like').

Storage periods

Short-term storage is preponderantly suited for brassicas which are sold as fresh vegetables or salads, such as cauliflowers, cabbage heads, baby leaves, Brussels sprouts, various forms of broccoli heads and their increasingly popular hybrids sold in bunches. These will typically be stored for 3–4 days at 0.5–10°C with 95% RH and in controlled atmospheres aiming at retaining maximum freshness and quality. Storage may be extended for up to 3 weeks when markets are oversupplied.

Long-term storage is suitable for swede (*B. napus* ssp. *rapifera*), turnip (*B. rapa* ssp. *rapifera*) and Dutch white head cabbage (*B. oleracea* var. *capitata*). The root crops are held for 5 months, or possibly longer, at 3°C and 90–95% RH. White cabbage can be stored for longer, up to 9–10 months, at 0°C and 95% RH. The commercial aspects of field heat removal by hydrocooling (for dense root crops), vacuum cooling and blast chilling are reviewed by Crowhurst (2018).

Storage controls

Commercial engineering companies are now installing very sophisticated computerized, automated, electronic systems which monitor storage environments and increase or reduce temperatures and atmospheric conditions in accordance with predetermined crop programmes. Real-time monitoring is relayed to producers' and packers' mobile platforms and warnings of impending malfunctions passed to engineers who then take remedial action, quickly minimizing disruption, downtime and produce deterioration. Recently, for example, JD Cooling Ltd (King's Lynn, UK) installed storage facilities for red, purple and French Breakfast radish (all *R. sativus* var. *longipinnatus*), destined for the pre-packed and bunched markets, creating an internal fogged atmosphere with ceiling-mounted coolers, condensing units, controls and humiDisk humidifiers for constant use and capable of handling 75% of the UK crop. They also provided Passive Up Flow systems suitable for stores containing swede (*B. napus* ssp. *rapifera*) crops.

Grading

Automated grading machines increase efficiency and substantially reduce requirements for manual labour. They provide greater uniformity of produce, hence increasing market value. They also provide accurate records, including the amount of waste, and contribute better traceability for food safety and security (Kondo, 2013). There are now more precise machine vision and near-infrared (NIR) systems which link with robots capable of handling delicate brassicas with great precision. Computerization much improves record keeping and traceability throughout the handling and the supply chain. Tong Engineering (www.tongengineering.com (accessed 17 August 2023)), for example, manufacture a blowing system which washes, dries and grades Dutch white cabbage (*B. oleracea* var. *capitata*) at the rate of 15 t/h. This company's lift roller graders sort swedes (*B. napus* ssp. *rapifera*) by width and diameter with smaller roots falling gently through the rollers first, followed by larger sizes. Tong Engineering equipment has auto-touch controls, intelligent diagnostics and proactive maintenance, minimizing downtime. The auto-touch controls are equipped with human–machine interface capabilities which provide centralized control units for manufacturing and processing lines and are equipped with data recipes, event logging, video feed and event triggering, which can be accessed at any time for any purpose.

Key to successful automation is the use of machine vision and NIR inspection systems (Kondo, 2010). This gives opportunities for measuring colour, size and shape, and for identifying external defects, sugar content and acidity.

The quality and storability of green broccoli (*B. oleracea* var. *italica*) could be rated using the normalized difference vegetation index (NDVI) and the inflection point of the red edge region where changes in spectral signature were identified (Kabakeris *et al.*, 2015b). Over 20 years, hyperspectral imaging technology has evolved into a powerful non-destructive tool for the assessment of vegetables, postharvesting. Imaging modes can include reflectance, transmittance, fluorescence and Raman spectroscopy for the assessment of produce quality and safety evaluation (Lu *et al.*, 2020). Near-infrared spectroscopy fulfils two functions. First, direct assessment of the chemical and physical status of produce in relation to postharvest quality and, second, predictions of storability (Walsh *et al.*, 2020). The identification of internal quality defects without physical sampling and dissection has considerable benefits. Internal brown heart or 'raan' in swedes, caused by boron deficiency, for example, cannot be identified by superficial visual inspection of root surfaces, but can be detected by ultrasound scanning, allowing the rejection of affected roots on a grading line (Holmes, 2015). Roots may then be individually sealed, preventing desiccation and deterioration.

Packaging

Packaging has up to now been a standard requirement for brassica crops in transit to the retail consumer. Effective packaging enhances the attractiveness of the product and retains quality characters for longer periods. But defective packaging accelerates deterioration and destroys quality. Deterioration is especially rapid where packaging allows the accumulation of toxic compounds, such as ethylene, that contribute towards accelerated deterioration. The formation of ethylene speeds up the processes of senescence and in brassica products, such as cauliflower, green broccoli and sprouting broccoli, leads to yellowing. Changing the concentrations of O_2 and CO_2 in small packs of brassica vegetables retains their fresh quality and extends shelf life. Lowering O_2 and increasing CO_2 concentrations around the produce blocks the synthesis of ethylene, slowing the ripening processes. The use of MAP started in the late 1940s when post-war plastic film polymers became widely available for civilian use. It is defined as 'an alteration in the composition of the gases surrounding fresh produce when such commodities are sealed in plastic films'. Overwrapping cabbages in polyvinylchloride (PVC) and controlled atmosphere treatment became especially popular because it was associated with reducing the pepper spotting physiological disorder (internal black necrotic flecking) by 50%. Additionally, the combination of 3% O_2 and 5% CO_2 with PVC film delayed tissue yellowing compared with ambient air treatments (Menniti *et al.*, 1997). The fresh produce industry is now having to reconsider packaging systems in order to avoid the use of plastic materials and to reduce their accumulation and consequent pollution.

Thin film plastics do, however, provide opportunities for the continued retention of quality by modifying the atmosphere within packages. Improvements in the design of polymeric films and other packaging materials allowed the expansion of cold-chain handling for supermarkets, displaying of fresh vegetables and salads, and even distribution through vending machines (Kader, 2010). Senescence is delayed by low-intensity visible light (Barcena *et al.*, 2020) as used in storage facilities, but accelerates in brightly lit retail outlets. Similarly, reducing ethylene levels around produce delays the senescence of fruit and vegetables which potentially reduces refrigeration requirements during transport and storage with consequent potential for energy savings. Postharvest life, as determined by consumer acceptance criteria of yellowing of pak choi (*B. rapa* var. *chinensis*) and green broccoli, increased as the temperature and ethylene concentrations decreased (Li *et al.*, 2017).

Following the upsurge of public concern regarding worldwide pollution caused by the accumulation of plastic packaging, manufacturers are developing biodegradable compounds. Alternatively, brassicas are being sold unpackaged as loose packs.

Food safety

The high frequency and incidence of food-borne outbreaks of disease related to fresh vegetable consumption is a major public health concern and an economic burden worldwide. Consequently, food safety is a key issue for the fresh-cut industry. Several outbreaks of food-borne illness caused by pathogens, such as *Escherichia coli* and *Salmonella* spp., among other enteric pathogens, have been widely reported in fruits and vegetables (Martinez-Hernandez *et al.*, 2018). Chlorine is the most widely used sanitizing agent for reducing pathogens on whole and fresh-cut produce, although it has some disadvantages that have led to research on new alternatives. Washing with selected neutral electrolysed water and a subsequent 6 kJ/m^2 UV-C treatment before packaging under high O_2 conditions offers a promising sanitizing treatment to reduce *E. coli* and *S. enteritidis* counts in fresh-cut Bimi® broccoli. Electrolysed water containing 10–80 ppm available chlorine is used in some commercial plants as an alternative disinfectant to sodium hypochlorite (Izumi and Inoue, 2018). Electrolysed water reduced the bacterial counts on fresh-cut vegetables, but the effectiveness on bacterial flora differed with the type of water and brassica involved. Pressurized precooling delayed the incidence of browning of fresh-cut purple cabbage, effectively suppressing the growth of microorganisms (total bacterial count and coliform group), and maintained its sensory quality (Xu *et al.*, 2019). Changes in chloroplast ultrastructure were inhibited and the degradation of chlorophyll delayed by micro-vacuum treatment (Sun and Li, 2017). The combined effect of ultrasonic treatment and MAP resulted in reductions of peroxidase and polyphenol oxidase activities – 10 min of ultrasonic exposure preserved pak choi (*B. rapa* var. *chinensis*) for 30 days (Zhang *et al.*, 2019).

Individual and combined application of ultrasound (40 kHz, 100 W) and ozone inactivated food-borne *E. coli* and *Salmonella* spp. (Traore *et al.*, 2020), maintaining the colour and vitamin C content of green cabbage (*B. oleracea* var. *capitata*). Low-quality storage and handling can lead to the evolution of hydrogen sulfide, a potent toxin for which propargylglycine is an inhibitor (Al Ubeed *et al.*, 2019).

Degradation

Traditional storage regimes allow preservation of cauliflower (*B. oleracea* var. *botrytis*) for up to 1 month. The main quality problems occurring during cauliflower storage are wilting, browning, yellowing of leaves and postharvest diseases (Roth *et al.*, 2012). Individual packaging of the curds, with an active film of suitable permeability, has increased the storage life of the product by reducing water loss and the magnitude of fungal rots. A film of medium-high permeability will maintain the quality of the fresh-cut product by reducing the O_2 content of packages, diminishing respiration and ethylene production, and repressing microbial and enzymic degradation. These processes start automatically with all brassicas from the instant they are harvested but storage at around 5°C reduces the speed of deterioration. The fresh, turgid appearance of produce is helped by increasing RH in the pack and results from reducing the temperature. The danger is that excessive moisture from brassicas accumulates inside the pack, especially when temperatures fluctuate. The higher humidity inside the pack condenses into water droplets when the film surface is cooler than the air inside the pack. A careful balance is required since produce such as green broccoli (*B. oleracea* var. *italica*), cauliflower and Chinese cabbage need a moist atmosphere as a means of avoiding desiccation.

Green broccoli, for example, is a highly perishable vegetable with limited shelf life. Hagen and Larsen (2020) demonstrated the importance of packaging to prevent weight loss and retain firmness in green broccoli, but none of the packages tested could prevent yellowing or decay development during the periods at room temperature, once purchased by the retail consumer. Countering the rapidity of this loss of quality and prolonging shelf life is best achieved by packaging and cold-chain handling during distribution and retail display. Green broccoli is now often overwrapped in polymeric films (with low water vapour transmission rates) or packaged in macro-perforated films, which results in a water vapour saturated headspace (around 100%) or sub-optimal RH (Caleb *et al.*, 2016). Packages incorporating a cellulose-based film window effectively prevented water vapour condensation on the film surface when compared to bi-axially oriented polypropylene and cling-wrapped commercial controls. Non-perforated packaging systems with cellulose-based film windows retained surface colour and maintained quality attributes at 10°C for 11 days, but at the expense of a significantly higher ($P < 0.05$) difference in the loss of product mass compared to control packaging.

Plastic film technology and packaging techniques have advanced substantially in the past 50 years and MAP is now a major marketing tool. Major difficulties remain such as the development of off-flavours and fermentation in green broccoli due to the innately high rates of respiration. But MAP is now a major marketing tool used for retaining fresh quality. Few consumers realise that opening a pack of brassica baby leaves releases a minute burst of CO_2 or nitrogen. The changed atmosphere is generated either by natural respiration or more recently by artificially adding nitrogen gas as the produce is sealed. There is a wealth of published information on MAP but a lack of systematic data providing information on which films to use for practical purposes and targets. Mahajan *et al.* (2007) developed two databases which include recommended gas composition for 38 products, 75 respiration rate models and permeability data for 27 polymeric films. The software was successfully tested for some products and, as an example, for Portuguese kale (*B. oleracea* var. *tronchuda*).

A key factor in MAP is finding films with a permeability that prevents an accumulation of damaging concentrations of either water vapour or CO_2, and excessive depletion of O_2. As an approximation, films are four to six times more permeable to CO_2 than they are to O_2. Transmission of water vapour is made independent of gas diffusion by using multi-layered plastics. One of the favoured newer products is 'linear, low-density polyethylene'. This has better structure compared with simpler films, making it tougher and more suited to heat sealing with greater impact resistance and tear strength. But the levels of CO_2 or O_2 take time to adjust inside the pack. Active adjustment by gas flushing speeds this process up and uses mixtures of CO_2, O_2 and nitrogen to purge the packages.

A low-O_2 controlled atmosphere is useful in maintaining freshness and qualitative traits of fresh-cut broccoli raab (*B. rapa* var. *ruvo*) during cold storage, while excessive accumulation of CO_2 (10–15%) should be avoided (Cefola *et al.*, 2016). Using combinations of polypropylene/polyethylene terephthalate and microperforated polypropylene/polyamide wrapping, and MAP using CO_2-absorbing sachets, produced equilibrium conditions that are near-optimal for broccoli raab (5% O_2 and 5% CO_2). Freshly harvested broccoli raab stored under these conditions for 8 days showed negligible reduction of appearance and odour scores, and was comparable in quality to the fresh samples.

The use of microperforated plastics and an ethylene retardant, such as 1-MCP, reduced senescence-related yellowing in green broccoli. A new generation of packaging materials uses environmentally responsive high-permeability films equipped with molecular temperature switches. These could help reduce the fermentation risks in MAP packs. This approach can be extended further with packaging that has an inbuilt ethanol sensor to identify the early stages of deterioration. Perforation-mediated MAP relies on the use of perforations (tubes) of different sizes to control the gas content of the package for both whole and fresh-cut brassicas. Designing efficient packages demands knowledge of gas exchange rates through the perforations.

In addition, circulating cooled air round the packages helps control the rate of gas exchange. A research aim for plastic manufacturers is to develop polymeric films with a selective barrier capable of matching the respiration of the produce. Alternatives are polypropylene, PVC and polyethylene films. For green broccoli, packing in polypropylene film increased storage life. Green broccoli stored in PVC deteriorated faster than when packaged in other materials. Green broccoli has a high respiration rate, is very sensitive to ethylene and loses water rapidly, and is consequently a difficult product to package. Using polypropylene extended shelf life to 3–4 weeks in air at 0°C compared with only 2–3 days at 20°C. The standard storage recommendations for green broccoli are 0°C with an RH of 95%, 1–2% CO_2 and 5–10% O_2. Modified atmosphere packaging and controlled atmosphere storage (hypoxia conditions) extends shelf lives of brassicas by depressing the O_2 uptake rate. Wang *et al.* (2017) reported that hypoxia may depress the expression of alternative oxidase activity and prolonged the shelf life. Aroma profiles can be related to storage conditions (temperature, time and packaging atmosphere) (Helland *et al.*, 2018) and are useful in controlling the storage of swede (*B. napus* ssp. *rapifera*) and turnip (*B. rapa* ssp. *rapifera*), which have been peeled and cut before packaging.

Modified packaging atmospheres containing 2% O_2 and 5% CO_2 provided environments which retained quality and freshness of pak choi (*B. rapa* var. *chinensis*), tatsoi (*B. rapa* var. *rosularis*), mizuna (*B. rapa* var. *nipposinica*), caisin (*B. rapa* var. *parachinensis*) and Chinese mustard (*Brassica juncea*) allowing easy mixing of these salad components with each other (O'Hare *et al.*, 2000). Similar packaging and preservation techniques are used for watercress (*N. officinale*), an increasingly popular western salad vegetable.

Green broccoli is one of the most popular brassica vegetables worldwide but highly perishable. Popularity is encouraged by its association with health benefits which potentially include retarding the onset of the diseases of affluence, such as cancers, strokes and heart failure. A crucial requirement for this vegetable is delaying the senescence of florets after harvest, which is associated with flower maturation and subsequent yellowing of foliage within the head. Green broccoli retained firmness and colour following hydrocooling and overwrapping with microperforated film and cold storage at 1°C, which allowed 5 days display at elevated retail temperatures (Toivonen, 1997). The senescence retardant 1-MCP markedly delays this process (Gomez-Lobato *et al.*, 2012). Controlled atmospheres extended the shelf life and maintained the quality of green broccoli (*B. oleracea* var. *italica* cv. Youxiu) florets. For example, atmospheres composed of 40% O_2 and 60% CO_2 prolonged the storage period of green broccoli heads to 17 days compared with 4 days in ambient air; and compared with 100% pO_2 which accelerated senescence and deterioration (Guo *et al.*, 2013).

Reducing waste resulting from postharvest processes, such as unused green broccoli leaves and stems, is now an environmental sustainability issue (Berndtsson *et al.*, 2020). Green broccoli production side products (leaves and stems), for example, can contribute health-promoting components for the human diet and also socio-economic and environmental benefits when marketed effectively, demonstrating their benefits for diets and contribution towards minimizing waste.

Alternative treatments

Exogenous applications of PSK-α, a signalling bioactive peptide, may delay senescence as demonstrated by applying at 150 nM to green broccoli (*B. oleracea* var. *italica*) florets stored at 4°C for 28 days (Aghdam and Luo, 2021). Fresh-cut green broccoli can be infected by microorganisms which is a persistent concern for growers and retailers alike. Contamination encourages deterioration and senescence of green broccoli and a range of prevention methods have been advocated including (Yu *et al.*, 2019):

- physical (decompression, microwave, packaging, ozone and UV-C);
- chemical (ethanol, fungicides, and 1-MCP); and
- biological (Chinese herbal extracts, edible film and essential oil) preservation techniques.

Treating green broccoli florets with humidified air has also been considered (Duarte-Sierra *et al.*, 2012) for delaying senescence. Also, treatment with chitin, a naturally occurring polysaccharide, claimed to be non-toxic and biodegradable, could be beneficial in retarding microbial activity (Duan and Zhang, 2013).

MARKETING SPECIFIC BRASSICAS

Head cabbage

Dense-headed white or Dutch autumn maturing cabbage (*B. oleracea* var. *capitata*) crops are stored for up to 10 months and ultimate sale year-round on to the fresh market, or for subsequent processing into increasingly popular coleslaw or as part of prepared salads. Such crops are grown extensively in north-west Europe (Germany, Netherlands, UK) and North America. Late maturing, dense-headed, red-headed types (*B. oleracea* var. *capitata* f. *rubra*) may be similarly stored for subsequent processing, especially in the Netherlands. Storage technology is increasingly sophisticated. Use is made of barns that are equipped with forced night air ventilation for cooling and dispersal of field heat. This permits preservation for 3–4 months until late February or early March. Methods providing closer control are required for longer term storage, involving purpose-built refrigerated stores in which cabbages are packed into palletized containers.

Maturation during the postharvest storage period is characterized by dry matter and sugar accumulation, and increasing firmness, with the highest quality in white cabbage developing with crops picked towards the end of the autumn harvesting period (Suojala, 2003) (see Fig. 8.13).

A precise refrigeration control system has over 2000 lines of IT instructions, catering for defrost, accurate maintenance of duct temperatures and monitoring to make the maximum use of low electricity tariffs. Precision is now increased by including touchscreen technology and digital control. There are large surface areas of crop in the store, so minimizing the machinery run-rates to make maximum use of low-cost electricity while achieving even air distribution is essential. Using 'defrost on demand' systems minimizes the amount of defrosting in the store.

White cabbage has been stored for extended periods under refrigeration in nitrogen atmospheres modified with 2–6% CO_2 and 1–5% O_2. These conditions result in fewer storage and trimming losses, longer retention of fresh colour, flavour and texture, and lower pathogen-induced spoilage. There may be indirect benefits from controlled atmosphere storage since reduced time is required in preparing the product for processing and marketing and the storage period can be extended to 10 months.

Cauliflower

Cauliflower (*B. oleracea* var. *botrytis*) is usually stored for short periods as a means of filling gaps in the supply chain caused by periods of hot dry weather. Under optimal conditions (0°C and 95% RH) cauliflower may be stored for up to 6 weeks. In practice, storage beyond 2–3 weeks is inadvisable. Cauliflower curd is very susceptible to damage during harvesting and even cryptic injuries will enlarge during storage into blemishes and discoloration associated with fungal and bacterial invasions. Spoilage can develop very quickly, leading to downgrading or outright rejection of the curds. Even during the normal time periods of distribution, curd discoloration develops and this is exacerbated where the heads are overwrapped with polyethylene film. *Alternaria* spp.

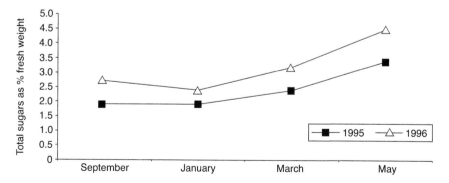

Fig. 8.13. White cabbage (*Brassica oleracea* var. *capitata*) core sugar content. (After Suojala, 2003).

causes leaf spots prior to harvest. These affect the wrapper leaves and can cause downgrading during cutting and grading.

Infection of the white curds is a source of much more serious wastage during storage and marketing. Small lesions (<5 mm) will spread and coalesce within 7 days of cutting, favoured by warm, moist conditions. Rapid cooling to remove field heat and cold-chain marketing at 4°C are recommended to inhibit *Alternaria* from inducing brown rotting of the curds. Since both *A. brassicae* and *A. brassicicola* are seed-borne pathogens, also capable of survival on soil debris, there are several avenues by which cauliflower crops may become infected. Transfer to the developing curds leads to postharvest sporulation and spoilage during storage, distribution and marketing.

Infection of cauliflower heads by downy mildew (*H. parasitica*) causes pale-greyish to brown discoloration on the curd surface. Inside the head, grey or black spotting and streaking extends through the bract tissues. Apparently healthy curds stored at 20°C and 70% RH are rapidly spoiled by downy mildew sporulation. Storage at 4°C will retard the expression of symptoms. Infection of seedlings may result in the pathogen entering a quiescent stage until the curd matures, which may then display brown wet lesions. Secondary invasion by soft rotting bacteria increases the rate of disintegration and deterioration of cauliflower heads. As with cabbage, ringspot (*M. brassicicola*), grey mould (*Botrytis cinerea*) and, in some instances, sooty moulds (*Cladosporium* spp.) have been associated with postharvest spoilage of cauliflower.

Several *Pseudomonas* spp. and *Erwinia carotovora* ssp. *carotovora* cause bacterial soft rots to cauliflower postharvest. Mechanical damage or bruising during harvesting and transit provides entry for soft rotting bacteria that may be spread between infected and healthy heads by the knives used for harvesting. Bacteria may also enter following primary infection by fungi such as *Alternaria*. Free water on cauliflower curds increases the rapidity of bacterial rotting, hence harvesting under wet conditions and overwrapping heads increases the likelihood of spoilage. Temperatures in excess of 10°C also encourage the rotting of cauliflower curds.

Brussels sprouts

Once sprout (*B. oleracea* var. *gemmifera*) buds have been harvested, they are either sent immediately to processing factories or packed into nets for the fresh market. Alternatively, the whole stem, with buds attached, may be marketed. This has become increasingly popular, especially in the run-up to the Christmas and New Year periods. Detached buds may be stored for a few days, but in ambient conditions the ends of sprout butts rapidly become yellowed or browned, reducing their acceptability to the consumer. Yellowing results from tissue senescence and this is accelerated by ethylene released from the crop or from microbes colonizing the sprouts. Storage for longer periods is achieved at 1–2°C and 95% RH; sprouts held in ice-bank coolers retain marketable quality for several weeks. A similar range of fungal and bacterial pathogens to those that reduce the postharvest quality of cabbage and cauliflower also spoil Brussels sprouts. In particular, the normally saprophytic fungus *Rhizopus stolonifera* (*R. nigricans*) will parasitize sprout buds, especially in warm, moist conditions. Infection of buds by powdery mildew (*Erysiphe cruciferarum*) in the field can result in dark, black lesions developing, especially during frosty periods.

Green broccoli and sprouting broccoli

Physiological senescence and yellowing are frequently followed by secondary fungal and bacterial rotting, which is promoted by high RH (>95%). Pathogens that damage broccoli (*B. oleracea* var. *italica*) heads include those affecting other brassicas. Each of these microbes is encouraged by moist, warm (>10°C) conditions. As with cauliflower (*B. oleracea* var. *botrytis*), modified and controlled atmosphere storage of these crops aids the retention of green coloration and delays senescence and the extension of flower buds. But there may be changes to texture and the development of off-odours. Reduced O_2 content (1–6%) failed to reduce losses caused by bacteria and while some fungi (*Alternaria* spp. and *Cladosporium* spp.) diminished the saprophytic *Mucor* spp. were encouraged. In California, USA, broccoli is harvested at an immature stage that extends its shelf life. The spears are shipped rapidly to Japan where they achieve a significant price that more than compensates for the sacrifice in yield. Green broccoli spears are especially susceptible to soft rot infection as they reach field maturity, caused by *Pseudomonas fluorescens* and other similar pathogens. Infections may render spears unharvestable or, where present at low levels, develop into unsightly and unacceptable lesions during cold-chain handling or when displayed for retail sale.

Chinese cabbage

The storage life of Chinese cabbage is relatively short, extending to only a few weeks at 0–1°C. Yellowing caused by tissue senescence is a major source of spoilage, even under these conditions. Leaf spotting (*Alternaria* spp.) will develop in stores at 0–1°C and 95–97% RH and may be accelerated at higher temperatures. Soft rotting (*E. carotovora* ssp. *carotovora*) is a major source of spoilage during transit and storage and blackening of the leaf veins is caused by *Xanthomonas campestris*.

MARKET DIVERSIFICATION

Arguably, the supermarkets have been at least partially instrumental in encouraging crop mechanization and subsequent roboticization. Recently, however, markets have become increasingly diversified. Direct delivery of 'fruit and vegetable boxes', especially of organically grown crops, from producer to consumer has expanded over the last decade. Each box is filled manually in accordance with customers' requirements. This does, however, add extra layers of complexity, cost and price. Smaller greengrocery shops, market stalls and direct doorstep deliveries have become increasingly popular as measurements of the carbon footprints of trading feature in consumer considerations. During the COVID-19 pandemic there was a very substantial increase in fresh produce sales through the wholesale markets (Dixon, 2021). Purchasing from local suppliers is now a greater feature of how people shop. Some of this has been accelerated as an outcome of the travel restrictions imposed as part of the COVID-19 pandemic.

PROCESSED BRASSICAS

Food processors accept outgraded and overripe produce. This is an expanding trade providing fillings for sandwiches, delicatessen portions of cauliflower (*B. oleracea* var. *botrytis*) or green broccoli (*B. oleracea* var. *italica*) florets and supplying the 'pre-prepared meals' trade for hotels, restaurants, airlines and events markets. These processors utilize both fresh brassicas, such as baby leaves, and cooked portions of Brussels sprouts, cauliflower and the swedes (*B. napus* ssp. *rapifera*) needed for Hogmanay and Burn's Night events. Processing also includes the preparation of European cabbage (*B. oleracea* var. *capitata*) and Chinese cabbage in sauerkraut and kimchi pickled produce. Also, various forms of mustard produced from crushed and purified seed have long been a condiment for meat dishes.

Market demand motivated by health considerations has transformed watercress (*N. officinale*) from a simple garnish to a major salad vegetable, consumed as fresh shoots and leaves or comminuted into hot or cold soup. Demand for wasabi (*Wasabia japonica*) expanded production from Japan into Australia, the UK and the USA. High-quality fresh roots and processed powders are now used worldwide in sushi dishes. The galaxy of Asian brassicas has been transplanted from Asia into Europe and North America, first satisfying the diaspora and then quickly taken up by their neighbours. Chinese cabbage is now a commonplace crop, used in similar ways to European cabbage, especially as a salad vegetable. Daikon radish (*R. sativus* ssp. *longipinnatus*) is increasingly popular as a salad constituent and has triggered a resurgence of interest and appreciation of European radish (*R. sativus*). This continues the trends for the consumption of vegetables internationally established during the colonization of the Americas, India, Australia and New Zealand, which resulted in cauliflower and other European brassicas being grown in new regions and subsequently local plant breeders developing cultivars more suitable for warmer climates.

The huge diversity of brassica vegetables and their use as fresh, prepared and processed food is increasingly appreciated because of their health-promoting properties and the continuous search for 'something new and novel' as required by the inquisitive younger generation. Changing social pressures place new requirements on the growers, increasing the use of mechanization, robotics and automation. Rapidly growing brassica crops, such as baby leaf salads, can be produced in urban factories – vertical farms using wholly artificial LED lighting and almost total automation. Urban or factory farming could become profitable once the problems of uneconomic energy footprints, the utilization of natural resources in the fabrication of buildings, and transport to and from the production sites are solved. Then novel means of food production could help resolve the dilemma of how to feed expanding populations and very imaginative automation and robotics will play increasing roles in making urban and factory farming more cost-effective.

REFERENCES

Aghdam, M.S. and Luo, Z.S. (2021) Exogenous application of phytosulfokine α (PSKα) delays senescence in broccoli florets during cold storage by ensuring intracellular ATP availability and avoiding intracellular ROS accumulation. *Scientia Horticulturae* 276, 109745. DOI: 10.1016/j.scienta.2020.109745.

Ahmad, W., Ayyub, C.M., Shehzad, M.A., Ziaf, K., Ijaz, M. *et al.* (2019) Supplemental potassium mediates antioxidant metabolism, physiological processes, and osmoregulation to confer salt stress tolerance in cabbage (*Brassica oleracea* L.). *Horticulture, Environment, and Biotechnology* 60(6), 853–859. DOI: 10.1007/s13580-019-00172-2.

Al Ubeed, H.M.S., Wills, R.B.H., Bowyer, M.C. and Golding, J.B. (2019) Interaction of the hydrogen sulphide inhibitor, propargylglycine (PAG), with hydrogen sulphide on postharvest changes of the green leafy vegetable, pak choy. *Postharvest Biology and Technology* 147, 54–58. DOI: 10.1016/j.postharvbio.2018.09.011.

Al-Bakheit, A. and Abu-Qatouseh, L. (2020) Sulforaphane from broccoli attenuates inflammatory hepcidin by reducing IL-6 secretion in human HepG2 cells. *Journal of Functional Foods* 75, 104210. DOI: 10.1016/j.jff.2020.104210.

Albornoz, K. and Cantwell, M.I. (2016) Fresh-cut kale quality and shelf-life in relation to leaf maturity and storage temperature. *Acta Horticulturae* 1141, 109–116. DOI: 10.17660/ActaHortic.2016.1141.11.

Anon (2015) *Crop Module 5377: Swedes, Turnips and Kohlrabi.* Red Tractor Assurance for Farms – Fresh Produce, Assured Food Standards, London.

Anon (2018) *Crop Module 6615: Brussels Sprouts.* Red Tractor Assurance for Farms – Fresh Produce, Assured Food Standards, London.

Babik, I., Rumpel, J. and Elkner, K. (1996) The influence of nitrogen fertilisation on yield, quality and senescence of Brussels sprouts. *Acta Horticulturae* 407, 353–359. DOI: 10.17660/ActaHortic.1996.407.45.

Bac, C.W., van Henten, E.J., Hemming, J. and Edan, Y. (2014) Harvesting robots for high-value crops: state-of-the-art review and challenges ahead. *Journal of Field Robotics* 31(6), 888–911. DOI: 10.1002/rob.21525.

Badea, C. and Basu, S.K. (2009) The effect of low temperature on metabolism of membrane lipids in plants and associated gene expression. *Plant Omics* 2(2), 78–84.

Baker, R.L., Leong, W.F., Brock, M.T., Markelz, R.J.C., Covington, M.F. *et al.* (2015) Modeling development and quantitative trait mapping reveal independent genetic modules for leaf size and shape. *The New Phytologist* 208(1), 257–268. DOI: 10.1111/nph.13509.

Barcena, A., Bahima, J.V., Casajús, V., Martínez, G., Lauff, D. *et al.* (2020) The degradation of chloroplast components during postharvest senescence of broccoli florets is delayed by low-intensity visible light pulses. *Postharvest Biology and Technology* 168, 111249. DOI: 10.1016/j.postharvbio.2020.111249.

Ben-Fadhel, Y., Ziane, N., Salmieri, S. and Lacroix, M. (2018) Combined post-harvest treatments for improving quality and extending shelf-life of minimally processed broccoli florets (*Brassica oleracea* var. *italica*). *Food and Bioprocess Technology* 11, 84–95. DOI: 10.1007/s11947-017-1992-2.

Berndtsson, E., Andersson, R., Johansson, E. and Olsson, M.E. (2020) Side streams of broccoli leaves: a climate smart and healthy food ingredient. *International Journal of Environmental Research and Public Health* 17(7), 2406. DOI: 10.3390/ijerph17072406.

Boonyakiat, D. and Boonprasom, P. (2012) Effect of active packaging on quality of Chinese kale. (Special issue: Agricultural and natural resources.) *Chiang Mai University Journal of Natural Sciences* 11(1 (Special)), 215–222.

Borkowski, J., Dyki, B., Oskiera, M., Machlanska, A. and Felczynska, A. (2016) The prevention of tipburn on chinese cabbage (*Brassica rapa* L. var. *pekinensis* (Lour.) Olson) with foliar fertilizers and biostimulators. *Journal of Horticultural Research* 24(1), 47–56. DOI: 10.1515/johr-2016-0006.

Cai, J.H., Cheng, S.C., Luo, F., Zhao, Y.B., Wei, B.D *et al.* (2019) Influence of ethylene on morphology and pigment changes in harvested broccoli. *Food and Bioprocess Technology* 12(5), 883–897. DOI: 10.1007/s11947-019-02267-1.

Caleb, O.J., Ilte, K., Fröhling, A., Geyer, M. and Mahajan, P.V. (2016) Integrated modified atmosphere and humidity package design for minimally processed broccoli (*Brassica oleracea* L. var. *italica*). *Postharvest Biology and Technology* 121, 87–100. DOI: 10.1016/j.postharvbio.2016.07.016.

Cantwell, M.I., Albornoz, K. and Hong, G. (2019) Ammonia accumulation is a useful indicator of the postharvest freshness and quality of spinach and kale. *Acta Horticulturae* 1256, 303–309. DOI: 10.17660/ActaHortic.2019.1256.42.

Cargill, B.F. (ed.) (1976) *The Potato Storage: Design, Construction, Handling and Environmental Control.* Michigan State University, East Lansing, MI and American Society for Agricultural Engineering, St Joseph, MI.

Cefola, M., Amodio, M.L. and Colelli, G. (2016) Design of the correct modified atmosphere packaging for fresh-cut broccoli raab. *Acta Horticulturae* 1141, 117–122. DOI: 10.17660/ActaHortic.2016.1141.12.

Chalupowicz, D., Osher, Y., Maurer, D., Ovadia-Sadeh, A., Lurie, S. *et al.* (2019) Characterizing the genetic potential of cabbage for summer storage. *Acta Horticulturae* 1256, 563–566. DOI: 10.17660/ActaHortic.2019.1256.80.

Chan, K.Y. and Heenan, D.P. (1996) The influence of crop rotation on soil structure and soil physical properties under conventional tillage. *Soil and Tillage Research* 37(2–3), 113–125. DOI: 10.1016/0167-1987(96)01008-2.

Conversa, G., Lazzizera, C., Bonasia, A. and Elia, A. (2020) Harvest season and genotype affect head quality and shelf-life of ready-to-use broccoli. *Agronomy* 10(4), 527. DOI: 10.3390/agronomy10040527.

Cox, D.N., Melo, L., Zabaras, D. and Delahunty, C.M. (2012) Acceptance of health-promoting *Brassica* vegetables: the influence of taste perception, information and attitudes. *Public Health Nutrition* 15(8), 1474–1482. DOI: 10.1017/S1368980011003442.

Cox, E.F. (1977) Pepper spot in white cabbage: a literature review. *Agricultural Development and Advisory Service (ADAS) Quarterly Review* 25, 81–86.

Crowhurst, R. (2018) The importance of removing field heat. *The Vegetable Farmer* September, 25–27.

Cubeta, M.A., Cody, B.R., Sugg, R.E. and Crozier, C.R. (2000) Influence of soil calcium, potassium, and pH on development of leaf tipburn of cabbage in eastern North Carolina. *Communications in Soil Science and Plant Analysis* 31(3–4), 259–275. DOI: 10.1080/00103620009370435.

Cunnington, A. (2019) *Store Managers' Guide*. Potato Council and Agricultural and Horticultural Development Board (ADHB), Stoneleigh, UK.

Cunnington, A., Mawson, K., Briddon, A. and Storey, R.M. (1992) Low temperature storage for quality crops. In: *Production and Protection of Potatoes. Conference Proceedings, 10–11 December 1992, Churchill College, Cambridge, UK. Aspects of Applied Biology Vol. 33*. Association of Applied Biologists, Warwick, UK, pp. 205–212.

Davies, A.C.W. and Wheeler, J.A. (1968) Effect of machine stripping on the yield and grade of Brussels sprouts. *Journal of Agricultural Engineering Research* 13(3), 241–244. DOI: 10.1016/0021-8634(68)90104-2.

Deane, C.R., Dix, P.J. and Fuller, M.P. (1996) Selection of hydroxyproline resistant proline accumulating mutants of cauliflower for improvement of frost resistance. *Acta Horticulturae* 407, 123–129. DOI: 10.17660/ActaHortic.1996.407.14.

DeJesus, J.M. and Venkatesh, S. (2020) Show or tell: children's learning about food from action vs verbal testimony. (Special issue: Parenting and the development of eating behaviors in the growing child.) *Pediatric Obesity* 15(10), e12719. DOI: 10.1111/ijpo.12719.

Díaz, M., de Haro, V., Muñoz, R. and Quiles, M.J. (2007) Chlororespiration is involved in the adaptation of Brassica plants to heat and high light intensity. *Plant, Cell & Environment* 30(12), 1578–1585. DOI: 10.1111/j.1365-3040.2007.01735.x.

Diego-Mas, J.A., Poveda-Bautista, R. and Garzon-Leal, D. (2017) Using RGB-D sensors and evolutionary algorithms for the optimization of workstation layouts. *Applied Ergonomics* 65, 530–540. DOI: 10.1016/j.apergo.2017.01.012.

Dixon, G.R. (2021) Thriving in changing markets. *The Vegetable Farmer* January, 20–21.

Dogan, A., Topcu, Y. and Erkan, M. (2018) UV-C illumination maintains post-harvest quality of minimally processed broccoli florets under modified atmosphere packaging. *Acta Horticulturae* 1194, 537–544. DOI: 10.17660/ActaHortic.2018.1194.78.

Downs, C.G., Somerfield, S.D. and Davey, M.C. (1997) Cytokinin treatment delays senescence but not sucrose loss in harvested broccoli. *Postharvest Biology and Technology* 11(2), 93–100. DOI: 10.1016/S0925-5214(97)01419-1.

Duan, J. and Zhang, S. (2013) Application of chitosan based coatings in fruit and vegetable preservation: a review. *Journal of Food Processing and Technology* 4(5), 271–280.

Duckett, T., Pearson, S., Blackmore, S. and Grieve, B. (2018). *Agricultural Robotics: The Future of Robotic Agriculture*. UK-RAS Network (Robotics and Autonomous Systems), London. Available at: www.ukras.org./wp-content/uploads/2021

/01/UKRASWP_AgriculturalRobotics2018_online.pdf (accessed 5 August 2023).

Duarte-Sierra, A., Corcuff, R., Angers, P. and Arul, J. (2012) Effect of heat treatment using humidified air on electrolyte leakage in broccoli florets: temperature-time relationships. *Acta Horticulturae* 945, 149–155. DOI: 10.17660/ActaHortic.2012.945.20.

Duarte-Sierra, A., Nadeau, F., Angers, P., Michaud, D. and Arul, J. (2019) UV-C hormesis in broccoli florets: preservation, phyto-compounds and gene expression. *Postharvest Biology and Technology* 157, 110965. DOI: 10.1016/j.postharvbio.2019.110965.

Duarte-Sierra, A., Munzoor Hasan, S.M., Angers, P. and Arul, J. (2020) UV-B radiation hormesis in broccoli florets: glucosinolates and hydroxy-cinnamates are enhanced by UV-B in florets during storage. *Postharvest Biology and Technology* 168, 111278. DOI: 10.1016/j.postharvbio.2020.111278.

de Vries, J., Evers, J.B., Dicke, M. and Poelman, E.H. (2019) Ecological interactions shape the adaptive value of plant defence: herbivore attack versus competition for light. *Functional Ecology* 33(1), 129–138. DOI: 10.1111/1365-2435.13234.

Du, D., Xie, L., Wang, J. and Deng, F. (2016) Development and tests of a self-propelled cabbage harvester in China. In: *Proceedings ASABE (American Society for Agricultural and Biological Engineers) Annual Meeting, Orlando, FL, USA, 17–20 July 2016.* DOI: 10.13031/aim.20162459786.

Ekman, J.H., Goldwater, A., Marques, J.R., Winley, E., Holford, P. *et al.* (2019) Preserving broccoli quality from harvest to retail: use of 1-MCP. *Acta Horticulturae* 1256, 23–30. DOI: 10.17660/ActaHortic.2019.1256.4.

El Didamony, M.I. and El Shal, A.M. (2020) Fabrication and evaluation of a cabbage harvester prototype. *Agriculture* 10(12), 631. DOI: 10.3390/agriculture10120631.

Everaarts, A.P. and Blom-Zandstra, M. (2001) Internal tipburn of cabbage (*Brassica oleracea* var. *capitata*). *The Journal of Horticultural Science and Biotechnology* 76(5), 515–521.

Fadhil, R., Agustina, R. and Hayati, R. (2021) Sensory assessment of sauerkraut using a non-numeric approach based on multi-criteria and multi-person aggregation. *Bulletin of the Transilvania University of Brasov, Series II: Forestry, Wood Industry, Agricultural Food Engineering* 13(62 Part 2), 111–118. DOI: 10.31926/but.fwiafe.2020.13.62.2.10.

Fisher, N.M., Gooderham, P.T. and Ingram, J. (1975) The effect on the yields of barley and kale of soil conditions induced by cultivation at high moisture content. *The Journal of Agricultural Science* 85(3), 385–393. DOI: 10.1017/S0021859600062250.

Frankowska, A., Jeswani, H.K. and Azapagic, A. (2019) Environmental impacts of vegetables consumption in the UK. *Science of the Total Environment* 682, 80–105. DOI: 10.1016/j.scitotenv.2019.04.424.

Gajewski, M., Smarz, M., Radzanowska, J. and Pudzianowska, M. (2015) The influence of storage conditions on quality parameters of head cabbage with conical heads. *Acta Horticulturae* 1099, 233–240. DOI: 10.17660/ActaHortic.2015.1099.25.

Gaskin, J.W., Cabrera, M.L., Kissel, D.E. and Hitchcock, R. (2020) Using the cover crop N calculator for adaptive nitrogen fertilizer management: a proof of concept. *Renewable Agriculture and Food Systems* 35(5), 550–560. DOI: 10.1017/S1742170519000152.

Geeson, J.D. (1983) Brassicas. In: Dennis, C. (ed.) *Post-Harvest Pathology of Fruits and Vegetables.* Academic Press, London, pp. 125–156.

Geeson, J.D. and Browne, K.M. (1980) Controlled atmosphere storage of winter white cabbage. *Annals of Applied Biology* 95(2), 267–272. DOI: 10.1111/j.1744-7348.1980.tb04746.x.

Gomez-Lobato, M.E., Hasperué, J.H., Civello, P.M., Chaves, A.R. and Martínez, G.A. (2012) Effect of 1-MCP on the expression of chlorophyll degrading genes during senescence of broccoli (*Brassica oleracea* L.). *Scientia Horticulturae* 144, 208–211. DOI: 10.1016/j.scienta.2012.07.017.

Gooderham, P.T. (1977) Some aspects of soil compaction, root growth and crop yield. *Agricultural Progress* 52, 33–44.

Grevsen, K. and Olesen, J.E. (1994) Modelling development and quality of cauliflower. *Acta Horticulturae* 371, 151–160. DOI: 10.17660/ActaHortic.1994.371.18.

Grieve, B.D., Duckett, T., Collison, M., Boyd, L., West, J. *et al.* (2019) The challenges posed by global broadacre crops in delivering smart agri-robotic solutions: a fundamental rethink is required. *Global Food Security* 23, 116–124. DOI: 10.1016/j.gfs.2019.04.011.

Guo, Y., Gao, Z., Li, L., Wang, Y., Zhao, H. *et al.* (2013) Effect of controlled atmospheres with varying O_2/CO_2 levels on the postharvest senescence and quality of broccoli (*Brassica oleracea* L. var. *italica*) florets. *European Food Research and Technology* 237(6), 943–950. DOI: 10.1007/s00217-013-2064-0.

Hagen, S.F. and Larsen, H. (2020) Shelf life of broccoli in different packages stored under two temperature regimes. *Acta Horticulturae* 1275, 329–336. DOI: 10.17660/ActaHortic.2020.1275.45.

Hagen, S.F., Borge, G.I.A., Solhaug, K.A. and Bengtsson, G.B. (2009) Effect of cold storage and harvest date on bioactive compounds in curly kale (*Brassica oleracea* L. var. *acephala*). *Postharvest Biology and Technology* 51(1), 36–42. DOI: 10.1016/j.postharvbio.2008.04.001.

Han, W., Huang, L. and Owojori, O.J. (2020) Foliar application of zinc alleviates the heat stress of pakchoi (*Brassica chinensis* L.). *Journal of Plant Nutrition* 43(2), 194–213. DOI: 10.1080/01904167.2019.1659350.

He, Q. and Xiao, K. (2018) Quality of broccoli (*Brassica oleracea* L. var. *italica*) in modified atmosphere packaging made by gas barrier-gas promoter blending materials. *Postharvest Biology and Technology* 144, 63–69. DOI: 10.1016/j.postharvbio.2018.05.013.

Heath, D.W., Earle, E.D. and Dickson, M.H. (1994) Introgressing cold-tolerant Ogura cytoplasm from rapeseed into pak choi and Chinese cabbage. *HortScience* 29(3), 202–203. DOI: 10.21273/HORTSCI.29.3.202.

Helland, H.S., Leufvén, A., Bengtsson, G.B., Pettersen, M.K., Lea, P. *et al.* (2016) Storage of fresh-cut swede and turnip: effect of temperature, including sub-zero temperature, and packaging material on sensory attributes, sugars and glucosinolates. *Postharvest Biology and Technology* 111, 370–379. DOI: 10.1016/j.postharvbio.2015.09.011.

Helland, H., Leufven, A., Bengtsson, G.B., Larsen, H., Nicolaisen, E.M. *et al.* (2018) Respiration rate and changes in composition of volatiles during short-term storage of minimally processed root vegetables. *Acta Horticulturae* 1209, 365–370. DOI: 10.17660/ActaHortic.2018.1209.54.

Holmes, M.J. (2015) *Investigation into Non-destructive Detection of Brown Heart in Swedes (Brassica napus) Using Ultrasound. Project FV 444 Final Report.* Agricultural and Horticultural Development Board (AHDB), Stoneleigh, UK, p. 32.

Holt, J.B. and Sharp, J.R. (1989) *A Simple Individual Cup Conveyor for Vegetable Harvesting. Divisional Note 1499.* AFRC Institute of Engineering Research, Silsoe, UK.

Hong, S.-J., Kim, B.-S., Park, N.-I. and Eum, H.-L. (2017) Influence of nitrogen fertilization on storability and the occurrence of black speck in spring kimchi cabbage. [Korean] *Horticultural Science & Technology* 35(6), 727–736. DOI: 10.12972/kjhst.20170077.

Hounsome, N., Hounsome, B., Tomos, D. and Edwards-Jones, G. (2009) Changes in antioxidant compounds in white cabbage during winter storage. *Postharvest Biology and Technology* 52(2), 173–179. DOI: 10.1016/j.postharvbio.2008.11.004.

Huang, Y.-C., Yang, Y.-H., Sridhar, K. and Tsai, P.-J. (2021) Synergies of modified atmosphere packaging and high-voltage electrostatic field to extend the shelf-life of fresh-cut cabbage and baby corn. (Special section: Emerging processing technologies to improve the safety and quality of foods.) *LWT – Food Science and Technology* 138, 110559. DOI: 10.1016/j.lwt.2020.110559.

Huffman, W.E. (2012) The status of labor-saving mechanisation in the U.S. fruit and vegetable harvesting. *Working Paper 12009.* Department of Economics, Iowa State University, Ames, IA, 25 pp.

Hunter, P.J., Jones, J.E. and Walsh, J.A. (2002) Involvement of Beet western yellows virus, Cauliflower mosaic virus, and Turnip mosaic virus in internal disorders of stored white cabbage. *Phytopathology* 92(8), 816–826. DOI: 10.1094/PHYTO.2002.92.8.816.

Hyun, J.-E., Bae, Y.-M., Yoon, J.-H. and Lee, S.-Y. (2015) Preservative effectiveness of essential oils in vapor phase combined with modified atmosphere packaging against spoilage bacteria on fresh cabbage. *Food Control* 51, 307–313. DOI: 10.1016/j.foodcont.2014.11.030.

Iizuka, M., Seyama, S., Suzuki, O. and Koizumi, T. (2013) Relationship between internal black spot of cabbage and cell death. [Japanese] *Horticultural Research* 12(1), 109–114. DOI: 10.2503/hrj.12.109.

Imahori, Y., Kodera, K., Endo, H., Onishi, T., Fujita, T. *et al.* (2016) The seasonal variation of redox status in komatsuna (*Brassica rapa* var. *perviridis*) leaves. *Scientia Horticulturae* 210, 49–56. DOI: 10.1016/j.scienta.2016.07.010.

Izumi, H. and Inoue, A. (2018) Bactericidal effect of four types of electrolyzed water on fresh-cut vegetables. *Acta Horticulturae* 1194, 1487–1494. DOI: 10.17660/ActaHortic.2018.1194.208.

Jenni, S., Dutilleul, P., Yamasaki, S. and Tremblay, N. (2001a) Brown bead of broccoli. I. Response of the physiological disorder to management practices. *HortScience* 36(7), 1224–1227. DOI: 10.21273/HORTSCI.36.7.1224.

Jenni, S., Dutilleul, P., Yamasaki, S. and Tremblay, N. (2001b) Brown bead of broccoli. II. Relationships of the physiological disorder with nutritional and meteorological variables. *HortScience* 36(7), 1228–1234. DOI: 10.21273/HORTSCI.36.7.1228.

Jenni, S., Dutilleul, P., Yamasaki, S. and Tremblay, N. (2017) Brown bead of broccoli. II. Relationships of the physiological disorder with nutritional and meteorological variables. *Journal of Agricultural and Food Chemistry* 65(39), 8538–8543.

Johnson, I.T. (2000) Brassica vegetables and human health: glucosinolates in the food chain. *Acta Horticulturae* 539, 39–44. DOI: 10.17660/ActaHortic.2000.539.3.

Johnson, J.R. (1991) Calcium accumulation, calcium distribution, and biomass partitioning in collards. *Journal of the American Society for Horticultural Science* 116(6), 991–994. DOI: 10.21273/JASHS.116.6.991.

Kabakeris, T. (2018) *Storability of Broccoli: Investigations of Optical Monitoring, Chlorophyll Degradation and Predetermination in the Field. Bornimer Agrartechnische Berichte.* Leibniz-Institut fur Agrartechnik und Biookonomie, Potsdam, Germany.

Kabakeris, T., Poth, A., Intreß, J., Schmidt, U. and Geyer, M. (2015a) Detection of postharvest quality loss in broccoli by means of non-colorimetric reflection spectroscopy and hyperspectral imaging. *Computers and Electronics in Agriculture* 118, 322–331. DOI: 10.1016/j.compag.2015.09.013.

Kabakeris, T., Ghadiri Khozroughi, A., Keiser, A., Geyer, M. and Huyskens-Keil, S. (2015b) The role of chlorophyll derivatives in postharvest quality determination of broccoli. *Acta Horticulturae* 1079, 201–205. DOI: 10.17660/ActaHortic.2015.1079.22.

Kader, A.A. (2010) Future of modified atmosphere research. *Acta Horticulturae* 857, 213–218. DOI: 10.17660/ActaHortic.2010.857.24.

Kanamitsu, M. and Yamamoto, K. (1996) Development of Chinese cabbage harvester. *Japanese Agricultural Research Quarterly* 30(1), 35–41.

Kasim, R., Kasim, M.U. and Erkal, S. (2007) The effect of packaging after 1-MCP treatment on color changes and chlorophyll degradation of broccoli (*Brassica oleracea* var. *italica* cv monopoly). *Journal of Food, Agriculture and Environment* 5(3/4), 48–51.

Kim, Y.C., Hussain, M., Anarjan, M.B. and Lee, S. (2020) Examination of glucoraphanin content in broccoli seedlings over growth and the impact of hormones and sulfur-containing compounds. *Plant Biotechnology Reports* 14(4), 491–496. DOI: 10.1007/s11816-020-00617-6.

Koide, S., Kumada, R., Hayakawa, K., Kawakami, I., Orikasa, T. *et al.* (2019) Survival of cut cabbage subjected to subzero temperatures. *Acta Horticulturae* 1256, 329–334. DOI: 10.17660/ActaHortic.2019.1256.46.

Koller, M., Vieweger, A., Total, R., Bauermeister, R., Suter, D. *et al.* (2008) Nitrogen effects of green manuring on organic cabbage crops. [German] *Agrarforschung* 15(6), 264–269.

Kondo, N. (2010) Automation on fruit and vegetable grading system and food traceability. *Trends in Food Science & Technology* 21(3), 145–152. DOI: 10.1016/j.tifs.2009.09.002.

Kondo, N. (2013) Robotics and automation in the fresh produce industry. In: Caldwell, D.G. (ed.) *Robotics and Automation in the Food Industry: Current and Future Technologies.* Woodhead, Cambridge, pp. 385–400.

Ku, K.M., Choi, J.H., Kim, H.S., Kushad, M.M., Jeffery, E.H. *et al.* (2013) Methyl jasmonate and 1-methylcyclopropene treatment effects on quinone reductase inducing activity and post-harvest quality of broccoli. *PLoS ONE* 8(10), e77127. DOI: 10.1371/journal.pone.0077127.

Ku, V.V.V. and Wills, R.B.H. (1999) Effect of 1-methylcyclopropene on the storage life of broccoli. *Postharvest Biology and Technology* 17(2), 127–132. DOI: 10.1016/S0925-5214(99)00042-3.

Kumar, S. and Kumar, M. (2007) Pre-cooling of horticultural produce. In: Goel, A.K., Kumar, R. and Mann, S.S. (eds) *Post Harvest Management and Value Addition.* Daya Publishing House, New Delhi, India, pp. 155–210.

Kusumam, K., Krajník, T., Pearson, S., Duckett, T. and Cielniak, G. (2017) 3D-vision based detection, localization, and sizing of broccoli heads in the field. *Journal of Field Robotics* 34(8), 1505–1518. DOI: 10.1002/rob.21726.

Kviesis, A. and Osadcuks, V. (2013) Development of robot manipulator and motion control using inverse kinematics for robotized vegetable harvesting. In: *Proceedings of the International Conference on Applied Information and Communication Technologies (AICT2013), 25–26 April 2013.* Jelgava, Latvia, pp. 137–146.

Lamichhane, J.R. and Venturi, V. (2015) Synergisms between microbial pathogens in plant disease complexes: a growing trend. *Frontiers in Plant Science* 6, 385. DOI: 10.3389/fpls.2015.00385.

Lee, Y.J., Ha, J.Y., Oh, J.E. and Cho, M.S. (2014) The effect of LED irradiation on the quality of cabbage stored at a low temperature. *Food Science and Biotechnology* 23(4), 1087–1093. DOI: 10.1007/s10068-014-0149-6.

Lema, M., Ali, M.Y. and Retuerto, R. (2019) Domestication influences morphological and physiological responses to salinity in *Brassica oleracea* seedlings. *AoB Plants* 11(5), plz046. DOI: 10.1093/aobpla/plz046.

Liao, J., Xu, Q., Xu, H. and Huang, D. (2019) Natural farming improves soil quality and alters microbial diversity in a cabbage field in Japan. *Sustainability* 11, 3131. DOI: 10.3390/su11113131.

Li, G.-Q., Xie, Z.-J. and Yao, X.-Q. (2013) Study on combining ability and heritability analysis of main economic characters for cauliflower. *Plant Science Journal* 31(2), 143–150. DOI: 10.3724/SP.J.1142.2013.20143.

Li, Y.-X., Wills, R.B.H. and Golding, J.B. (2017) Interaction of ethylene concentration and storage temperature on postharvest life of the green vegetables pak choi, broccoli, mint, and green bean. *The Journal of Horticultural Science and Biotechnology* 92(3), 288–293. DOI: 10.1080/14620316.2016.1263545.

Lin, C.-W., Fu, S.-F., Liu, Y.-J., Chen, C.-C., Chang, C.-H. *et al.* (2019) Analysis of ambient temperature-responsive transcriptome in shoot apical meristem of heat-tolerant and heat-sensitive broccoli inbred lines during floral head formation. *BMC Plant Biology* 19(1), 3. DOI: 10.1186/s12870-018-1613-x.

Liu, Y. and Darmency, H. (2019) Morphological differences among *Raphanus raphanistrum* populations and their relationship to related crops. *Plant Breeding* 138(6), 907–915. DOI: 10.1111/pbr.12729.

Lougheed, E.C. (1983) 'Fresh' Ontario-grown vegetables in winter: controlled atmosphere and 'ethylene-free' storage. *Highlights of Agricultural Research in Ontario* 6(1), 7–9.

Loughton, A. and Riekels, J.W. (1988) Black speck in cauliflower. *Canadian Journal of Plant Science* 68(1), 291–294. DOI: 10.4141/cjps88-037.

Lu, Y., Saeys, W., Kim, M., Peng, Y. and Lu, R. (2020) Hyperspectral imaging technology for quality and safety evaluation of horticultural products: a review and celebration of the past 20-year progress. *Postharvest Biology and Technology* 170, 111318. DOI: 10.1016/j.postharvbio.2020.111318.

Magnusson, M. (2002) Mineral fertilisers and green mulch in Chinese cabbage (*Brassica pekinsensis* (Lour.) Rupr.): effect on nutrient uptake, yield and internal tipburn. *Acta Agriculturae Scandinavica, Section B – Soil & Plant Science* 52(1), 25–35. DOI: 10.1080/090647102320260017.

Mahajan, P.V., Oliveira, F.A.R., Montanez, J.C. and Frias, J. (2007) Development of user-friendly software for design of modified atmosphere packaging for fresh and fresh-cut produce. *Innovative Food Science & Emerging Technologies* 8(1), 84–92. DOI: 10.1016/j.ifset.2006.07.005.

Makhlouf, J., Castaigne, F., Arul, J., Willemot, C. and Gosselin, A. (1989) Long-term storage of broccoli under controlled atmosphere. *HortScience* 24(4), 637–639. DOI: 10.21273/HORTSCI.24.4.637.

Makino, Y. and Amino, G. (2020) Digitization of broccoli freshness integrating external color and mass loss. *Foods* 9(9), 1305. DOI: 10.3390/foods9091305.

Makino, Y., Inoue, J., Wang, H.-W., Yoshimura, M., Maejima, K. *et al.* (2020) Induction of terminal oxidases of electron transport chain in broccoli heads under controlled atmosphere storage. *Foods* 9(4), 380. DOI: 10.3390/foods9040380.

Manolopoulou, E. and Varzakas, T. (2011) Effect of storage conditions on the sensory quality, colour and texture of fresh-cut minimally processed cabbage with the addition of ascorbic acid, citric acid and calcium chloride. *Food and Nutrition Sciences* -2(-9), 956–963. DOI: 10.4236/fns.2011.29130.

Martínez-Hernández, G.B., Formica, A.C., Falagán, N., Artés, F., Artés-Hernández, F. *et al.* (2013) Extending the shelf life of the new Bimi® broccoli by controlled atmosphere storage. *Acta Horticulturae* 1012, 925–932. DOI: 10.17660/ActaHortic.2013.1012.124.

Martinez-Hernandez, G.B., Navarro-Rico, J., Gomez, P.A., Artes, F. and Artes-Hernandez, F. (2018) Emerging sanitizing techniques on Inoculated fresh-cut Bimi® broccoli. *Acta Horticulturae* 1209, 353–358. DOI: 10.17660/ActaHortic.2018.1209.52.

Martinez-Hernandez, G.B., Venzke-Klug, T., Carrion-Monteagudo, M.D.M., Artés Calero, F., López-Nicolás, J.M. *et al.* (2019) Effects of α-, β- and maltosyl-β-cyclodextrins use on the glucoraphanin-sulforaphane system of broccoli juice. *Journal of the Science of Food and Agriculture* 99(2), 941–946. DOI: 10.1002/jsfa.9269.

Mashabela, M.N., Maboko, M.M., Soundy, P. and Sivakumar, D. (2018) Variety specific responses of cauliflower varieties (*Brassica oleracea* var. botrytis) to different N application rates on yield, colour and ascorbic acid content at harvest. *Acta Agriculturae Scandinavica, Section B – Soil & Plant Science* 68(6), 541–545. DOI: 10.1080/09064710.2018.1440002.

Mayberry, K.S., Bradford, K.J. and Rubatzky, V.E. (1991) Yellow cotyledon: a seedling disorder of broccoli. *HortScience* 26(1), 21–23. DOI: 10.21273/HORTSCI.26.1.21.

Maynard, D.N., Warner, D.C. and Howell, J.C. (1981) Cauliflower leaf tipburn: a calcium deficiency disorder. *HortScience* 16(2), 193–195. DOI: 10.21273/HORTSCI.16.2.193.

Mazza, G. (2004) Diet and human health: functional foods to reduce disease risks. *Acta Horticulturae* 642, 161–172. DOI: 10.17660/ActaHortic.2004.642.18.

Melander, B. and Kristensen, J.K. (2011) Soil steaming effects on weed seedling emergence under the influence of soil type, soil moisture, soil structure and heat duration. *Annals of Applied Biology* 158(2), 194–203. DOI: 10.1111/j.1744-7348.2010.00453.x.

Menniti, A.M., Maccaferri, M. and Folchi, A. (1997) Physio-pathological responses of cabbage stored under controlled atmospheres. *Postharvest Biology and Technology* 10(3), 207–212. DOI: 10.1016/S0925-5214(97)01415-4.

Miceli, A., Vetrano, F. and Romano, C. (2018) Quality and shelf-life of minimally processed cauliflower. *Acta Horticulturae* 1209, 277–280. DOI: 10.17660/ActaHortic.2018.1209.40.

Michell, K.A., Isweiri, H., Newman, S.E., Bunning, M., Bellows, L.L. *et al.* (2020) Microgreens: consumer sensory perception and acceptance of an emerging functional food crop. *Journal of Food Science* 85(4), 926–935. DOI: 10.1111/1750-3841.15075.

Miret, J.A., Munné-Bosch, S. and Dijkwel, P.P. (2018) ABA signalling manipulation suppresses senescence of a leafy vegetable stored at room temperature. *Plant Biotechnology Journal* 16(2), 530–544. DOI: 10.1111/pbi.12793.

Moon, E.-W., Yang, J.-S., Yoon, S.-R. and Ha, J.-H. (2020) Application of colorimetric indicators to predict the fermentation stage of kimchi. *Journal of Food Science* 85(12), 4170–4179. DOI: 10.1111/1750-3841.15532.

Nandini, D.B., Rao, R.S., Deepak, B.S. and Reddy, P.B. (2020) Sulforaphane in broccoli: the green chemoprevention!! Role in cancer prevention and therapy. *Journal of Oral and Maxillofacial Pathology* 24(2), 405. DOI: 10.4103/jomfp. JOMFP_126_19.

Nestle, M. (1998) Broccoli sprouts in cancer prevention. *Nutrition Reviews* 56(4), 127–130. DOI: 10.1111/j.1753-4887.1998.tb01725.x.

Nett, L., Sradnick, A., Fuß, R., Flessa, H. and Fink, M. (2016) Emissions of nitrous oxide and ammonia after cauliflower harvest are influenced by soil type and crop residue management. *Nutrient Cycling in Agroecosystems* 106(2), 217–231. DOI: 10.1007/s10705-016-9801-2.

Nishio, T. (2017) Economic and academic importance of radish. In: Nishio, T. and Kitashiba, H. (eds) *The Radish Genome (Compendium of Plant Genomes Series)*. Springer, Cham, Switzerland, pp. 1–10. DOI: 10.1007/978-3-319-59253-4.

Noman, M., Aysha, J., Ketehouli, T., Yang, J., Du, L. *et al.* (2021) Calmodulin binding transcription activators: an interplay between calcium signalling and plant stress tolerance. *Journal of Plant Physiology* 256, 153327. DOI: 10.1016/j. jplph.2020.153327.

Nowbuth, R.D. and Pearson, S. (1998) The effect of temperature and shade on curd initiation in temperate and tropical cauliflower. *Acta Horticulturae* 459, 79–88. DOI: 10.17660/ActaHortic.1998.459.7.

Ogura, H. (1968) Studies on the new male sterility in Japanese radish, with special reference to the utilization of this sterility towards practical raising of hybrid seeds. *Memoirs of the Faculty of Agriculture, Kagoshima University* 6(2), 39–78.

O'Hare, T.J., Wong, L.S. and Prasad, A. (2000) Atmosphere modification extends the postharvest shelf life of fresh-cut leafy Asian brassicas. *Acta Horticulturae* 539, 103–107. DOI: 10.17660/ActaHortic.2000.539.12.

O'Hare, T.J., Pyke, M., Scheelings, P., Eaglesham, G., Wong, L. *et al.* (2012) Impact of low temperature storage on active and storage forms of folate in choy sum (*Brassica rapa* subsp. *parachinensis*). *Postharvest Biology and Technology* 74, 85–90. DOI: 10.1016/j.postharvbio.2012.06.020.

Pekkeriet, E.J., van Henten, E.J. and Campen, J.B. (2015) Contribution of innovative technologies to new developments in horticulture. *Acta Horticulturae* 1099, 45–54. DOI: 10.17660/ActaHortic.2015.1099.1.

Pellegrino, R., Wheeler, J., Sams, C.E. and Luckett, C.R. (2019) Storage time and temperature on the sensory properties broccoli. *Foods* 8(5), 162. DOI: 10.3390/ foods8050162.

Perez-Balibrea, S., Moreno, D.A. and García-Viguera, C. (2011) Genotypic effects on the phytochemical quality of seeds and sprouts from commercial broccoli cultivars. *Food Chemistry* 125(2), 348–354. DOI: 10.1016/j.foodchem.2010.09.004.

Perkins, S. and Tiffin, D. (2008) *Comparative Study of Refrigerated Storage for UK Horticulture*. Project TF 187 for the Horticultural Development Council (HDC), now the Agricultural and Horticultural Development Board (AHDB), Stoneleigh, UK.

Pierscianowski, J., Popović, V., Biancaniello, M., Bissonnette, S., Zhu, Y. *et al.* (2021) Continuous-flow UV-C processing of kale juice for the inactivation of *E. coli* and assessment of quality parameters. *Food Research International* 140, 110085. DOI: 10.1016/j.foodres.2020.110085.

Pintos, F.M., Hasperué, J.H., Vicente, A.R. and Rodoni, L.M. (2020) Role of white light intensity and photoperiod during retail in broccoli shelf-life. *Postharvest Biology and Technology* 163, 111121. DOI: 10.1016/j.postharvbio.2020.111121.

Pliakoni, E.D., Deltsidis, A.I., Huber, D.J., Sargent, S.A. and Brecht, J.K. (2015) Physical and biochemical changes in broccoli that may assist in decision-making related to international marine transport in air or CA/MA. *Acta Horticulturae* 1071, 951–957. DOI: 10.17660/ActaHortic.2015.1071.86.

Porter, K., Collins, G. and Klieber, A. (2004) Effect of postharvest mechanical stress on quality and storage life of Chinese cabbage cv. Yuki. *Australian Journal of Experimental Agriculture* 44(6), 629. DOI: 10.1071/EA03056.

Pringle, B., Bishop, C. and Clayton, R. (2009) *Potatoes Postharvest*. CAB International, Walllingford, UK. DOI: 10.1079/9780851995021.0000.

Rihan, H.Z., Al-Issawi, M., Al-Shamari, M., Elmahrouk, M. and Fuller, M.P. (2015) Plant abiotic stress tolerance analysis in cauliflower using a curd micropropagation system. *Acta Horticulturae* 1083, 43–52. DOI: 10.17660/ActaHortic.2015.1083.3.

Rihan, H.Z., Al-Issawi, M. and Fuller, M.P. (2017) An analysis of the development of cauliflower seed as a model to improve the molecular mechanism of abiotic stress tolerance in cauliflower artificial seeds. *Plant Physiology and Biochemistry* 116, 91–105. DOI: 10.1016/j.plaphy.2017.05.011.

Roth, E., Gomez, A., Barriobero, J., Ozcoz, B., Mir-Bel, J. *et al.* (2012) Optimisation of the postharvest chain for whole and fresh-cut cauliflower. *Acta Horticulturae* 934, 1261–1267. DOI: 10.17660/ActaHortic.2012.934.171.

Ryall, A.L. and Lipton, W.J. (1972) *Handling, Transportation and Storage of Fruits and Vegetables: Volume 1, Vegetables and Melons*. AVI Publishing, Westport, CT.

Saldana, N., Cabrera, J.M., Serwatowski, R.J. and Gracia, C. (2006) Yield mapping system for vegetables picked up with a tractor-pulled platform. *Spanish Journal of Agricultural Research* 4(2), 130. DOI: 10.5424/sjar/2006042-185.

Samec, D., Kruk, V. and Ivanisevic, P. (2019) Influence of seed origin on morphological characteristics and phytochemicals levels in *Brassica oleracea* var. *acephala*. *Agronomy* 9(9), 502. DOI: 10.3390/agronomy9090502.

Satheesh, N., Workneh Fanta, S. and Yildiz, F. (2020) Kale: review on nutritional composition, bio-active compounds, anti-nutritional factors, health beneficial properties and value-added products. *Cogent Food & Agriculture* 6(1), 1811048. DOI: 10.1080/23311932.2020.1811048.

Sato, F. and Okada, K. (2014) Daily red LED illumination improves the quality of cabbage plug seedlings during low-temperature storage. *The Journal of Horticultural Science and Biotechnology* 89(2), 179–184. DOI: 10.1080/14620316.2014.11513066.

Schouten, R.E., Zhang, X., Tijskens, L.M.M. and van Kooten, O. (2008) The propagation of variation in glucosinolate levels as effected by controlled atmosphere and temperature in a broccoli batch. *Acta Horticulturae* 802, 241–247. DOI: 10.17660/ActaHortic.2008.802.30.

Shikano, I. and Cory, J.S. (2015) Impact of environmental variation on host performance differs with pathogen identity: implications for host-pathogen interactions in a changing climate. *Scientific Reports* 5(1), 15351. DOI: 10.1038/srep15351.

Shin, D.J., Yoo, S.J., Hong, J.K., Weon, H.Y., Song, J. *et al.* (2019) Effect of *Bacillus aryabhattai* H26-2 and *B. siamensis* H30-3 on growth promotion and alleviation of heat and drought stresses in Chinese cabbage. *The Plant Pathology Journal* 35(2), 178–187. DOI: 10.5423/PPJ.NT.08.2018.0159.

Shipway, M.R. (1978) Winter white cabbage: evaluation of "pepper-spot" resistant varieties. In: *Kirton Experimental Horticulture Station Annual Report for 1978.* pp. 62–63.

Sivaranjani, A., Senthilrani, S., Ashok kumar, B. and Senthil Murugan, A. (2022) An overview of various computer vision-based grading system for various agricultural products. *The Journal of Horticultural Science and Biotechnology* 97(2), 137–159. DOI: 10.1080/14620316.2021.1970631.

Sohail, M., Wills, R.B.H., Bowyer, M.C. and Pristijono, P. (2021) Beneficial impact of exogenous arginine, cysteine and methionine on postharvest senescence of broccoli. *Food Chemistry* 338, 128055. DOI: 10.1016/j.foodchem.2020.128055.

Song, H., Yi, H., Han, C.-T., Park, J.-I. and Hur, Y. (2018) Allelic variation in *Brassica oleracea CIRCADIAN CLOCK ASSOCIATED 1 (BoCCA1)* is associated with freezing tolerance. *Horticulture, Environment, and Biotechnology* 59(3), 423–434. DOI: 10.1007/s13580-018-0045-8.

Song, H., Lee, M., Hwang, B.-H., Han, C.-T., Park, J.-I. *et al.* (2020) Development and application of a PCR-based molecular marker for the identification of high temperature tolerant cabbage (*Brassica oleracea* var. *capitata*) genotypes. *Agronomy* 10(1), 116. DOI: 10.3390/agronomy10010116.

Soriano, F., Claudio, M.T.R. and Cardoso, A.I.I. (2019) Germination of broccoli organic seeds treated with thermotherapy. *Acta Horticulturae* 1249, 73–78. DOI: 10.17660/ActaHortic.2019.1249.14.

Sorin, C., Leport, L., Cambert, M., Bouchereau, A., Mariette, F. *et al.* (2016) Nitrogen deficiency impacts on leaf cell and tissue structure with consequences for senescence associated processes in *Brassica napus*. *Botanical Studies* 57(1), 11. DOI: 10.1186/s40529-016-0125-y.

Stansell, Z., Farnham, M. and Björkman, T. (2019) Complex horticultural quality traits in broccoli are illuminated by evaluation of the immortal *BolTBDH* mapping population. *Frontiers in Plant Science* 10, 1104. DOI: 10.3389/fpls.2019.01104.

Steindal, A.L.H., Rødven, R., Hansen, E. and Mølmann, J. (2015) Effects of photoperiod, growth temperature and cold acclimatisation on glucosinolates, sugars and fatty acids in kale. *Food Chemistry* 174, 44–51. DOI: 10.1016/j.foodchem.2014.10.129.

Strandberg, J.O., Darby, J.F., Walker, J.C. and Williams, P.H. (1969) Black speck, a non-parasitic disease of cabbage. *Phytopathology* 59(12), 1879–1883.

Su, T., Li, P., Wang, H., Wang, W., Zhao, X. *et al.* (2019) Natural variation in a cal-reticulin gene causes reduced resistance to Ca^{2+} deficiency-induced tipburn in Chinese cabbage (*Brassica rapa* ssp. *pekinensis*). *Plant, Cell & Environment* 42(11), 3044–3060. DOI: 10.1111/pce.13612.

Sun, B., Jiang, M., Zheng, H., Jian, Y., Huang, W.-L. *et al.* (2020) Color-related chlorophyll and carotenoid concentrations of Chinese kale can be altered through CRISPR/Cas9 targeted editing of the carotenoid isomerase gene *BoaCRTISO*. *Horticulture Research* 7(1), 161. DOI: 10.1038/s41438-020-00379-w.

Sun, X., Luo, S., Luo, L., Wang, X., Chen, X. *et al.* (2018) Genetic analysis of Chinese cabbage reveals correlation between rosette leaf and leafy head variation. *Frontiers in Plant Science* 9, 1455. DOI: 10.3389/fpls.2018.01455.

Sun, X.-X., Basnet, R.K., Yan, Z., Bucher, J., Cai, C. *et al.* (2019) Genome-wide transcriptome analysis reveals molecular pathways involved in leafy head formation of Chinese cabbage (*Brassica rapa*). *Horticulture Research* 6(1), 130. DOI: 10.1038/s41438-019-0212-9.

Sun, Y. and Li, W. (2017) Effects the mechanism of micro-vacuum storage on broccoli chlorophyll degradation and builds prediction model of chlorophyll content based on the color parameter changes. *Scientia Horticulturae* 224, 206–214. DOI: 10.1016/j.scienta.2017.06.040.

Suojala, T. (2003) Compositional and quality changes in white cabbage during harvest period and storage. *The Journal of Horticultural Science and Biotechnology* 78(6), 821–827. DOI: 10.1080/14620316.2003.11511704.

Swegarden, H., Stelick, A., Dando, R. and Griffiths, P.D. (2019) Bridging sensory evaluation and consumer research for strategic leafy brassica (*Brassica oleracea*) improvement. *Journal of Food Science* 84(12), 3746–3762. DOI: 10.1111/1750-3841.14831.

Terzo, M.N., Russo, A., Ficili, B., Pezzino, F.M., Tribulato, A. *et al.* (2020) Neglected sicilian landraces of black broccoli (*Brassica oleracea* var. *italica* Plenck) and health benefits: an *in vivo* study. *Acta Horticulturae* 1267, 91–96. DOI: 10.17660/ActaHortic.2020.1267.15.

Thompson, A.K. (1998) *Controlled Atmosphere Storage of Fruits and Vegetables.* CAB International, Wallingford, UK.

Tian, M.S., Davies, L., Downs, C.G., Liu, X.F. and Lill, R.E. (1995) Effects of floret maturity, cytokinin and ethylene on broccoli yellowing after harvest. *Postharvest Biology and Technology* 6(1–2), 29–40. DOI: 10.1016/0925-5214(94)00047-V.

Toivonen, P.M.A. (1997) The effects of storage temperature, storage duration, hydrocooling, and micro-perforated wrap on shelf life of broccoli (*Brassica oleracea* L., Italica group). *Postharvest Biology and Technology* 10(1), 59–65. DOI: 10.1016/S0925-5214(97)87275-4.

Toivonen, P.M.A. and DeEll, J.R. (1998) Differences in chlorophyll fluorescence and chlorophyll content of broccoli associated with maturity and sampling section. *Postharvest Biology and Technology* 14(1), 61–64. DOI: 10.1016/S0925-5214(98)00022-2.

Traore, M.B., Sun, A., Gan, Z., Senou, H., Togo, J. *et al.* (2020) Antimicrobial capacity of ultrasound and ozone for enhancing bacterial safety on inoculated shredded green cabbage (*Brassica oleracea* var. *capitata*). *Canadian Journal of Microbiology* 66(2), 125–137. DOI: 10.1139/cjm-2019-0313.

Tyc, O., Putra, R., Gols, R., Harvey, J.A. and Garbeva, P. (2020) The ecological role of bacterial seed endophytes associated with wild cabbage in the United Kingdom. *MicrobiologyOpen* 9(1), e00954. DOI: 10.1002/mbo3.954.

Valida, A., Rivera, F.R., Salabao, A., Benitez, M., Sudaria, E. *et al.* (2018) Cabbage (*Brassica oleraceae* var. *capitata*) quality grading at two storage temperatures. *Acta Horticulturae* 1213, 129–133. DOI: 10.17660/ActaHortic.2018.1213.17.

van Stokkom, V.L., Teo, P.S., Mars, M., de Graaf, C., van Kooten, O. *et al.* (2016) Taste intensities of ten vegetables commonly consumed in the Netherlands. *Food Research International* 87, 34–41. DOI: 10.1016/j.foodres.2016.06.016.

Vijayakumar, H., Thamilarasan, S., Shanmugam, A., Natarajan, S., Jung, H.-J. *et al.* (2016) Glutathione transferases superfamily: cold-inducible expression of distinct GST genes in *Brassica oleracea. International Journal of Molecular Sciences* 17(8), 1211. DOI: 10.3390/ijms17081211.

Walkey, D.G.A. and Webb, M.J.W. (1978) Internal necrosis in stored white cabbage caused by turnip mosaic virus. *Annals of Applied Biology* 89(3), 435–441. DOI: 10.1111/j.1744-7348.1978.tb05971.x.

Walkey, D.G.A. and Neely, H.A. (1980) Necrosis in white cabbage II. Resistance to pepper-spot. *Tests of Agrochemicals and Cultivars (A Supplement to Annals of Applied Biology)* 94(1), 18–19.

Walsh, J. and Hunter, P. (2004) Pest and disease management: sunken and disorderly. *The Grower* 15(1), 14–15. DOI: 10.1564/15feb07.

Walsh, J., Hunter, P. and MacDonald, N. (2004) *Internal Disorders of Stored White Cabbage. Factsheet No. 11/04*. Horticultural Development Council, Coventry, UK, p. 8.

Walsh, K.B., McGlone, V.A. and Han, D.H. (2020) The uses of near infra-red spectroscopy in postharvest decision support: a review. *Postharvest Biology and Technology* 163, 111139. DOI: 10.1016/j.postharvbio.2020.111139.

Wang, A., Hu, J., Gao, C., Chen, G., Wang, B. *et al.* (2019a) Genome-wide analysis of long non-coding RNAs unveils the regulatory roles in the heat tolerance of Chinese cabbage (*Brassica rapa* ssp. *chinensis*). *Scientific Reports* 9(1), 5002. DOI: 10.1038/s41598-019-41428-2.

Wang, C., Yang, S., Liu, Y., Zhang, X. and Wang, R. (2014) Effects of water stress and nutrient deficiency on ethylene synthesis in harvested broccoli. *Journal of Food, Agriculture and Environment* 12(2), 725–730.

Wang, H.-W., Makino, Y., Inoue, J., Maejima, K., Funayama-Noguchi, S. *et al.* (2017) Influence of a modified atmosphere on the induction and activity of respiratory enzymes in broccoli florets during the early stage of postharvest storage. *Journal of Agricultural and Food Chemistry* 65(39), 8538–8543. DOI: 10.1021/acs.jafc.7b02318.

Wang, J., Huang, F., You, X. and Hou, X. (2019b) Identification and functional characterization of a cold-related protein, BcHHP5 (*Brassica rapa* ssp. *chinensis*). *International Journal of Molecular Sciences* 20(1), 93. DOI: 10.3390/ijms20010093.

Wang, L., Zhang, Y., Chen, Y., Liu, S., Yun, L. *et al.* (2019c) Investigating the relationship between volatile components and differentially expressed proteins in broccoli heads during storage in high CO_2 atmospheres. *Postharvest Biology and Technology* 153, 43–51. DOI: 10.1016/j.postharvbio.2019.03.015.

Wieczorek, M.N., Walczak, M., Skrzypczak-Zielińska, M. and Jeleń, H.H. (2018) Bitter taste of Brassica vegetables: the role of genetic factors, receptors, isothiocyanates, glucosinolates, and flavor context. *Critical Reviews in Food Science and Nutrition* 58(18), 3130–3140. DOI: 10.1080/10408398.2017.1353478.

Wilhoit, J.H. and Vaughan, D.H. (1991) A powered cutting device for selectively harvesting broccoli. *Applied Engineering in Agriculture* 7(1), 14–20. DOI: 10.13031/2013.26206.

Wilson, R.A., Gupta, S., Sangha, M.K. and Kaur, G. (2019) Effect of heat stress on enzymatic and non-enzymatic antioxidants in *Brassica rapa. Journal of Environmental Biology* 40(1), 119–124. DOI: 10.22438/jeb/40/1/MRN-877.

Wydrzynski, T.J., Chow, W.S. and Badger, M.R. (eds) (1995) Chlorophyll fluorescence: origins, measurements, interpretations and applications. *Australian Journal of Plant Physiology* 22(2), 1–355.

Xu, F., Wang, H., Tang, Y., Dong, S., Qiao, X. *et al.* (2016) Effect of 1-methylcyclopropene on senescence and sugar metabolism in harvested broccoli florets. *Postharvest Biology and Technology* 116, 45–49. DOI: 10.1016/j.postharvbio.2016.01.004.

Xu, Q., Wang, R., Wang, L., Xing, Y., Li, W. *et al.* (2019) Effects of different pre-cooling methods on quality changes in fresh-cut purple cabbage during cold-chain transportation. *Food and Fermentation Industries* 45(7), 135–143. DOI: 10.13995/j. cnki.11-1802/ts.019049.

Yamauchi, N., Harada, K. and Watada, A.E. (1997) *In vitro* chlorophyll degradation in stored broccoli (*Brassica oleracea* L. var. *italica* Plen.) florets. *Postharvest Biology and Technology* 12(3), 239–245. DOI: 10.1016/S0925-5214(97)00063-X.

Yin, X. and Huang, Z. (2020) Effects of polylactic acid antimicrobial films on preservation of Chinese rape. *Packaging Technology and Science* 33(11), 461–468. DOI: 10.1002/pts.2529.

Yu, J.-X., Hu, W.-Z., Zhao, M.-R., Guan, Y.-G., Hao, K.-X. *et al.* (2019) Research progress on preservation technology for fresh-cut broccoli. *Food and Fermentation Industries* 45(15), 288–293.

Yu, Y.J., Zhao, X.Y. and Xu, J.B. (2004) Screening method for resistance to tipburn in Chinese cabbage. *Acta Horticulturae* 637, 189–193. DOI: 10.17660/ ActaHortic.2004.637.22.

Yuan, L., Liu, S., Zhu, S., Chen, G., Liu, F. *et al.* (2016) Comparative response of two wucai (*Brassica campestris* L.) genotypes to heat stress on antioxidative system and cell ultrastructure in root. *Acta Physiologiae Plantarum* 38(9), 223. DOI: 10.1007/ s11738-016-2246-z.

Zhang, J., Xiao, H., Yao, S., Song, Z., Jin, Y. *et al.* (2020) Physical and mechanical properties of cabbage. *International Agricultural Engineering Journal* 29(1).

Zhang, X., Su, Y., Liu, Y., Fang, Z., Yang, L. *et al.* (2016) Genetic analysis and QTL mapping of traits related to head shape in cabbage (*Brassica oleracea* var. *capitata* L.). *Scientia Horticulturae* 207, 82–88. DOI: 10.1016/j.scienta.2016.05.015.

Zhang, X., Zhang, M., Devahastin, S. and Guo, Z. (2019) Effect of combined ultrasonication and modified atmosphere packaging on storage quality of pakchoi (*Brassica chinensis* L.). *Food and Bioprocess Technology* 12(9), 1573–1583. DOI: 10.1007/ s11947-019-02316-9.

Zhao, Y.-S., Gong, L., Huang, Y.-X. and Liu, C.-L. (2016) A review of key techniques of vision-based control for harvesting robot. *Computers and Electronics in Agriculture* 127, 311–323. DOI: 10.1016/j.compag.2016.06.022.

Zhao, Z., Sheng, X., Yu, H., Wang, J., Shen, Y. *et al.* (2020) Identification of candidate genes involved in curd riceyness in cauliflower. *International Journal of Molecular Sciences* 21(6), 1999. DOI: 10.3390/ijms21061999.

Zhou, F., Zuo, J., Xu, D., Gao, L., Wang, Q. *et al.* (2020) Low intensity white light-emitting diodes (LED) application to delay senescence and maintain quality of postharvest pakchoi (*Brassica campestris* L. ssp. *chinensis* (L.) Makino var. *communis* Tsen et Lee). *Scientia Horticulturae* 262, 109060. DOI: 10.1016/j.scienta.2019.109060.

Zong, R.J., Morris, L.L., Ahrens, M.J., Rubatzky, V. and Cantwell, M.I. (1998) Postharvest physiology and quality of gai-Lan (*Brassica oleracea* var. *alboglabra*) and choi-sum (*Brassica rapa* subsp. *parachinensis*). *Acta Horticulturae* 467, 349–356. DOI: 10.17660/ActaHortic.1998.467.39.

Index

Page numbers in *italics* refer to figures; Page numbers in **bold** refer to tables.

CABI – who we are and what we do

This book is published by **CABI**, an international not-for-profit organisation that improves people's lives worldwide by providing information and applying scientific expertise to solve problems in agriculture and the environment.

CABI is also a global publisher producing key scientific publications, including world renowned databases, as well as compendia, books, ebooks and full text electronic resources. We publish content in a wide range of subject areas including: agriculture and crop science / animal and veterinary sciences / ecology and conservation / environmental science / horticulture and plant sciences / human health, food science and nutrition / international development / leisure and tourism.

The profits from CABI's publishing activities enable us to work with farming communities around the world, supporting them as they battle with poor soil, invasive species and pests and diseases, to improve their livelihoods and help provide food for an ever growing population.

CABI is an international intergovernmental organisation, and we gratefully acknowledge the core financial support from our member countries (and lead agencies) including:

Ministry of Agriculture People's Republic of China

Australian Government
Australian Centre for International Agricultural Research

Agriculture and Agri-Food Canada

Ministry of Foreign Affairs of the Netherlands

Schweizerische Eidgenossenschaft
Confédération suisse
Confederazione Svizzera
Confederaziun svizra
Swiss Agency for Development and Cooperation SDC

Discover more

To read more about CABI's work, please visit: **www.cabi.org**

Browse our books at: **www.cabi.org/bookshop**, or explore our online products at: **www.cabi.org/publishing-products**

Interested in writing for CABI? Find our author guidelines here: **www.cabi.org/publishing-products/information-for-authors/**